The Rise of Science in Islam and the West

This is a study of science in Muslim society from its rise in the 8th century to the efforts of 19th-century Muslim thinkers and reformers to regain the lost ethos that had given birth to the rich scientific heritage of earlier Muslim civilization. The volume is organized in four parts; the rise of science in Muslim society in its historical setting of political and intellectual expansion; the Muslim creative achievement and original discoveries; proponents and opponents of science in a religiously oriented society; and finally the complex factors that account for the end of the 500-year Muslim renaissance.

The book brings together and treats in depth, using primary and secondary sources in Arabic, Turkish and European languages, subjects that are lightly and uncritically brushed over in non-specialized literature, such as the question of what can be considered to be purely original scientific advancement in Muslim civilization over and above what was inherited from the Greco–Syriac and Indian traditions; what was the place of science in a religious society; and the question of the curious demise of the Muslim scientific renaissance after centuries of creativity. The book also interprets the history of the rise, achievement and decline of scientific study in light of the religious temper and of the political and socio-economic vicissitudes across Islamdom for over a millennium and integrates the Muslim legacy with the history of Latin/European accomplishments. It sets the stage for the next momentous transmission of science: from the West back to the Arabic-speaking world of Islam, from the last half of the 19th century to the early 21st century, the subject of a second volume.

John W. Livingston is Associate Professor of History at the William Paterson University of New Jersey, USA.

The Rise of Science in Islam and the West

From Shared Heritage to Parting of the Ways, 8th to 19th Centuries

John W. Livingston

Routledge
Taylor & Francis Group

LONDON AND NEW YORK

First published 2018 by Routledge

2 Park Square, Milton Park, Abingdon, Oxon OX14 4RN
605 Third Avenue, New York, NY 10017

Routledge is an imprint of the Taylor & Francis Group, an informa business

First issued in paperback 2021

British Library Cataloguing-in-Publication Data
A catalogue record for this book is available from the British Library

Library of Congress Cataloging-in-Publication Data
Names: Livingston, John W. (John William), 1932–
Title: The rise of science in Islam and the West : from shared heritage to
    parting of the ways, 8th to 19th centuries / John W. Livingston.
Description: Abingdon, Oxon ; New York, NY : Routledge, 2018. |
    Includes index.
Identifiers: LCCN 2017019716 | ISBN 9781472447333
    (hardback : alk. paper) | ISBN 9781315101507 (ebook)
Subjects: LCSH: Science, Medieval. | Science—Islamic countries—History. |
    Islam and science—History.
Classification: LCC Q124.97 .L58 2018 | DDC 509/.02—dc23
LC record available at https://lccn.loc.gov/2017019716

ISBN: 978-1-4724-4733-3 (hbk)
ISBN: 978-0-367-34942-4 (pbk)

Typeset in Times New Roman
by Apex CoVantage, LLC

I dedicate this book to my long deceased professor and friend, Martin Dickson. Martin was a professor of Persian language and Iranian history, specializing in the Safavid period and Islamic mysticism, at Princeton University, from 1958 to 1991, the year of his death. He was a mentor and inspiration in learning and life, to me and to many other of his students and those who were fortunate enough to become his friend. His memory lives on, carried by his former students who now teach in universities across the country. Like the mystical poets he studied, and emulated, Martin had a knack of turning the prism to see the universe revealed in unexpected lights. It was my good fortune, my best fortune I should say, to have had him as a friend. With him always in mind, I wrote this book.

Contents

Illustrations

Plates

Figures

Preface

The lands of Islam were once graced with proud cities in a high civilization that for over half a millennium led the world in intellectual productivity. From Cordova in Spain to Samarqand in Central Asia, with Fez, Marakkesh, Cairo, Damascus, Baghdad, Nishapur, Shiraz and Bukhara in between, a legion of creative men of science contributed over the centuries to building a body of scientific, medical and philosophical knowledge that gave Islamdom a precedence in civilization openly recognized by its neighbors to the east and west. Between roughly 800 and 1500 AD, Islamic civilization was accepted as the world's uncontested leader in science and its affiliated branches. Until 1600, military success against the West confirmed the Quran's scriptural assurance of Muslims being the best of all people. As for the moral and mental superiority of Muslims over Westerners, and over everybody else, that went without question. Muslims were God's favored community. Their holy book said so. Victory in arms over the infidels was proof. Superiority of mind, as in war and religion, could be taken for granted.

Around 1130, European scholars began translating Arabic scientific, medical and philosophical texts to Latin. Sometime between 1400 and 1500, the impetus of scientific study in the civilization of Latin Christendom reached a level comparable to that of its traditional Muslim rival. But there was a difference. Whereas the interest and investment in science in the West were on the rise, in the East they were in decline, and had been for centuries. The science and technology that through the early Arab conquests had been inherited and translated during Islam's youth and then had been assimilated and expanded during the civilization's mature years, were in the 12th and 13th centuries well along the way of losing social standing and being replaced by both mystical devotion and studies more stringently oriented to a narrow interpretation of religious orthodoxy, and this at the very time that, in Spain, Italy and Sicily, Arabic science, medicine and philosophy were being translated to Latin and assimilated. With the fading importance of this scientific and philosophical heritage in the lands of Islam went not only the creative spirit of curiosity regarding physical nature, but also the splendid technique of contriving mechanical devices in which the Muslims had been so adept, and which amazed visitors to the caliph's court. With it went that ease of mental boundaries, that flexibility of adaptation, of borrowing, of learning and improving on inventions made beyond the domain of Islamdom by non-Muslims that had earlier characterized the genius of Muslim civilization and made it what it was.

This inversion of mental openness and creativity did not bring about an immediate eclipse of Muslim military glory, for a spark of that same flexibility and ease of borrowing, of adapting and improving, was exhibited again by another young Muslim power, the Ottomans, in the 14th and 15th centuries, when artillery was taken from the West and advanced models of it turned on the traditional foe, Constantinople, which was taken in 1453 and Vienna, which was almost taken in the next century. While Muslim armies under Ottoman sultans continued Islam's triumph over the West for most of the 16th century and vast stretches of Christian territory in eastern and central Europe fell under Ottoman sway, giving fresh proof of God's abiding favoritism, the expansion of science in the West, originally fueled by the Latin absorption of Arabic–Muslim learning in the 12th, 13th and 14th centuries, continued quietly, steadily and far from the battlefields, building western Europe's human potential, in spite of the internal discord of political and religious strife in the 16th and 17th centuries and the Church's humiliation of Galileo. It was not until the late 17th century, when the forward progress of Muslim arms had for a second time come to a halt at the gates of Vienna and then suffered a series of shattering reversals, that a few highly placed officials in the sultan's court were at last able to perceive, and admit to, a shift in military power. It did not take long before it also dawned upon them that the shift had something to do with technological expertise and scientific knowledge. This was a grave revelation accepted with great reluctance and hesitation. There was no end of resistance to admitting it. Only a brave few dared admit that those infidel Frankish louts of Abbasid and Crusader times had overtaken God's community in the methods and weapons of war and the mind. No one wanted to listen. It would take until the early 18th century, with the shock of more defeat and lost territory, each loss greater than the previous, before a wider circle of courtly elites were forced to recognize that western military power could overcome anything the Ottomans were able to put up in resistance to it. Further Ottoman conquest was out of the question. It was now a matter of defense; soon it would be of survival.

Frightened by this shift that was so out of order in the affairs of the world, a world designed and controlled by God, the God of Islam and the Quran, the sultan and his ministers sent a mission to the West to see what it was that made these Frankish ruffians so strong after having been so weak for so long. It was discovered that military power was the product of better technology, which in turn was the product of a technique of social and political organization, of planning and execution surpassing anything Muslims possessed. Power had to do with factories and education and the printing press and science. Henceforth, the Ottoman ruling elite, rather, the party in the elite urging innovative reform, became increasingly more curious about the West. But for most Muslims it was too traumatic to contemplate. Steeped in denial, they claimed it was not Frankish strength but Muslim weakness that caused the change. Muslims had to become better Muslims. It was a question of morality. Weak morality made for weak institutions, which opened the gates to corruption and a loss of spiritual vigor.

Many in high places refused to accept the observations and conclusions of the mission and rejected with religious passion the reformist party's judgment that the empire must accept Western methods and inventions or perish. The idea of Muslims as inferior in anything bordered on heresy. Not only did it mean admitting an end, or even worse, a reversal in the millennial experience of Muslim expansion, it threw the value of venerable social tradition and practice into question and, ultimately, the value of Muslim education. In the Muslim logic of life under God during a thousand years of historical experience, inferiority to the West on the battlefield ultimately put to question the fundamentals of religion. There were things that could not be questioned, or should not be, and it took a century of not just Ottoman defeat, but defeat and occupation of Muslim lands everywhere, the Balkans, North Africa, Egypt, the Caucasus, Central Asia and India, before Muslims could admit something was not right in the community of God's religion. Whatever it was, it had to be set right or the whole of Islamdom would be swept up by the relentless onslaught of Britain, France, Austria and Russia.

Reforming the old ways and introducing new ones came with great difficulty and demanded a high price, socially, culturally and materially. The idea of innovation and change was seen to be the first step to heresy. Reformers who tried to revive the old heritage of science and introduce the new science of the West in order to face the existential challenge were accused by both religious authorities and political rivals of destroying Islam and of being agents of the West. To avoid the charge, reformers insinuated the new into the old, hoping to make change palatable to at least religious sensitivities. Reformers faced the daunting task of reviving the lost spirit of scientific creativity that had given birth to what became the Islamic scientific heritage, which the Latin West had centuries earlier translated into its own cultural traditions and then transformed into something original in the Scientific Revolution of the 17th century, with a final revolutionary twist provided by Darwin in the 19th century. It meant learning from the West, that contemptible enemy of a lower religion and civilization that was now, strange as it seemed, threatening Islam.

Learning from the West meant, in addition to its being perceived as betrayal of religion, the horror of bringing into Islam a slew of repugnant ideas: an earth that moved, a set of universal natural laws that competed with God's unimpeded will and rivaled divine law, that threw into question Quranic truth and Prophetic tradition; and horror of horrors, it meant embracing the idea of God's glorious creation, man, having a common ancestor with an ape.

These dedicated reformers did not enjoy a large or enthusiastic following. Unlike the experience of 19th century Japan, young men did not flock to the reformers and pay their way to the West to study science and engineering. Even in those courtly circles where the impetus to reform was born, reformers faced murderous opposition from men and powerful families with vested interests in maintaining the status quo. Opposition came from every direction: the highest ranks of government, the bureaucracy, the military and educational and religious institutions. Innovative reform had faithful enemies across the social spectrum. With illiteracy hovering around 90 percent or more, ignorance and superstition offered plentiful recruits for forces defending religion and tradition. Popularity was assured to those who called the West satanic and everything that came from it inventions of the devil.

Two centuries of reform did not save the Ottoman Empire. The first World War that threatened to destroy Europe destroyed the Ottomans. The Arab provinces of the empire fell under European domination. The Turks managed to escape this humiliation. From the ashes of the Ottoman Empire a secular Turkish republic dedicated to westernized reform emerged. To the east, Iran was under British economic control and partial occupation. The Muslims of India had for more than a century been under direct British control, with those of Central Asia under first the tsars, then their Soviet successors. European powers carved new states out of the conquered Arab lands and drew the borders as they wanted. It appeared that Islam's humiliation was complete; but not quite. The rest of the 20th century and the start of the 21st would be far more unkind to the Arabs and peoples of the wider Islamic world in their relations and interactions with the West.

As the mounting friction of 19th century nationalism, imperialism, militarism and economic rivalry boiled into the Great War between the European powers, the gap in scientific knowledge and technological expertise between Islamdom and the West continued to widen at an accelerated pace and never stopped. Secularized Muslim autocrats were either force-drafting ill-conceived reforms and half-baked programs of modernization or abolishing those begun by their predecessors. The men of reform themselves were hamstrung between a frightened and insecure despotism and a disjointed, mostly illiterate population whose attitudes were shaped by spokesmen of populist religion, the ulema, who looked to the holy law of the Shari'a, the Quran, the Prophetic traditions and the ways of the pious ancestors of early Islam as an authentic way back to the pristine past in order to confront the present, men for whom change in any form other than going back to that idyllic past when Muslims were truly Muslims was a path to heresy and unbelief. The flawed programs of reform and the opposition to them opened a rift between reformers and conservatives that weakened Islamic society all the more. The problems reformers faced were truly staggering, but social regeneration was not an impossible task. The Japanese, a people who no less than Muslims cherished tradition, faced the same challenge at the same time and transformed their society. Their success both shamed and inspired Middle Eastern reformers: shame for having failed so far, and inspiration to succeed.

Disaster following disaster drained away the will and energy that illumine inspiration: the collapse of the Ottomans and Anglo–French occupation of the Fertile Crescent, the destruction of Palestine and the creation of Israel, the losses of wars with Israel, the American destruction of Iraq and the attempt to dominate the oil-producing countries. One might expect that such a history would drive Muslim states to do what it took to defend themselves and run their own economies and control their own natural resources.[1] Yet Middle Eastern governments have clearly failed to adopt the science, technology and industry that self-defense demands in the modern world. They prefer to buy it off the shelf from industrialized countries, and as a result, they are at their mercy. This is truly remarkable in a community of Muslim societies that prides itself on its past technical and scientific achievement. How is it to be explained? For that matter, how is it that after centuries of brilliant scientific achievement the tradition of science all but disappeared in medieval Muslim society? Is there any connection between the medieval loss of the tradition and the inability to regain it? Religious opposition is often given as the reason that the Islamic spirit of scientific inquiry was smothered and never revived. But in that case, why did science, medicine, philosophy and mathematics flourish as they did, at one place or another from Spain to Central Asia, between the 8th and 16th centuries? If religion was indeed the cause of decline, then how does one account for the rise of science in that same religious society? Islam is a religion that originates in a book of revelations, the Quran, whose verses over and over again direct the believer to know God through nature.

Do the unbelievers not realize that
the heavens and the earth used to be
one solid mass that
we exploded into existence?
And from water we made all living things.
(21:30)

He created seven universes in layers.
You do not see any imperfection
In the creation by the Most Gracious.
Keep looking; do you see any flaw?
(67:3)

The vagaries of science in Muslim society give rise to many questions. Did Arabs and Muslims really create science or just borrow and imitate? If they did create, then what original science came from them to deserve the praise that their civilization has received from those versed in the history of science? What, if anything, did thinkers and experimenters in Islamdom accomplish in advancing the frontiers of scientific knowledge beyond what they inherited and translated from their Greek, Iranian and Indian predecessors? Why would Muslims, supremely self-confident and complacent in believing themselves possessors of the eternal truth of God's final revelation in the holy Quran, have bothered with science in the first place? What need did they have of science? And what was it that drove Muslims, as well as the Christians, Jews and residual pagans who joined them in their scientific enterprise, to cultivate natural philosophy and mathematics? What precisely was the position that religion took in regard to their pursuit of science and philosophy?

These are but some of the many questions that beg for answers in the minds of those who delve into the dramatic rise and fall of scientific creativity in the Islamic experience and of the difficulties modern Muslim states have had in reviving that creative impulse.

There seems to be no end to the questions, once the first is asked. A legion of them link the Muslim East to the Christian West. How was it that Muslim knowledge was transmitted from the Arabs in Spain, Italy and Sicily to Latin scholars of the 12th and 13th centuries? What drove the Latins to acquire this knowledge and what did religion have to say about it? Do the differences in religious response say anything about essential differences between the two civilizations? Can the Western experience throw light on the Eastern? Is there any parallelism here between the translation movement of Arabic to Latin and the Muslim translation and assimilation of Greek knowledge three centuries earlier? And how was it that the descendants of the Latin translators were able in the 16th and 17th centuries to rework and revolutionize the Greco–Muslim legacy into a new science? Why was there a scientific revolution in Christian Europe and not in Islamic civilization, where Muslims possessed the mathematical tools, the observational apparatus, the planetary tables and the theoretical models of heavenly motion that were available to Copernicus, Brahe, Kepler, Galileo and Descartes? Muslim astronomers were just as plagued as their European counterparts by the same critical doubts about the physics of Aristotle and the geometric sleight of hand of Ptolemaic astronomy that flawed the planetary models as mirrors of reality. The esthetics and astronomy of the Pythagorean tradition excited Muslim astronomers and cosmologists as much as they did Copernicus and Kepler. What were Muslim scientists missing?

Sustaining these hefty questions is a drama, both uplifting and depressing, but always astonishing. With respect to Islam, it would be hard to exaggerate the poignancy of a civilization, a world leader in science, medicine and military power for close to a millennium, that so completely, as it would appear, lost the critical spirit of scientific inquiry and then failed to recapture that energy even as its millennial enemy (Latin Christendom transforming to modern Europe and imperial America) relentlessly advanced on the Muslim heartlands and hallowed sanctuaries, swallowing up land after Islamic land, as if Muslims had destined themselves to succumb through one colossal act of negation in order to remain heroically faithful to a fantasized past of purity, greatness and glory. Instead of acting to put science and technology in the service of the state for survival as defeat followed defeat and economies floundered, as they do to this day, many Muslims and Muslim states ignored or rejected modern science, as the West continued to advance. The vacuum created by the defeat, neglect and ineptitude of state leaders opened the way to political Islam. There is no reason to believe Islamist government would be any less repressive than the dictatorships that have been overthrown or are presently being challenged.

For anyone familiar with the achievement of the men who contributed to science in Islamic civilization, the fall from grace and subsequent failure of their modern descendants to recapture the contemporary version of this lost heritage, and the creative spirit that produced it, is tragic. In the great movements of history, Islam's scientific burnout is comparable to the abysmal fall of the ancient Egyptians to 2,500 years of foreign occupation, to the fall of the Roman Empire in the West, or the falls of the Incas, Mayans and Aztecs to a handful of horsemen from over the sea.

Trying to determine when and how the loss of creative genius began in the long trajectory of Muslim civilization is like gathering mercury. What adheres in one place escapes in another. Decline in one region of the vast world of Islamdom was often set off by scientific renaissance in another. Decline did at last set in everywhere, but in regard to the commonly held views on the history of science in

Muslim civilization, it came surprisingly late. In fact, it was not long after the last candle of scientific brilliance went out that Muslims began coming face to face with the new science from the West, that old Greco–Arabic science that had been revolutionized beyond recognition in the century and a half between Copernicus and Newton. Here begins the wrenching sequel to medieval creativity and decline. To appreciate the hard experience of Islam's continuing centuries-long grappling with modern science and modernity after its rich scientific heritage had withered away, a word must be said about how Muslims perceive themselves as products of history and religion.

The cradle of Muslim civilization was Arabia, more precisely, western Arabia, in the cities of Mecca and Medina, the former living on international trade, the second on local agriculture. In this relatively primitive environment, the Prophet Muhammad received the intermittent revelations that comprise the holy book of Islam, the Quran. It was here that the earliest ethos of Muslim society evolved, its genetic strands being formed by trade, by Arab tribal organization and customs, and, inasmuch as the nascent Muslim community was perceived as a threat to the commercial and religious status quo, by defensive warfare against those who attacked the Prophet and his followers, followed by offensive warfare. This earliest ethos was reinforced by a century of Arab tribal conquest of those higher civilizations to the north, east and west and was then modified and enfolded within a more universal ethos whose cultural and intellectual fabric came from the norms and traditions of the conquered civilizations, whereby primitive Islam transcended but never quite lost its Arab roots to become a religion, a law and an organizing principle of a world civilization. The primitive ethos lived on within the universal ethos, nurtured by mythologized legends and histories of the early days of struggle and glory, when the Prophet and a small band of trusted followers triumphed in the fight for God against the forces of evil.

In the broadest sense, classical Islamic civilization was a synthesis of elements from two worlds: Quranic dicta, Arab tribal norms idealized and sanctified in the body of transmitted and collected traditions recording what the Prophet purportedly had said and done relevant to life's myriad problems, and of the social, political, and intellectual traditions of the higher civilizations. The Arabs brought into this universalizing synthesis the Prophet Muhammad, the Quranic word of God, and the Arabic language that gave divinity human expression. Conjoined to these are the earliest histories that record the life of Muhammad: his struggle in Mecca and ultimate triumph in Medina, and the proud century of conquest by those Arab tribes the Prophet had united, all written long after the events.

The forms of higher civilization were absent in even the most advanced of pre-Islamic Arabian cities. (The earlier civilizations centered in Yemen had disappeared before the advent of Islam.) The Arabs of the 7th century knew very little of literature and astronomy and nothing of philosophy and mathematics. Their medical knowledge was primitive. Politically they were unschooled. The high culture of Muslim literature, music, science, medicine, mathematics, philosophy, geography, historiography and architecture arose later in the conquered territories of the Byzantines and Sassanians, in Syria, Egypt, Mesopotamia, Iran and Central

Asia, where civilizations extended back to the Bronze Age. The continuing traditions of these pre-Islamic civilizations became the inheritance from which arose the literary, artistic and intellectual greatness of Islamdom. The one exception to this may perhaps be the formulation of Islamic law and the science of jurisprudence, the purely "Arab science." But even here powerful influences from Syria, Mesopotamia and Egypt can be seen. Except for the judge and jurisprudent Malik ibn al-Anas, who did his work in Medina, and Ahmad ibn Hanbal, whose legal thought manifested the most radical adherence to the Medinian traditions of Arab social custom, the greatest legal thinking was accomplished by Arabs more influenced by traditions and legal thinking outside of Arabia: abu Hanifa and al-Shafi'i, the former being the most liberal of the founders of the four schools of law and the latter the most widely traveled, the most philosophical, and the most rigorously logical in his method of jurisprudence.

Islamic civilization was an enterprise shared by all Muslims, whether Arabs or non-Arab converts to Islam, and by non-Muslims as well. Science, philosophy and medicine, along with the trade and agriculture that went into supporting their practitioners through royal patronage, were professions open to all: Jews, Christians, and even the star-worshipping pagans of al-Harran in northern Iraq. During the first two centuries following the conquests, before the mass conversions to Islam when Muslims were still a ruling minority, the emerging civilization was admirably egalitarian in its intellectual openness. By the time intellectual enterprise had first begun to blossom during the generation between the late 8th and early 9th centuries, the twin processes of Arabization (in North Africa, Egypt, Syria and Mesopotamia) and Islamization (in Iran, India and Central Asia) had proceeded far enough that the former ruling caste of Arabs, having lost their political dominance in 750, were well on their way to melting into the vastly more numerous local populations. Iran and its central Asian cultural annex west of the Oxus River resisted Arabization beyond adopting Arabic as the language of religion, science, literature and government; though eventually a revived Persian language would find broad acceptance in the region for literature, history and politics.

In the regions of Iranian cultural ascendance, Islamization transcended Arabization, and from the 10th century on, a distinct Persian style of Islamic civilization evolved in those lands. Signs that Iran would go its own way became evident soon after Arab military and political power had been broken and the capital of the Muslim empire shifted from Arab Damascus to the new city of Baghdad on the Tigris, which at the time was part of the cultural heartland of Iran. Iranian converts to Islam had helped greatly in bringing the new dynasty to power. Its chief propagandizer, organizer and recruiter had been an Iranian convert. Chafing at their empire being overrun and their proud civilization destroyed, the Iranians did much in shaping the style of the new dynasty, which they served at the highest levels. An Iranian-driven literary movement in Baghdad at the height of the dynasty's power boasted of the superiority of the Iranian character, virtue and cultural excellence over the Arab. That the Iranians were writing this in Arabic was an irony that would be changed in the 10th century when Persian, written in Arabic script, began coming into its own.

The Iranians had reason to boast. They played a vital part in Abbasid courtly life as viziers, governors, generals, essayists, poets, scribes, physicians, astronomers and drinking companions of the caliph and his princes. Khorasanians from northeast Iran were, in the early decades of the new caliphate, the mainstay of the empire's military organization. Iranian converts had been enlisted into Arab political service because of their administrative and political experience in running an empire. Having mastered the language of their conquerors, Iranians by the middle of the 8th century figured among the greatest composers of Arabic poetry and belles lettres and contributors to Arabic historiography, science, philosophy, medicine. Iranians were the first to analyze Arabic constructions to formulate a logically systematized set of grammatical rules to facilitate learning the language by non-Arabs. They wrote the first manuals for the training of government scribes, clerks and courtiers, establishing the forms of Abbasid absolutism, courtly life and bureaucracy on Sassanian models. Iranians proved to be masters in Arabic, outdoing the Arabs themselves in both the elegance of prose and the lyricism and eroticism of poetry. Some, it was believed, must have lived years with Arab bedouin tribes to have known the language so well. But Iran's pride of civilization, which went back more than a millennium to Cyrus the Great, had been too deeply rooted to succumb to Arabization; roughly three centuries after the Arab conquest, Persian was revived as a literary language, beginning naturally with poetry, the soul of a language's beauty. Arabic nonetheless endured in the Iranian region and everywhere else as the civilization's language of religion, science, philosophy and medicine.

The Iranian contribution was fundamental to the evolution of classical Islamic civilization in every facet of its genius, but it should be stressed that members of all ethnic and religious groups within that wide empire conquered by the Arabs participated in the grand enterprise of intellectual renaissance, of refining and expanding the scientific, philosophical and medical heritage that had come with the conquests and that formed the cerebral core of the high period of Islamic civilization, just as Muslims of all ethnic origins, Arab and non-Arab converts and their descendants contributed to the evolution of Islam as a religion and comprehensive way of life through interpretation of the Quran and the analysis and evaluation of the Prophetic Traditions, equally with legal commentary and articulation of theological and mystical understandings of the Prophetic message.

The corpus embodying the intellectual traditions of high civilization was first put into Arabic by non-Muslims, Christians mainly, and then integrated into an Islamic world view whose nucleus was the Quran, the Prophetic Tradition, and the legal embodiment of Quran and tradition, namely the holy law, Shari'a.

Not all literate Muslims accepted the rational knowledge of non-Arab, pre-Islamic provenance. Paralleling the religio-political divide that separated Muslims into Sunnis, Shi'is and a number of minor sects was another divide, one that separated Muslims into those who valued reason and the rational sciences as a legitimate path to religious understanding and those who did not, who believed instead that the Quran, the Prophetic Tradition and the Holy Law were complete in themselves and that all rival sources should be suppressed as being alien and non-religious.

The studies of the science, philosophy, medicine and mathematics that were patronized by Muslim rulers, and that were later woven into a religio-rational world view, opened the way to the Muslim world's Christians and Jews, whose own particular religious views were framed in the same rational paradigms that derived ultimately from Aristotelian and Neoplatonic sources. On this level the three religions were truly sisters, which made for an easy collegiality. Jews and Christians, as well as some pagans, shared in the endeavors of intellectual life with their Muslim counterparts, with whom they were in close association, often in a teacher–student relationship.

In the lands of North Africa, Egypt, Syria and Mesopotamia, Arabization had reduced Syriac and Coptic to liturgical languages of the surviving Christian churches, the majority of whose members, having accepted Arabic, had converted to Islam by the end of the 10th century. The Christian Egyptians, Syrians and Mesopotamians had few regrets for the demise of Byzantine rule, with its coercive Greek rulers and state-enforced orthodoxy. In terms of religious toleration and taxation, the Arab conquest had come almost as a liberation. With time, the natives of these regions came to regard themselves as Arabs, even those Christians who resisted conversion. In the Arab world, Muslims and Christians today live side by side as Arabs, with some Maronite exceptions in modern Lebanon. Though the Jews were as much as the Christians a part of Arab–Islamic society and civilization in language, lifestyle, and cultural and intellectual orientation, those who remained faithful to their mother religion retained a communal ethnic identity that stopped short of accepting Arabism. In Spain, where Arabism never went as far as it did on the other side of Gibraltar and where Islamization never went as far as it did in Iran, Latin continued as the literary language of the large Christian majority under Arab rule. Many Christians in Spain and Portugal chose to write in Arabic, which they knew far better than Latin, eliciting a bitter complaint from at least one 9th century Spanish bishop.[2] But eventually Arabic faded from the Iberian peninsula with the destruction of Arab political power and the expulsion of the Muslims (and with them the Jews), though not without their leaving behind a hefty intellectual legacy, as seen in the many words of Arabic origin in modern Spanish.

Religion was the primal bond of commitment that held these diverse peoples together across the lands of Islam, whether Arab, Iranian, Turk, Armenian, Kurd, Baluchi, or Berber, and whether they were converts from Christianity, Judaism, astral paganism or Hinduism. As did science in its own realm of endeavor, Islam transcended ethnic and linguistic differences. Arabic, the language of the Quran, of the angels and of the mind of God, was the scriptural and intellectual thread that stitched the many peoples together in mental patterns that, with religion underlying them all, gave unity within a legion of cultural variations.

Notes

1 Iran, having dispossessed itself of the American-supported shah, appears at the moment to be doing exactly this but in face of the most dire American–Israeli threats of war. As will be brought out in the latter part of this study, any move by a Middle Eastern country toward effective self-defense is taken by the West as a threat to its interests.

2 W. Montgomery Watt and P. Cachia, *A History of Islamic Spain*, Edinburgh University Press, Edinburgh, 1965, p. 56.

Acknowledgements

In addition to Martin Dickson, to whom this book is dedicated, I would like to thank William Paterson University for two decades of financial assistance in my research for this book; and also the library, whose interlibrary loan librarians were able to locate and provide me with hard-to-find, centuries-old books in Arabic and Ottoman Turkish.

I should also like to thank the American Philosophical Society for a research grant I was awarded, which helped cover expenses for a year in the National Library of Egypt and the library of the American University at Cairo. I also wish to thank Dr. John McClain, formerly in the Department of Physics at the American University of Beirut, who meticulously went over every sentence I wrote, correcting grammar, punctuation, word choice and errors in the science. Professor George Robb of the History Department of William Paterson University is as well to be thanked for the time and effort he graciously put into reading the book and giving critical and insightful recommendations in organization and in deleting what could be deleted and what could be written in shorter form.

Lastly, I would like to thank Malissa Williams for her uncanny ability to put into type a most messy, illegible, hand-written copy covered with crossed-out lines, a dozen arrows stretching across the scrawl-crowded page indicating where to insert a wild artistry of words within bubbles that might have won a prize in a Jackson Pollack imitation contest. How she did it is a mystery, and I am forever indebted to her for it.

Part I
Islam in ascendance

1 The historical setting of the great age

For a thousand years Muslim arms triumphed over Christendom. Palestine, Syria and Egypt were torn from Byzantine Christian rule within a decade of the first earnest Arab raid in 634. In the east, the Zoroastrian Empire of the Sassanids was conquered. The Byzantine capital was besieged several times, but Constantinople's strong walls withstood the Arab army and navy. A civil war over the caliphate a quarter century after the death of the Prophet Muhammad halted conquests for a few years, in addition to sowing the seeds of the Sunni–Shi'i split in the Community of Believers. In North Africa the mountain Berbers had, for a period, checked the resumed Arab advance but were eventually subdued. Many Berber tribes converted and were enlisted as contingents in the Arab military, making the conquest of western North Africa and Spain a joint Arab–Berber venture.

To later generations of Muslims it would appear that even tribal strife among the Arabs could not stop the banner of Islam from advancing. The triumph of God's soldiers fighting for the true religion was the Divine Plan unfolding in history.

The downside of conquest was the social problems that came with the land. Indigenous converts to Islam were denied equality to Arab Muslims, as if the factious tribes were not problems enough for the rulers of the expanding empire. An Arab family ruling a vast empire in which non-Arab Muslims came to outnumber the favored Arab elite created a condition too fragile to last. The Arab dynasty of Umayyad caliph-kings succumbed in less than a century. The upside of the conquest was the gift of high civilization that came with the territory and that would transform the primitive traditions of the conquerors and their new religion. The era of conquest came to a halt with the fall of the Arab kingdom. The revolution that overthrew it began in Iran.

The new ruling dynasty, named Abbasid after a descendant of an undistinguished uncle of the Prophet, Abbas, who never accepted Islam, was a bitter disappointment to those non-Arab Muslims and anti-Umayyad Arab tribesmen who had joined the revolution. Arab tribal pretension was submerged in a religious universalism of Muslim equality, but the Abbasid promise of revolutionary egalitarianism suffered much in the daily practice of absolutist government. In effect, the Arab aristocracy had simply been replaced by an Irano–Arab despotism. The Abbasid caliphate encountered a long series of major and minor revolts, many deriving from the unresolved social problems that existed before the Arab conquests but in their Muslim setting expressing discontent through Shi'i opposition to Sunni rule. Beyond the issue of political legitimacy, there was not a great deal of religious difference between Shi'i and Sunni; the Sunni community would have been equally happy with a leader from the Prophet's family, if one with the proper qualifications were to gain power.[1]

As for the Abbasid caliphs, rather than setting off to win new conquests for the greater glory of Islam and the dynasty, they were kept busy fending off the endemic uprisings of rebels from every quarter. For glory of a gentler kind, when not exerted to containing revolt, the new dynasty focused on internal development and the peaceful pursuits that went into making a flourishing civilization. A new Abbasid capital was built at Baghdad, where the Tigris comes closest to the Euphrates, a stone's throw from the former Iranian imperial capital Ctesiphon. It was in Iran that the revolution against the Arab Umayyads had begun, and it was in large part Iranians who organized, supported and fought to overthrow them, though it should be emphasized that there were Arab tribes settled in Iran who were as hostile to the dynasty as were Iranian and other non-Arab Muslims and who fought against it just as fiercely.

The Arab caliphate had, to a large extent, modeled its imperial rule along the lines of Byzantine monarchy and administration. The Umayyad capital Damascus had been a Byzantine provincial capital, and Syrian locals who had served in the Byzantine bureaucracy were in turn employed by the new rulers to administer civil affairs for the Arabs, most of whom were off fighting and who, in any case, had little experience in statecraft. The Abbasids, on the other hand, took the Sassanid structure as their model for imperial autocracy and administration.[2]

Baghdad was deliberately chosen for its geographical location. The rich soil between the rivers offered abundant agriculture; the waters around it promised a lucrative long distance trade. Served by rivers and seas: Tigris, Euphrates, Persian Gulf, Black Sea, Caspian, Aral, Mediterranean, Red Sea, the new capital became the hub of international commerce from China to Spain.[3] Baghdad had two names, both symbolic: The Abode of Peace (Dar al-Salam) for the peace and stability the new dynasty would restore to the Muslim community; and The Round City (al-Madina al-Mudawwara) for the geographical and cosmic importance of the new capital. The city's circular shape represented in two dimensions the spherical embrace of the universe. As the universe was ruled by God, its center, the earth, was presided over by the caliph, who was now styled as God's shadow on earth and caliph of God, instead of caliph of the Prophet of God, which was used by the earlier caliphs. Baghdad was the center of the caliph's empire, the center of the earth, the center of God's universe and the most powerful and prestigious city on earth.[4]

Mansur, the founding caliph of Baghdad, made a wise choice, and Baghdad prospered. The caliphs had riches to patronize the arts and sciences, and what came to be known as Classic Islamic Civilization evolved and flourished, giving new life and modes of expression to art, architecture, literature, poetry, music, theology, philosophy, mathematics, astronomy, optics, mechanics, medicine and mysticism. Within a generation of its founding Baghdad had come to embody the collective intellectual spirit of Athens, Alexandria and Persepolis. Gifted Jews, Christians,

Harranian star worshippers and Muslims, wherever they existed in the vast lands of Islamdom and whatever their ethnic origins, were drawn to the new capital, bringing with them the inherited intellectual traditions of Greek, Syriac, Iranian and Indian civilizations. By the middle of the 10th century, these traditions, articulated in Arabic, had been elaborated, advanced and redressed in accordance with the spiritual essentials of Islam, in which form they were transmitted from Baghdad to a constellation of provincial capitals, many of them by then having become autonomous capitals in themselves: Bukhara, Tus, Shiraz, Samarqand, Nishapur, Cairo, Palermo, Cordova, Seville and Toledo, all in their unique way striving to emulate the mother city whose style and magnificence became legendary in its own time of greatness, where in Islamic garb, Greek philosophy thrived alongside Alexandrian and Indian astronomy, mathematics, medicine and the other natural sciences (see Plate 1).

Andalusian Spain became home to a renaissance in the sciences, medicine, philosophy, poetry, music, and architecture. From its center in the Guadalquivir basin, Cordova radiated its brilliance to the cities of the Iberian provinces, as had Baghdad to the Asian provincial cities a century earlier. Toledo, situated at the geographical center of the Iberian Peninsula, would be the rich source from which western scholars traveling over the Pyrenees slaked their thirst for Muslim knowledge in the 12th century.

In the east, the Abbasid caliphs began losing control of their outer provinces in North Africa and central Asia within the first half century of the dynasty's founding, not because of conquest from without but rather because of the lack of principled institutions that could curb the caliph's will when personal whim overtook sound political sense. The empire that was won by revolution from the Umayyads of Damascus, and centralized by the first four of the Abbasid caliphs and their Iranian ministers and bureaucrats, suffered from self-inflicted wounds by some of those rulers who followed. First, the slackening of the caliph's firm hand over his governors frayed the power of central government and its ability to collect provincial taxes. The most famous of the caliphs, Harun al-Rashid (786–809), allowed a favorite general to administer Tunisia as a personal fief, and toward the end of his life arbitrarily divided the empire between two of his sons, a personal act that led to years of destructive civil war, until the victorious son, Ma'mun, at last reunified the empire.[5]

Ma'mun realized that the caliphate, the one and only institution that held the empire together politically, had become unstable, and he searched for a supporting pillar, initially in Shi'ism, then in rationalist theology. Neither came to fruition. His youngest son and successor, Mu'tasim, found it in soldiers, Turkish slaves and a royal bodyguard of thousands. This proved to be almost fatal to caliphal power. Within less than a decade (861–869), the Turkish commanders had seated and unseated four caliphs, murdering two.

The political power of Abbasid Baghdad dimmed precipitously as agriculture and trade suffered under the internecine warfare of the rival chieftains. Shops were looted by soldiers, while the dams, ditches, sluices and canals that irrigated the rich land between the Tigris and Euphrates fell into disrepair. Agricultural output declined drastically, trade was interrupted, taxes stopped coming in from the provinces, and as fiscal shortfall contracted the reach of central government, provinces were lost to governors, generals, Turkish favorites and powerful local families who established their own dynastic rule. Taking advantage of the troubles at the center, provincial revolts of protest by lower levels of society threatened the very existence of the imploding caliphate. At times, the caliph and his Turkish masters ruled no more than Baghdad, Samarra and their environs. A resurgence of caliphal power lasted little more than a generation (869–908) before falling back into the hands of exogenous military chieftains.

This was the price of not building political institutions that could have curbed both the frivolous whims of absolutist rulers and the voracious appetites of military commanders turned politicians. Turkish military domination of what was left of the Abbasid empire became official in 935 when the enfeebled caliph, putting the best face he could on his impotence, conferred upon the leading Turkish chieftain the title *amir al-umara* or commander of commanders, whose office was to "protect" the caliph and the truncated empire. Ten years later, Baghdad fell to a northern Iranian dynasty of Shi'i orientation that took the place of the Turkish chieftains as protectors of the caliph. Caliphal power had been squandered. The Sunni political rulers of the Islamic community had failed in their mission of social and religious cohesion.

The fragmentation of the empire had its upside in the provinces. Weakening at the center allowed taxes previously sent to Baghdad to be locally invested. The affluence that had been Baghdad's for a century was now diffused throughout the provinces. With affluence came the intellectual and artistic creativity of high civilization as new dynasts emulated the courtly life and patronage that had been established by the early Abbasid caliphs. The new cities that were built and the older ones that were expanded became centers of civilization from which learning and cultural refinement radiated outwards. As an example, Tunis under the Aghlabid dynasty, begun by a governor appointed by Harun al-Rashid, was the starting point for the expansion of Muslim arms and civilization into Sicily and southern Italy, with Palermo and Salerno in turn becoming important centers of high civilization. To the west, in Morocco, Fez was built by the breakaway Idrisid dynasty; to the east, in central Asia, the Samanid dynasty of aristocratic Iranian origins arose with its capital in Bukhara, north of the Oxus basin, where the lush valleys of Oxus and Jaxartes made for rich agriculture, the revenues of which were supplemented by the lucrative trade carried over the silk route that traversed Transoxiana.

It was in Bukhara and the lands of the Samanids that the remembered traditions of the old civilization of Iran were revived. Situated between Iran and the lands of the Turks, the Samanids, recalling with pride the civilized traditions of Sassanid times, initiated a Persian literary revival, which, during the course of the 10th and 11th centuries, evolved into a rich strain of Islamic literature in Persian, alongside Arabic. In the Samanid court of Bukhara, the poet Firdawsi began composing Iran's great national epic, *The Book of Kings*, or *Shahname*. It is not the kind of poem devout Muslims would like. Though written in Arabic script and three times as long as *The Iliad*, Firdawsi's epic contains no Arabic loan words that had worked their way into the spoken Persian of the time, nor is Islam ever mentioned. Greatness, glory and heroism belong to the Sassanid kings fighting to save Iranian civilization from the barbarous Turks of the plains. Cast in the drama of the light of Iran, land of the Aryans, battling the darkness of Turan, land of the Turks, Firdawsi's classic epic renders the Turko–Iranian struggle in the dualistic hues of Iranian Zoroastrianism.

Owing to courtly patronage of the Samanids and their successor dynasties in Iran and central Asia, Persian became the second Islamic language of literature and learning. Concomitant with the literary revival, a Persianate form of Muslim civilization evolved, apparent in the style of script, architecture, dress and cuisine as quite distinct from the Arabic form to the west.

In 1055, the Iranian Shi'i "protectors" of the shadow caliph in Baghdad were replaced by a powerful Sunni Turkish tribal ruler who restored the political dignity of Sunni Islam and was given the high sounding title of "sultan," protector and military commander of the caliph and the Sunni community. The caliphate remained Abbasid in name and religious in function, while the ruling dynastic sultanate of Turks, called Seljuk, took military and administrative command of the empire that they created by conquest, extending from Central Asia to Syria. From then on, military and political power east of the Nile would rest in the hands of various Turkish dynasts, who played the part in the central and eastern lands of Islam that the Berbers played in Morocco and Andalusia.

While at the western end of the Muslim world Toledo, Seville, Saragossa and Cordova were falling to the conquistadors during the 13th century, in the east, the brilliant cities of Shiraz, Nishapur, Bukhara, Khiva, Samarqand and Baghdad were being leveled and depopulated by an incursion of pastoral tribal armies from the central Asian steppes, the Mongols. In 1258, Hulagu Khan brought an end to the once-glorious Abbasid caliphate by having the last of the line trampled under the hooves of his cavalry.

The destruction of Baghdad has been taken to mark the end of the great period of scientific study in Muslim civilization: the Golden Age. In terms of number of scientists, originality, productivity and breadth of interest, this putative age in fact ended a century and a half before the destruction of Baghdad; during the centuries following the Mongol incursion, a considerable number of highly creative men produced brilliant original science, limited primarily but not exclusively to astronomy and mathematics. Great astronomer–mathematicians appeared sporadically in Islamdom right up to the 16th century. It could be argued that the astronomy and mathematics produced after the Mongol destruction of Baghdad amounted to somewhat of a renaissance, an after-glow silver age of the earlier golden one.

Across this broad historical tapestry the Muslim experience in scientific originality is written. The period 770 to 1100 was one of the great scientific and philosophical renaissances prior to modern times. Mongol and post-Mongol rule in Iran and Central Asia gave Muslim civilization other moments of high scientific creativity. There is strong evidence that the influences of the innovative astronomy and mathematics patronized by those post-Mongol rulers reached right to Copernicus and the Scientific Revolution in the West.

Assimilating the scientific traditions

The question is, how did high science come into this society whose ethos originated in a cultural fusion of primitive tribal desert warriors and semi-urbanized merchants bound in the expansive energy of religion and conquest? Why would science be so attractive to a community of tribal merchants and warriors suddenly turned caliphs, conquerors and tax collectors that they would actively appropriate, naturalize and expand the science, medicine and mathematics of their classical Greek, Hellenistic and Byzantine predecessors?[6]

It would seem obvious that the most immediate reason was the practical benefits the knowledge could offer. Indications are that medicine, astrology and alchemy were indeed the earliest sciences to be explored: the first to attend to the health of the ruler and his family, the second to establish a propitious relationship with the stars in planning military campaigns, and the third to set up an enterprise whose economic benefits could supplement the royal treasury's gold supply. The founder of the Umayyad caliphate had a physician and an astrologer, both Christians learned in the Greco–Syriac traditions of their respective professions. An Umayyad prince is claimed to have studied alchemy in Egypt and promoted the translations of Greek texts in the late 7th century.[7]

The translations of medical, astrological and alchemical texts made during the Umayyad period were few, sporadic and poorly made. This was to change when the Abbasids came to power and established Baghdad as the new capital.

A bibliographical work of the 10th century contains a story, possibly apocryphal, that reveals an early connection between the Abbasids and Greco–Syriac science. At the heart of it is the caliph's health. Mansur, founder of Baghdad, is said to have complained of a stomach ache that his doctors at court could not relieve. He was told of a Christian physician of great fame working and teaching in a hospital in a place called Jundi Shahpur, near the old Sassanian capital, Ctesiphon, not far from Baghdad. The physician's name was Jirjis Bakht Ishu' (George Joy of Christ). Jirjis, a Nestorian Christian of Syrian origin who had translated several medical works of Galen into Syriac, the local language of learning that is a close kin to Arabic and written in a very similar script, was chief physician of the Jundi Shahpur Hospital.[8]

Founded by the Sassanian ruler Shahpur (241–271 AD), Jundi Shahpur had by the 6th century reign of Anushirvan the Great become known for its hospital and medical training center. The hospital was staffed by learned Nestorian physicians who had fled religious repression by state-enforced Byzantine Orthodoxy and been offered refuge in Iran. These Nestorian physicians tended to the royal family's health and made Persian translations of Greek scientific and medical texts that were used to train Iranian physicians. As physicians, their studies included logic and natural philosophy. Often their translations were from Syriac versions of Greek originals, Greek and Syriac being the languages of scholarship in Syria and Mesopotamia. Jirjis Bakht Ishu' was director of Jundi Shahpur when called to Baghdad to cure the caliph's stomach ailment. Satisfied with his treatment, Mansur made Jirjis his personal physician. Bakht Ishu's Nestorian descendants would be physicians of the caliphs for the next 150 years.

The philosophical knowledge of classical Athens and science and medicine of Hellenistic Alexandria made it to Baghdad in a roundabout way, via Nestorian scholars who had sought refuge in Iran from Byzantine religious persecution.[9]

Before the advent of Islam, science and philosophy had nearly expired from Roman neglect and Christian persecution. Romans were indeed not much for science, and fervent Christians burned books, destroyed pagan temples, smashed statues and persecuted philosophers, but the Hellenistic tradition in medicine, logic, music, mathematics, optics, astronomy, along with Aristotelian natural philosophy and Neoplatonic metaphysics, continued to be actively pursued in the Byzantine empire, regardless of the aggressive efforts of Christian emperors, patriarchs and theologians to curtail intellectual exploration in the cause of political and religious unity. Disregarding the inhospitable intellectual climate, men like Proclus, Johannas Philoponus, Themistius and a host of others in the 5th and 6th centuries carried on their scientific and philosophical studies, though some suffered the ultimate punishment, such as the Neoplatonist Hypatia, who was dragged from her home and torn apart in the streets of Alexandria by a crazed mob of Christian fanatics, making her the first woman to join Socrates in the hall of philosophical martyrdom.

By aggressively fostering religious unity in the hope of solidifying political unity, Byzantine emperors tolerated only the Orthodox version of Christianity. It had not worked for Constantine the Great, nor for those rulers who followed him. Those Christian communities outside the Orthodox persuasion, such as Monophysites, Nestorians, Arians and other regional sects in Syria, Egypt and North Africa who resented their Greek-speaking Roman rulers, were pressed in various ways to conform. The forceful effort for unity coming from Constantinople caused severe collateral damage to the old schools of philosophy. Because of the philosophical penchant of asking questions and coming to resolutions that were more in harmony with the principles of Aristotle, Plato and Plotinus than with the theology that came out of the Councils of Nicea and Chalcedon, the emperor Zeno closed the school in Athens, and Justinian closed the one in Alexandria. Religious and intellectual intolerance served to push natural philosophers to regions further afield, eastwards from the centers of imperial power. They brought with them what they could of their libraries of Greek and Syriac manuscripts and set up schools to continue the intellectual traditions their forefathers had inherited from the Greeks. Adept in logic, optics, mathematics, music, metaphysics and astronomy, the migrating scholars were a virtual Byzo–Syriac college of arts and sciences.

As seen by the emperors and patriarchs, the chief threat to the Orthodox belief system, the hinge of imperial unity, was metaphysics. Because of its speculations on the divine, the metaphysical culprit had to go. Suppressing metaphysics meant suppressing philosophy in general. Metaphysics was the crown of philosophy, the control center as it were. The principles of metaphysics gave place, purpose and unity to the universal system that related nature, man and soul under God, the first cause and universal agent that presided over the cosmic whole. Metaphysics provided the unifying paradigms and was inseparably joined to natural philosophy and medicine, as head to body, an organic and systemic one. Philosophy, logic and medicine were sides of the same prism. As a physician, the philosopher healed the physical body; as a logician, he disciplined the mind; and as a metaphysician, the soul. Body, mind and soul were states of being in a unified system of knowledge. To prohibit metaphysics was to cripple the intellectual and spiritual system that made sense of nature, man, God, the universe, and the relationships uniting them. To attack metaphysics was to attack the soul of philosophy. The Arabic word *hakim*, one possessing wisdom, even today conveys the dual sense of physician and philosopher.[10]

Knowledge was perceived as a whole. Being well-versed in the principles of the whole gave moral, esthetic and philosophical, and hence ethical and religious, meaning to knowledge, thereby elevating knowledge to wisdom. Otherwise, the parts fell into isolated pieces, and as such failed to ennoble and enlighten, regardless of how thoroughly any one part of the whole might be mastered. Virtue resided in wholeness, in the integrity of knowledge. Without metaphysics there could be no natural philosophy; the various sciences would become bodies, hollow shells, devoid of spirit, direction and intellectual substance. Not until the century of Galileo, Newton and the secularizing revolution of knowledge did natural philosophy cut loose from its Greek-given metaphysical foundations to exist on its own.

In protest to the religious imperialism of Constantinople, Syrian Nestorians claimed a vestige of cultural autonomy by clinging to their version of Christianity. Resent though they did their Greco–Byzantine rulers, Syrian scholars nonetheless cherished the intellectual heritage of their pagan ancestors: Plato, Aristotle, Archimedes, Euclid, Hippocrates, Galen, Ptolemy and Appolonius, to name but the major figures. Their works were studied, commented on, copied and translated into Syriac by Monophysite and Nestorian bishops and monks who were actively formulating a rationally structured theology, independent of Constantinople. As Byzantine pressure came more and more to curtail their studies, moving them eastwards into northern Syria and the highlands of upper Mesopotamia, they founded monastery schools to educate their young and perpetuate their religious and intellectual heritage. Edessa in northeast Anatolia became an important center of Nestorian learning. A medical school was founded there, where the study of non-Orthodox philosophy and theology accompanied medicine.

But it was not long before the reach of imperial officers forced the Edessa school to close, causing the learned Nestorian community to flee to upper Mesopotamia where they resettled in Nisibis, which soon became a center for philosophical and medical studies, their scholars busily translating Greek texts to Syriac right up to the 7th century and the Arab conquests. Edessa and Nisibis were the principal centers of Nestorian learning, with lesser ones in Syrian Seleucia, south Anatolian Antalya and Amida. The farther east of Constantinople the Nestorian scholars settled, the weaker the hand of imperial control over them. Jundi Shahpur in Sassanian Iran was a welcome refuge.

Bordering the Sassanian Empire to the east was India, home of another high civilization with a tradition of science, medicine and mathematics, whose influences reached Jundi Shahpur and the Nestorian scholars residing there. It was in Jundi Shahpur that the collection of Indian folk stories known as the *Pancatantra* was translated by Nestorian scholars into Syriac, thanks to the royal patronage of the Sassanian shah. The stories were later translated from Syriac to Arabic as *Kalila wa Dimna* when Baghdad had become the world's leading center of learning. The *Pancatantra* was also translated into Greek, where it was known as the *Bedpai Fables*. Originally composed in Sanskrit, it had gone in turn to Syriac, Greek and Arabic. Books of Aristotle's natural philosophy were also translated to Syriac at Jundi Shahpur or retranslated.

The scientific, philosophical, medical and mathematical studies and translations done by Nestorians and Monophysites, joined by some Syrian theologians of the Greek Church in Damascus, formed the earliest channel through which Greek learning, much of it Christianized and in Syriac, entered Muslim society of the Umayyad period. Three of the most important (of the many) Syrian Christians who contributed to this intellectual enterprise were a Monophysite, a Nestorian and a Greek Orthodox.

The Monophysite was a bishop, George, called "Bishop of the Arabs" (d. 724) who made translations from Greek into Syriac, and who wrote a commentary on Aristotle's *Organon* at his bishopric in Mesopotamia. The Nestorian was Sebokht, bishop of Nisibis, who did the same for Aristotle's *Analytics*, and who also wrote on astronomy and the Indian number system, which through him was transmitted to mathematicians living in Syria and Iraq during the early Umayyad period. The Greek Orthodox was John of Damascus (d. 750), a theologian of the Church and a trained logician who critically debated religion with pious Muslims who were at the time barely beginning to acquaint themselves with logic and metaphysics.

The interaction introduced religious-minded Arabs – those who were devoted to studying the Quran and prophetic tradition and had not joined the conquests to get rich – to the logic of theology and provoked them into thinking through a theological defense in answer to the critical and often embarrassing questions John and other Christian theologians posed to their Arab hosts in their debates at the Umayyad court in Damascus and elsewhere. With Greek learning having been thoroughly diffused throughout Syria and Iraq a century and more before the Umayyads established themselves in Damascus, a Christian–Muslim theological confrontation was inevitable.

The early intellectual interactions could on occasion be abrasive. Arab interlocutors, knowing little of Aristotelian logic and metaphysics with its intricacies of essences, substances and qualities, were at a disadvantage. Faring poorly against their theologically trained Christian antagonists, pious Arabs suffered bruised sensitivities. In early Abbasid times, the third caliph, Mahdi, acutely sensing the need for logic, had the Nestorian patriarch Timothy I translate into Arabic Aristotle's *Topics*. The caliph studied it well and engaged Timothy in a cordial debate over the Christian belief of God having a son.[11] While Muslims considered Christians borderline polytheists, Christians considered Muslims to be anti-trinitarian Ishmaelites, that is, Judaized–Christian heretics. Out of such debates and somewhat misinformed branding, sometimes antagonistic, sometimes cordial, arose early in Islamic theology.

Christians had been welcome at the caliph's court since the early Umayyads. The dynasty's founder, Mu'awiyya, had a physician, a court poet, an astrologer and a favorite wife, all Christians. He employed Christians experienced in state administration. Christian theologians frequented the court. One esteemed court visitor would become an Orthodox saint, the aforementioned John of Damascus. The tolerant atmosphere of Umayyad rule that allowed Christians in Syria the freedom to build churches (but not each and every Umayyad Caliph was so tolerant), also allowed Christian–Muslim interaction at the highest level.

Could this theological interaction have reached into natural philosophy? The question of Arab scientific interest during Umayyad times is problematic. An Umayyad prince, Khalid ibn Yazid (d. 704), grandson of Mu'awiyya and son of the second Umayyad caliph, was claimed to have encouraged translations of scientific works from Greek to Arabic and to have lived in Egypt where he studied alchemy.[12] Any Arab interest in alchemy may have been related to the dynasty's innovation of minting gold coins, one of the reforms of the caliph abd al-Malik (d. 705). Until then, the Arabs had used Byzantine and Sassanian coins. Rather than in alchemy for itself, Arab interest may have been in the metallurgy of detecting alloys of gold,[13] but recent scholarship offers no support of Khalid's being involved in alchemy or any other of the sciences.

Another outside figure, Jabir ibn Hayyan (the famed Geber of Latin tradition), who lived shortly after Khalid ibn Yazid and whose fame as an alchemist would become as great in the West as in the East, is considered the father of alchemy in the Muslim tradition. Both figures, however, Khalid and Jabir, are steeped in legend and are as illusive to the lens of history as the alchemical marvels they are said to have achieved are to rational comprehension.

According to the generally accepted interpretation of Umayyad history, the germ of intellectual curiosity neither included natural science nor reached up to the ruling class as patrons. Umayyad rulers are thought, for the most part, to have been far too occupied with conquest, keeping the tribes together, administering and taxing the expanding empire – in addition to applying themselves to the diversions of slave girls, music, poetry, drinking, feasting, hunting and building palaces in the desert – to have had any time or inclination to study or patronize science. According to this interpretation, it was not until the Umayyads were overthrown, the focus on conquest was brought to an end, and Baghdad was built that the caliphs began to give serious attention to Sassanid-style royal patronage.

But close analysis questions this broad treatment of Arab intellectual genesis. Every river has a source, and the flood of science translation that characterized Abbasid times may well have its origins in the Arabization reforms of abd al-Malik. According to a recent interpretation,[14] the Greek-speaking Syrian Christians who had lost their positions in the bureaucracy to Arabs because of the reforms set themselves to regaining their status by training their children in Arabic and the fine art of translation. The scientific, medical and philosophical translations of 9th century Abbasid Baghdad followed a century of preparation and were the work of those Syrian descendants who had staffed the Umayyad ministries prior to 700. How else to explain the scientific and linguistic expertise found in the early Abbasid period translations, exemplified by the precise technical terminology in the early 9th century Arabic version of Ptolemy's *Almagest*? Going from Greek to

Arabic, the Umayyads provided a century-long training period in the art of translation, the fruits of which came to maturity in early Abbasid times.

According to this interpretation, called by its author "an Alternative Narrative" (as opposed to the generally accepted "Classical Narrative"), the high positions held by the earlier generations of Syriac-speaking Christians who knew Greek and monopolized the Umayyad ministries (*diwans*) were regained by their more educated descendants, competing among themselves to become the physicians, astronomers and translators serving the dominant Arab elite. The spirit of competition that demanded accuracy and expertise of ministerial translation under the Umayyad rulers during and following abd al-Malik's reorganization was transmitted over the course of a century to the educated elite of the Greco–Syriac community, from which came the talent required to render Greek science in Arabic translation. Otherwise, the fine translation made of the *Almagest* by the Syriac Christian Hajjaj ibn Matar would hardly have been possible. Competitive demands of refinement in the art of translation were at the same time accompanied by a competitive demand for increased exactitude in the science being translated, as seen in the more precise re-measurements of values given by Ptolemy for precession, solar apogee and the inclination of the ecliptic, which had been made during the first half of the 9th century. More precise methods of measurement were also devised.

The translation of Greek science was only a part of this Muslim scientific renaissance in the making. Concomitant with the Jundi Shahpur and Baghdad connection occasioned by Mansur's apocryphal stomach ache was the serendipitous arrival of an Indian embassy in Baghdad bearing gifts for the caliph, one of which was a copy of the book of astronomy called the *Surya Siddhanta*, known also as the Sindhind, written by the 7th century Indian astronomer Brahmagupta. Mansur had his Iranian court astrologer/astronomer Muhammad al-Fazari learn Sanskrit to translate Brahmagupta's astronomy. Since Greek astronomy was still relatively unknown to Muslims, the Indian system became the basis of early Abbasid astronomy.

The confluence of texts from East and West in their different languages made for wide and often enriching diversity in early Baghdad's scientific activity. At the same time that astronomers in Mansur's court were basing their horoscopes and celestial predictions on Indian methods, the Jundi Shahpur connection was giving Muslim medical interest a Greek orientation. With Indian astronomy came the Indian mathematics of Aryabhata (5th century AD) and Brahmagupta, solidifying Indian scientific influence in Islam for a century. Not until the beginning of the 10th century, by which time the great bulk of extant Greek science had been translated into Arabic and a generation of Muslim astronomers had come to maturity possessing the critical knowledge to appreciate the power and elegance of Ptolemaic astronomy compared to the Indian systems, did the latter give way to the former.

For roughly two generations, from the time of Mansur in the 760s and through the reigns of his successors, Hadi, Mahdi and Harun al-Rashid, Indian, Greek and Syriac scientific texts were being translated into Arabic, though in a flawed and desultory manner. Translations were most usually commissioned by the caliph, though the Barmakid dynasty of princely viziers who served the caliphs from Mansur to Harun, and who rivaled them in wealth and emulated them in patronage, also commissioned translations. One of the Barmakid viziers had Euclid's *Elements* translated. He also commissioned an Iranian Jew, Sahl Rabban al-Tabari, to translate the *Almagest*, "the Greatest," the abbreviated Arabization by which Ptolemy's great work became known in the Muslim East and later in the Latin West.

Translation work in Baghdad was with few exceptions done by Christians, as it had been in Jundi Shahpur. The Christian communities in the Fertile Crescent had been translating Greek texts for centuries in the Nestorian and Monophysite bishoprics and monasteries and were the midwives of scientific transmission when the Arabs took over. There was no pressure on Christians or Jews to convert to the religion the Arabs brought with them, though some did. Sahl Rabban al-Tabari's son did, taking the name Ali, while continuing in his father's career translating scientific and medical texts to Arabic for the Abbasid court. Conversion was in fact an important entry through which scientists first started coming into Islam, though most men of science who served the caliphs held to the faith of their ancestors, be they Christians, Jews or pagans. Ayyub (Job) of Edessa was a Christian philosopher, physician and naturalist who taught in Baghdad in the early 9th century and translated many Galenic texts into Syriac and Arabic for the Abbasid court. Hunayn ibn Ishaq, his son Ishaq, the Bakht Ishu' family and Qusta ibn Luqa, to name but a few, were Christian translators and writers of scientific treatises patronized by the caliphs. Thabit ibn Qura was a star-worshipping pagan from Mesopotamian Harran. The caliphs and their ministers were more concerned with good translations than conversions, just as the Umayyads before them had been more concerned with having high bracket tax payers than converts.

Being provided with good translations was a central challenge to the delivery of science from Greek and Syriac to Arabic. By the early decades of the 9th century, as scientific interest was gaining in courtly circles and demands for greater precision in translations grew, it became evident that the early translations left much to be desired. Many of the earliest translations were defective. There were various reasons for this: poor manuscripts to work from, lack of a uniform methodology, translators who were not adept in the subject they were translating, and lack of scientific terminology in Arabic. Nestorian, Jacobite and Monophysite physicians and scientists of the early 9th century were dissatisfied with the flawed Arabic translations they had to work with and desired better ones. The problem was answered when the caliph Ma'mun (813–833) founded his House of Wisdom.[15]

Ma'mun, the seventh Abbasid caliph, had fought his way to power with the support of Iranian forces under his Iranian general, Tahir, after the caliph's father, Harun al-Rashid, had divided the empire between his sons Amin and Ma'mun. Toward the end of his reign Ma'mun founded his House of Wisdom, one of the greatest centers of scientific and medical study in ancient and medieval times, comparable to the museum and library of Ptolemaic Alexandria and the Sassanid medical center at Jundi Shahpur. (Something like the House of Wisdom had existed in Sassanid times, a royal library the name of which in Arabic was exactly that,

House of Wisdom.) Ma'mun's version was not precisely an institution of translation; it was much more than that, but the great translators of Greek material into
Arabic were closely associated with it and did their translation work in accordance
with the demands of that climate.[16] By the time of Ma'mun, the best of Jundi
Shahpur's Nestorian physicians had followed Bakht Ishu' to Baghdad, in particular
abu Zakariyya Yuhanna ibn Masawayh, who translated several medical works into
Arabic and as the first director of Ma'mun's institute, directed scientific interest
toward the Greek tradition.

How the House of Wisdom came to be is anecdotally explained by the 10th century bio-bibliographer ibn al-Nadim (d. 995). Ma'mun dreamt that Aristotle visited
him in his bedchamber. The philosopher introduced himself, then he and the caliph
had an uplifting exchange about "The Good." It was this dream that resolved
Ma'mun to devote himself to the good, in pursuit of which he founded The House
of Wisdom.[17] If Mansur's stomach ache and the Indian embassy that came to his
court promoted an interest in translating medicine and astronomy, Ma'mun's
dream put the pursuit of translation and scientific study on an institutional basis.
What the story of the dream conveys historically is that translation and the study
of science were now in full flower and spreading.[18]

Owing to both the continuing intellectual vitality of the Christian community
and the liberality of its Muslim rulers who brought learned Christians into their
service to elevate Arabic as a language of learning, Syriac translations and the
Syriac tradition of scholarship continued for centuries after the Arab conquests.
Succeeding Masawayh as director of the House of Wisdom was another Nestorian
physician, Hunayn ibn Ishaq. Hunayn translated, or directed Arabic translations
of, hundreds of Galenic texts from Greek and Syriac. He himself is credited with
translating 95 treatises on the eye from Greek to Syriac and 39 from those languages to Arabic. He and his associates made the classics of Greek medicine available to Arabic readers. In addition to translating, Hunayn wrote medical treatises
of his own. Of the many Christians who contributed to making translations and at
the same time composed original scientific works in Arabic, Hunayn was the most
prodigious. Besides his translations and original treatises on optics, the structure
of the eye and the diseases that afflict it, Hunayn established a critical method that
rendered Arabic translations more faithful to the original Greek or Syriac.[19]

Making a sound translation of a text was no trivial undertaking. Knowing that
to make a good translation required having a good copy to translate from, Hunayn
would gather as many copies of a text as possible and select from the pool the four
or five he considered to be the best and most complete.[20] Having multiple copies
was necessary. Copyists made errors, they garbled passages, left out words, lines,
pages and even chapters of a book. Since a copyist often had no understanding of
the subject he was copying, all kinds of confusion arose. Having multiple copies
of a text was absolutely necessary for a decent translation.

In order to obtain the copies, and to find books that were known to exist but were
not found in the House of Wisdom collection, scientists and government agents
were sent to the older centers of learning to buy manuscripts. The Christian scientist and translator Qusta ibn Luqa is said to have gone as far as Constantinople and

Armenia in search of texts. To ensure the accuracy of the text to be translated, Hunayn's method was followed. From the collected pool of copies, the very best were selected. A copy would be given to each member of the team and together they would read it out word by word filling in gaps, correcting copyist errors and deciding which reading made the most sense when the copies failed to agree. Once a sound Greek or Syriac copy had been constructed, it would be translated into Arabic.

This introduced another problem: finding a word in Arabic that conveyed the sense of a scientific or technical word in Greek. The kinship between Arabic and Syriac helped provide Arabic equivalent to a Greek terms. Since Syriac translations of Greek works had been going on for generations, Syriac texts were rich sources in bringing the Greek corpus into Arabic. But often Hunayn had to create Arabic equivalents to the Greek, which demanded a thorough knowledge of the latter language. With Hunayn at the lead, the scholarly corps of Christians serving their Muslim patrons enabled Arabic, a language entirely innocent of science, medicine and philosophy until well over a century after the advent of Islam, to express the corpus of rational knowledge that had been built up by Greeks, Syrians, Iranians and Indians over the course of a millennium. The work of translation and philology was a long process. Hunayn's work as translator and director of the House of Wisdom was continued by his son, and then his grandson, paralleling the Christian Bakht Ishu' dynasty of physicians in Abbasid service.

Perfection was an abiding virtue of the Nestorians and other scholars who joined in the work of expressing science, philosophy, medicine and mathematics in Arabic. Works were translated and retranslated as the methodology improved along with the construction of Arabic nomenclature. Euclid's *Elements*, for example, was retranslated by Hunayn and was worked on again by the Harranian astronomer Thabit ibn Qurra. Ptolemy's *Almagest* was also retranslated, receiving a total of four Arabic translations and recensions in the 9th century.

In the 9th and early 10th centuries, scientific translation by Christians and assimilation by Muslims, as well as by non-Muslims, progressed hand in hand. More and better translations of Greek and Syriac texts were made available to a new generation of science enthusiasts who came from all the religious communities. The scientific culture taking shape in Baghdad was an enterprise that transcended communal identity. Among the early scientists were Arabs, but more were converts or descendants of converts from the subject peoples, and a good number of them were Iranians. Jews and pagan Harranians were also creative participants.

The enthusiasm for science in the early Abbasid period extended beyond the circle of royal patronage. Emulating the caliph, wealthy viziers sponsored translations. Some had their own observatories built, such as the Barmakid family, who served as viziers during the first half century of the Abbasid Caliphate. The wealthy Iranian family called the Banu Shakir, sons of Shakir (actually two sons and their father) funded translations. Practicing astronomers, they figured prominently in the formative period of Muslim science.

Within half a century of the founding of the House of Wisdom, the greatest part of the extant corpus of science, medicine and philosophy in Greek and Syriac had been translated into Arabic. Not all the achievements of Hellenistic science had survived. Several important treatises by Archimedes had not, nor had many of those by lesser scientists. Some works not found in the centers of learning that had come under Muslim rule had been preserved in Constantinople and so were unavailable to the translators (though they would become available to the Latins in the 15th century, as the Ottoman conquests were bringing Byzantine civilization to an end).

The Greek legacy that Muslims inherited via Syriac formed the backbone of what was to become a scientific renaissance.[21] Coming a millennium after the great burst of Hellenistic creativity, the Muslim renaissance in science is often referred to as Arabic, just as Hellenistic science is Greek. The language of science in Muslim civilization was indeed Arabic, but peoples other than Arabs contributed to the achievement. To call it Muslim is also a misnomer, as it would exclude the contribution made by Christians, Jews and pagan Harranians. Science transcended race and religion. This tremendous creative undertaking, which endured for several centuries, employed scholars of all faiths and peoples from all the far-flung lands under Muslim rule. Within the scientific community there were no second class members. Rank and rivalry were expressed in terms of systems, texts, ideas and personalities. The different cultures, religions and languages from which the scientific sources came, and the reverence in which their authors were held, made for tolerance.

It has been mentioned how the Indian astronomy of Brahmagupta's *Surya Siddhanta* was introduced to the Abbasid court at the time of Mansur. For a century the Indian system and its planetary models formed the basis upon which Abbasid astronomers learned and practiced their profession. It took that long before most, but not all, astronomers discovered the superiority of Ptolemaic models and methods of computation and discarded the Indian system. Only gradually did the Ptolemaic system come into its own, and even more gradually did Indian methods fade out. The translations of Greek astronomical works commissioned by Harun al-Rashid and the Barmakid viziers, followed by the more rigorous translations made in the House of Wisdom and culminating in the work of the Harranian astronomer al-Battani (d. 926), firmly established Ptolemy as the new basis of Muslim astronomy.[22]

Battani is a good case in point of the non-racial and non-denominational character of science. He was neither Arab, Muslim, Christian nor Jew, but like his earlier coreligionist Thabit ibn Qurra, a pagan hailing from Harran in upper Mesopotamia, a traditional center of astronomical study, even before the days of Edessa and Nisibis. Like any Jew or Christian, Battani was free to adhere to his ancestral astral religion with roots in Babylonian times.

The relationship between religion and the starry heavens made for an enduring tradition in Harranian astronomy. When Islam came on the scene, the pagans of Harran could have been confronted with the dire choice that all polytheists faced: conversion, expulsion or death. This was more of a religious fiction than historical fact. The Arabs were more interested in living tax payers than dead pagans. In addition, the astrology that was the spiritual core of Harranian astronomy was of great interest to the new rulers, enough so that a pious fiction was adopted to resolve the dilemma. The pagan star worshippers were given the name of a people mentioned in the Quran, Sabians, who like Jews and Christians were accepted as monotheists, "People of the Book." As simple as that.

The importance of the Harranian astronomical tradition in the rise of astronomy in Islamdom can be traced back to Thabit ibn Qurra, who, like his contemporary Hunayn ibn Ishaq, was a primary figure in the legion of non-Muslim scholars who built the Arabic scientific corpus. A prevailing tolerance allowed Christians, Jews and other non-Muslims to live their lives and prosper, son following father in the family's career line. As the son of Hunayn followed in his father's profession and became director of the House of Wisdom, and the son of Jirjis Bakht Ishu' did the same as the caliph's personal physician, so too did Thabit's son, Sinan, follow in the footsteps of his father as a Sabian and a scientist, all of them following the family tradition of religion and profession while serving the caliph's scientific enterprise in Baghdad.

And if there was room for pagan star worshippers in the scientific community, there was also room for professed atheists. The famous Iranian medical authority abu Bakr Zakariyya al-Razi (d. 925) was born a Muslim, his family having converted from the ancestral religion when conversion to the dominant religion was becoming more and more the trend among the urban strata of society after almost three centuries of Muslim rule. But al-Razi left the scriptural belief of Islam for a belief system expressed in science and reason and made no secret of it. He did not have to, because he felt no threat. The civilization was intellectually rich, self-confident and wide-minded enough that one could ignore religion, even pass it off as a crutch for the mentally dependent and live without fear of *fatwas* of heresy, as long as one was not too publicly antagonistic and insistently or overly defiant in challenging authority.

Being under the wing of royal patronage, most ranking scientists, physicians and philosophers were, it is true, somewhat insulated from mainstream public prejudices, but the tolerance of rulers had a softening effect on society in general. Tolerance at the highest level of government opened the way to assimilation of the intellectual traditions of high civilization. Note has been made of the caliph Mu'awiyya's Christian physician, astrologer and court poet. A century later, a Jew, Ma-Sha' Allah, was one of the astrologer-astronomers in Mansur's court to cast horoscopes in determining the most propitious date to break ground for the new Abbasid capital of Baghdad. A Greek Christian from Baalbak, Qusta ibn Luqa, a primary scientist and translator who has been mentioned as midwifing Greek science into Arabic, was commissioned by the Abbasid court to seek out and purchase medical and scientific manuscripts.

Some Muslims resented the favored position the caliphs bestowed upon the scientific and medical consortium of Baghdad's Nestorians, Jacobites, Greeks, Jews, Harranians and Monophysites. A plaintive joke among Arabs in the 9th century was that to be a doctor one had to be a non-Muslim and speak Arabic with a foreign accent.[23]

In addition to their translations and original scientific writing, Eastern Christians served as personal teachers to Muslims, usually converts from the subject peoples who had been gradually going over to the dominant religion. The philosopher and theorist of music al-Farabi (d. 950), known in Islam honorifically as "The Second Teacher" (Aristotle was the first) studied under the Nestorian abu Bishr Matta, a chief logician of his day and famed for his commentaries on Aristotle. Matta had been educated by Jacobite monks at Dayr al-Qunna, a Nestorian monastery and center of learning southwest of Baghdad that was functioning long into Islamic times. Yuhanna ibn Halan had been another of Farabi's Christian teachers. When he had imbibed and digested the philosophy and sciences of the Greeks from his Christian tutors and made a name for himself with his writings, Farabi in turn became the teacher of a Jacobite Christian from Takrit in central Iraq, Yahya ibn 'Adi (d. 975), who would gain high scholarly prestige in his own time as a logician and translator from Syriac to Arabic. Yahya ibn 'Adi taught the famous Muslim historian of religions and science, Mas'udi (d. 957), author of the Arabic classic *Muruj al- Dhahab* (*Fields of Gold.*) Christian–Muslim teacher–student relationships were not restricted to scholarly circles. Ibn 'Adi also taught one of the sons of a famous Abbasid vizier, Ali ibn 'Isa (d. 1001). Another of his students was the Iranian Mu'tazilite theologian abu Hayyan al-Tawhidi (d. 1010). The chain of logical studies intellectually interlinking Judaism, Christianity and Islam through teacher–student relationships continued for generations. One of the teachers of the famous physician and philosopher ibn Sina (Avicenna) was the Christian physician and logician abu 'Isa ibn Yahya (d. 1010).

The lineage of Christian physician–philosophers as teachers to Muslims lasted right up to the end of the 11th century, over four centuries after the Arab conquests. One of the sons of the Bakht Ishu' family of Baghdadi physicians, Gabriel, was chief physician and medical teacher at the al-Adud Hospital at the end of the 10th century. Named after one of the rulers of the Buwayhid dynasty, the hospital was well staffed with Christian doctors. One of them was a Greek, known by his honorific name Nazif al-Nafs, Clean Soul, who translated medical texts from Greek. More well-known was the Christian physician, scientist and philosopher ibn Butlan (d. 1068).

A sign that the tradition of high Christian scholarship was coming to an end in the late 11th century is the career of Yahya ibn 'Isa (John son of Jesus, d. 1100). As a youth he wanted to study logic but could find neither a Muslim nor Christian teacher in the whole of Baghdad to instruct him. The Muslim tradition of logic and philosophy in Arabic had been going strong since the early 10th century but was losing ground by 1100. Yahya ibn 'Isa at last found a Mu'tazilite theologian who agreed to teach him logic, and under his teacher's influence, he converted to Islam taking the name of Ali. The story of Yahya ibn 'Isa could be taken as a metaphor of both the social pressure put on non-Muslims to convert to the dominant religion and of the coming to its end of the great period of Christian scholarship in medieval Islamdom after three centuries of productive vigor.

Emblematic that the Christian midwives had been effective in delivering Greek science, medicine and philosophy to Muslim civilization is the career of abu Ya'qub Yusuf al-Kindi. Because he was the earliest full-blooded Arab who contributed significantly as a scientific commentator and original thinker, al-Kindi was known as "The Philosopher of the Arabs." Along with the Harranian astronomer al-Battani, he signifies the maturing of an authentic scientific tradition in Muslim society. Al-Kindi's keen interest in all the sciences – mechanics, mathematics, optics, astronomy, physics and medicine – and his prolific literary output in them, present a brilliant preface to the sustained intellectual creativity of Muslims in the 10th and 11th centuries, the golden age of their scientific renaissance.

The House of Wisdom was central to the transformative process of translation flowering into original creation. Baghdad and Ma'mun's institute were together the magnet to which scientific, medical, philosophical and mathematical talent throughout the empire was attracted, and then later the mirror from which courtly life, patronage and scientific study were projected out to the provincial capitals, when they were being ruled by autonomous governors turned emulating dynasts. All over the Muslim world, independent and autonomous rulers were patterning themselves on the Abbasid style: Aghlabids in Tunis and Sicily, Idrisids in Morocco, Fatimids in Egypt, Hamdanids in Syria, Umayyads in Spain, Tahirids in Iran, Samanids in Central Asia. The Abbasids had become shadow caliphs, but what they had accomplished in the 8th and 9th centuries became an enduring legacy to the empire's provincial successors.

The broad embrace of Islamdom had through tribal conquest brought all the old centers of learning within a single political organization; then, with the building of Baghdad and the House of Wisdom, it brought all the scientific traditions together, concentrated in an academic center in a thriving commercial city where Greek, Syro–Mesopotamian, Iranian and Indian knowledge was pooled, translated, digested, expanded and synthesized within an Islamic context and then transmitted throughout the breadth of the former empire, in whose many capital cities of the emergent dynasts science was further expanded and refined. From Spain to Samarqand, the capital cities of a dozen major provinces opened opportunities of patronage to a legion of physicians and astronomers. Where one city went into economic depression, political instability and religious rigidity, another flourished. People were always on the move: scholars, merchants, religious leaders, mystic preachers, proselytizers. Relocation was the norm. Travel was in the cultural genes. The commercial ethos of Islam's civilization born in the merchant community of 6th and 7th century Mecca made for a mobile society of merchants who took their caravans and ships from city to city and port to port across the lands of Islam. Owing to the yearly hajj pilgrimage, religion reinforced the urge to journey. The proliferation of dynasties and patronage created a wide market for talent. Like the merchants who felt at home in the familiarity of Muslim cities where they traded their goods, religious scholars, poets, musicians, physicians and astronomers were able as well to take to the road to sell their skills to the ruler who was buying. If there was no employment in one place, there would be in another. An astronomer or physician with a name had his choice of place. The Spanish Umayyad court welcomed with open arms any talent that chose to leave Baghdad for Cordova.

Owing in great part to the linguistic, religious and cultural unity of the Muslim world, relocating was no big problem. The aforementioned 10th century philosopher and theoretician of music who wrote on the agreement between Plato and Aristotle, abu Nasr al-Farabi (d. 950), was a Turk from Central Asia whose patron was a Syrian Arab dynast, the Hamdanid in Aleppo at the far edge of the Asian continent. Such cases were not out of the ordinary. In the 10th and 11th centuries, opportunities abounded for those who had made their names. The physician–philosopher ibn Sina spent much of his time traveling from court to court, enough for him to have quipped that a good portion of his literary output was written on the back of a camel. If a scholar had to leave a city for political or religious reasons, there were places he could go where the intellectual climate was favorable, and where the climate was favorable there was royal patronage for anyone of some genius.

The answer to the question as to why the caliphs and the elites at court chose to subsidize science and philosophy is found in great part in the reports of Umayyad patronage of alchemy, astrology and medicine. The anecdote of Mansur's stomach problem expressed the need of medicine for bodily well-being. But then, with an eye to practicality, what did Ma'mun's dream of Aristotle have to do with his patronizing the sciences and philosophy? Everything. The paradigmatic interconnectedness of the branches of natural philosophy, medicine and metaphysics has been touched upon. The sciences did not come separately. They were interdependent, and together presented a whole vision of reality, from rocks to angelic intelligences, with man bodily and spiritually in between, and crowning it all, God, made comprehensible by a set of rational principles, metaphysics. At the crown of philosophy, metaphysics complemented the order, structure, motion and purpose of the heavenly bodies while accounting for man's soul and its existential place between physical creation and God. Aristotelian physics gave Galenic medicine its central paradigm. The practicality of the latter necessitated the former. Physics would have been meaningless without its metaphysical infrastructure of cause, effect, purpose, and God's thinking mind embracing the parts in a universal whole and imparting motion to the system of concentric spheres, from the firmament to earth's cyclical seasons. There was no picking and choosing from the integrated corpus of natural and metaphysical philosophy inherited from the Greeks. One branch was meaningless if cut from the body of the whole. The leaf withered when detached from the trunk.

Another compelling reason why Muslim dynasts patronized science and philosophy was inherited tradition. The tradition of Sassanid courtly patronage was communicated to the Abbasids and Samanids through Jundi Shahpur and the high-born Iranian viziers who served them, such as the Barmakids, who at their own expense commissioned translations of Greek mathematics and astronomy. Royal patronage became an important part of courtly life and prestige in all but the narrowest and religiously restrictive dynasties. Rulers competed for great astronomers or physicians to grace their courts as much as they did for famous poets, singers and architects. When the dour Sultan Mahmud of the Ghaznavid Turks defeated the Samanid ruler, he took with him to his capital Ghazna the well-known astronomer–mathematician al-Biruni and the poet Firdawsi as part of his spoils of war.

Neither scientist nor poet desired to leave the civilized comforts of Bukhara for the strict religious atmosphere of Ghazna, but the choice was not theirs. They were the rewards of conquest. Like the riches Mahmud would bring back from his plunder of India, Biruni and Firdawsi were jewels that would add intellectual glitter and prestige to his court. Caring little for astronomy or poetry, Mahmud nonetheless honored his precious captives and supported their work. An uneducated, rough Turkish warrior, Mahmud as a ruler appreciated the style and forms of civilized Muslim dynasts, and just as the Samanid court in Bukhara had been modeled on the Abbasid's in Baghdad, Mahmud of Ghazna modeled his on Samanid Bukhara. Courtly patronage ennobled a ruler's name, dignified his court and engendered a sense of dynastic legitimacy and prestige in the community and among other Muslim dynasts. Patronage of scholarship and literature was the badge of civilization. Mahmud looted and burned Hindu temples and shrines, but supported his scientist Biruni, who learned Sanskrit and wrote one of the great medieval texts on Hindi religion, customs and civilization, which his Turkish patron was savaging.

Strict Sunni that he was, Sultan Mahmud detested Firdawsi's epic *Book of Kings* that rejected Arabic loan words and ennobled Sassanid shahs as if Islam never existed, but he patronized him nonetheless. It was the civilized thing to do, and this even with the poet being Shi'i and a rationalist Mu'tazilite in his theology!

Mahmud's brilliant scientist also troubled him. He suspected that Biruni's astronomy harbored beliefs incompatible with religion. Biruni, on the other hand, feared Mahmud for his simple-minded religious fervor, shown as much by his character as his merciless ravaging of Hindustan. He refrained from talking science with Mahmud, lest the sultan mistake it as an attack on religion. But on one occasion the issue between Biruni's astronomy and Mahmud's religion came out. An ambassador from the Volga Turks, who had trade relations with people near the polar regions, came to Ghazna on a diplomatic mission and in conversation happened to mention to the sultan that if you went far enough north you would reach a point where the sun never set. In some places, he said, it was six months of continuous daylight followed by six months of dark night. Mahmud could not believe what he heard. It was contrary to religion. How could a Muslim know the time to do his prayers in such conditions? What about fasting during Ramadan? One would die of starvation before the sun went down. God could not have made such a world as to thwart Islam and God's own Quran and Holy Law! The merchant assured him it was true. Convinced the merchant was wrong, Mahmud had his astronomer settle the issue. Biruni must have feared that the astronomical truth would surely bring the sultan's wrath down on him, but he could not deny what he knew to be true. The scientist bravely explained as simply as he could that since the plane of the orbit of the sun around the earth was tilted approximately some 23 ½ degrees to the plane of the celestial equator, which rotated daily on its polar axis, the resulting solar motion as viewed from earth was a helix, and that during the winter season, if one was very far north, say 50 or 60 degrees from the equator, the sun would be seen not to rise above the earth's horizon for as much as half the year, and then not set for the other half. It was a question of astronomy, not religion. So the merchant was right.

How Mahmud resolved this contradiction between what his religion told him and what his distinguished astronomer told him is not known, but, confident enough in his power as ruler, he continued to support Biruni in his work and never brought the matter up again. In return, the name of Mahmud's son and successor, Mas'ud, was immortalized in the annals of science when Biruni dedicated to him his *Mas'udic Canon*, one of the great works of Muslim observational astronomy.[24]

Only under the severest conditions of religious threat and public disapprobation did rulers disengage from the royal tradition of supporting scholars of the non-religious sciences. It was a courtly tradition that even religiously narrow rulers like Mahmud respected. The Berber Unitarians, or Muwahhidun, untutored north African tribesmen who invaded Spain as fervent religious reformers, were within a generation patronizing naturalists, logicians, mathematicians, astronomers and, horror of horrors, metaphysicians. The dynasty they established in Andalusia presided over a scientific and philosophical efflorescence rivaling even that of the Umayyad caliphate in Cordova before them. The Berber Muwahhidun and the Turkish Ghaznavids, though somewhat akin in religious spirit to today's Saudi Wahhabis, conducted their royal court and offered patronage according to the civilized norms of their times.

Another consideration in understanding the tradition of royal patronage of science was the practical religious and economic benefits to be gained, benefits which might soften a people's suspicion and resentment of an untutored warrior who fought his way to power and whose authority as ruler was of questionable legitimacy. Scientists themselves argued these benefits, though perhaps for reasons not wholly devoid of self-interest. Science served religion. Biruni called astronomy the handmaiden of religion. Astronomy and its instruments were able accurately to determine the times of prayer: a sundial or gnomon for daylight prayers, an astrolabe for evening prayers. It required astronomy to determine the *qibla*, the niche built in the mosque wall that faced the direction of the holy Ka'ba in Mecca. Astronomy could foretell the new moon and the beginning of religious feasts, the beginning and end of Ramadan, the month of fast, and when the daily fast began and ended. It was also important in other affairs of daily life. Astronomy could predict lunar and solar eclipses, give directions to navigators on the high seas and merchants leading camel caravans overland. Astronomy could determine the locations of cities on the earth's surface and determine, within minutes, the beginning of the seasons. It was a boon to long distance trade on land and sea and helped merchants make fortunes as much as it helped agriculturists calculate the times to sow and harvest. Life, civilization and religion required an accurate calendar, and it was astronomy that made one possible. Religion, commerce, travel, navigation and agriculture, there was no end to the benefits of astronomy. Above all, the moving portrait of the heavens with their beauty, harmony and perfection of design and movement provided the believer with a glimpse into the cosmic magnificence of God's mind. Astronomy could not but deepen faith and enrich life both materially and spiritually.

Unmentioned of course, but much in any astronomer's or ruler's mind, and in everyone's mind who could afford having a horoscope cast, was astrology. Who did not want to know if the stars were for or against? If the heavens governed the tides and seasons, why could they not influence human events as part of the cosmic unity of existence? Even the rigorous scientist Biruni, who claimed to disbelieve in the vulgar kind of astrology that claimed to foretell events, wrote a well-known book on its principles and methods.[25] What ruler could afford not to keep an astrologer on hand who could tell him what the stars had in store? With the astrologer came astronomer and mathematician, just as with the physician and his medical kit came the natural philosopher and metaphysician.

Scientists were not shy to let rulers know of the many marvelous benefits science gave to religion and rulers. Mathematics, for example, like astronomy, was a boon to civilization and mankind. Algebra, as explained by the algebraist al-Khwarizmi, helped simplify the computations in the complex inheritance laws laid down by the Quran. Geometry and trigonometry helped agriculturists survey their fields, determine areas of regular and irregular shapes and plan irrigation flows. It allowed the calculation of surface areas and volumes for builders. Support of science was of great benefit to society and a badge of honor for a noble, caring and religious ruler. It was an important part of what rulers did for political legitimacy.[26]

Rare were the Muslim dynasts who failed to buy into it.

Notes

1   In fact, one of the Abbasid caliphs, Ma'mun (813–833), had seriously considered transferring the caliphate to a member of Ali's family, but the candidate he had in mind was too advanced in age to win much support among the caliph's Iranian advisors. H. Kennedy, *The Prophet and the Age of the Caliphates*, Longman, New York, 1986, pp. 153–154.

   What began as a power struggle between Ali, who was caliph in the capital Medina, and Mu'awiyyah, the governor of Syria, following the murder of the third caliph in 656, developed over the generations into a religious split, one sect paralleling the other in its religious and theological evolution, each possessing its own books of law, canons of Prophetic Tradition, mosques, shrines, days of religious celebration, Islamic heroes, holy men and mystical orders. The Shi'ites damned the first three caliphs for having usurped power from Ali and passionately celebrated the martyrdom of his son Husayn, who in 680 was killed at Karbala in the Iraqi desert in an abortive attempt to unseat the Umayyads who were ruling from Damascus. In later years that bloody event was taken to be the central drama establishing the Shi'i apotheosis of martyrdom.

   The Shi'ites divided into many different sub-sects in the 8th and 9th centuries, the majority the results of revolutionary movements aiming to accomplish what had been hoped for in the Abbasid revolution. The mainstream Shi'ite community was content to live peacefully with their Sunni coreligionists and unobtrusively practice their own form of Islamic worship in their own mosques. The two sects considered each other bona fide Muslims but perhaps just slightly out of step, but not at all enough to deter joint business ventures or intermarriage. On the political level, however, Abbasid rulers kept a close eye on the descendants of Ali, a number of whom emulated the martyrdom of Husayn but not always by choice. Claiming the Shi'i Imams their leaders and justice their cause, revolutionaries of any hue could tap into latent sources of religious and social discontent and, if gifted with charisma and organizing ability, mold and move it under the banner of Shi'i legitimism.

2   J.J. Saunders, *A History of Medieval Islam*, Routledge and Kegan Paul, London, 1972, p. 103.

3 Bernard Lewis, *The Arabs in History*, Oxford University Press, Oxford, 1993, pp. 87–88, 97.

4 Philip Hitti, *History of the Arabs*, Palgrave, London, 1990, pp. 304–305.

5 Realizing the war had cost the office of the caliph much of its prestige and religious dignity, the caliph searched for a unifying principle that might anchor the ruling institution to religion. Ma'mun first considered appointing as his successor a descendant of the revered family of Ali, a Shi'i Imam, whereby the rule of the succeeding caliphs would be seen to embody the prophetic light of Muhammad, a form of Sassanian religio-political absolutism. This, Ma'mun hoped, would render the caliphal office inviolable. The idea was abandoned not only because of the opposition of Ma'mun's ministers but because a descendant of Ali who was willing and up to the task was not to be found. Ma'mun then decided that the basis of the ruling institution's religious legitimacy would be the state supported doctrine of a school of rationalist theologians (called Mu'tazilites by their opponents, the name by which they are commonly known since they lost out in the theological scuffle for dominance), whose religious beliefs were based on logic, free will, divine justice and the principle that the Quran had been created by God at a moment in time and was not co-eternal with God. These so-called Mu'tazilites called themselves The People of Justice and Oneness of God.

6 That the Arabs dynamically appropriated science and philosophy to make them their own rather than just passively receiving the intellectual traditions, see A.H. Sabra, "The Appropriation and Subsequent Naturalization of Greek Science in Medieval Islam," *History of Science*, XXV, 1987, pp. 223–243.

7 However, this has been effectively refuted by Manfred Ullman, "Halid ibn Yazid und die Alchemie: Eine Legende," *Der Islam*, 55, 1978, pp. 181–218.

8 P. Hitti, *History of the Arabs*, MacMillan, New York, 1970, p. 309. According to some accounts this Royal Military Hospital was thought to have originated from a Sassanian prisoner of war camp for captured Roman soldiers, among whom were medically trained Greeks that the Sassanians employed as military physicians. Recent scholarship, however, has cast some doubt on the hospital of Jundi Shahpur having been a prisoner of war camp and a subsequent model for Muslim hospitals. Michael Dols, "The Origins of the Islamic Hospital: Myth and Reality," *Bulletin of the History of Medicine*, 61, 1987, pp. 367–390.

9 The route of transmission is one of the more curious detours in the history of science. Max Meyerhof, "From Alexandria to Baghdad: A Study of the History of Philosophical and Medical Teaching Among the Arabs," *Sitzungsberichte der Preussischen Akademie der Wissenschaften*, 23, 1930, pp. 389–429. For more recent, detailed and incisively critical examination of the transfer of Greek science and rise of science in Muslim society, see Dimitri Gutas, *Greek Thought, Arab Culture: The Graeco-Arabic Translation Movement in Baghdad and Early Abbasid Society*, Routledge, London, 1998. J.L. Berggren, "Islamic Acquisition of the Foreign Sciences: A Cultural Perspective," in *Tradition, Transmission, Transformation*, edited by F.J. Ragel and S.P. Ragep, Brill, Leiden, 1992, pp. 310–324. Also George Saliba, *Islamic Science and the Making of the European Renaissance*, MIT Press, Cambridge, MA, 2007, pp. 1–72.

10 Philosophy and medicine had also been conjoined by the Greeks of Galen's treatise that bears the title "That the Best Doctor Is Also a Philosopher." John Freely, *Aladdin's Lamp: How Greek Science Came to Europe Through The Islamic World*, Knopf Doubleday, New York, 2004, p. 56.

11 Dimitri Gutas, *Greek Thought, Arab Culture: The Graeco-Arabic Translation Movement in Baghdad and Early Abbasid Society*, Routledge, London, 1988, pp. 67–69.

12 *Fihrist*, ibn al-Nadim, translated and edited by Bayard Dodge, vol. II, pp. 581–582. As noted earlier, Manfred Ullman has fairly convincingly shown this claim not to reflect historical reality.

13 Saliba, *Islamic Science*, p. 151.

14 Saliba, *Islamic Science*, pp. 68–83.

15 *Bayt al-Hikma* in Arabic. For its Sassanian precedents and caliphal precursors of Ma'mun see Ghutas, *Greek Thought, Arab Culture*, pp. 53–60.
16 Gutas, *Greek Thought, Arab Culture*, pp. 58–59.
17 The story is found in ibn al-Nadim's *Fihrist*, 2 vols., edited and translated by Bayard Dodge, vol. II, Columbia University Press, 1970, pp. 583–584.
18 Saliba, *Islamic Science*, p. 52.
19 Hitti, *History of the Arabs*, pp. 312–313; G. Strohmaier, "Hunayn ibn Ishaq," in *Encyclopedia of Islam*, 2nd edition, edited by P. Berman, Th. Bianquis, C.E. Bosworth, E. van Donzel, W.P. Heinrichs, Brill, Leiden, the Netherlands. vol. 3, pp. 578–580. Also abd al-Hamid al-Sabra, "The Exact Sciences," in *Genius of Arabic Cicilization: Source of Renaissance*, MIT Press, Cambridge, MA, 1978.
20 Manfred Ullman, *Islamic Medicine* (*Islamic Surveys II*), Edinburgh University Press, Edinburgh, 1978, p. 9.
21 A.I. Sabra, "Situating Arabic Science: Locality Versus Essence," *Isis*, 87, no. 4. December, 1996, pp. 654–670.
22 W. Hartner, "Battani," in *Dictionary of Scientific Biography*, vol. I, edited by C. Coulson, Gillispie, Scribner and Sons, Detroit, 1970–1980, pp. 507–516.
23 G. Saliba, *A History of Arabic Astronomy*, NYU Press, New York, 1994, p. 52 citing al-Jahiz, *Kitab al-Bukhala* and Max Meyerhoff, *from Alexandria to Baghdad*.
24 E.S. Kennedy, "Biruni," in *Dictionary of Scientific Biography*, vol. II, edited by Charles Coulson, Gillispie, Scribner and Sons, Detroit, 1970–1980, p. 150.
25 *Tafhim al-Nujum*, edited and translated by R. Ramsey Wright, Luzac, London, 1934.
26 Dimitri Gutas, *Greek Thought, Arabic Culture*, Routledge, New York, 1998, pp. 29–52, for the Abbasid Caliph Mansur's legitimizing the new dynasty by appropriating Sassanid royal policy of patronage.

2 The record of original achievement

Medicine

Since it was medicine that began Muslim interest in the sciences, it is only fitting that any exploration of Muslim scientific achievements begin with it.[1]

The hospital was to medicine what the observatory was to astronomy. Workshops of the profession, the hospital and the observatory were the institutional bases for the study, refinement and advancement of their respective crafts, supported by royal patronage. A hospital, however, brought a ruler greater social prestige than an observatory. Closer to the common people than the stars, the hospital administered to their needs more directly than the observatory. A hospital that bore the founding ruler's name proclaimed his piety, humanity and, above all, his cognizance of the people's welfare. A hospital was good politics in public relations, a badge of honor. It established a ruler's credentials as a good Muslim and caring governor. The grander the hospital, the more devoted was its founder perceived to be in serving religion and community. The Buwayhid rulers in Baghdad, Shi'i Muslims posing as protectors of the Sunni Caliph, founded a hospital in 982 that employed 25 doctors of various specialties and was described by a traveler as an immense palace. The poor were treated free of charge.

The Sassanian hospital staffed by Syriac Christians at Jundishahpur has been taken by historians as the model for the earliest hospitals built by Muslim rulers in Baghdad, Damascus, Shiraz and Nishapur. However, a critical review of the early Syriac and Arabic sources questions the existence of a major and truly important hospital at Jundishahpur, comparable to the ones in Nisibis and Susa, just to the west of Jundishahpur.[2] What is certain is the relation between the Syriac medical tradition that migrated to the western region of the Sassanian realm, where it received Iranian patronage, and the rise of medicine in Muslim civilization. Sassanian rulers patronized Syriac Christian physicians and founded public hospitals that served the poor free of charge. Those who could afford it were obliged to pay. The larger hospitals served as medical schools where students studied and practiced under a single physician for a number of years. The hospital schools were not organized in terms of a graduated curriculum with scheduled classes and exams, though teaching physicians did give lectures and had the responsibility of accrediting their students as having mastered the authoritative texts.

Christian physicians such as Bukht Ishu' and Masawayh, who were trained in Greek medicine, were brought from former Sassanian hospitals at Nisibis, Susa and Jundi Shahpur to serve the Abbasid caliphs in Baghdad. This began the transferal of Greek medical practice to the Arabs. The Iranian contribution to the Greek-based Muslim medical tradition was perpetuated by several great Iranian physicians of the 10th and 11th centuries, foremost among them being al-Razi, al-Tabari, al-Majusi and ibn Sina. All four of them composed encyclopedic works that collectively dominated the teaching and practice of medicine. Al-Tabari's *Firdaws al-Hikma* (*Paradise of Wisdom*) was the first of these authorities. His more famous student's monumental compendium, al-Razi's *al-Hawi* (*Continens* in the Latin tradition), remained an authoritative medical text for over seven centuries in both the Muslim East and Latin West. Al-Razi's work was followed first by al-Majusi's medical encyclopedia, then a generation later by ibn Sina's *Qanun al-Tibb* (*Canon of Medicine*). Al-Razi's *al-Hawi* and ibn Sina's *Qanun* endured as the twin fundaments of medical knowledge in Islamdom until the late 18th century and until the 17th century in the Christian West. More monumental works would be written by Muslim physicians in later centuries, but none would surpass in prestige those written by al-Razi, al-Majusi and especially ibn Sina, whose *Qanun* would remain at the pinnacle of the civilization's medical excellence.

Composing voluminous comprehensive works was a medical tradition across the lands of Islam. During the high point of Spanish Umayyad civilization in the 10th century, abu al-Qasim al-Zahrawi composed a medical encyclopedia that is usually referred to by its abbreviated title, *al-Tasrif*. The 30th volume of the work, devoted to surgery, bonesetting and surgical instruments, gained prominence as the authority in surgical practice, circulating as an independent book throughout the Islamic world. Translated to Latin in the 13th century, it gained similar prominence in Europe.[3] Another encyclopedic work of the Muslim medical corpus was by ibn Rushd, more famous as a philosopher but also a leading physician and Maliki judge. His *Kulliyat at Tibb*, or *Universal Medicine*, called *Colliget* in its Latin translation, ranks in authority with those compendiums of his encyclopedic predecessors. Yet another medical encyclopedia was composed late in the next century, this by ibn al-Nafis, a teaching physician in the famous Nuri Hospital in Damascus.

The comprehensiveness and great clinical detail of these voluminous works gave them authoritative status as the final word in medicine. But excellent as they were as general texts and handy reference books for all things medical, there may possibly have been a downside to the production and popularity of such ponderous compendiums, in that their sheer size and exhaustive treatment marginalized interest in composing smaller and more specialized treatises, the diminishing number of which impeded the continuation of the empirical tradition of Muslim medicine. One made one's mark not in small research studies but producing massive encyclopedic texts containing all that needed to be known in medicine. Yet before this encyclopedic mind-set had signaled the path to fame, physicians of all religions in Muslim society did well in extending the frontiers of medical knowledge and experience, especially in clinical diagnostics, ophthalmology, surgery, bonesetting, tracheotomy, amputations, cataract removal, treatment of trachoma and other eye diseases.

Having through their early Nestorian mentors inherited the traditions of Hippocrates, Galen and Dioscorides, the medical scholars of Muslim civilization advanced beyond the medicine they inherited. But they never lost respect for their Greek predecessors. They always regarded Greek medical knowledge with the same deep admiration that Muslim astronomers held for Ptolemy, paralleling the critical reverence that Muslim thinkers held for Plato and Aristotle. Yet in spite of the Muslim scientist's worshipful admiration of his Greek forerunners, there was abundant criticism, reorganization and innovative reform of what the Greek authorities had accomplished, as much in medicine as in astronomy and philosophy. Ibn al-Haytham's *Doubts Concerning Ptolemy* and Aflah's *Reform of the Almagest* had their medical counterpart in Razi's *Doubts Concerning Galen*. Though their underlying principles were considered sound, Greek systems and techniques were seen to be flawed in certain details, leaving room for expansion and improved systematization. Accordingly, the advances Muslims made were within the general framework of Greek principles. Razi's certitude that Galen, though not incorrect in principle, had fallen short on clinical analysis, motivated him to study the eye from a physiological point of view. Razi's exhaustive descriptions of the symptoms and development of smallpox and measles derived from his dissatisfaction with Galen's lack of detail in his clinical analysis of common diseases. His studies continued to be considered definitive for over half a millennium, giving Razi's work an enduring fame that Majusi's better organized and more systematically critical compendium failed to achieve. The critical spirit evinced by Razi continued into the 13th century, when the physician 'abd al-Latif al-Baghdadi cast his own doubts on Galen. During his studies of skeletons of bodies laid low by a plague in Egypt, he discovered misconceptions in Galen's anatomy, as well as in other parts of Galen's medical system. Observation was sounder than theory, Baghdadi concluded.[4]

Indian and Iranian influences accompanied Greek in the earliest formation of Muslim medical practice and theory. This was a general pattern in the Muslim assimilation of science. The Christian convert to Islam, Ali ibn al-Rabban (d. 855), who was a medical authority during the formative period of Muslim medicine, combined Iranian and Indian with Greek medical ideas. Paralleling the development of Muslim astronomy as it moved from the Indian-based system to the Ptolemaic system, medical science shed its early Irano–Indian elements to emerge essentially Greek in theory and practice.

The fundamentals of Muslim medicine came to be based on Galen's system of the four humors: blood, phlegm, yellow and black bile, which were derived from Aristotle's four basic elements: earth, air, fire and water. The elements, like the four humors, corresponded to the four basic qualities that defined the physical composition of natural objects: cold, hot, wet and dry. Each humor was a coupling of two qualities. Blood was hot and wet; its dominant elements, fire and water, were perceived to produce the heat the living body required. Phlegm was cold and wet; yellow bile was hot and dry; and black bile was cold and dry. From this derived the four basic dispositions of personality: sanguine, phlegmatic, choleric and melancholic. Health was preserved by maintaining a balance among the bodily humors; nothing in excess: Aristotle's golden mean as a medical paradigm. The Muslim contribution to the advancement of medical science was not in finding new principles but rather in its refinement and systematization and also in the clinical descriptions and diagnoses made by its great physicians, al-Razi and ibn Sina being the most notable in this respect. Both were Neoplatonic in their epistemology and philosophies of the physical world. Accordingly, their methodical organization and categorization of the causal relationships governing the vast body of medical knowledge that had accumulated over the millennium were framed in physio-spiritual paradigms, where the soul, a non-physical organ of the body, was seen as the balance that determined bodily health. Soul and spirit were not abstracts but determinants of bodily health and well-being.

The apogee of medical productivity was roughly from 900 to 1100 and centered in Baghdad, Iran and Andalusia. Following that period, the tradition was then continued in Egypt and Syria, supported first by the royal patronage of the Ayyubid sultanate established by Salah al-Din, then by its successor, the slave sultanate of the Turkish Mamluks. Ayyubid and Mamluk emulation of Abbasid and Fatimid traditions of patronage produced new hospitals in Cairo and Damascus, which became the principal centers of medical study in the 12th and 13th centuries. One of the most important of these centers was the Nuri Hospital founded in the 12th century by the Turkish amir Nur al-Din al-Zangi, who ruled Damascus as an appannage of the Seljuq sultanate in Anatolia.[5] In the century after the hospital's foundation, two renowned physicians were practicing there, ibn abi 'Usaybia (d. 1270), who wrote an important medical biographical dictionary, *Uyun al-Anba fi Tabaqat al-Utibba*, and ibn al-Nafis (d. 1288), who planned on immortalizing his name by outdoing all his encyclopedic forerunners with an immense medical compendium of 300 volumes. Only eight were completed.

Ibn al-Nafis achieved posthumous fame in the Latin West for his discovery of the lesser pulmonary circulation of the blood. The discovery was made apparently without recourse to dissection of cadavers, something which was strictly prohibited by Islam and which undoubtedly impeded further anatomical discoveries.[6] Ibn Nafis' description of the pulmonary system is found in the anatomical section of his commentary on ibn Sina's *Qanun*, which for all its fame is an organizational shambles. To give unity and coherence to the *Qanun*, ibn al-Nafis collected all the bits of information and narrative on a subject scattered through the work's hundreds and hundreds of pages and organized them in one section. It was in the course of his reordering the *Qanun* that he made this discovery, one of the most important anatomical discoveries in ancient and medieval medicine. Perhaps postulation is more accurate than discovery to describe this contribution by ibn al-Nafis. By simply reflecting on Galen's insistence that blood passed through one ventricle of the heart to the other by an invisible corridor, he realized something was wrong about it, and in a thought experiment, he decided blood had to circulate from the heart through the lungs before passing through the heart. Four centuries later William Harvey, aided by his dissection of cadavers, would provide a description of the full circulatory process.

That ibn al-Nafis's description of the pulmonary circulation went unnoticed was one of many instances in the history of Muslim science of a brilliant insight being left unexplored.

The Nuri Hospital was the last glow of Muslim medical achievement. After ibn Nafis and ibn abi 'Usaybia (and their two students, ibn al-Quff, a Christian, and Muhammad Akfani) the medical tradition reached an uncreative plateau. Grand hospitals would continue to be founded in the 15th, 16th and 17th centuries by Ottomans, by Safavids in Iran and by Moghul Timurids in India, but the authorities would continue to be the works of Razi, Majusi, ibn Sina, ibn Rushd and ibn al-Nafis, with Galen, for all the criticism he received from Muslim physicians, remaining the master of their medical principles.

Mathematics

Building on Greek, Babylonian and Indian precedents, Muslims developed the branch of mathematics that came to be called algebra. The name comes from the title of an early 9th century book on the subject by the Muslim mathematician–astronomer abu Musa al-Khwarizmi. He was one of Ma'mun's court astronomers from Khiva in Central Asia. Khwarizmi titled his book (in its shortened form) *al-jabr wa'l muqabala*, which conveys the meaning and operational technique of "setting and confronting" the terms of an algebraic equation. Mathematically it means setting the equation equal to zero by transposing all the elements to one side, that is, having them confront each other, and then simplifying the equation by canceling out the plusses and minuses. The word *al-jabr*, a medical term meaning to set bones straight, mathematically conveyed the idea of setting and forcing the equation into the straightest and simplest form possible. From *al-jabr* came the Latinized algebra and from al-Khwarizmi's algorism, a method of iterative computation he used to obtain closer and closer approximations of an unknown value in an equation.[7] Khwarizmi's form of algorism was later modified by another scientist from the Turko–Iranian sphere of the Muslim world, al-Biruni, who devised an ingenious trigonometric method of constructing algorisms to determine sines, chords and trisection of angles.

Al-Khwarizmi established six basic algebraic models corresponding to all forms of quadratic algebraic equations and gave solutions for them, the purpose being to systematize the equations and provide methods to find the roots of the unknown in order to solve everyday problems. Among the practical applications he lists for algebra are: determining the portioning out of a family inheritance among its various members; the width, depth and length in constructing canals; and land surveying. To that one could add formulating corporate business transactions to calculate running costs and profits, and computing commercial insurance rates based on statistical averages of goods damaged during transport measured against the distance of transport – the word average comes from Arabic, *'awwariyya*, meaning damaged. Khwarizmi's book on algebra helped spread the use of Indian numerals based on the decimal system, which was a blessing for practical numerical operations, even simple ones like division and multiplication. One of the purely theoretical problems that fascinated him was to find two cubes whose sum is another cube: $X^3 + Y^3 = Z^3$, a particular case of Fermat's last theorem which states that $X^n + Y^n = Z^n$ has no solution, a proof of which was produced in 1993 by an Englishman, Andrew Wiles, after years of intensive solitary work on the problem at Princeton.

Like the Christian physician–translators who were with him in Ma'mun's House of Wisdom, al-Khwarizmi was productive in a broad range of related sciences and left a large legacy as an astronomer and geographer. His astronomical book on planetary observations, or *zij*, as these tables were called, was dedicated to his patron, the caliph Ma'mun. The basis of his astronomy, like elements in his algebra, clearly comes from Indian sources. His method of using algebra to solve geometric problems was further expanded by the Egyptian abu Kamil al-Shuja' in the 10th century. By then Muslim scientists, mathematicians and physicians were coming into their own as masters of their professions alongside their Christian colleagues, upon whom they were becoming less and less dependent. In the same decade that the mathematician abu Kamil was writing in Egypt on algebraic analysis, the Greek Christian Qusta ibn Luqa was in Baghdad translating the algebraic *Arithmetica* of Diophantus. Not long after that, when the great bulk of the extant Greek texts had been translated, science in Islamic civilization was in full bloom and being advanced by both Muslims and non-Muslims.

Algebra occupied a central place in the Muslim restructuring and organizing of Hellenistic mathematics. Of the new methods Muslim mathematicians applied to algebra, one of the most fruitful was lexicographical analysis, that is, determining frequency of occurrence, arrangements and combinations of words and letters in a text, a branch of algebra called combinatorial analysis. Al-Kindi developed and applied it to cryptography and cryptanalysis.[8]

The contribution of non-Muslim scientists continued through the 10th century and into the generations of most intensive creativity in the 11th century. The polymath and translator (of the nominally monotheist Saba'ian sect in the Harran region of upper Mesopotamia) Thabit ibn Qurra, resuscitated Archimedes' long lost method of computing by the integral sums of infinitesimals. Archimedes had demonstrated the method of infinitesimal calculus in his treatise *The Sand Reckoner*, but no Hellenistic mathematician continued Archimedes' work and the treatise was eventually lost. Inspired by Archimedes' other works, such as *The Sphere and The Cylinder*, Thabit wrote his *The Sections of a Cylinder and Their Surface*, in which he demonstrated the method of integral sums by computing areas under parabolic, hyperbolic and elliptical curves sliced into infinitesimal strips. Thabit's son Sinan continued his work. By refining and generalizing the method of Thabit's geometrical calculus of infinitesimals, Sinan's son Ibrahim achieved more fame as a mathematician than both his father and grandfather.

Experimental application of geometric infinitesimals as a method of computation was advanced in the 11th century. The Basra-born Arab physicist and astronomer ibn al-Haytham (d. 1037) applied the method to volumetrics by rotating a section of a parabola around the central axis to generate a paraboloid and computed the resulting volume by integral sums of the infinitesimals. This was a monumental work of mathematics, even more so because it was accomplished without the simplifying shorthand apparatus of signs, notations and symbols that is so familiar today and makes advanced mathematics possible.

Ibrahim ibn Sinan's and al-Biruni's work on geometric transformations of generating conic sections from a circle was used in making the stereographic projections employed in astrolabes. Advancing along the line of mathematical exploration established by ibn al-Haytham and Umar al-Khayyam in the 11th century, an early 15th century scientist in Samarqand, Ghiyath al-Din al-Kashi, discovered a method of approximate integration to calculate dimensions of arches and vaults, the area of stalactites, and the volume contained in architectural shapes, such as in the hollow cupola.

Parallel to the refinement of geometric computational methods, algebraic methods were also further developed over the centuries. Abu Kamil's reworking of Khwarizmi's algebra combined with Qusta ibn Luqa's reworking of the equations of Diophantus laid the groundwork for advanced methods of computation and analysis. Following a few decades after the work of Qusta ibn Luqa and abu Kamil, two exceptional Iranian Muslim astronomer–mathematicians in Baghdad, Habash al-Hasib and al-Karaji, made further advances in the method of algebraic solutions. Habash devised a convergent iterative algorithm to solve an algebraic equation involving the sum of an unknown and its sine function, $a = x + b \sin(x)$, also known as Kepler's equation in the West. In the next century, ibn al-Haytham demonstrated a method of solving a complex problem involving a cubic equation, namely determining the angle of a beam of light reflected from a spherical mirror to an object at a given location. The problem became known as Alhazen's Problem in medieval Latin mathematics because of the algebraic technique he used to solve it.

Haytham was a mathematical master in many fields of study. In addition to his work in optics, he derived methods of algebraic, trigonometric and geometric solutions for solving complex problems in astronomy and physics. He was also a master at solving problems that were purely mathematical with no practical application to scientific problems. Indeed, Muslims pursued pure mathematics no less than their Greek predecessors had. Beginning with al-Kindi, who based his own work on several centuries of Hellenistic and post-Hellenistic mathematicians, Muslim scientists played with mathematics for the pure joy of discovering and proving its beauty, such as in isoperimetry (finding the largest possible volume given a specified surface area), where a circle is found to be the largest area per given perimeter, a sphere per given surface area, and an equilateral triangle for any configured triangle of a given perimeter.

Greek geometers had been obsessed with squaring the circle, that is finding a method to transpose the area of a circle into a square using whole numbers. Ibn al-Haytham recast the problem in trigonometric form, approaching it according to what the Greeks referred to as quadrature of the lunules, literally little moons, a simple and neat construction produced by a triangle and three intersecting circles where each leg of the triangle is a diameter of a circle, the three lunules being the lunar shaped curves inside the triangle. In this and other problems Haytham attacked, as in isoperimetry, he in effect married geometry to trigonometry, quite as al-Khwarizmi, abu Kamil and their algebraist successors married algebra to geometry. Muslim mathematicians would bring all three branches together before the 13th century.

Living at the height of Muslim scientific productivity in terms of creative quality and numbers of men engaged in scientific activity, Haytham, Biruni and Khayyam represent that generation of the 11th century whose work was the culmination of 200 years of continuous scientific and mathematical advances and would be continued for centuries more by scientists such as Nasir al-Din al-Tusi, Kamal al-Din al-Farisi and Ghiyath al-Din al-Kashi. In the early 14th century, the Iranian physicist and mathematician Kamal al-Din al-Farisi was applying algebraic analysis to that old problem of Khwarizmi's, finding two numbers, the sum of whose cubes equals the cube of another number ($a^3 + b^3 + c^3$). The problem was still being pursued in the 17th century, around the time of Fermat, by the Iranian mathematician popularly known as al-Yazdi.

In the purest sense of creative originality, it is not exactly historically accurate to say Muslim mathematicians invented algebra. Elements of what was to become algebra existed in the sources of their Greek and Indian predecessors. What Muslims accomplished was to bring diverse elements together in producing methods of solution. The same can be said for trigonometry, where the Indian system of sines of central angles in a circle was married to the Greek system of chords and arcs of a circle. Greek astronomers had calculated chord tables relating the measure of arc on a circle to its corresponding chord, with the line inside the circle connecting the ends of the arc. Indian astronomers had calculated sine tables, which related the degrees of a central angle (the angle formed by two radii) to the half-chord in the form of a pure fraction. Indians and Greeks applied their trigonometric functions to astronomy, as did their counterparts in Islamic civilization, but there were Muslim astronomers who explored mathematics for itself, for its beauty of perfection and elegance of simplicity. Following centuries of incremental development in which the functions of sine, cosine, tangent, secant and so forth were elaborated by a succession of mathematicians from Musa al-Khwarizmi in the 9th century to Nasir al-Din al-Tusi in the 13th, trigonometry was at last liberated from being a computational tool in the service of astronomy and elevated to the status of an independent branch of mathematics. In this sense trigonometry was an invention of Muslim mathematicians.

Precision and simplicity were the guiding principles in the advance of Muslim mathematics. This went as well for astronomy, in whose service mathematics was given its greatest attention. Hand in hand with the construction of larger and finer observatories and observational equipment, which enabled observers to track the heavens more accurately and give more precision to Ptolemaic parameters of planetary models, Muslims excelled in developing simpler and more powerful computational methods, a mathematical tradition that endured at least until Ghiyath al-Din al-Kashi and the 15th century re-efflorescence of astronomy in Timurid Samarqand.

Muslims refined and simplified methods for extracting square and cubic roots whose origins went back to Indian mathematics. Khwarizmi, for instance, developed the algorithm for extracting square roots: $\sqrt{n} = a + \frac{r}{2a}$, where n is the number whose square root is to be found, a is the closest whole integer to the root and r = n − a². The mathematician al-Uqlidisi (d. 952) improved this to $\sqrt{n} = a + \frac{r}{2a+1}$. Al-Baghdadi found an algorithm for approximating cube roots: $\sqrt[3]{n} = a + \frac{4}{3a^2 + 3a + 1}$, where r is the difference between n and a^3. Based on the work of al-Biruni, Umar Khayyam and another mathematician, al-Samaw'al, Muslim astronomers used an algorithm, known in the West as the Ruffini–Horner method, to extract the *nth* root of a sexagesimal integer.

Muslims used sexagesimals as well as other systems of numerical notation. Besides the system based on 60 inherited from ancient Mesopotamia, a system of alphabetic letters was inherited from the Greeks, and decimal symbols from the Indians. They were all used by mathematicians and astronomers, and often the systems were mixed, with letters of the Arabic alphabet standing for sexagesimal values. Sometimes the sexagesimal and decimal systems were used together. The decimal place-value system of numerical notation was discussed by al-Khwarizmi in a manual on Indian reckoning. In time the system was adopted and generalized and supplied with a symbol to represent nothing, the zero, *sifr* in Arabic, origin of the word cypher. Conceiving of a symbol that stood for nothing, an emptiness, took tremendous intuitive imagination. It simplified mathematical operations to no end but was anything but simple to think of rationally. The comprehension of the idea of an absolute nothing was as mind boggling as comprehending the idea of infinity.

What is equally impressive is that all this was done without the simplifying benefit of mathematical symbols. All numerical operations – equalizing, dividing, multiplying, adding, subtracting, squaring, cubing, taking square and cube roots – all these operations, along with all the numbers involved, were written out in terms of their names, and the numbers were all written out as well. One marvels at how mathematicians and astronomers of the time could follow page after mind-numbing page of complex mathematical narrative, devoid of the signs, symbols and operational notations now taken for granted, while keeping it all in their minds.

The lack of a mathematical shorthand did not impede Muslim mathematicians from making important breakthroughs in the simplification of computational methods. Greek astronomers had used a cumbersome and time consuming method called the Menelaus Theorem in making computations on a sphere. Based on a spherical quadrilateral, the Menelaus Theorem involved laborious dogwork involving four arcs and as many angles. Three Iranian astronomers in Khorasan discovered a much easier method based on a spherical triangle formed by three intersecting great circles. Called the Law of Sines today, the method is based on the equality of the ratios of the sines of the angles and their corresponding arcs, expressed in modern form, where a, b and c are interior angles and A, B and C are the corresponding arcs. The three mathematicians abu al-Wafa al-Buzjani (d. 998), abu Muhammad al-Khujandi (d. 1000) and abu Nasir Mansur ibn Iraq (d. 1036) claimed prior discovery, medieval forerunners of the squabble between Newton and Leibnitz over the calculus of infinitesimals. The beauty of the discovery was the great simplification of computation that again brought geometry and trigonometry together in a powerful technique of determining earthly and astronomical distances and angles.

Their discovery of this marvelous tool was, like all scientific and mathematical discoveries, a step by step, scientist by scientist process. The Menelaus Theorem, as used by Ptolemy, was improved on by an ethnic mix of Muslim astronomers and mathematicians, first by the Turk al-Khwarizimi, and then, to cite only the most famous, the Arab al-Kindi, the Turk al-Farabi, the Iranian abu al-Wafa, the Arab ibn al-Haytham, and finally the Iranian from Iraq, abu Nasr ibn Iraq, teacher of al-Biruni. The other two Iranian co-discoverers, Buzjani and Khujandi, followed a parallel line of development to the sine theorem. Simplification from the old Greek quadrilateral method to the triangular sine method did not end the use of the Menelaus theorem. Rather, the old and the new existed side by side. As late as the 13th century, Nasir al-Din al-Tusi was attempting to reduce the Menelaus method to a simpler form of computation, an example of science being no more immune to old traditions that refuse to die than any other discipline, even when a better way has clearly been shown.

Astronomy and mathematics were like Siamese twins joined at the head. Advances that looked to be purely mathematical were most often the result of dissatisfaction with Ptolemaic astronomy. Dissatisfaction was centered on the system's parameters, not its primary principles. Muslim scientists accepted the latter and revised the former. In his book *Islah al-Majisti* (*Reform of Ptolemy's Almagest*), the 12th century Andalusian astronomer Jabir ibn Aflah derived a trigonometric relationship of arcs of great circles that was most probably based on Thabit ibn Qurra's treatise on the Menelaus Theorem and independent of abu'l Wafa's work. Because Jabir ibn Aflah was translated into Latin and considered an authority on spherical trigonometry, his form of the Law of Sines became known in the Latin world as Geber's Theorem. One of the most remarkable revisions was Biruni's expansion of the Greek system of two-dimensional rectangular and polar coordinates to a system of spatial coordinates.

Muslim mathematicians did not stop at three dimensions. A geometer of Biruni's and Haytham's generation, abu Sa'id al-Sizji (d. 1025), added a 4th dimension, a spherical hyper-dimension used to explore spatial problems of high mathematical complexity, as explained in his book *Measuring Spheres by Spheres*. The 3-dimensional universe was, as it were, a stereographic projection of a 4-dimensional universe, in the same way a Mercator map is a 2-dimensional projection of a 3-dimensional world. Medieval Arabic did not have a word for hyperspace, but geometrically the concept was present. Muslim geometers took off on this, working with equations up to the 5th and 6th degree, as if to imply there existed that many dimensions of hypergeometric space. Al-Farabi wrote a book called *Introduction to the Imaginary Geometry*, which has not survived but appears to have been a geometric exposition of a universe beyond 3-dimensions. It failed to survive perhaps because it was too far out for anyone at the time to bother making a copy of it. Dimensions beyond three were not considered in Europe until the 16th century, with the so-called "Christoff's Cube," which was simply abu Sa'id al-Sizji's division of a cube into two cubes and six parallelepipeds.[9]

One wonders what frontiers these imaginative minds might have reached had they devised a symbolism to sweep away the cumbersome clutter of words that would have given mental space for their continuing exploration into the universe of pure mathematics.

The high point of mathematical and scientific creativity was reached in the middle of the 11th century; but creativity continued several centuries more, especially, if not solely, in astronomy and the language in which it was expressed: mathematics. Though the period after 1100 brought forth few polymaths the stature of Biruni and Haytham, and though the instances of brilliance and originality became less frequent as the centuries advanced, the instances when they did occur were as brilliant and creative as the work of any astronomer or mathematician of the 10th and 11th centuries. Two of the most creative in the late period were Umar al-Khayyam (d. 1131), an Iranian in Seljuq patronage, and Ghiyath al-Din Jamshid al-Kashi (c. 1420), a Turko–Iranian in the patronage of the Timurid ruler Ulug Beg, grandson of Timur and himself an astronomer. More famous in the West for Fitzgerald's translation of his Persian quatrains than for his astronomy and mathematics, Khayyam accomplished for algebraic cubic equations what Khwarizmi and abu Kamil did for quadratics. By solving cubic equations representing three-dimensional space through the intersection of conics, that is, curves produced by passing a plane through a cone at critical angles of inclination, namely the parabola, ellipse and hyperbola, each of which has a defining quadratic algebraic equation, Khayyam married algebra to three-dimensional geometry, anticipating by half a millennium the analytic geometry of Fermat and Descartes. The geometrical beauty of these intersectional curves as defined by their algebraic equations was for Umar Khayyam equivalent to a divine revelation. Anticipating Galileo, Kepler and Descartes, he called mathematics the purest form of philosophy, the first step up the ladder to knowledge of the true essence of being. In less cerebral moments, when quatrains replaced quadratics, the mathematician seemed to find this as well in wine, women and poetry.

Khayyam's work on cubics was continued by another Iranian, Sharaf al-Din al-Tusi (d. 1170), who analyzed different forms of 3rd degree equations by intersecting 2nd degree curves. Sharaf al-Din's analytical geometry marks the frontier of mathematical sophistication reached by Muslims. Further progress in the field would not be made until centuries later when Descartes, Fermat and their successors devised a language of mathematical symbolism that simplified complex expressions.[10]

The high tradition of Muslim mathematics drew to an end with the passing of Ghiyath al-Din Jamshid al-Kashi in the 15th century. Director of the monumental observatory built by the Timurid ruler Ulugh Beg in Samarqand, al-Kashi's departure marked the last bloom of Muslim scientific creativity. With a passion for precision, he had the observatory's huge mural quadrant graduated to seconds of degrees to attain the highest observational accuracy he could. To this end he originated an ingenious algorithm that he used to calculate π to the 16th decimal place. In his *Key to Computation* (*Miftah al-Hisab*) he introduced several signs of mathematical notation and further simplified mathematical expression by using decimal fractions in his computations. Decimal fractions had been used and discussed centuries before Kashi. Khwarizmi had written on them, and in the 10th century a Muslim mathematician renowned for his work on Euclid, and hence known as Uqlidisi, also wrote on decimal fractions. But Kashi used them consistently and systematically.

Fractions had generally been expressed in Babylonian sexagesimals, meaning that what had been expressed as 50/60 + 6/3,600 + 10/216,000, Kashi could write as 0.83472 in Arabic numerals. It would be another two centuries before decimal fractions found their way to Europe.

In addition to establishing the basis of the decimal fraction system, Kashi demonstrated its utility by computing areas, volumes and other geometric problems that had no neat solution in whole numbers. He also worked out conversion tables between sexagesimal and decimal fractions and formulated simple rules to go from sexagesimals to decimals for those trigonometric functions that were not tabulated or when a conversion table was not at hand. Possibly his vigorous use of decimal fractions was influenced by Chinese astronomy. The Chinese method of computing new moons involved measuring time by dividing the day into 10,000 units, each unit being 1/10,000 of a day. The Mongol period had brought China and Islamdom into close contact, most strikingly seen in the Chinese influence on eastern Islamic art and Persian miniatures but also visible in astronomy. There are indications that Muslim astronomers visited or were employed by the Mongol court in China. Kashi in Samarqand was well within the sphere of Chinese influences, whether through Muslim astronomers visiting Mongol China or Chinese astronomers visiting Samarqand.

Rather than Kashi's genius being the dawn of another brilliant renaissance in Muslim astronomy and mathematics, it was the setting sun.

Optics

Early Muslim optical studies were based on the Hellenistic accomplishments of Euclid, Hero, Ptolemy and Theon; most works were translated into Arabic and elaborated by two of the leading Christian translator–physicians, Hunayn ibn Ishaq and Qusta ibn Luqa. The science of optics had three branches: the physics of light, which like astronomy and music was expressed mathematically; the study of vision, which was related to the physiology of the eye; and ophthalmology, which combined vision, physiology and medicine. Whereas in the Greek tradition of optics the three branches were pursued independently, in the research of the founders of optics in Muslim civilization they often converged.

Hunayn's optical works focused on vision and physiology, while Qusta combined Euclid's geometrical optics with Galen's ophthalmology to work out a curious explanation of vision called the theory of emission. According to Euclid, sight was produced by rays of light coming from a point at the center of the eye and falling on the object, forming a conic field of vision with the apex at the eye and the cone's base as the object. Euclid's theory produced more problems than it solved. It implied that each eye acted like a flashlight, emitting rays of vision, the result of which would be two intersecting cones of light. It also implied that objects upon which the eyes focused would be visible in the dark. Modifications of the theory were more promising. One propounded by the early Arab scientist al-Kindi working from Arabic translations of optical studies reversed Euclid's theory and had the rays of light going from a luminous body to the eye. Light rays traveled from every point on an object's surface to the eye, forming Euclid's cone of vision. Al-Kindi's theory of intromission became the fundamental framework for optical studies from the 10th century on, first in Islamdom and then later in the Latin West. The theory was not entirely original. Aristotle had theorized that the image of an illuminated or luminous object was transmitted to the eye, but neither Aristotle nor Euclid went into the mathematics, graphics, physiology and anatomy of the eye in explaining the mechanics and power of vision, as did al-Kindi. By the same token, the physicians Galen and Herophilus limited their optical studies to anatomy, physiology and ophthalmology. Muslim scientists, on the other hand, integrated the branches of optics as far as they could, building the theory of intromission into a unifying science.

Al-Kindi's writings covered optics in all aspects: color, reflection, refraction, concave and convex lenses, eye diseases, why the sky is blue and burning mirrors, the last being a field of study pioneered by Archimedes and Anthemius of Tralles and one that intrigued al-Kindi for its military applications.[11]

Conical geometrics applied to the study of vision led al-Kindi, Qusta ibn Luqa and the Muslim physicist ibn Sahl to study the behavior of light reflected from concave and convex mirrors. Ibn Sahl experimented with hyperbolic mirrors; another experimenter, the astronomer–mathematician abu al-Wafa, investigated parabolic mirrors. Ibn Sahl's student ibn al-Haytham extended his teacher's studies to spherical, cylindrical and conical mirrors. Ibn al-Haytham's 3rd degree algebraic equation relating the position of objects and reflected images from geometrically curved mirrors has already been alluded to. The high point of mathematical optical analysis came with the work of ibn Sahl and ibn al-Haytham, who built on the earlier achievements of Hunayn, Qusta and al-Kindi. Studying the angle by which light bends when passing through crystals, ibn Sahl discovered the geometrical equivalent to what in 18th century Europe would be called Snell's Law, the correlation between the sine of the angle of refraction and the change in the speed of light as it passes from one medium to another. Ibn Sahl realized that light traveled at an immense speed, but neither he nor his more famous student conceived of a way to measure it. It was not until the last half of the 17th century that the Danish astronomer Roemer was able to measure it with the aid of a telescope.

Continuing his teacher's research, ibn al-Haytham established a new frontier in optics with his experimental work on the rainbow and the mathematical relationships he established for reflections by parabolic and hyperbolic mirrors.[12]

His monumental work on optics, *Kitab al-Manazir*, or *Book of Optics*, a two-volume encyclopedia of optical knowledge whose thorough and systematic examination of light and vision made it the single most authoritative text on the subject for half a millennium in both Islamdom and Latin Christendom, combined the mathematical and experimental work of Qusta, al-Kindi and ibn Sahl with the physiological and anatomical work of Hunayn[13](see Plate 2).

In his application of geometry and physiology to vision, ibn al-Haytham reached a new understanding of optics and sight. According to this, vision is produced by innumerable rays of light being reflected from the surface of an object and striking the retina, upon which an image of the object is formed in accordance with the point-to-point correspondence of light-ray stimuli of the retina and the geometrical principles of optics. All the point-stimuli of the light rays impinging on the eye at the apex of the visual cone are integrated and transmitted to the optic nerve and then transmitted to the brain, where a coherent image of the object is formed. Ibn al-Haytham theorized that through refraction at the surface of the retina all the point-to-point rays conveying the object's disintegrated image (disintegrated into as many points as there were rays carrying the image to the eye) were canceled out, except for the ray that impinges perpendicularly on the retina, that is, the normal. This normal ray enters the retina without being refracted and carries with it the image of the object framed at the base of the visual cone.

Ibn al-Haytham's pointillistic intromission theory, along with his work on the psycho-physiology of vision and experimental analysis of the rainbow, defined the problems that future scientists of optics would investigate, such as image inversion, or the upright perception of an object whose image is projected upside down by point-to-point correspondence of light rays from object to eye; unity of perception, that is, the binocular fusion of an image that is received by two eyes so that we see only one image and not two; and depth perception, that is, the projection of a 3rd dimension onto a 2-dimensional image.[14]

Al-Haytham's analysis of the rainbow is a classic example of experimental physics. He made hollow spheres of clear glass, filled them with water, then sealed and suspended them in a dark room to simulate falling raindrops. Placing a screen some distance behind them, he directed a beam of light onto them in the dark room and succeeded in projecting a rainbow onto the screen. Analyzing his observations, he concluded that the rainbow was the result of a combination of reflections and refractions; some light rays were reflected off the smooth surface of the drops, which acted like mirrors, and reached the screen only after being reflected from one drop to another, and other rays passing through the drops and undergoing two refractions before being projected onto the screen. He concluded the primary rainbow was produced by a reflection and two refractions, while the secondary rainbow was generated by two of each. (The critical experiment of Newton's prism proving white light to be composed of the color spectrum was still 6 ½ centuries in the future.)

All physical aspects of light fascinated Haytham. Observing the bend of light as it passed from one medium to another, he concluded that the rays took the easiest and quickest path, anticipating Fermat's principle of least time. Light, he theorized, moved at great speed, existed independent of vision and, in intensity, varied inversely with distance. His analysis of reflections from geometrically curved mirrors produced the mathematical equation that in the Latin tradition, as mentioned previously, would be given his name, Hazen's Problem: Given the location of any two points some distance in front of a mirror (convex, concave, spherical, parabolic, and so forth.), find the solution for determining a point on the mirror's surface where the incident ray is a line from one of the two given points to the point on the mirror, and the reflected ray is a line to the second given point.

Muslim research in optics did not go beyond the level to which Haytham advanced it. It was 2 ½ centuries after his death before his book on optics seriously engaged another Muslim scientist, an indication that the momentum of interest in the physical sciences was already, by the middle of the 11th century, beginning to slacken. The short-lived resuscitation of ibn al-Haytham's work after being so long dormant was the work of the Iranian scientist Kamal al-Din al-Farisi (d. 1320). Al-Farisi was a student of Qutb al-Din al-Shirazi, an important Neoplatonic illuminationist and revisionist astronomer attached to the Mongol observatory at the capital Maragha in northern Iran, near Azerbaijan. Shirazi's teacher had been the polymath Nasir al-Din al-Tusi, astronomer, mathematician, ethicist, gnostic philosopher, Mongol advisor and political manipulator. The illuminationist symbolism of light that was central to Nasir al-Din's and Qutb al-Din's cosmological beliefs uniting science and religion appears to have been a vital factor motivating Kamal al-Din to pursue the long-neglected study of optics. It also took some encouragement. This was provided by his teacher Qutb al-Din. The result was a detailed rewriting of al-Haytham's encyclopedic book on the science of light.[15]

After studying the refraction of light passing through glass and then water, Farisi simplified Haytham's experiment by assuming that the difference in the degrees of refraction of the two mediums could be disregarded, allowing him to disregard the effect of the glass envelope around the droplet.

> He examined the propagation of rays inside the sphere between two refractions and also treated the different types of reflection. . . . He knew, moreover, that the rays refracted in the drop of water after one or several reflections in its interior are not sent equally in all directions but produce a mass of rays in certain regions of space. . . . He expressed these ideas in a complicated language of "cones" of rays that have been refracted after having undergone one or two reflections in the interior of the sphere.[16]

Though adding little to optical knowledge that had not already been presented in Haytham's work, Farisi did recycle in fresh form what had been accomplished. However, this fine recension of al-Haytham's work failed to spark a revival, since interest in optics for its own sake as an important field in the physical sciences had atrophied beyond recovery. This was at the moment optical study in the Latin West was on the rise. While Kamal al-Din was experimenting on light, analyzing the rainbow and rewriting Haytham, a Latin scientist, Theodoric of Freiburg, supplied with a Latin translation of Haytham's book, was doing the same and coming to the same conclusion as al-Farisi. The coincidence indicates the proximity of scientific levels prevailing in the civilizations of Islam and the West in the early 14th century. The critical difference is that scientific interest in the West was on the ascent, while in Islam it was on the descent. Muslims were altogether unaware of the scientific efflorescence to the northwest, while becoming more neglectful of their own.

With the death of al-Farisi, optics again fell into abeyance for another 2 ½ centuries, until revived by Taqi al-Din al-Ma'ruf (d. 1574). His book on optics neither added anything new nor succeeded in arousing interest in optical studies. Neglected for centuries, Ma'ruf's book was unknown to exist until a manuscript of it was discovered in 1987. The promising foundation in optical studies established by Qusta ibn Luqa, Hunayn ibn Ishaq and al-Kindi, brilliantly continued by ibn Sahl and ibn al-Haytham, after whom it was then left in neglect for centuries, came at last to a sputtering end with al-Farisi and al-Ma'ruf. In the West, until the publication of Descartes's original research on optics in the 17th century, ibn al-Haytham's work lived on as a primary authority, translated and printed over and over again.

Mechanics and physics

In the creative centuries of their civilization, Muslims adopted useful technology without feeling obliged to justify it by searching the Quran or Prophetic Tradition. The use of paper and the technology of producing it were early adoptions. Paper making began in China around 100 AD and came to the attention of the Arabs in 751 in Samarqand, where some Chinese technicians who had fallen prisoner after the Arab victory over the Chinese army at a Battle of the Talas River taught their captors how to make it. Paper manufacturing spread westwards through Islamdom during the 8th and 9th centuries, eventually reaching Europe from Spain and Sicily. The cheapness of paper compared to vellum made books much more accessible to a larger section of the population and helped spread learning, first in the lands of Islam and later in the Latin West, where the first paper mill was built in Italy in 1276.

Mechanical contraptions fascinated Muslims, though the art of cleverly crafted machines was not in principle something that fascinated scientists themselves. Greek philosophers and scientists tended to regard mechanics and building of machines as a lower category of thought, the way some theoretical physicists today might regard engineering. Theory, the work of a pure mind, was nobler than the practical application of it, the work of hands. This high-minded attitude of thinking being nobler than doing, elevated to philosophical dogma by Aristotle, was communicated to those Hellenistic Greeks who took time to think of machines but rarely stooped to making them, however ingenious. The joy was in the conception and design, mental experiments rather than the gritty labor of physical construction. The reality in idea was enough. The abundance of slave labor in Hellenistic and Roman society made labor saving technology unnecessary. Economic exigency rarely challenged philosophical prejudice, leaving the machines of Archimedes, Hero and Euclid as drawing board marvels.

With few exceptions, the old adage that necessity is the mother of invention holds true. Would Archimedes have constructed his complex balance had the king of Syracuse not given him the crown to measure for its content of gold and silver? Archimedes was himself a rare exception: a mathematical, scientific and mechanical genius. To his balance could be added his design and construction of war machines to defend Syracuse against the Roman siege and the system of pulleys that enabled him to drag a fully charged sailing vessel across dry land with ease.[17]

Muslims were more keen on actually constructing the machines that for the Hellenistic Greeks were for the most part an amusement in thought and sketch. When the treatises of the Greek mechanicians had been translated, a tradition in mechanics developed in Muslim society that focused on the combined application of weights, levers, balances, pulleys, screws, hydrostatics and air pressure. Mechanicians delighted in constructing machines that awed the beholder.

In the Muslim world, the art of contriving ingenious mechanical devices was a science in itself, called '*ilm al-hiyal al-handasa*, '*ilm* meaning science, *hiyal* conveying the idea of cunning, crafty, deceptive, and *handasa* geometric engineering. Taken together, the name stood for the science of creating awe and wonder through the cunning construction of machines whose connected inner workings were hidden from the viewers who saw only the outward final effect. The caliphs gave patronage to those who could build them thrones that, like Justinian's, appeared to float upwards, or who could decorate their courts with fountains adorned with mechanical trees whose leaves moved and birds whose wings flapped as they sang to the rhythm of the water playing up and down the fountains. A 13th century Iraqi mechanician, al-Jazari, wrote a fine book with illustrations on the construction of these mechanical devices, which in 1974 was translated to English by a British engineer, Donald Hill, as *The Book of Ingenious Mechanical Devices*. In it Jazari described and drew in detail constructions of 60 different types of machines, including water clocks, pumps and washing machines.

Jazari's work was the culmination of a long tradition of applying scientific principles to mechanics to measure time, weigh objects, determine specific gravity, do useful work and solve physical problems. While Archimedes related mechanical principles to geometrical figures as an exercise in the pure joy of mental gymnastics, Ibn Sina, al-Khazini, ibn al-Haytham, al-Biruni and Umar al-Khayyam conceptually investigated the physics of mechanics in terms of weight (*wazn*) and heaviness (*thiql*). Their work in this, particularly concerning the concept of *thiql*, was later conceptualized as the effect of the earth's gravity pulling the center of an object toward the center of the earth, the force (*quwwa*) being proportional to the distance between the center of the object and the center of the universe, which was of course the earth's center.[18] It was hypothesized that as an object approached the center of the earth, the gravitational pull diminished until totally vanishing at the center, meaning that an object no longer possessed any "tendency" (*mayl*) to be in motion by virtue of its "weightiness" (*thiql*), since it had reached its ultimate place, the universe's center.

"Tendency" or "impetus" (or "inclination," *inclinatio* as Latin scholars later translated it literally from the Arabic) was the inherent power of an object to move by its own weight toward the center of the universe. It was theorized that because the universe was spherical and the earth was a sphere located at the center the universe, the force of gravity therefore acted spherically on objects. All parts of an object were acted on by the force of gravity in a spherical field. This was demonstrated by a thought experiment. At sea level the surface of the water in a hemispherical bowl has a curvature equal to that of the earth's surface. When the bowl is taken to a mountain top, the curvature of the water's surface decreases in proportion to its height above sea level. At an infinite distance above the earth, there is no curvature, while below the earth's surface the curvature increases in relation to the distance between the center of the bowl and the earth's center. As the bowl descends toward the center of the earth the curvature of the water approaches sphericity. At the center, where there is no gravity, the water in the bowl is in the form of a sphere, its center coinciding with the earth's.

In another example involving a hemispherical bowl, this one actually carried out by experiment, two balls of different weight and diameter were rolled down the sides of a bowl that rested on a horizontal plane. When the balls came to rest side by side, the heavier one was of course closer to the bottom of the bowl than the lighter. The experiment proved that the center of gravity of the combined balls was their point of tangency. Two balls of equal weight would come to rest such that a vertical line through their point of tangency was coincident with a radius of the bowl, while balls of unequal weight would come to rest in such a configuration that their combined center of gravity would be displaced from a vertical line through the center of the bowl by a distance inversely proportional to their respective weights. This is the closest Muslim physicists came to Galileo's experiments of dropping balls from towers or rolling them down inclined planes.

Muslim scientists made the leap from statics to the study of objects in motion, but it was purely theoretical and innocent of mathematization. The kinetics of distance measured over time in the motion of freely falling bodies of different weights was a curiosity that never broke through the Aristotelian conviction that heavier objects fell faster. Muslims limited their mechanical studies to the static relationship between weight and gravity. Kinetics did not become a subject of investigation until the 14th century, when scholars at the universities of Paris and Oxford, building on the previous century's Latin translations of Arabic scientific texts, initiated the long process of applying geometrical methods to relate distance, time, speed and acceleration.

Muslim physicists did not challenge the Aristotelian theory of the separateness of heavenly and earthly motion. Heavenly motion was divine: circular and uniform. The celestial orbits could therefore undergo no change: no speeding up, no slowing down. Heavenly motion was the turning of the complex clockwork of spheres carrying the luminaries and planets in their periodic orbits. The spheres were real but not physical: real in the sense that mind, thought, intelligence, angels and God were real; that is, they were endowed with essence and substance but had no corporeal, tangible materiality. The heavens above the sphere of the moon were pure and unchangeable, devoid of the change and corruption of the physical realm. Hence there could be no physics of the heavens, only metaphysics.

Astronomy and physics, therefore existed in separate realms of the universe, one upper, the other lower. Physics pertained to earthly matter where motion was impressed, changeable and caused by a force acting on an object. Once impelled to motion by an external force, an object continued moving by, as Aristotle had explained it, the force of the air that rushed from the front of the object to its rear. The air displaced by the forward motion of the object acted as a propellant by pushing the object from behind. The theory presented many problems. One was that it failed to explain how a moving object would ever stop moving unless it hit something, or entered a vacuum where there would be no air to be displaced to the rear for forward propulsion.

Aristotle's theory was criticized by the 6th century Alexandrian philosopher John Philoponus, known in the Arabic tradition as Yahya al-Lughawi. His revision was taken up and developed by scientific thinkers in Islamdom, with motion now conceptualized as the effect of an impressed force. The impressed force was termed *mayl*, which conveyed the idea of a tendency or inclination toward motion that had been imparted to an object.

Three leading scientists advanced the theory of *mayl*, the Iranian physician–philosopher ibn Sina in the 11th century, the Iraqi Jewish philosopher–physician abu al-Barakat al-Baghdadi (d. 1153) and the Andalusian philosopher–physician ibn Baja (d. 1135). *Mayl* could be impressed upon or imparted to an object when it was thrown, struck, catapulted, shot from a bow or slung from a sling. Motion impressed by *mayl* was opposed to that inherent in a freely falling body whose motion resulted from the effect of gravity and whose descent was vertically toward the center of the earth. An external or impressed *mayl* imparted a tendency of motion to an object, which would remain in motion until the instant the *mayl* was depleted, upon which the object would abruptly fall to earth. Late medieval sketches of projectile trajectories made by Europeans, to whom the theory of *mayl* was transmitted, depict abrupt, cliff-like drops rather than gradual descents.

Muslim physicists did not attempt to mathematize the concept of motion; nor did they, in their analysis of it, combine an object's externally imparted tendential motion with the physical pull of vertical gravitational attraction, dual forces that produce a parabolic trajectory, a curve Muslim scientists were quite familiar with from their work in geometry and algebra. Rather than combining, synthesizing and mathematizing their studies of gravity, weight and motion, as ibn al-Haytham did in applying cubic equations to light, Muslim physicists treated each force separately, and they missed the parabola invisibly inscribed across the air by every hurled object. The closest that physicists in the Muslim world ever came to the mechanical dynamics developed by Galileo, Kepler, Descartes, Hooke and Newton was in an exposition on freely falling bodies by abu Barakat al-Baghdadi, who wrote that a constantly applied force would cause a moving body to accelerate. But he attempted no quantitative measurements and the insight, left uninvestigated, was promptly forgotten: another example of the lack of continuity as a stumbling block to a Muslim scientific renaissance progressing to a revolution.

Physicists in Islamdom excelled in problems of statics, such as determining specific weights and centers of gravity of rigidly joined systems of bodies. Through geometrical analysis they determined the loads acting on a beam. They wrote on the mechanics of the lever, pulley and screw, the hydrostatics of floating and submersible bodies, and the "Archimedes water balance" (consisting of three pans and a calibrated beam with weights) to determine specific weights and the proportions of specific metals in alloys. Many of the scientists who first mastered and then advanced beyond the mechanics of Archimedes and Hero have been encountered before – Khwarizmi, al-Kindi, Thabit ibn Qurra, ibn Sina and Umar ibn al-Khayyam, with Biruni and Khazini at the frontier. Working with actual physical objects rather than the geometric shapes upon which Archimedes based his mechanics and employing algebra and computational techniques that they originated on their own, Muslim scientists elevated statics to a level beyond that to which the Hellenistic scientists had taken it. Encyclopedic compendiums of mechanical knowledge were written by ibn Sina (*Mi'yar al-Aql*: Measure of the Mind), Umar Khayyam (*Mizan al-Hikam*: Balance of Principles), and Khazini (*Mizan al-Hikma*: Balance of Wisdom). Employing the relationships between volume, water displacement and weight of a submerged body, Biruni constructed a 3-pan hydrological balance to determine specific gravities of metals to a high degree of accuracy, and he perfected a method of finding the percentages of metallic composition in a bi-metallic alloy, as had Archimedes.

One construction project considered by a Muslim scientist, which would have required unsurpassed mechanical ingenuity for the time and would have rivaled not only the building of the Pyramids but also the Hanging Gardens of Babylon and the rest of the Seven Wonders, was in the end not undertaken; but that it was even considered is a mark of the tremendous self-assurance, if not arrogance, of Muslim scientists and engineers in their prime.

This story is worth telling. Ibn al-Haytham had once boasted when in Syria that he could solve Egypt's irrigation and flood-control problems by having a huge dam built up river around Aswan. Word of this boast reached the Fatimid ruler of Egypt and Syria, Hakim bi-Amr Allah, who invited the scientist to come and build it. Haytham blithely accepted and the caliph provided a yacht and sent him up the Nile to scout out a site for this Aswan dam. Reaching Upper Egypt and seeing in astonishment the mammoth statues, palaces and temples left by the ancient Egyptians, the scientist realized that if a dam were at all possible these master builders of antiquity would have surely built it. Ibn al-Haytham admitted defeat to himself but not to the Fatimid caliph, for Hakim had a reputation of being violently unpredictable and on the mad side. He did strange things, such as riding a donkey into the desert at night by himself or dressing as a beggar and wandering through the streets of Cairo at night, always alone. He had killed several of his viziers, without apparent reason, had demolished Christian churches, including the Church of the Holy Sepulcher in Jerusalem, had forced Christians and Jews to wear black robes (this in spite of his chief vizier being a Christian) and issued a raft of quixotic laws that convinced the populace he was truly unhinged. Ibn al-Haytham was said to be terrorized with fear at what this cruel, mentally afflicted ruler might do to him for having failed to build the dam. To save himself he feigned to have been stricken dumb in Upper Egypt and cloaked himself in madness and epilepsy until Hakim's mysterious disappearance years later. Perhaps anecdotal, the story nonetheless delivered its humbling message to Muslims who would boast beyond their abilities.

Astronomy

Much of the work done by scientists in the early period of assimilation was a refining reassessment of what the Greeks had done. Astronomical observations made from new observatories with better crafted equipment enabled astronomers to record more accurate values for planetary positions. The caliph Ma'mun commissioned his astronomers to calculate the distance of one degree on the earth's surface to determine the earth's circumference. Eratosthenes of Alexandria had computed the earth's circumference almost a thousand years earlier. He had seen that at noon on the day of the summer solstice in Aswan in Upper Egypt the sun was directly overhead since its rays fell vertically down a deep well to the surface of the water. This established a vertical line from a known point on the earth's surface to the sun at a precise moment, giving him the beginning of a right triangle. All he needed to complete it in order to compute the earth's circumference was a side and an angle of the triangle. By measuring the altitude of the sun at Alexandria at precisely noon on the day of the summer solstice, and knowing the distance between Aswan and Alexandria, he had them. Or almost. The value Eratosthenes used for the distance between the two cities was not accurate and Aswan is not directly south of Alexandria, so there was some error. Nonetheless, his method is considered one of the ten greatest experiments in the history of science, ranking with Galileo's rolling balls down inclined planes and dropping objects from the leaning tower of Pisa, and Newton's directing sunbeams through a prism to discover the composite nature of light.

Ma'mun's astronomers used a method different from that of Eratosthenes. Using a rope of known length, they counted the number of lengths it took to cover the distance between two locations in a north–south direction, and with one astronomer in one location and another in the other location, they measured the respective solar altitudes at the midpoint of a solar eclipse. The eclipse, clearly apparent to both observers at the same time, made it possible to make simultaneous measurements, giving them a triangle of two known angles and a side. The value they calculated for the earth's circumference was an improvement over that determined by Eratosthenes.

In geodesic studies, astronomers and geographers produced maps giving the latitude and longitude of cities, or rather the longitudinal degrees between them. The caliph Ma'mun is reported to have assembled 70 experts to map the empire. As a member of Ma'mun's mapping commission, al-Khwarizmi computed the latitude and longitude of 545 cities, along with those of mountains, rivers and seas.

Such work was of great importance in observational astronomy, as it was vital to know the geodesic coordinates of the place where heavenly observations were made. In his geodesic study *Tahdid al-Amakin* (*Determining Latitudes and Longitudes of Places*), al-Biruni developed a method for determining longitudinal differences between two locations. One of his methods for determining longitude was for observers in two different places to simultaneously measure the position of a heavenly body, a mutually observed eclipse allowing simultaneity. With this information, and measuring accurately the distance between the two places, the longitudinal difference between them could be computed by simple trigonometry. By employing a theorem of Ptolemy on inscribed quadrilaterals, Biruni worked out a more accurate method of longitudinal determination. He also computed shadow tables that made it easy to find the time of day. Knowing the latitude, one simply measured the angle of the sun from the horizon and referred to the tables to find the hour of the day, a particularly handy thing to have in a culture whose religion prescribed daytime prayers at certain times in relation to the sun's position overhead.

Known as '*ilm al-hay'a*, science of the structure of the universe,[19] astronomy was queen of the natural sciences, the crowning glory that metaphysics was to philosophy. Astronomy revealed heaven's map of the divine mind. Many of the advances made in mathematics and in methods of computation were spin-offs from astronomy, the science whose structure and content was most mathematically imbued.

The framework of principles governing the heavens as established by Ptolemy was eventually adopted by Muslim astronomers. Previously, Muslim astronomy had been dominated by Indian influences. Indian methods had been adopted in the late 8th century. It was not until the beginning of the 10th century and the work of one of Islamdom's greatest astronomers, al-Battani, that the century-long process of clearing the field of Indian influences, coming mainly from Brahmagupta's *Surya Siddhanta* that had been translated from the Sanskrit early in the Abbasid period, that Ptolemy and the Almagest replaced Brahmagupta and the Surya Siddhanta. In deference to its geometrically more powerful Greek rival, the Indian system was gradually superseded, at which point a number of Muslim astronomers exerted a considerable amount of mental energy in their effort to liberate Ptolemy's system from what they considered to be its annoying wrinkles, if not flagrant flaws.

Muslim astronomers were puritanical fundamentalists when it came to the particulars of theory. They consistently criticized Ptolemy for his less than strict adherence to the divine principles that ruled the universe, namely circular and uniform motion of the heavens around a central point, the earth. Much geometric tinkering went into re-engineering Ptolemy's cluttered machinery to reach the cherished goal of a system that ran strictly according to principle and at the same time faithfully represented the actual motions of the heavens as recorded by observation. Ptolemaic reform and improved mathematical techniques in performing computational operations on a sphere were the dynamics driving Muslim advances in astronomy.

Planetary models that represented observed celestial motion had to conform to cosmic reality as defined by principle. Another goal Muslim astronomers strived for, this having to do with observation rather than theory, was to perfect the precision of the geometric clockwork so that the planetary models came as close as possible to the observed motions. Some of Ptolemy's parameters did not do the job to the satisfaction of observational astronomers. The larger Muslim observatories, with their more finely designed and accurate equipment, allowed for more precise observations of the heavenly bodies than had been made by the Babylonians, Greeks and Indians. Increased precision of observation required tighter values for the parameters governing planetary motions represented by the geometric models. Improvements in one demanded improvements in the other. Ptolemy's modular constructions were a superb feat of geometrical ingenuity, but they were not quite good enough for Muslim astronomers.

The planetary models with their wheels, cranks, wheels on wheels and little circular mechanisms to impart latitudinal motion to the planets, and the modular parameters that governed the distance and timing of the moving parts, approximated to a remarkable degree what the astronomer observed when tracking the heavenly orbits over the years. The whole thing was a geometric masterpiece of brilliant imagination and a cunning sleight of hand that fooled no astronomer worth his salt. It took all that to capture in geometry the God-given system of the heavens, since in reality Ptolemy's (and before him Aristotle's) divine trinity of principles were all three wrong: Earth is not at the center; the motion of the orbiting bodies is not uniform; and the motion is not circular. Considering the formidable task of capturing the motions of the heavenly bodies in a systematic geometrical structure that presumed those erroneous principles, Ptolemy's system came remarkably close to accomplishing an impossible task, at the price of sacrificing simplicity. It could not have been otherwise. But that it survived 1500 years, in spite of flaws, artifices and the criticisms it suffered from first Muslim and then European astronomers, says something about the quality.

The system worked like this. A large circle (called the deferent) carried another circle (called the epicycle) that carried the planet. The center of a given planet was on the circumference of the epicycle, whose center was on the circumference of the deferent circle. The center of the deferent represented the center of the universe, where earth stood motionless. The deferent and epicycle synchronously completed a full 360 degree rotation, representing a solar year. In effect, the planet circled a point that was at the same time circling another point. The composite planetary motion produced by the deferent–epicycle arrangement was able to approximate the retrograde motion of the outer planets, that is, that the outer planets, Mars, Jupiter and Saturn, as observed from a supposedly stationary earth, appear to slow down, halt, go backward and then start forward again, since in reality they are moving slower than the earth in their solar orbits.

But that was not all. Yet another geometric device was required, one to keep the planetary orbits to a circular path, as principle demanded. Their actual elliptical orbits are close to circular but not quite, and the non-circularity was apparent to astronomers. Unable to question principle by considering the orbits to be other than

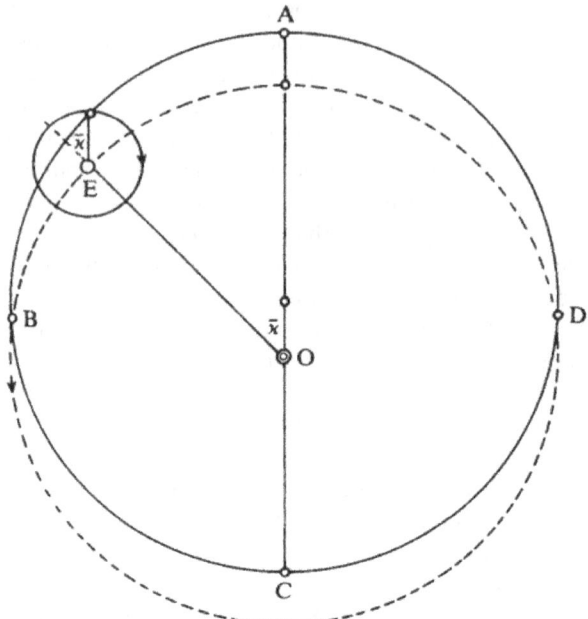

Figure 2.1 Ptolemy's solar model (Figure 3.1 of George Saliba's *Arabic Planetary Theories in Encyclopedia of the History of Arabic Science*).

ABCD is the eccentric sphere; DEB the solar orbit or ecliptic, with the earthly observer at O, the center of the universe. The center of the epicycle is on the ecliptic at E and rotates at the same angular velocity as the ecliptic but in the opposite direction. The center of the (imaginary) eccentric sphere is a distance x from the observer. The eccentric and the motion of the epicycle could account mathematically for the change in motion and apparent size of the moon in its orbit around the earth at apogee, perigee and quadrature.

circular, the models had to be adjusted to preserve circularity and at the same time impart planetary motions that agreed with observations. To correct for the difference between circular and observed motion, Ptolemy put the center of the planetary orbits off-center from the center of the deferent, where the earth was placed. The distance between the center of the deferent circle and the center of the planetary orbit was called the "eccentric." It wa successful. If observed from that point, the planet would appear to orbit the earth in perfect circles. And there he had it. By constructing a circle around the eccentric point, giving that circle the same radius as the deferent and placing the center of the epicycle on the circumference of the eccentric circle, Ptolemy came impressively close to crunching the heavenly system into his tight-fisted geometry of spinning spheres. A marvel of simplicity it was not. The planet was carried along on the turning circumference of the epicycle whose center was carried along on the oppositely turning circumference of the eccentric circle whose center was to be thought of as the center of the universe, center of the earth, which it was not. There was more.

The planets also have latitudinal motion as they orbit the earth from east to west, that is, their orbits are tilted slightly in a north–south direction. This required more geometric finesse. To capture the latitudinal motion of the outer planets, Mars, Jupiter and Saturn, Ptolemy introduced a circular seesaw mechanism.

All in all, it was elegant considering the restraints imposed upon it. Supplied with the right parametric values of angular displacements, diameters and eccentrics, the eccentric-epicyclic models came tolerably close to what astronomers observed. The agreement between observation and heavenly geometry made it possible for some scientists to accept the constructions as being real, similar somewhat to the way modern scientists consider the mathematical constructs interrelating light, mass, energy, gravity, force, distance and time to express a level of reality in nature, or more precisely, the way some theoretical physicists accept Superstring theory to be real on the sub-atomic level of nature, even though the reality only makes sense in terms of mathematical constructs. Ptolemy's *Planetary Hypotheses*, translated into Arabic at Ma'mun's House of Wisdom, described the heavenly spheres to be real. That Aristotle accepted the heavenly reality of the concentric

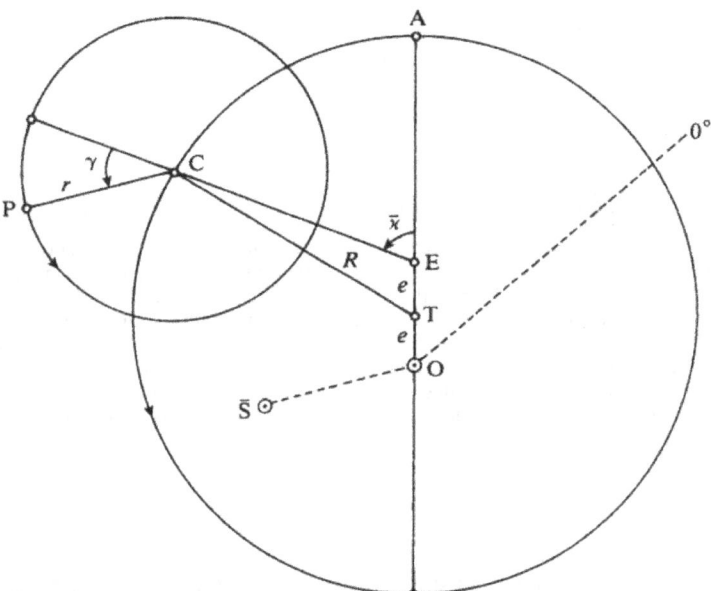

Figure 2.2 Ptolemy's models for the upper planets, those beyond the solar orbit: Mars, Jupiter, Saturn (Figure 3.3 of Saliba's *Planetary Theories*).

The motions of the upper planets were observed to be simple enough for representation in a self-explanatory diagram. T is the center of the deferent carrying the epicycle, center C, which in turn carries the planet at P. The equant is at point E; O the center of the universe, the observer. The distances marked by e represent the distance of the equant and observer from the center of the deferent.

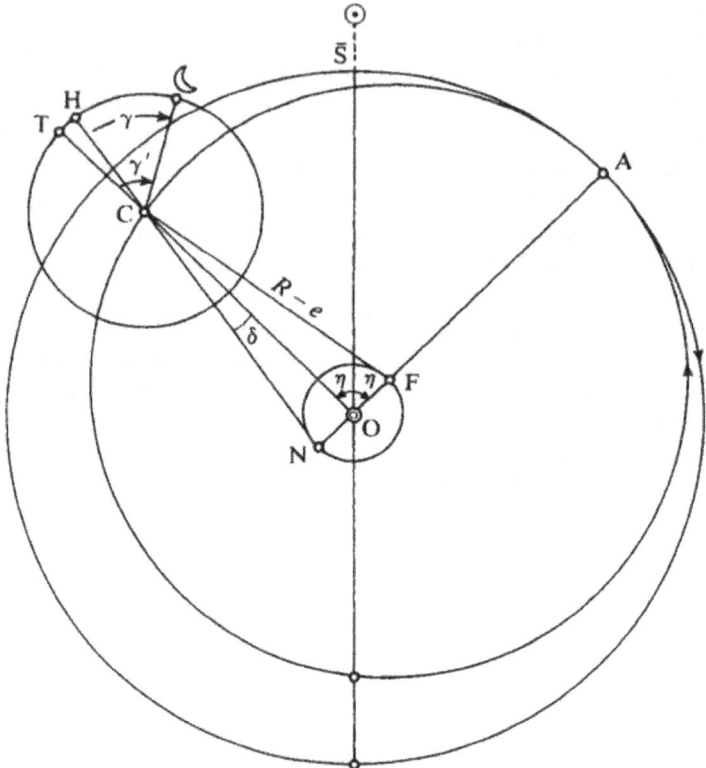

Figure 2.3 Ptolemy's lunar model (Figure 3.2 of Saliba's *Planetary Theories*).

As evident in the diagram, lunar motion was more complex than solar. Ptolemy added a tiny encompassing sphere around the earthly observer, called sphere of the nodes (diameter NF in the diagram), whose center is the center of the ecliptic. As the sphere of the nodes moved, it carried the apogee of the deferent A. The center of the epicycle carrying the moon is at C on the ecliptic. The mean lunar apogee is at H on the epicycle. Ptolemy's lunar model confronted Muslim astronomers with inconsistencies that contradicted Ptolemaic principles, leading to modular reconfiguration by Muslims. With respect to the moon, Muslims saw Ptolemy's great defection from principle as being his failure to conceive of a model that had the spheres moving the moon uniformly around a common center.

spheres in his own planetary system was proof enough of their existence in Ptolemy's system, and they were accepted as such by some physics-oriented Muslim astronomers.

There was, however, one geometric device Ptolemy introduced that Muslim astronomers found repugnant and wanted to be rid of. This was the equant, an imaginary point that equalized a planet's angular motion as observed from earth. It was obvious by observation that planets did not move in uniform motion around either earth or the eccentric point. They sped up and slowed down. But if viewed from this imaginary point, they appeared to conform to the principle of uniformity. The equant would be the system's aggravating thorn that Muslim scientists could not suffer. Deferents, epicycles, eccentrics, latitudinal seesaw circles and equants, to mention only the most obvious elements in the Ptolemaic modular edifices, composed the geometry of a universe constructed in accordance with the principles of circle, uniform motion and geocentricity, when in fact the heavenly architecture of planetary motion is, as came to be known in the 17th century, patterned on a sun-centered ellipse, with the planetary bodies undergoing varying motion and distance from the sun during the period of their solar orbits. But of all the geometric paraphernalia that went into making the system run in close approximation to the observed clockwork of the heavens, it was that little equant that Muslim astronomers found impossible to swallow. It was cheating. The equant too obviously violated the principle of circular motion. Eccentrics, epicycles and latitudinal crank devices were no problem, they were essential parts of the machine, but the equant was just stuck there, an imaginary point not moving anything; an ignoble blemish standing shamelessly naked defiantly doing nothing: a cheeky barefaced scandal to the beauty and elegance of God's creation. It topped the list, by far, of criticisms Muslim astronomers leveled against Ptolemy's system. Like a woman disgraced, it had to be banished.

The Muslim endeavor to square the reality of the heavens as seen by observations to the reality of Aristotle's and Ptolemy's holy trinity produced a new branch of theoretical astronomy, *'ilm al-hay'a*, the science of the structure and system of the heavens. Its scientists strove to achieve an astronomy of inner consistency, and in so striving produced a series of revisionist models of Ptolemy that found their way to the Copernican models, when European astronomers were, in their turn, searching for coherence.

Criticism of Ptolemy has a long history in Muslim astronomy. Ibn al-Haytham and the astronomer abu 'Ubayd al Juzjani (d. 1070) reacted with the same critical sense to what they regarded as Ptolemy's violation of both philosophical principle and physical reality. It contradicted common sense and reality to have a planet orbiting in a circle if the axis of the sphere carrying the planet on its perimeter did not pass through the center of the orbit. Ibn al-Haytham's *Doubts Concerning Ptolemy (Shukuk 'ala Batlamayus)*, which criticizes not just the *Almagest* but Ptolemy's *Optics* and *Planetary Hypotheses* as well, attacks his planetary models for their complexity and superfluity. As nonsensical as the equant that gave non-uniform motion to the planet was Ptolemy's clumsy construction that was supposed to account for the motion in latitude of Venus and Mercury. That also violated the principle of uniform circular motion around the center of the earth. The epicyclic mechanism of his lunar model was another infraction of the principle.[20]

Ibn al-Haytham focused on the logic of physical reality of the geometric constructions underlying heavenly motion. Just as Ptolemy in his *Planetary Hypotheses* accepted the physical reality of the spheres carrying the planets in their heavenly orbits around the earth, so too did ibn al-Haytham; but, Haytham rejected the imaginary points and planes that Ptolemy employed to make his planetary

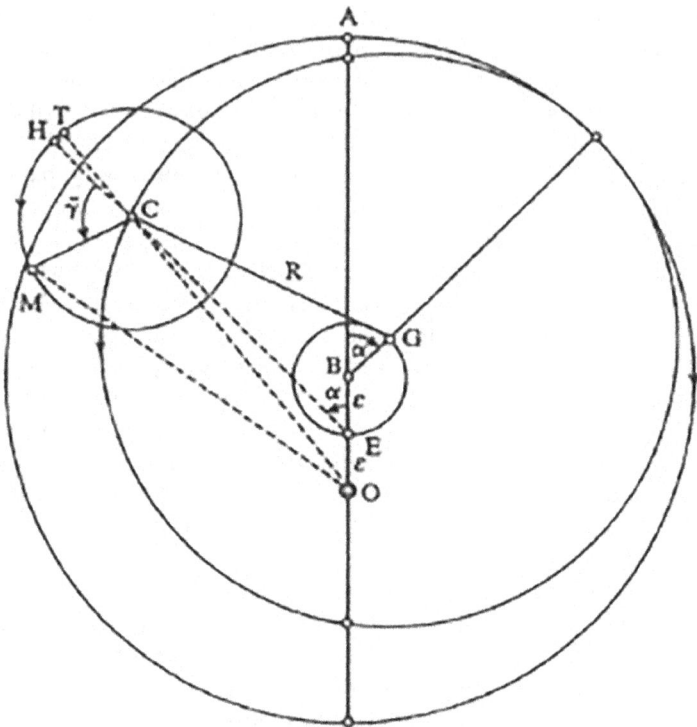

Figure 2.4 Ptolemy's model for Mercury (Figure 3.4 of Saliba's *Planetary Theories*).

Mercury's motion was observed to be complex, compared to the sun and upper planets. To help account for it, Ptolemy added a small sphere with center B inside the spheres of the deferent and ecliptic, as he had in his lunar model with the sphere of nodes. The center of the deferent at G is on the inner sphere. The deferent radius is GC. As the inner sphere rotates, G is carried by its motion. The extension of BG to the deferent circumference represents the apogee. This inner sphere, analo- gous to the mechanism Ptolemy employed in his lunar model, acted similar to an epicycle whose sphere moves another sphere with its center fixed on its surface. Both deferent and epicycle (MHT) move in the same direction opposite to the zodiac. E is the equant point, O the center of the universe and e the distance from the equant to the observer and to the center of sphere carrying the center of the deferent.

models work. According to Haytham, these had to be mental geometric constructs that Ptolemy had made up in order to create a modular system for tracking plan- etary motion and not something meant to be physically real. Haytham wanted a system without imaginary constructs, a structure of the heavens whose geometry represented the existence of a physically real system of spheres rotating uniformly around a point at the center of the earth, with all motions accounted for as simply as possible: In other words, nothing less than a geometric construct of heavenly reality that coincided with physical reality. By maintaining that there existed, in reality, a geometrically structured system of the orbiting planets free of all the contradictions and impossibilities that riddled the Ptolemaic system, he came the closest of all Muslim scientists to bringing a mathematical physics to bear on astronomy.

Ibn al-Haytham's quest for a flawless system of a physically structured astronomy, like his optical theory, found no followers to explore the possibilities of his original ideas. Not even ibn al-Haytham himself, with all his doubts about the Ptolemaic models, was driven to search for a simpler system that might better correspond to observations, though he had no doubt such a system existed, however complex the motions:

> The truth that leaves no room for doubt is that there are correct configurations for the movements of the planets, which exist, are systematic, and entail none of these impossibilities and contradictions, but they are different from the ones established by Ptolemy.[21]

Applying observational data to find a simpler and better system was more than a half millennium in the future, and though Muslim astronomy prepared the way for it, it was not to happen in Islam. As far as concerned theory, observational data were used only to sharpen Ptolemaic parameters. The Ptolemaic system itself and its principles stood.

Nonetheless, the many observatories Muslim dynasts built from Cordova to Samaqand over the course of seven centuries immensely enriched astronomy through the thousands of individual observations recorded column by column, page after page in a hundred and more bound tables, or zijes. The zijes, the collective product of centuries of royal patronage, formed the bedrock of Muslim astronomy.[22] New observational centers, such as those built in Fatimid Cairo and Umayyad Cordova, affluent cities where competing caliphates were founded around the same time in the 10th century, added to the impetus of astronomical study that had begun in Baghdad. In Fatimid Egypt, the reputedly mad Caliph Hakim bi-Amr Allah (c. 1010) founded his *Dar al-Hikma* (House of Wisdom) in the dynasty's new capital city. Patterned on Ma'mun's *Bayt al-Hikma* in Baghdad, Hakim's *Dar al-Hikma* was centered around the institute's primary component, the observatory, which was erected by the caliph's chief astronomer, ibn Yunis. Ibn Yunis dedicated to his patron the set of tables he had composed (the *Zij al-Hakimi*) while directing the observatory. Ibn Yunis's zij was one of the chief tables in the long tradition of Muslim observational astronomy. These observations enabled astronomers to determine more accurate values for a planet's latitudinal motion and orbital cycle, as well as a more precise value for precession, that is, the degree of wobble of the earth's axis of rotation.[23]

Muslim expertise in building better and larger instruments of observation enhanced precision of measurement, and hence an appreciable sharpening of Ptolemaic parameters. Muslim astronomers were famous for their precisely crafted mural quadrants, astrolabes, sundials, celestial spheres and armillaries. Improved equipment and methods of observation allowed for more accurate determinations of star positions, such as those recorded in abd al-Rahman al-Sufi's *Star Catalog*; and, as alluded to in the above paragraph, for more precise observations of the latitudinal motions of the Moon, Mercury and Venus, which in turn allowed astronomers to more finely tune Ptolemy's planetary models in the perennial quest of conformity with observational astronomy.

Observation, the practical side of astronomy, was considered by its royal patrons more important than reforming the theory. Reform was what astronomers did in their minds, thinking in circles turning on circles, when they were not busy observing. It could be done with pen and paper. Observation required expensive observatories and equipment, hence the need of royal patronage or support of wealthy court officials who wanted precise observations of heavenly bodies, whether for the glory of astronomy, their own glory (of course) or for the sake of having more precise forecasts made by astronomy's somewhat disreputable but widely popular sister, astrology. Theoretical astronomy was the heady plaything of critical discontent, and however deep the discontent, theoretical astronomy remained more or less grudgingly faithful to Ptolemaic principles and modular patterns until late in the Muslim scientific experience, when there at last arose two attempts to break through the Ptolemaic mold, the first in 12th century Andalusia, the second in 13th and 14th century Iran.

Andalusia

Muslim Spain was a latecomer to scientific productivity. Not until the end of the 10th century did astronomy there approach the level of work progressing in Egypt, Syria, Iraq and Iran. The breakup of the caliphate and subsequent dispersion of science to the provincial capitals contributed to the efflorescence. Around 1000, just as the prestigious Umayyad caliphate in Andalusia was about to go down, Andalusia's entry into major league science was signaled by the publication of al-Majriti's corrected edition of Khwarizmi's zij. Toward the end of the century the astronomer al-Zarqali, in cooperation with a Jewish astronomer, abu Ishaq Ibrahim ibn Yahya, produced a table of observations made in Toledo just before it fell to Christendom in 1085. Zarqali's improved astrolabe, whose design and construction gained fame throughout Islamdom and then later in the West, signaled Muslim Spain's sophistication in crafting the complex plates of inscribed stereographic projections of stellar positions at different latitudes that, among other components, go into making an astrolabe. Andalusian astronomers had by then mastered the science enough to start reforming theoretical astronomy to rid it of the imperfections Ptolemy's models had laid on it.

A century after ibn al-Haytham wrote his criticism of Ptolemy, the influence of which had reached Andalusian astronomers, Jabir ibn Aflah wrote his *Reform of the Almagest* (*Islah al-Majisti*), listing his own problems with the system. A major criticism was the values Ptolemy had for the distances of Venus and Mercury from the center of the earth. If the values were anywhere near accurate, the two planets would be beyond the solar orbit, somewhere out there with Mars, Jupiter and Saturn. It was a perplexing problem. All astronomers agreed that the planets circled earth with the sun placed after Mercury and Venus in order of proximity to earth. But observations showed that Mercury and Venus were, during a part of the year, closer to earth and the other part more distant from it than the sun, as indeed they are, a fact that poses no problem in a sun-centered system.

The problem was not Ptolemy's values but the system itself. It could give no explanation why a solar parallax of three minutes could be measured but no parallax at all could be seen for the inner planets part of the year (when they were on the other side of the sun and too distant for parallax to be observed by the naked eye).[24] The obvious conclusion would be that Mercury and Venus were, with respect to earth, on one side of the sun part of the year and on the other side the other part. By putting the sun at the center and the earth where the sun was thought to be, the confusion could have been resolved. Muslim astronomers had in fact considered heliocentricity, among them al-Biruni and Nasir al-Din al-Tusi, but the idea was not seriously pursued, neither as a Pythagorean esthetic in the manner of Copernicus, nor as a Platonic idea substantiated by observation married to mathematics in the manner of Kepler. Like the break of Copernicus and Kepler with Ptolemaic astronomy, the 12th century Andalusian cleavage was philosophy-driven: not Pythagorean or Platonic, but Aristotelian.

Three vizier–scholars in the Berber Muwahhidun court, ibn Baja (c. 1150), ibn Tufayl (d. 1185) and ibn Rushd (d. 1198), one the student of the other in succession, were all three so passionately devoted to Aristotle (ibn Rushd going so far as to call Aristotle "divine") that even his astronomy with all its weaknesses was embraced. Consequently, a system burdened with major contradictions was adopted to replace a system burdened with contradictions hardly of a lesser order. It was nonetheless a beautiful and coherent system.

Aristotle's configuration of the heavenly spheres was so elegant and geometrically perfect that any omniscient deity would have had no choice but to construct the heavens accordingly. A series of hollow contiguous spheres were configured concentrically nested around the earth, each sphere having its own uniform motion with respect to the others within and without it. Everything was spherical, concentric, harmonious, uniform and pure: no equants, no epicycles, no eccentrics, no crank and swing mechanisms for latitudinal motion. It was a majestic system consistent within itself. The problem was, it required some 55 spheres. And even with that, it was geometrically hopeless as a model patterned on the observed periodic motions of the heavens. Philosophically pure, it was a shamble with respect to observational astronomy. The difficulties Aristotle's system presented produced a critical reaction on the part of Muslim astronomers as sharp as their critique of Ptolemaic astronomy.[25]

The second of this trio of Muwahhidun greats, ibn Tufayl, author of the famous philosophical romance on self-enlightenment, *Hayy ibn Yaqzan* (*Living, Son of Consciousness*), produced nothing in the way of astronomical literature supporting Aristotle's system of nested spheres, but according to his student, the astronomer al-Bitruji (d. 1204), he had intended to find a system that was faithful to

Ptolemy's principles, avoiding eccentric and epicyclic spheres.[26] Ibn Tufayl's other student, ibn Rushd, a much greater physician and metaphysician than astronomer, believed that astronomy had become an unreal jumble of contradictory motions whose only reality was mathematical computation.[27] Ibn Rushd's grand aspiration was to abandon Ptolemy altogether and prove the Aristotelian system of concentric spheres to be the real one that reigned in the heavens. He never quite got around to it:

> In my youth I hoped I should be able to complete such an investigation. But now in my old age, having been hindered from doing this by various obstacles, I have lost the hope of accomplishing it. But these words may help excite some investigators to inquire into these things, since in our times the science of astronomy is nonexistent in that it conforms to opinion rather than to reality.

But how, without resorting to epicycles and eccentrics, to explain the apparent change in orbital speed, retrogression and distances of the planets in a geocentric system? Like Aristotle, his mortal god, ibn Rushd had an answer for everything. Optical illusion. The play of light in the clouds and the mist veiling the heavens from earth confused and distorted observations. Rather than admit it was an impossible task, ibn Rushd excused himself with the face-saving gesture of infirmity and heavenly obfuscation. Nonetheless, the quest to liberate astronomy from the improbable baggage with which the Ptolemaic system burdened the heavens was a brave gesture. It signaled the urge to break out of the Ptolemaic straightjacket and reach a simpler system. Possibly, with continuity of improved equipment and more exact observations conjoined to theoretical exploration, Andalusian reform astronomy might have reached a Copernican conclusion, once a rebellious generation of young Andalusian astronomers had rejected the rationalist devotion to Aristotle of their fathers.

Given the metaphysical constraints imposed on any system accounting for planetary motion, it was a choice between, on the one side, unlikely complexity with geometric accuracy accompanied by some contradictions in principle, and on the other, principled simplicity but in the face of unswallowable contradiction to observable phenomena. Ibn Tufayl and ibn Rushd were competent but not creative astronomers. They were, above all, physicians and philosophers who, like any of the educated elite in the rational tradition, had sound training in astronomy, as well as in mathematics, music, optics and all the branches of natural philosophy. They were competent astronomers whose reformist astronomy was a consequence of their absolute devotion to Aristotle as much as it was a rejection of Ptolemy's geometric complexity and transgression of principle. They found it difficult to believe that anyone who could have divined the workings of physical nature and hidden metaphysical principles of the universe and brought them all together in a tidy, well-structured and harmonious system as tightly and elegantly as had Aristotle, from starry firmament to frogs and acorns, could not but be right about heavenly structure and motion.

A more fertile break with the Ptolemaic dead end that had challenged over two centuries of Muslim astronomers and would-be reformers was al-Bitruji's astronomical theory. Realizing that the Aristotelian system was as hopeless as the Ptolemaic system, al-Bitruji developed a hybrid system in an attempt to avoid the flaws of both. Striving to marry the philosophical purity of Aristotle's concentric spheres to the geometric accuracy provided by Ptolemy's eccentric, epicycle and equant, Bitruji devised a complex system of rotating concentric spheres that carried the planets in their orbits around the earth.[28] This was purely Aristotelian. To gain the accuracy of Ptolemy, and account for retrograde motion, Bitruji configured the moving spheres in such a way that they imparted a spiraling or helical motion (*lawlabiyya*) to the orbiting planets.

His *lawlabiyya* system sacrificed mathematical power for philosophical purity but was still a daring departure from astronomical tradition in pursuit of a fitting system for the heavens, a medieval Ptolemaic–Aristotelian hybrid. On the order of Tycho Brahe's Ptolemaic–Copernican hybrid, where the earth orbits the sun while the other planets orbit the earth, it was a timid step away from Ptolemaic authority and could have been a fruitful lead to a sun-centered system, had there been other astronomers to follow it up. There is no telling what Bitruji's system might have led to if he or a few inspired followers had, like Kepler centuries later, observed Mars closely with the best equipment available at the time – and had they been assisted by a colleague with the mathematical power of an ibn al-Haytham, Umar al-Khayyam or al-Kashi to fill the role that Kepler played for Tycho Brahe. It should be noted that, had it not been for the continuity, connectedness and relatively speedy intellectual exchange that webbed European scientists into a community, loose as it might appear to have been when viewed from the present, the Copernican and Tychonian systems might well have shared the fateful oblivion, at least for some period, that Bitruji's *lawlabiyya* spiral was destined to suffer: a potentially fruitful escape hatch from the Ptolemaic cul-de-sac left to die.

One of the primary differences between the Muslim and Western scientific experiences, typified by Bitruji's innovation in astronomy or ibn Baja's theory of *mayl* in the earthly physics of moving bodies, is that nobody in Cordova or other Muslim capitals came along to think on and explore ibn Baja's or Bitruji's scientific breakaways. Bitruji's helix attracted no fellow astronomer in the Muwahiddun court, no student or visiting astronomer from Egypt, Syria, Iraq, Iran or Samarqand to keep alive and transmit a possibly fruitful insight that could have served as a springboard to other orbital possibilities.

In the miniscule nub of Latin Europe, laced as it was with universities in the 16th and 17th centuries, the seminal constructions of Copernicus were carried on and modified by Brahe and Kepler, with Mersenne, who knew Galileo and Descartes, busily communicating new ideas through his wide correspondence with scientists and philosophers across Europe. Communication kept the European scientific community growing. Scientific thinking explored possibilities outside the established systems existed in Muslim civilization, but the scientific community lacked a supporting institutional infrastructure that connected its members into what was going on in their world.[29] After 1100, Islam became, with respect to science at least, a civilization of solitary genius. Umar Khayyam and al-Bitruji in the 12th century, Nasir al-Din al-Tusi in the 13th, al-'Urdi in the 14th, Ghiyath al-Din

al-Kashi in the 15th; the greats came and went. Their post-12th century efflorescences were short and sweet.

With but one significant exception, there were no continuities of advance, with
one generation building on the work of its predecessor, as had existed from the 8th
to the end of the 11th century in Islamdom. The exception was the more than
century-long tradition of reformist astronomy initiated by Nasir al-Din al-Tusi (d.
1274), director of the Mongol Il-Khanid observatory at Maragha in northern Iran,
whose variant versions of Ptolemaic models were advanced by several first-rate
astronomers – the Syrians Mu'ayyid al-Din al-Urdi (d. 1266) and ibn al-Shatir (d.
1350) and the Iranian Qutb al-Din al-Shirazi (d. 1375), among other bright lights.

Maragha and Samarqand

Between 800 and 1050, the area of highest concentration of observational astronomy fell between the Nile and Tigris Rivers. Thereafter, the area of highest concentration shifted east to northern Iran and central Asia, to observatories in
Nishapur, Maragha, Bukhara and Samarqand. The eastward shift more or less
parallels the decline of Abbasid caliphal wealth and authority and patronage as
powerful provincial dynasties arose in the east.

Creative astronomy continued long after the period that is generally considered
the golden age of Muslim science. One obvious reason for this is that astronomy was
in many ways useful to religion, as in determining times and direction of prayer,
lunar visibility for beginning and ending Ramadan and other religious feasts, such
as the Hajj pilgrimage. Muslim theologians "saved," as it were, astronomy by remov-
ing it from its Hellenistic philosophical moorings and making it Islamic and declaring it to be a manifestation of God's glory, perfection and wisdom, thereby stripping
astronomy and its heavenly motions of all causal autonomy. Even in the 12th century
and after, after al-Ghazali, "more and more men of science were men of religion."[30]
More accurately stated, those men of science Professor Saliba refers to, who were
more and more men of religion, were in fact men of religion who practiced astronomy only in its time-reckoning branch, to determine times of prayer, new moons,
religious feats and orientations of mosques. This is known as '*ilm al-tawqit*: Its
practitioners were men of astronomy but not of science in general.

Accordingly, astronomers of the 13th and 14th centuries and after have been
termed legist-astronomers, to distinguish them from the Hellenizing philosopher–
scientists of the earlier period, before the Hellenistic autonomy of science and
philosophy had been "tamed" by *kalam*, or Muslim scholastic theology.[31] For this
reason, the scientific efflorescences that appeared in Islamdom into the 16th century were essentially in astronomy, the finest of them occurring in Iran following
the Mongol incursion.

The several astronomers who followed Nasir al-Din al-Tusi's lead in reforming
Ptolemy's system have been grouped together by contemporary historians of Muslim
astronomy as "The Maragha School." The reform astronomy was centered on the
fine observatory built in Maragha by Hulagu Khan at Nasir al-Din's urging, following the destruction of Baghdad.[32] The Maragha observatory had a staff of 20

astronomers from various regions of the Muslim and Mongol world, among them a Chinese mathematician. Generously funded by the Mongol ruler, the observatory covered a large area and was provided with an extensive library and workshops for constructing observational equipment, whose size and precision mark an appreciable advance in science.[33] Appointed by Hulagu as director of the observatory, Nasir al-Din, as deft in his astronomy as he was in his political machinations, obliterated the offensive equant with the same decisiveness with which he is reported to have counseled Hulagu to obliterate the Sunni Abbasid caliphate: one an offense to his sense of heavenly propriety, the other to his earthly religious persuasion.[34]

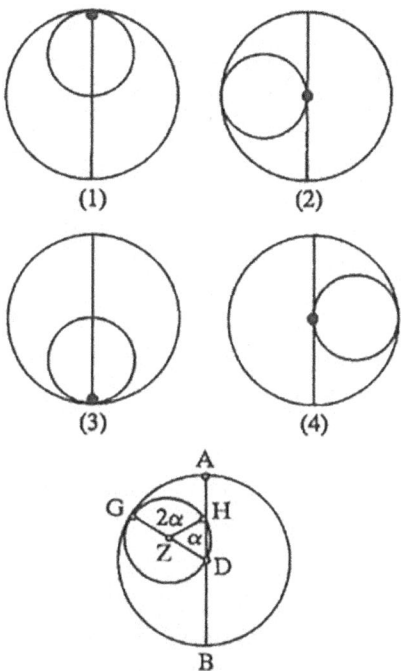

Figure 2.5 Nasir al-Din's al-Tusi's double epicycle, known as the 'Tusi Couple' (Figure 3.11 of Saliba's *Planetary Theories*).

The brilliance of the double epicycle is that it renders two circular motions into a linear motion and accounts for observed planetary motion without having to introduce Ptolemy's arbitrary equant. Marking an advance in planetary modeling, the double epicycle may have also added, in a passive way, to theory and practice by serving to exhaust the imaginative power of later astronomers in 16th century Europe, who endeavored to compress the observed motions of the planets into conforming to Ptolemaic principles in a geocentric universe. The inner sphere is half the diameter of the encompassing one. They move in opposite directions, the smaller sphere at twice the angular velocity as the larger. A point on a circumference of the inner sphere will trace out a linear trajectory, AHDB, of the bottom figure. Tusi applied it to the moon and the upper planets and was able to do away with the crank mechanism Ptolemy introduced in his lunar model to account for the changing distances of the moon as observed on earth during the lunar orbit.

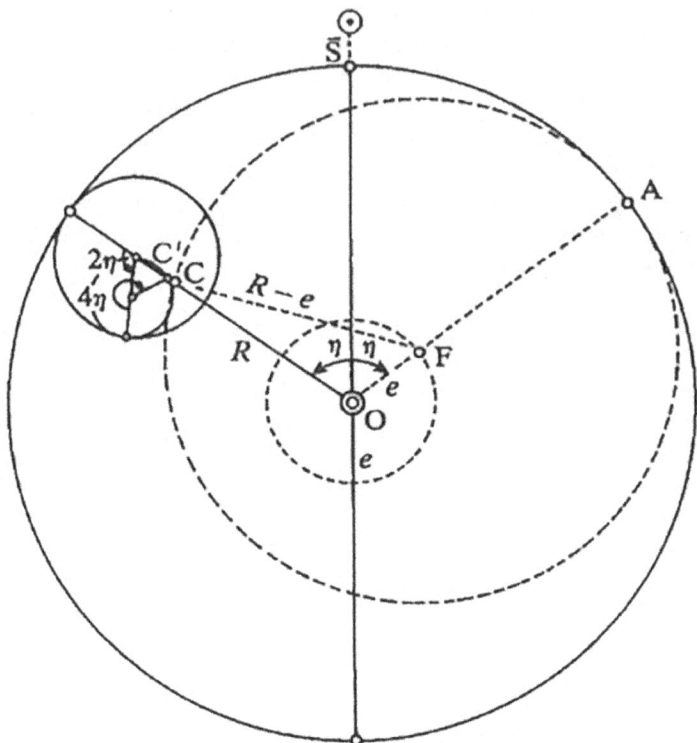

Figure 2.6   Nasir al-Din's al-Tusi's lunar model with double epicycle (Figure 3.12 of Saliba's
Planetary Theories).

The double epicycle has the deferent moving uniformly around the center of the universe and also brings
the moon closer to the earth at quadratures and takes it further away at conjunction and opposition.

Nasir al-Din was not the first Muslim astronomer who played around with epicycles to be rid of the equant, though he usually gets the credit for it. Earlier efforts
had been made, some of them showing imaginative innovation, the most effective
of which added an outer secondary epicycle to the existing primary one. This
system placed the planet on the circumference of the outer epicycle, whose center
turned on the circumference of the primary epicycle. With the correct angular
direction and velocity, the extra epicycle, whose radius was equal to the value of
the deferent circle's eccentric, provided somewhat of an improvement in creating
the appearance of uniform planetary motion. But the discrepancy between modular
and observed motion of the planet was still too great to be ignored.

A variant of the extra epicycle model was made by a contemporary of Nasir
al-Din, Mu'ayyid al-Din al-Urdi, a Damascene astronomer credited with designing
the Maragha observatory complex and its equipment. Al-Urdi added to the Ptolemaic model an epicycle of radius half the value of the eccentric. This proved to

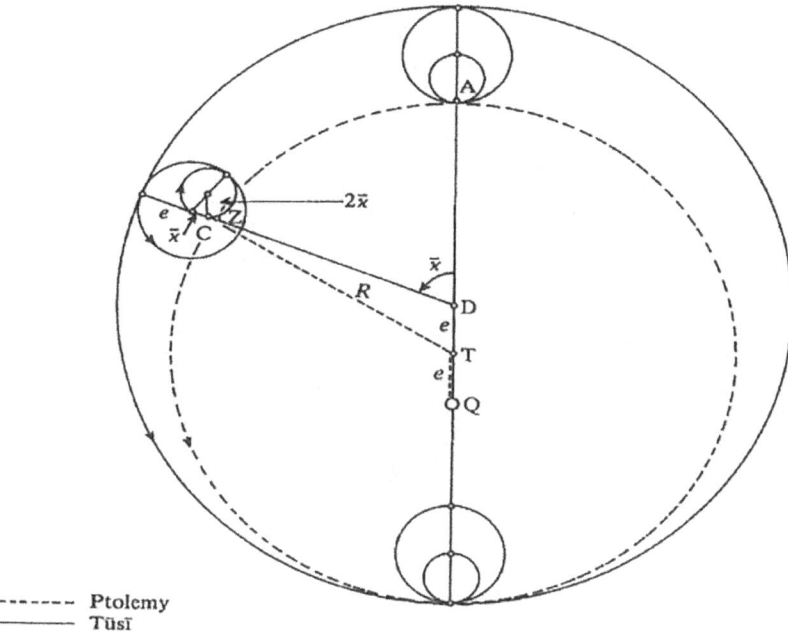

------- Ptolemy
———— Tūsī

Figure 2.7 Tusi's planetary model with double epicycle (Figure 3.24 of Saliba's *Planetary Theories*).

The double epicycle, or as it is often called, the Tusi Couple, did away with Ptolemy's imaginary equant, which was an affront to principle that could not go unchallenged and eradicated to Muslim astronomers. A sphere rolls within a larger sphere of double the diameter so that a point on the circumference of the inner sphere traces out a straight line trajectory, transforming circular motion into linear. D on the diagram is the center of Tusi's deferent sphere. The broken line and circle on the diagram relate to Ptolemy's model, and as can be seen, Tusi's model has annihilated the equant point at T.

be more successful than earlier variants and was adopted by other astronomers who felt obliged to squeeze the curves of observation into the girdle of principle. Still, Urdi's models failed to produce perfect circles for the planets. In reworking Ptolemy's lunar model to produce one that conformed to principle, Urdi wrote that if a planet was moved by a sphere that rotated on its axis and at times sped up and slowed down,

> there would be no need for all the efforts expended in regard to this astronomy, and the final quest would then be the knowledge of the equations to be applied to the motions, even if those were based on false notions.[35]

This was a rich insight. It was not followed up until three centuries later when Kepler, working on the orbit of Mars and led by false assumptions, stumbled on the ellipse and the mathematics of non-uniform planetary motion.

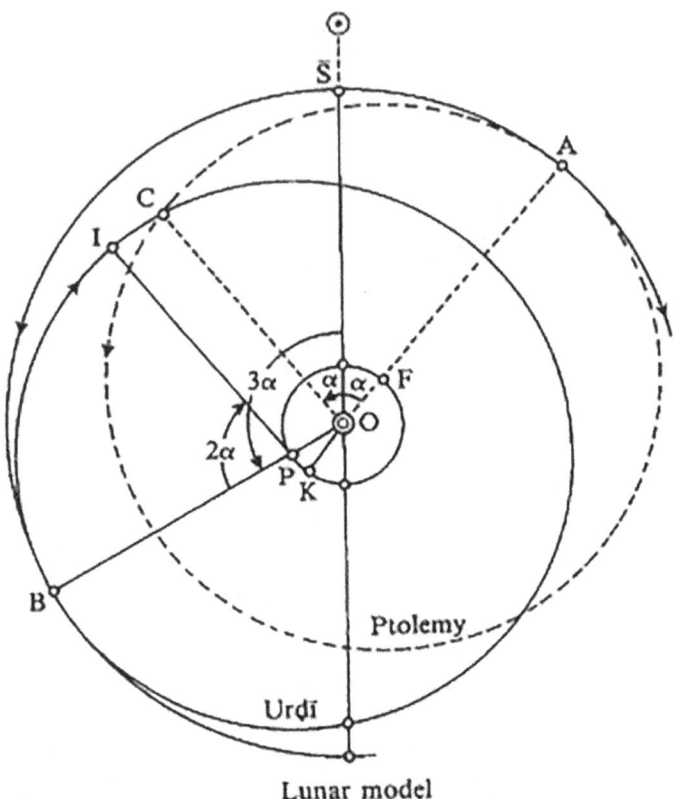

Lunar model

Figure 2.8 Al-'Urdi's lunar model compared to Ptolemy's (Figure 3.10 of Saliba's *Planetary Theories*).

Al-'Urdi's lunar model represents a great improvement over Ptolemy's in that all spheres move in uniform motion around their own centers. Al-'Urdi introduced a small encompassing sphere around the center of the universe that acted as an epicycle in carrying the deferent sphere around their common center (the point of observation) at uniform motion. As can be seen from the diagram, Ptolemy has the deferent move around its own center, which is not at the point of observation or center of the universe, an egregious contradiction to principle. According to al-'Urdi's innovative model, the deferent moves in the same direction as the celestial sphere; a, 2a and 3a in the diagram indicate that al-'Urdi had the deferent sphere moving three times faster than the value of Ptolemy's angular velocity. In other words, angle SOB is three times larger than angle SOA. Al-'Urdi was pleased with his model, not for being more accurate regarding observations, but because it preserved the principle of uniform motion around a common center.

The more the Muslim quest for philosophical purity led to complexity, the more the frustration and yearning for simplicity. It never occurred to astronomers that the principles might be at fault. A sun-centered universe was on occasion considered but not taken too seriously. Nasir al-Din al-Tusi himself considered it, but then fell back on his double epicycle, believing it was more philosophically correct to go the way of complicating the machinery than abandoning the elegant simplicity of principle.

Tusi's genius was to take the secondary epicycle of al-'Urdi from the circumference of the primary epicycle and place it inside the primary epicycle, the inner sphere being half the diameter of the outer.[36] The geometric construct annihilated the detested equant.[37]

The modifications introduced by Tusi were then further modified by another Iranian astronomer working at the Maragha observatory, Qutb al-Din al-Shirazi. Again, sacrificing simplicity for purity of principle, Qutb al-Din added even more spheres to Tusi's models.

Here, al-Shirazi replaces Ptolemy's crank mechanism and 'Urdi's small, earth-centered encompassing sphere with an epicycle. The model corrects the Ptolemaic flaw of the deferent going around its own center and not the center of the universe. Shirazi considers 'Urdi's lunar model a vast improvement over Ptolemy's model, and to enhance the model even more, he adds an epicycle that has the eccentric sphere (DHK) and epicycle move around the center of the universe. Al-Shirazi's eccentric sphere has half the eccentricity of Ptolemy's eccentric (ABG) and moves twice as fast in the direction of the zodiac. The epicycle, centered on H, has a radius of half the Ptolemaic eccentricity and moves uniformly in the same direction as the eccentric sphere at half the angular velocity. The eccentric sphere moves uniformly around its center at F.

Ibn al-Shatir, who like al-'Urdi was a Damascene Arab, was the last of these first-rank astronomers who redesigned the basic configuration of Tusi's epicyclic innovation. Al-Shatir's prime target was the Moon, the planetary body whose

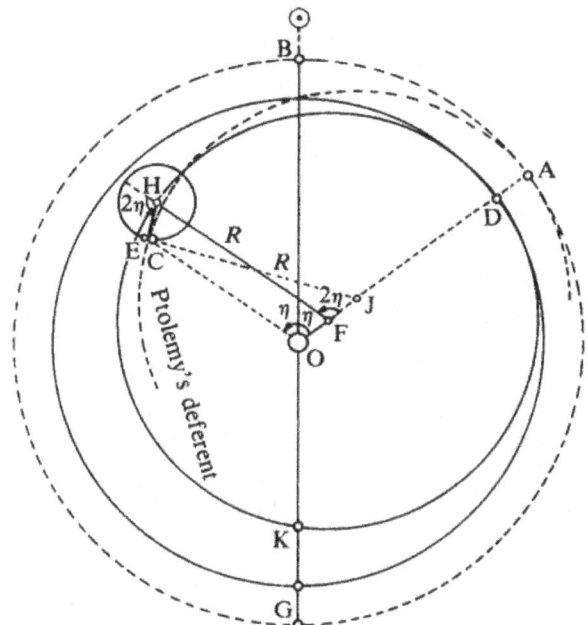

Figure 2.9 Qutb al-Din al-Shirazi's lunar model compared to Ptolemy's lunar model with its deferent and eccentric spheres (Figure 3.13 of Saliba's *Planetary Theories*).

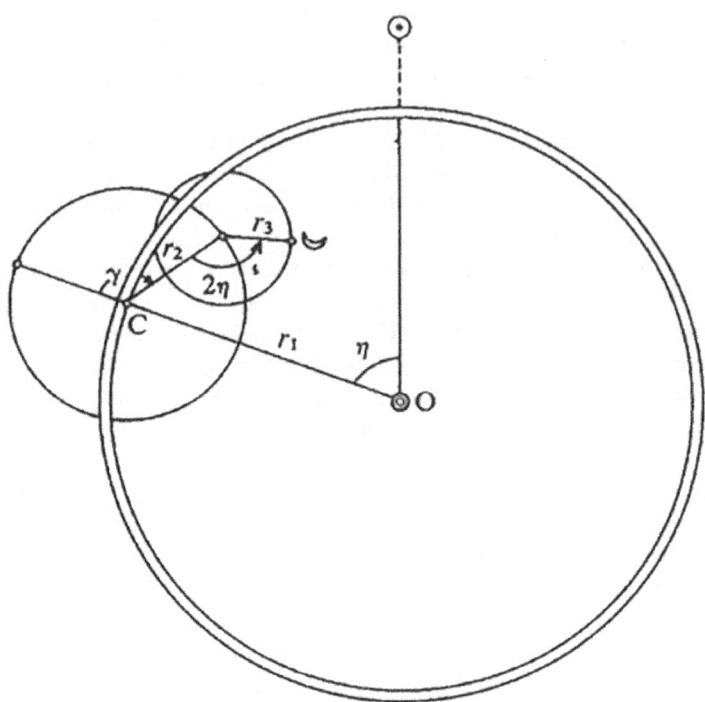

Figure 2.10 Ibn al-Shatir's lunar model (Figure 3.15 of Saliba's *Planetary Theories*).

Ibn al-Shatir wanted to be rid of Ptolemy's eccentric sphere and the inner encompassing crank-like sphere that other Muslim astronomers had avoided out of principle, even though it did account well for lunar motion. Ibn al-Shatir introduced a new sphere which he called al-Mumaththal, or parecliptic, whose center was the center of the universe, and so concentric to the sphere of the fixed stars but moving in the opposite direction. He added another concentric sphere, accounting for lunar motion in latitude. The inclined sphere moved in the direction of the celestial sphere. A third sphere (radius r_2), whose center is on the inclined sphere, acts as an epicycle and moves opposite to the celestial sphere. A fourth sphere is added, with radius r_3 and its center on the epicycle. This is called the al-mudir sphere, as its radius is the mudir, meaning director. This fourth sphere is carried by the epicycle. The moon is carried by the director, as seen in the diagram, and for this reason, the sphere and its radius are named director.

Marking an advance in planetary modeling as all of these added spheres certainly did, the double epicycle and additional spheres may have also added in a passive way to theory and practice by serving to exhaust the imaginative power of later astronomers in 16th century Europe who endeavored to compress the observed motions of the planets to conform to Ptolemaic principles in a geocentric universe. An improvement as it indeed was over Ptolemaic modeling constrained to adhere to the principles of astronomical motion, the geometric complexity introduced by Muslim astronomers was, in a practical sense, unhelpful in that the multiplicity of circular motions made for a dynamic of spheres spinning on spheres spinning on spheres that lost all simplicity and ended in a cul-de-sac. Perhaps the one great contribution of this obsessive fixation that led to dizzying complexity was that it convinced Copernicus, who was familiar with Muslim accomplishments and the Nasirian tradition of latter-day Muslim astronomy but who apparently had problems grasping the complexity of the twirling dynamics, to turn away from what appeared to have become a dead end and think in another direction, one that several Muslim astronomers had over the centuries suggested as a possibility but never pursued as a viable system representing physical reality, for reasons of sense perception and for their devotion to the classical principles of heavenly motion that reflected the divine mind.

unruly motion most singularly abused the divine principles of heavenly motion.[38] No more cumbersome than Ptolemy's crank-centered model, ibn al-Shatir's version above all preserved principle, which for Muslim astronomers was as holy for the law of heavenly motion as was the Quran for Shari'a law.

Ibn al-Shatir's lunar model was a masterpiece of simplicity compared to that of his immediate predecessor and fellow Damascene, Ubaydallah ibn Umar Sadr al-Shari'ah. Working on precedents set by Nasir al-Din, Ubaydallah produced a lunar model requiring five spheres, among them a trinity of epicycles, one moving on the other, and all without doing away with the Ptolemaic crank.

The challenge of taking observations seriously and imposing them on Ptolemaic principles was driving Muslim astronomers to ever further geometric extremes, narrative explanations of which can deaden the reader's attention as much as geometric sketches of them can dazzle the eye. But not to give these Maragha astronomers their due does no service to any study endeavoring to capture the monumental effort Muslim scientists exerted in attempting to break through the mind-bending twilight zone they found themselves in, imprisoned between the fading day of their Greek inheritance and the unbreaking dawn of their advance to new frontiers – a

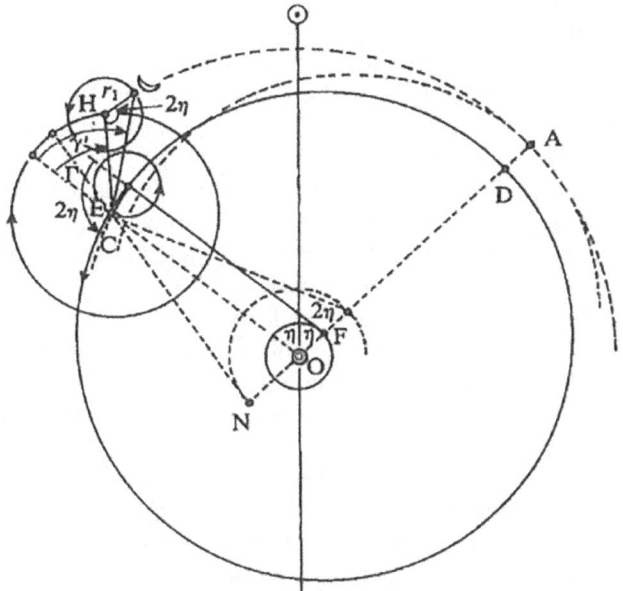

Figure 2.11 Lunar model of Ubaydallah ibn Umar Sadr al-Shari'ah (Figure 3.14 of Saliba's *Planetary Theories*).

Qutb al-Din al-Shirazi's lunar model was simplicity itself compared to an astronomer later in the 14th century, Sadr al-Shari'ah (d.1346/47), who endeavored to improve al-Shirazi's lunar model, specifically by accounting for the moon's 'prosneusis point.' The prosneusis is what Ptolemy called the discrepancy between lunar apogee This he did by keeping al-'Urdi's small, universe-centered encompassing sphere that was placed inside the deferent and adding a third epicycle.

dawn that glimmered but whose sun never quite lifted over the horizon to illuminate those truths that would become crystal clear to Copernicus, whose insight is mirrored in the astronomy of Maragha.

Ibn al-Shatir's model represented a large improvement in accuracy over that of Ptolemy's in accounting for lunar motion and differences in the Moon's distance from earth and its change in apparent size at syzygies and quadrature. His solar model seems to have combined philosophical theory and observation, an exception to the rule that observational astronomy existed apart from mathematical and theoretical astronomy.[39] Ibn al-Shatir refined his solar model by adding an encompassing sphere and a smaller one, the parecliptic, that was concentric to it and centered on the earth's center. The parecliptic carried the deferent that now served as an epicycle. This arrangement accounted for the observable change in the diameter of the solar disk, something that Ptolemy's model had simply ignored. Ibn al-Shatir's solar modeling was marvelously in agreement with observations and principles, but it rendered the geometric clockwork so complicated that the trade-off may not have been all that acceptable to other astronomers. His reconfiguring of the complex model that accounted for Mercury's problematic motion is of particular interest because of the implications it suggests concerning Muslim influence on the Copernican remodeling of Ptolemaic astronomy.[40] Ibn al-Shatir's ingenious incorporation of his advanced variations of Urdi's and Tusi's innovations is of especial importance in that his model of Saturn's motion appears to have been used by Copernicus in his own attempt, a century and a half later, to address that conundrum voiced by ibn al-Haytham in the 11th century, ibn Tufayl, ibn Rushd and al-Bitruji in the 12th, and Urdi in the 13th, that a simpler, non-Ptolemaic astronomy was out there somewhere, elusive but waiting to be found.

Not long after ibn al-Shatir passed from the scene, the Maragha School died out. As far as the present state of research in Muslim astronomy has gone, ibn al-Shatir appears to have been the last of the creative minds in that tradition of astronomy. The Ptolemaic epicyclic system continued on into the late 16th century, during which time several astronomers appeared who were comparable in greatness to Urdi, Tusi, Shirazi and Shatir, but they offered no creative innovation; merely the continuation of an old tradition with improved parametrical values. Perhaps like mathematics, which seemed to have reached a limit of development for lack of a system of symbolic notation, Maragha-style astronomy reached its limit with models so complex they could go no further within the iron-clad constraints of their traditional principles. Could those Maragha astronomers have felt anything akin to today's particle and string theory physicists who find that the deeper they delve into nature's nuclear innards the more complex nature appears? The Maraghans piled on geometrical spheres the way string theorists pile on mathematical dimensions and modules without coming any closer to understanding anything but the mathematics of it all, which becomes an end in itself.

Ironically, as the models of planetary motion became increasingly more complex, the mathematics of computational spherical astronomy and its methods became simpler, such as, for example, the transition from the Menelaus Theorem

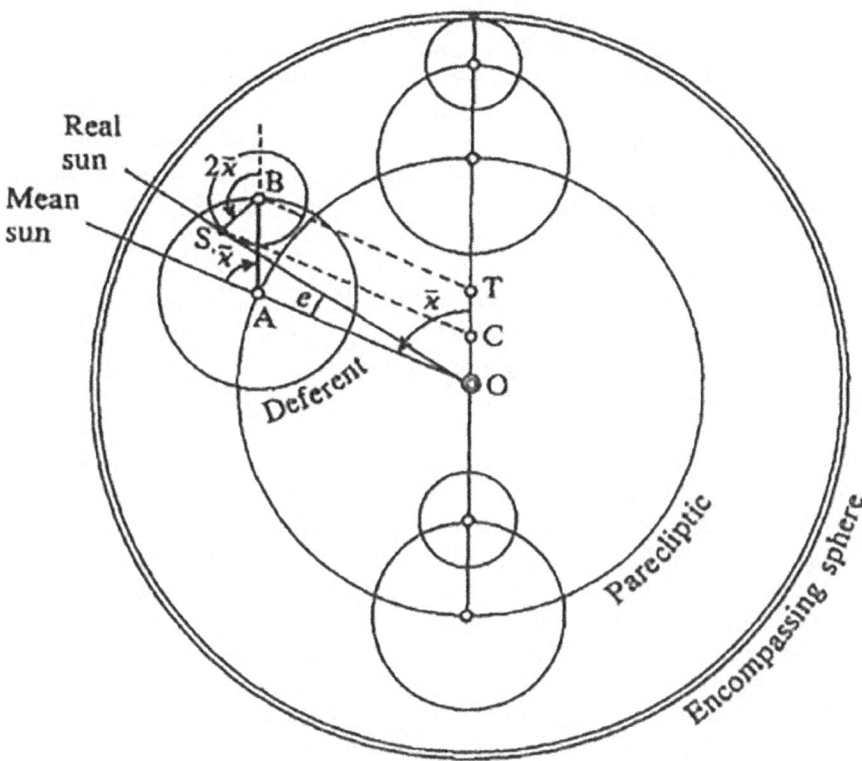

Figure 2.12 Ibn al-Shatir's solar model (Figure 3.9 of Saliba's *Planetary Theories*).

Here ibn al-Shatir adds an encompassing inner sphere he names the parecliptic. The deferent is the epicycle whose center is carried by the parecliptic of radius OA, O being the center of the universe and A being the center of the major epicycle, of radius AB. A smaller secondary epicycle of radius SB (half of AB), which ibn al-Shatir calls the director (al-mudir), is centered on the surface of the deferent sphere. The sun is at S on the surface of the secondary epicycle. The deferent and epicycles move at the same angular velocity. The observer at point O is the center of the universe. The parecliptic moves in the direction of the sun at the same angular motion. The deferent moves opposite the parecliptic, so that line AB is always parallel to OCT. The director moves opposite the parecliptic at twice the angular rate. The complex construction of sphere moving sphere moving sphere at different rates and directions accounts for the apparent change in size of the solar disk at apogee, mean distance and perigee. Ptolemy had maintained that through the seasons there was no difference in the size of the solar disk, an obvious contradiction to observation. Ibn al-Shatir and other Muslim astronomers had measured the differences, and Muslim astronomers could tolerate contradictions between theory and observation as little as theologians could between metaphysics and unrestrained divine power.

to the Law of Sines. By the end of the 14th century Muslim astronomers had exhausted the possibilities of Ptolemaic methods without achieving satisfaction. Al-Urdi's lament for the simplicity to be gained by a sphere of non-uniform motion must have expressed the frustration of many astronomers. By the time of ibn al-Shatir, much of the work accomplished by Muslim astronomy was already in the process of being passed on to the West in Latin translation. All it took was a dash of Pythagorean and Platonic cosmology from the creative imagination of Copernicus for the inherited achievement of late Muslim astronomy to break through the constraints of Aristotelian–Ptolemaic principle. In the words of one of the greatest historians of Muslim astronomy:

> Indeed, it has been shown that most of the Copernican planetary models are duplicates of these exhibited either by the Maragha scientists or by ibn al-Shatir. All that was left to Copernicus was the philosophically important reintroduction of a heliostatic universe.[41]

Creative astronomy in Islamdom did not end with the eclipse of the Maragha school in the late 14th century. Other rulers of the Turko–Mongol successor dynasties that emerged from the body of the great Mongol empire in the 15th and 16th centuries followed Hulagu Khan as patrons of astronomy. Ulug Beg, grandson of Timur, who conquered an empire stretching from the borders of China to Iran, Iraq and Anatolia, built an observatory in 1420 in the Timurid capital Samarqand that was even grander than the Maragha observatory. Housed in a three-story cylindrical structure 150 feet in diameter, the observatory was equipped with a mural quadrant with a radius of 130 feet, the arc of the quadrant being over 200 feet in length.[42] The quadrant was oriented to observe the altitude of bodies at the meridian and built underground to improve measurements. Reputed to have been the largest and finest ever built, the instrument was graduated in degrees subdivided into minutes, providing a high degree of accuracy.

One a lineal descendant of the other, these two crowning glories of Muslim observational astronomy, Maragha and Samarqand, had much in common. Both observatories were built by rulers descended from great conquerors, Genghis and Timur, the latter claiming to be related by blood to the former and thereby laying claim to, and continuing, Mongol military glory and the tradition of high civilization. Both observatories were monumental constructions with workshops and craftsmen who made precision equipment, and both had master astronomers as directors: Nasir al-Din al-Tusi in the court of Hulagu and Ghiyath al-Din Jamshid al-Kashi in the court of Ulug Beg. Writing in Persian, the courtly language of literature and learning in Mongol Maragha and Timurid Samarqand, al-Kashi pays his respects to Nasir al-Din al-Tusi in the preface of his table of observations, which is, according to its title, a supplement to the *Zij al-Ilkhani*, Nasir al-Din's table of observations he dedicated to Hulagu Khan.

Though lineally related, the astronomy practiced at Samarqand differed in style from that of its Maragha predecessor. Al-Kashi did not continue the line of development initiated by the scientist he so highly respected, al-Tusi. The Samarqand astronomers knew of and discussed the reform astronomy of the Maragha School, but, perhaps in view of ibn al-Shatir's refinements of al-Tusi's double epicycle, they figured that development along that particular line of planetary remodeling had gone as far as it could in complexity, leaving nothing more to be gained. A superb astronomer and mathematician, al-Kashi followed the traditional pattern established by pre-Maraghan astronomy, going back to Abbasid, Buwayhid and Seljuq times.

Where the Maraghan tradition of astronomers honed their skills by developing unique planetary models girded in faithfulness to principle, the Samarqand tradition excelled in refinement of observation and mathematical precision. Ulug Beg himself was a practicing astronomer and stickler for precision; he recomputed sine tables for each minute of arc, 1/3,600 of a circle, to the accuracy of one part in 6 million, while his chief astronomer al-Kashi, in a treatise devoted entirely to π, the ratio between circumference and diameter of a circle, computed its value using an algorithm he devised based on a regular polygon inscribed in a circle,

> to such a degree of precision that if it is used to calculate the circumference of a circle with a diameter of 600,000 earth-diameters, the error involved will be less than the smallest linear unit known to him – the breadth of a horsehair.[43]

Kashi's major contribution was in numerical analysis and in the elegance and precision of his groundbreaking computational algorithms, which were not to be surpassed until over a century later in the West, where, in Kashi's own time, toward the middle of the 15th century, scientists were well on their way to reaching the level of their Muslim counterparts. It would nonetheless take some time before Western mathematicians and astronomers reached the level al-Kashi achieved in generalizing a system of numerical notation that tremendously simplified mathematical computation. It will be recalled that Muslims used a mixed system in expressing numbers and fractions: integers written in Hindi decimal symbols and fractions in Babylonian sexagesimals using the Greek system of alphabetic letters to represent numbers. Sometimes pure sexagesimals were used for integers and fractions. In his *Key to Computation* (*Miftah al-Hisab*), al-Kashi generalizes the use of Hindi decimals for integers and fractions, demonstrating how much simpler the system was for multiplication, division, extracting roots and computing exponential powers. It was not until almost two centuries later that the system was introduced to Europe.

Scientists in Timurid Iran and Central Asia neglected the analytical studies in optics, algebraic equations, mechanics and medicine that earlier Muslim scientists had taken to new frontiers. These sciences were not forgotten. They were commented on and glossed, but no longer emulated in originality. The same might be said of the empires that succeeded in power and prestige the Il-khanid and Timurid in Iran and Central Asia, namely the Ottoman, Safavid and Mogul empires. Of these, the Ottomans are the most relevant to the history of Muslim science. Patterning courtly culture on precedents set by the Mongol Ilkhanids, Ottoman sultans assumed their predecessors' tradition of royal patronage that had been essential in Muslim courts since the earliest Abbasid Caliphs. Accordingly, Ottoman astronomy was closely linked to that astronomy patronized by the Ilkhanids and Timurids. The astronomer Qushji left Samarqand for Istanbul after it had fallen to the Ottomans. In the late 16th century the Ottoman successors to Timurid power in the Muslim world built an observatory in Istanbul. It was planned on a large scale in the monumental style of Maragha and Samarqand at the urging of Sultan Murad III's (1574–1595) energetic grand vizier Sokollu. Its chief astronomer, Taqi al-Din al-Dimashqi, had equipment made modeled on that of its grand predecessors.[44] But before the observatory had been fully constructed and was operational, a rival party at court led by the Shaykh al-Islam, head of the ulema hierarchy in the Ottoman state, prevailed upon the sultan to have it destroyed, claiming it would anger God and bring misfortune because of these pride-swollen astronomers who dared arrogate to themselves the ability to understand God's ways. Three years in constructing, the new observatory was in 1580 torn down in a few days. Insecure rulers in Islamdom had always been vulnerable to religious pressures. In 12th century Fatimid Egypt, an observatory had been destroyed for the same reason that would bring down the one in Istanbul. Weak or illegitimate rulers in need of religious support did whatever they had to do to gain it.

From the time the rational sciences arose in Muslim society, scientists and their ruling dynastic patrons had, at various times, varying degrees of difficulty with the conservative ulema and their followers. The destruction of the Istanbul observatory, however, bears particular significance, even fateful, for at the time it was being leveled, Tycho Brahe's grand observatory on the island of Hven with its many astronomers and mathematicians, among them Johannes Kepler, was in full swing tracking the heavenly bodies at the dawn of what would later be seen to be a scientific revolution (see Plate 3).

Starkly marked by the works of Copernicus, Brahe and Kepler on the one side, and the 1580 destruction of the Istanbul observatory on the other, a more poignant expression of the contrary trajectories of scientific activity in the West and Islamdom would be hard to find. It was to be a bitter pill for Muslims to swallow. Long before Copernicus, Tyche Brahe and Kepler, the Muslim scientific achievement had assembled the fundamental components that would in European hands be put together and turned to a new way of looking at nature and the heavens. The scientific expertise that Tycho Brahe assembled at Uranoborg was matched by the astronomers of Maragha and Samarqand. The descriptions of instruments used by al-'Urdi at Maragha and al-Kashi's description of the observational equipment at Samarqand indicate that Muslim astronomers had at their disposal equipment comparable to that used by Brahe and his team of astronomers.[45] Mathematically, in the collective expertise of Haytham, Biruni, Khayyam, Tusi and Kashi. Muslims possessed the analytical techniques achieved by Kepler and Galileo. Muslim dissatisfaction with Ptolemaic astronomy, and the certitude among some Muslim astronomers that there did exist a discoverable system of the heavens, pushed Muslim scientists to find it no less than it pushed Western scientists in the 16th and 17th centuries.

What Muslim scientists lacked was the sustained funding and connected community of science that made for continuity and institutional autonomy to protect, preserve and advance their work. Lacking this support, the vigor and number of scientists dwindled after the 12th century, and so too did the probability that one of them would have the conviction that one or more of the holy trinity of circularity, uniformity and geocentricity was wrong. Indeed, up until the time of Copernicus and Brahe, Muslims were busily carrying on the tradition of perfecting theoretical astronomy in accordance to the primary cosmological principles laid down by Aristotle and confirmed by Ptolemy, though they were not happy with them. The recently uncovered work of Shams al-Din al-Khafri (d. 1550), a first-rank Iranian astronomer/mathematician writing in Arabic, expresses it in the title he gave his book: *Hall ma la Yanhall* (*Solving What Cannot Be Solved*). Muslim scientists were alive and kicking even into the mid-16th century.[46]

Unable to wrench themselves from those principles, Muslim astronomers loaded the heavens with spheres turning on spheres that turned on spheres. A sun-centered system of planets harking back to Pythagoras and Aristarchus had been considered by Biruni, Tusi and other Muslim astronomers, only to be discarded without further exploration of an alternate system's possibilities that could have led to a consideration of non-uniformity and non-circularity.[47] Theoretical and observational astronomy lived to a large extent in their own separate spheres, but what also short-circuited the Muslim possibility of scientific advance was lack of continuity and support, summed up by the demolition of the Istanbul observatory.[48]

This brings us to the place of the sciences in Muslim society.

Notes

1 For the general works on Muslim medicine see Manfred Ullman, *Islamic Medicine*, Edinburgh University Press, Edinburgh, 1978; and Peter Pormann and Emile Savage-Smith, *Medieval Islamic Medicine*, Edinburgh University Press, Edinburgh, p. 2007.
2 Michael Dols, "The Origin of the Islamic Hospital: Myth and Reality," *Bulletin of the History of Medicine*, 61, 1987, pp. 367–390.
3 Ullman, *Islamic Medicine*, pp. 44–45.
4 Ullman, *Islamic Medicine*, p. 71.
5 Nur al-Din's father, Imad al-Din, had initiated the counter-crusades by conquering Edessa around 1140. Years later, Nur al-Din's son sent a military force to help the Fatimid caliph put down a rebellion which started a chain of events resulting in the demise of the Fatimids and the rise of Salah al-Din in Egypt who advanced the work Imad had started by permanently crippling the Crusaders.
6 The practice was strenuously resisted, and even when the storm of modernity hit with its full force in the 19th century and modern hospitals and medical schools were established, cutting up cadavers for anatomy classes had to be done circumspectly if not secretly.
7 A common mistake is the linkage often made between logarithm and al-Khwarizmi. Logarithm was coined by the 18th century Scottish engineer Napier who invented both logarithms and the slide rule and has nothing to do with al-Khwarizmi.
8 Roshdi Rashed, *Encyclopedia of the History of Arabic Science*, vol. 2, edited by R. Roshdi and R. Morelon, Routledge, New York, 1996, p. 328.

 9 Rosenfeld and Youschkevitch, *Encyclopedia of the History of Arabic Science*, vol. 2, p. 492.
10 Roshdi Rashed, *Encyclopedia of the History of Arabic Science*, vol. 2, p. 363.
11 Archimedes was believed to have designed a giant parabolic mirror of highly polished metal which, by focusing the sun's rays, he used to destroy a Roman fleet attacking Syracuse. He did not in fact build such a weapons system, but the invading Romans killed him anyway.
12 A delightful little monograph on this greatest scientist of light in Muslim civilization is available for the general reader in Bradley Steffens' *Ibn al-Haytham: First Scientist*, Morgan Reynolds, Greensboro, NC, 2007.
13 The book includes detailed chapters on physical and mathematical analyses of refraction; reflections from conical, cylindrical, spherical and parabolic mirrors; burning mirrors; the light of the moon and its phases; the light of stars; the rainbow and halo; solar and lunar eclipses; propagation of light; theory of vision; physiology and anatomy of the eye and their function in vision; and the psychology of perception.
14 Gil Russell, "The Emergence of Physiological Optics," in *Encyclopedia of the History of Arabic Science*, vol. 2, pp. 703–704; also David Lindberg in the same volume, p. 721.
15 Rather than a simple commentary, al-Farisi produced an extensive revision of it, in which he went beyond Haytham's optical theory. He had his own ideas on the behavior of light and in several areas of his study diverged from the great authority, being more systematic in his application of mathematics to reflected and refracted rays impinging on a glass sphere. This is seen, for example, in his criticism of Haytham's method of experimentally analyzing the rainbow. Hollow spheres filled with water, he theorized, would not duplicate drops of water, as Haytham had assumed in his famous experiment where water-filled glass spheres simulated droplets of moisture in the atmosphere. Believing Haytham's physical and geometric analysis to be lacking, al-Farisi held that the rainbow was the result of a combination of one or more reflections and two refractions. A ray of light passing through a water-filled glass sphere, unlike the droplet of water which has no glass surface around it, would undergo four refractions, not two as Haytham had surmised.
16 Rashid Roshdi, *Dictionary of Scientific Biography*, vol. VII, edited by Charles Coulson, Gillispie, Scribner and Sons, Detroit, 1970–1980, pp. 212–218.
17 A frivolous application of the ingenious mechanics and machines conceived by Archimedes and the Alexandrians came centuries after the Hellenistic period, in 6th century Constantinople, when Justinian had a hydraulic system built and concealed beneath his throne to awe visiting embassies with his godlike power as he levitated above them.
18 Rozhouskaya, *Encyclopedia of the History of Arabic Science*, vol. 2, pp. 620–623.
19 Also as *'ilm al-falak*, science of the heavenly bodies. On this, see Regis Morelon, "General Survey of Arabic Astronomy," in *Encyclopedia of the History of Arabic Science*, edited by Roshdi Rashed, 1996, vol. I, pp. 1–57.
20 A.I. Sabra, "An 11th century Refutation of Ptolemy's Planetary Theory," in *Science and History: Studies in Honor of Edward Rosen* (Studia Copernica XVI). Polish Academy of Sciences Press, Warsaw, 1978, pp. 117–131. For ibn al-Haytham's rejection of Ptolemy's equant because it gave non-uniform motion to the deferent sphere carrying the epicycle upon whose circumference the planet was fixed see Sabra, "Ibn al-Haytham's Treatise: Solution of Difficulties Concerning the Movement of Iltifaf," *Journal of History of Arabic Science*, 3, 1979, pp. 388–392.
21 G. Saliba, "Arabic Planetary Theories," in *Encyclopedia of the History of Arabic Science*, vol. 2, p. 82, citing abd al-Hamid al-Sabra, Nabil Shahaby, ibn al-Haytham's *Shukuk 'ala Batlamayus*, Cairo National Library Press, Cairo, 1971, p. 64.
22 E.S. Kennedy, *A Survey of Islamic Astronomical Tables*, *Transaction of the American Philosophical Society*, Philadelphia, 1956.
23 The earth spins unstably, like a top, its axis describing a full circle once every 26,000 years. Taking the earth as stationary, astronomers projected the wobble onto the axis of the celestial sphere carrying the fixed stars, the firmament. Working on a problem related

to precession, the Andalusian astronomer al-Zarqali (d. 1100) is said to have spent 25 years observing the sun to determine the movement of the solar apogee, which he found to be one degree every 299 years.

24 The larger the measurable parallax, the closer the observed body, parallax being the angular difference of a heavenly body observed from two different points on earth, or at different times during the year as the celestial spheres orbited the earth. An extremely distant object, the fixed stars for example, would show no parallax. Ptolemy's models for Mercury and Venus gave them six minutes of parallax for the whole year.

25 G. Saliba, *Islamic Science and the Making of the European Renaissance*, MIT Press, Cambridge, MA, 2007 p. 185.

26 A.M. Sabra, "The Andalusian Revolt Against Ptolemaic Astronomy: Averroes and al-Bitruji," in *Tradition in the Sciences: Essays in Honor of I.B. Cohen*, edited by E. Mendelsohn, Cambridge University Press, Cambridge, 1984, p. 147 note 7.

27 *Tafsir ma ba'd al-tabi'a*, edited by M. Bouyges, Beirut, 1942, p. 1664.

28 A.I. Sabra, "The Andalusian Revolt Against Ptolemaic Astronomy: Averroes and al-Bitruji," pp. 133–153.

29 On this, Toby Huff, *The Rise of Early Modern Science: Islam, China and the West*, Cambridge University Press, Cambridge, 1993, pp. 62–107.

30 Saliba, *Islamic Science and the Making of the European Renaissance*, MIT Press, Cambridge, MA, 2007, p. 243. Professor Saliba sees theoretical astronomy (*'ilm al-hay'a*) as a Muslim category of astronomy without Greek precedents. Having survived religious resistance because of its many practical uses, it went on to be accepted into the religious corpus of study. As a point of clarification, the Muslim study of theoretical astronomy had no Greek precedents as a category but was in content steeped in Greek astronomy. *Islamic Science*, pp. 127–128.

31 A.H. Sabra, "Science and Philosophy in Medieval Islamic Theology," *Zeitschrift fur Geschichte der Arabische-Islamischen Wissenschaften*, 9, 1994, p. 40.

32 Nasir al-Din, obliged to accept Hulagu's patronage when the Mongols seized the Ismaili (Assassin) fortress of Alamut where he was chief astronomer, is reported to have saved 400,000 of the city's manuscripts from being destroyed. Saliba, *Islamic Science*, p. 243.

33 S.H. Nasr, "Nasir al-Din al-Tusi," in *Dictionary of Scientific Biography*, vol. XIII, edited by Charles Coulson, Gillispie, Scribner and Sons, Detroit, 1970–1980, p. 509.

34 The centerpiece of Tusi's Ptolemaic reform was a cunning geometric device that had a smaller epicycle rolling along the inner circumference of a larger one, making a double-epicycle, referred to by contemporary scholars as "the Tusi couple." The couple was formed by giving the inner epicycle a diameter half of the outer. The epicyclic coupling was an elegant achievement of geometric imagination that transformed a combination of circular motions into linear motion, outdoing Ptolemy himself in what could be done with circles on circles. It was this ingenious device of al-Tusi's that gave distinctive identity to those astronomers who adopted it and successively refined it, collectively forming what could be called a school or tradition in astronomy within the larger Ptolemaic tradition of Muslim scientists who continued to employ the discredited equant.

35 G. Saliba, "Arabic Planetary Theories," p. 93.

36 The inner sphere rolls like a ball bearing along a circumference of the encompassing sphere. Both spheres are in motion, the inner at twice the angular velocity of the outer, as the double-sphered epicycle is carried along on the circumference of the turning deferent. The beauty of this roller bearing configuration is that any given point on the circumference of the rolling inner sphere will trace out a diameter of the larger sphere, in effect producing a straight line from two purely circular motions. It is this periodic shortening and lengthening of the diameter of the planet's orbit around Earth that produces the effect of uniform motion as observed from the eccentric. By giving the proper values to the radii and angular velocities, the resultant motion of the double-epicycle arrangement adequately approximates the observed motion of the planets. In al-Tusi's models, the outer planets, Mars, Jupiter, Saturn, were carried on the circumference of

the smaller sphere, the larger sphere being the epicycle whose center was on the circumference of the deferent whose center was at the eccentric point.

37 In addition to remodeling the outer planets, Tusi reworked Ptolemy's lunar model, which was particularly offensive because of a circular crank-like mechanism that made the center of the deferent orbit the point that represented the center of the universe. The mechanism worked well enough in making a lunar model, but it was astronomically and philosophically hideous: a universal center that moved! The center of the deferent was supposed to be the unmoving earthly center of the universe. But lunar motion had been too much of a challenge because of motion in latitude and because of the large differences in lunar distance and size at syzygies (when the moon or any planet is in conjunction with, or in opposition to the sun) and quadrature (when the orbital points midway between the two syzgies of conjunction and opposition). Ptolemy had the moon twice as large at the latter than at the former, and so resorted to having the center of the deferent move to obtain reasonable correspondence between observation and the working of his lunar model. Tusi attacked the problem with his double epicycle, transforming non-circular to circular motions, thus saving the divine principle of heavenly motion.

38 Put simply, ibn al-Shatir transferred the effect of Ptolemy's circular crank that moved the center of the universe in a small circle to a small outer sphere whose center was carried by the extremity of the radius of the deferent's epicycle. To account for the Moon's latitudinal motion, the plane of the deferent was slightly inclined to the plane of the ecliptic. This modification required an additional outer concentric sphere which ibn al-Shatir termed the "parecliptic." All in all, ibn al-Shatir required four spheres for his lunar model: the parecliptic, the inclined deferent, the epicycle, and the epicycle of the epicycle.

39 Saliba, "Arabic Planetary Theories," pp. 86–87.

40 Shatir's model of Mercury consisted of a combination of the innovations made by the astronomers Urdi and Tusi. The same was true for his thoroughly restructured model for Saturn. Shatir did away with both equant and eccentric by using improved variations of Urdi's and Tusi's innovations, the effect of which had Saturn orbiting in purely circular motion around Earth, with an inclined sphere accounting for latitudinal motion. This was achieved by using five spheres: one representing the zodiacal signs; one inclined sphere for latitude; the third being the deferent sphere with center on the periphery of the inclined sphere and so acting as an epicycle; the fourth sphere, having its center on the deferent, called the "director," acted as a second epicycle; and the fifth sphere, centered on the periphery of the fourth sphere, was in effect a third epicycle. The planet was carried along the periphery of the fifth sphere, that is, the third epicycle. The deferent acting as epicycle was given a radius 1½ times larger than the deposed eccentric; the "director" epicycle's radius was ⅓ that of the deferent. The same method was applied to ibn al-Shatir's models for Mars and Jupiter. By this configuration of spheres and epicycles, with their distinctive deferent and director radii set as ratios to annihilate the eccentric, and having the inclined sphere, along with the epicycles, move in the direction of the sphere of the zodiac and the deferent sphere in the opposite direction, and with each sphere of this imposing structure given the right value, models were provided for Mars, Jupiter and Saturn that accounted for longitudinal and latitudinal motion around the earth, a triumph perhaps more of geometrical imagination than astronomical genius.

41 E.S. Kennedy, "The Exact Sciences in Iran under the Saljuqs and Mongols," in *Cambridge History of Iran*, vol. 4, Cambridge University Press, Cambridge, 1975, p. 670. See also George Saliba, "Arabic Astronomy and Copernicus," *Zeitschrift fur Geschichte der Arabischen-Islamischen Wissenschaft*, I, 1985, pp. 225–229.

42 Kennedy, "The Exact Sciences in Timurid Iran," in *Cambridge History of Islam*, vol. 4, Cambridge University Press, Cambridge, 1975, pp. 375–395.

43 Kennedy, "The Exact Sciences in Timurid Iran."

44 Ahmad Y. al-Hassan, "Factors Behind The Decline of Islamic Science after the 16th century," *Islam and the Challenge of Modernity*, edited by Sharifa al-Attas, International Institute of Islamic Thought And Civilization, Kuala Lumpur, 1996, pp. 353–354.
45 E.S. Kennedy, *Cambridge History of Iran*, p. 672.
46 Saliba, *Renaissance*, p. 242.
47 For Muslim astronomers who speculated on a moving earth see: Eilhard Wiedemann, "Zu den Anschauuugen der Araber uber die Bewegung der Erde," *Geschichte der Medezin und der Wissenschaften*, no. 30, vol. III, Nr. 1, 1909, pp. 1–3.
48 Professor Saliba would rephrase the comparative dynamics suggested by the terms. "Muslim scientific decline" and "Western progress" by describing the West as "outpacing" its rival. This has the virtue of preserving the historical fact that the work of science at no time disappeared from the face of the Muslim world.

3 Science in a religious society

The place of science in Muslim society at large, aside from the haven it received in the elitist circles of the royal court and wealthy viziers attached to it, was, to some degree, a function of the relationship between science and religion and religion's place in society. By establishing the relation of science to religion, or rather the regard science had among the accredited religious authorities who monopolized the prerogative of interpreting the holy sources and determining from them what was acceptable to religious belief and what was not, an idea can be gained of the social space that science was able to achieve for itself, which in turn will help clarify at least one of the leading factors underlying the changing fortunes of science in Muslim society. This is not to say there was at all times an interaction between science and religion. With the odd exception of an irrepressible conservative purist imposing himself into the otherwise gently passive debate, science and religion more often than not existed in a felicitous state of neutrality, mutually oblivious to one another. This is especially true in the earliest centuries of Muslim history, when a Muslim society and religion was forming and lines of self-definition and demarcation had not yet been set.[1]

Since the Muslim state evolved as a result of the nascent community of believers struggling to defend itself and then, having done that successfully in Medina, expanding to crush its Meccan enemies, all in the name of Islam, religion gave birth to the state. The idea of separation of one from the other was alien, repugnant and historically contradictory. The oneness of religion and state was embodied in Muhammad's career, first in Mecca then in Medina, as prophet, statesman, warrior, diplomat and lawgiver. The state was in its origins to be the political and institutional embodiment of religion, as it was in Muhammad's lifetime. When after his death it did not work out that way because of the power struggles and consequent rise of monarchy, the men of religion, those who believed themselves to be the true followers of Muhammad and protectors of his mission and the Quranic message, chose to distance themselves from the political rulers of the state. So powerful was the communal memory of the Prophet and of what he and his small band of companions had accomplished in a brief span of a generation, that the pious of the community could not but hold reservations about the legitimacy of their dynastic rulers who acted more like Byzantine or Sassanian emperors than leaders governing the Muslim community by the law of the Quran and the Prophetic Tradition.

The purely religious core of the Islamic cultural ethos evolved from several concomitant and interrelated pursuits: interpretation of the Quran; commentary on the Hadith, that is, the prophetic model of Muhammad as preserved in collected traditions of the Prophet's words and actions and believed to represent human perfection; and the working out of the holy law, or Shari'a, perceived to be rooted in both Quran and Hadith. Pre-Islamic Arab custom was tacitly sanctioned through the collections of Prophetic Tradition, which were used to delineate a Prophetic Sunna, or way of the Prophet, which portrayed the idealized figure of Muhammad living his daily life as an Arab man, the best of them, the best of all possible men: the Perfect Man. The Sunna, the idealized way of the Prophet, how he lived, in every detail, became the communal model that all good Muslims should follow.

Of course, a great deal that was outside Arab custom and the Quran entered Islamic civilization through the conquests and was authenticated by reports claiming to be Prophetic Tradition. Any custom, idea, style or mode of public thought or conduct that was obviously contradictory in principle or spirit to them was rejected. Any such sort that might be cloaked in a tradition would be categorized as repugnant and ignored. Those ideas and practices that were not contradictory were accordingly modified, restated or recast in Islamic dress. What did not appear to negate or in any way vitiate Islamic law, public custom and Quranic revelation was considered permissible, if not by the stricter interpreters of public propriety, the ulema, then by the communal majority of the *umma*, by consensus – as determined or sanctified by the men learned in religion whose judgments were tuned to public sentiments.

The elitist culture of the caliph and his courtiers diverged from the norms of expected religious behavior, sometimes sharply. The spatial and intellectual distance between the palace at one pole of urban life and the marketplace and public mosque at the other made for two different cultures. At the center of social activity in the mosque and marketplace, the pressure of conformity was not easily avoided, and anyone who did dare avoid it had to pay a price commensurate to the degree of deviation. The ulema, the moral conscience as it were, those guardians of communal propriety learned in Shari'a, Quran and Prophetic Sunna, held the whip of public censure.

The ulema were dedicated to maintaining religious autonomy from the state and its rulers, upon whom they looked with undisguised disdain for their extravagant and irreligious courtly life, drinking, licentious poetry, music, dancing girls and extrajudicial executions. The ulema perceived themselves as society's teachers and spiritual protectors. They protected the community from all seductions – appetitive, intellectual and pseudo-religious. They interpreted, preserved and executed the religious law. And this is as they were seen by the *umma*. It was not the state and its rulers that Muslims looked to for moral guidance or education. The rulers had betrayed the simplicity of Muhammad and his companions. They had betrayed the Quranic injunction of an egalitarian society of all Muslims equal before God and God's Law. If the guardians of religion and public morality could not rein in a caliph's arbitrary behavior, they could in their mosques at least guide lesser placed Muslims in the rightly directed life. As teachers they molded the young and preserved the message of Islam as contained in the Quran, Hadith and Shari'a. Education was a religious monopoly of the ulema. Anything outside of religion taught by anyone outside the ulema could be resented and, at least passively, resisted. The secular literature of the court and especially the ruler's patronage of the rational disciplines were regarded by some ulema with suspicion or hostility as trapdoors leading to heresy's gates, or at best were dismissed as vain pursuits. Most of the ulema simply ignored non-religious studies so long as they made no pretensions to deliberate on subjects within the sphere of religion, a sphere not so easy to avoid for students of philosophy.

Baghdad was for a century and more of its founding free of religious and political constructs not its own making. In the words of a contemporary witness:

> The good thing about Baghdad . . . is that the rulers can feel secure against any head of (religious) party winning the upper hand, as the 'Alids and the Shi'ites frequently do over the people of Kufa. In Baghdad opponents of the Shi'ites live together with the Shi'ites, opponents of the Mu'tazilites together with the Mu'tazilites, and opponents of the Kharijites together with the Kharijites; each group holds the other one in check and prevents it from setting itself up as leader.[2]

For the same reason of religious autonomy that the ulema kept their distance from the ruler, they equally refused to be organized into an institutional hierarchy with a chief at their head leading the way in scriptural and legal interpretation. Islam evolved without acquiring church organization or papal director. Organized Christianity became patterned along the lines of the imperial Roman political and administrative environment it had been born into, its church an empire, its pope an emperor, its bishops and archbishops standing as high provincial officers and governors. The church structurally replaced the empire it evolved in. In the early medieval Latin West, when the state was weak or hardly existed, the pope and Church did indeed assume the part of emperor and empire. Islam, on the other hand, patterned itself along the lines of a totally different environment. Quranic scholars and legal thinkers reflected the individuality of the Arab tribesmen and autonomy of the tribe. Religious scholars cherished their independence, free of the state and any overriding religious institution that could call councils and impose doctrine. Authority resided in a scholar's learning and piety, in his acuity of legal and Quranic interpretation, in his books and sermons and probity of personal life, in the number of people who accepted him as truly pious in deed and character, a person whose spoken and written words were taken to exemplify the highest values of Islam. Religious authority resided in the individual, not, as in Christendom, an institution of assembled scholars under a roof arguing and deciding doctrine.

Rather than a canonical law as in the Latin or Byzantine church, four schools of law named after individual legal scholars arose in the Muslim community: Hanafi, Maliki, Shafi'i and Hanbali. Six separate collections of Prophetic Tradition made by Arab and Iranian religious scholars across the Muslim world came to be accepted as authoritative. No institution or authority existed to compound the six collections or the four schools into one. This pluralism of legal schools and Hadith was an expression of individual, regional and religious autonomy that was no doubt necessary in a far-flung society that perceived religion to include everything human beings did or wanted to do. The closest those early scholars of the 8th and 9th centuries collectively came to institutionalizing authority within the disparate body of the ulema is seen in a tradition that reports the Prophet as having said that his community would never agree on an error, most probably a posthumous creation to give prophetic credence to the principle of communal consensus, *ijma'*, a source of jurisprudence used liberally in the Hanafi and Shafi'i schools of law to legitimize local custom by projecting it into individual Hadith, which was then used in formulating sacred law.

By avoiding government association and institutionalization, the learned spokesmen of Islam believed themselves to be acting in harmony with the prophetic spirit of Muhammad and the Quranic message. Muhammad and Islam came to liberate the Arabs from superstitious fears, polytheism and arbitrary human authority so that they obey one true God while treating their fellow Muslims with honesty, fairness, kindness, charity and justice. The Islamic revolution that overturned the clay and wooden gods of pagan Arabia housed in the Ka'aba of Mecca, and that, for a time, unseated the pagan Meccan aristocracy whose commercial interests were inextricably tied to those gods and the Ka'aba, meant for those early Muslims an egalitarian revolution whereby all men, low and high, rich and poor, were equal before God and the holy law of the Quran. Superstitious fears were replaced by fear of God, fear here serving the cause of justice and charity. It was a bloodless triumph, a glorious moment of divine guidance and human obedience, when pagan Mecca fell in 630 AD without a fight after having fiercely resisted Muhammad's monotheists for years: a moment that would shine ever brighter in the minds of Muslims through the centuries.

The Umayyad caliphate, descended from the powerful family that opposed Muhammad right up to the fall of Mecca, appeared to the pious to be starkly political, "secular," consumed in conquest, power, wealth and self-indulgence. This left the ulema, as they themselves understood their obligation, to preserve and propagate the simple and pure spirit that had lived in Medina during the time of Muhammad, when the Quranic message lived in the souls of men through their proximity to the living Prophet. Confronted by political rulers who were seen to betray the spirit of Medina, the ulema became antagonistically aloof of the rulers and their government, producing a situation somewhat resembling the present divorce between Middle Eastern dictators and the less radical and violent movements of political Islam, which in their opposition to what they call misguided, irresponsible, irreligious and corrupt government show themselves to be protectors of the repressed and dispossessed.

The ulema had little patience for, and less appreciation of, the exigencies of an Umayyad government ruling and holding together an expanding empire whose complexity of daily operations outpaced the formulation of Holy Law, which, at the time was in its earliest stages and had, as yet, little concern for central and provincial administration and the legal organization of the conquered lands and their non-Muslim inhabitants. Umayyad insistence on taxing non-Arab converts to Islam at a higher rate than that of Arab Muslims distanced the egalitarian-minded ulema all the more from the purely political rule of a leader who went from being a simple follower of Muhammad to imitating an emperor or shah, the very arrogance the Quran condemned. Monarchy was anathema to the pious spirit of the young religion; it was repellant to the memory of Muhammad's life and the egalitarian message of the Quran. Arab tribal independence reinforced communal distaste. Umar refused the title "caliph" as being an imperial pretension, preferring to be called simply *amir al-mu'minin*, commander of the faithful.

The pious looked back to the pristine days of Muhammad in Medina and to his humble successors who ruled after him, abu Bakr, Umar, Uthman and Ali. (The Shi'i community would eventually come to reject the legitimacy of the first three.). The Abbasid dynasty of caliphs that supplanted the Umayyads was no better in the eyes of the devout. Religious hypocrisy, as some of them saw the turnover from one family to another, had taken the place of godless politics. To the puritanical members of the ulema, the liberal patronage that Abbasid caliphs provided the Christians and Jews who were engaged in bringing the rational learning of the pagan Greeks into Islam was proof enough of the hypocrisy that lay barely hidden by their outer display of piety. Were it not for them, the ungodly works of the Greeks would have remained unknown to Muslims.

The pressure of the ulema's occasionally expressed antipathy regarding the rational sciences was sufficient to deny science and philosophy any permanent anchorage in the hearth of Muslim society. Public space within the community was denied. Cultivation of the Greek and Indian heritage was limited to the royal court, to the patronage of the ruler and his high viziers.[3] What few institutions came into being for scientific study were products of high patronage. Indeed, without support from the rulers there would have been no scientific renaissance in Muslim civiliza-tion, or at least a much lesser one, since far fewer bright young men would have devoted their lives to a dead-end career. The ruler's court was the only source of funding for such work, patronage coming direct from the caliph himself or a wealthy minister of the court. Muslim society provided no public ground for sci-entific clubs, meeting places or institutions. Science did not penetrate into the heart of society. It never grew into becoming anything near an accepted mainstream intellectual pursuit, even during the reigns of the most patronizing of rulers. Mosque schools did not include mathematics, logic or the rational sciences in their corpus of studies. There were no Muslim equivalents to Plato's academy, Aristo-tle's Lyceum or Zeno's Stoa. The closest thing may have been the caliph Ma'mun's House of Wisdom or the Pythagorean-inspired Brethren of Purity, the *Ikhwan al-Safa*, anonymous authors of a collection of Gnostic Isma'ili-oriented philosophical and scientific treatises, more Neoplatonic theosophy than science, popular in Fatimid Egypt and among the Shi'i of Iraq.[4]

Why it was that institutions fostering the study of science did not spread roots from the royal court to society at large is an intriguing question. The royal courts were, it is true, set some distance apart from the society of the *umma*, in space as well as culture, but for all that, the ruler's court was still a part of society, the ruling part. Influential people came and went from the palace to the busy commercial centers. There were conduits connecting the polar world of ruled and rulers through which influences could flow. Wealthy merchants were no strangers to courtly officials. Learned ulema visited the royal court. Some were invited and some went. Some allowed themselves to be attached to it, most notably abu Yusuf, the renowned judge, scholar and disciple of abu Hanifa, founder of the school of jurisprudence bearing his name. The presence of abu Yusuf added to the Abbasid court that glimmer of legitimacy rulers eagerly sought and purchased through patronage when possible.

Scholars who have studied the rise of educational institutions in Muslim society, namely the *madrasa*, offer an explanation of why natural science did not take institutional root. In regard to Muslim education, they consider the individualistic, personalist nature of the *ijaza* system: A single scholar giving a certificate to a single student who had satisfactorily mastered a single book under his tutelage, meaning the student had memorized the book or a good part of it – enough to paraphrase it. The *ijaza* documented the holder to teach that book, in the name of the issuer. The *madrasa* offered no curriculum, no graduated years and levels of study, no degree, nor graduate school for a doctorate. The student studied under various scholars and collected certificates, one for each book. One could study in a madrasa year after year collecting certificates until running through all the attending scholars and then move on. This could mean a lot of traveling, depending on the size and fame of the madrasa. A student who persevered and collected certificates and made a name for himself over the years by sermons and writing his own books might attract students of his own and begin issuing *ijaza* certificates himself. No council of scholars controlled courses or curriculums. No high authority decided who could teach what at what level of instruction. *Ijazas* were licenses to teach a certain book, not a particular subject, and were invested with no more authority than the esteem the issuer held among his fellow scholars, the closest the system came to tenure or peer review.

The individualistic nature of the *ijaza* system obviated structure and control. A student of the natural sciences would be obliged to do a lot more traveling than a religious one. He would need to be consummately devoted to the subject or to the scholar teaching it. There were such students, though too few to inform society sufficiently to blunt prejudices fed by religious antipathy and popular superstition. Also mitigating against the rise of private institutions teaching the natural sciences was the pattern of extended kinship forming self-reliant groups in dealing with the outside world. "On the most general level, the overwhelming dominance of the extended family was a major factor behind the fact that in medieval Islam there arose no corporate municipal institutions."[5]

Consequently, scholarly guilds or other autonomous professional groups were not there to provide the seeds from which could sprout institutions capable of sustaining scientific inquiry. In addition, by not recognizing corporate entities, Islamic law reinforced the traditional, pre-Islamic extended family pattern, further impeding any rise of corporate personalities or legally autonomous institutions. It would be quite the opposite in the Latin West.

Being so dependent on the ruler's predilections, science in Muslim civilization was always vulnerable to becoming a sometime thing. If for any reason rulers found it impolitic to patronize the rational sciences, the well of production went dry. Scientists had nowhere else to turn. Accordingly, the names of the great scientists are associated with rulers: Bakht Ishu' and Fazari with the Caliph Mansur, Khwarizmi with Caliph Ma'mun, al-Farabi with the Hamdanid ruler of Syria, ibn Sina with the Samanid Nuh ibn Mansur, Biruni with Sultan Mahmud of Ghazna, ibn Baja, ibn Tufayl and ibn Rushd with the Berber Muwahhidun, Umar Khayyam with the Seljuqs, Nasir al-Din al-Tusi with the Mongol Hulagu Khan, and Kashi with the Timurid Ulug Bey in Samarqand. So too for the many tables of astronomical observations named in honor of ruler–patrons.

It could be said, in fact, that science existed on two simultaneous levels, one the upper level of large scale science sustained by royal patronage, that is, the science of great libraries, observatories and hospitals with their well-rewarded court astronomers and physicians, and a small-scale lower level of private individuals far from the ruler's court doing their own in keeping the traditions alive by collecting, studying, copying and discussing scientific texts in the privacy of their homes with a few trusted friends, informal groups of like-minded men meeting discretely, more or less out of sight of the ulema, with scientific knowledge being passed on from father to son and master to student. The odd religious scholar or judge might circumspectly sit in to satisfy his curiosity or indulge an intellectual craving he was unable to satisfy in the religious texts or schools. The brightest and most accomplished young men of the lower level would be the ones to seek out private libraries and, if able to afford it, private lessons from an astronomer, physician or mathematician.[6] The exceptional few would find royal patronage as court physician or astronomer.

Al-Farabi (d. 950) hailed from humble village origins beyond the Oxus River. Led by his father to study medicine under a Christian physician, he learned what he could from his teacher and then began studying on his own by reading books on science and philosophy whenever he found one, convinced that natural knowledge and philosophy deepened religious understanding. A self-educated polymath like all the great students of science and philosophy, he excelled in medicine, wrote a book reconciling the philosophies of Plato and Aristotle, and then a monumental encyclopedia on the science of music. By the time al-Farabi traveled to Syria and was appointed physician to the Hamdanid ruler in Aleppo, he was a famous scholar, destined to be known as *al-Mu'allim al-Thani*, the second teacher, the first being Aristotle.

Another example of how a bright lad might become a scientist is offered in ibn Sina's autobiographical account of his rise into the Samanid court from his undistinguished village origins outside of Bukhara. Fathers and medicine would seem to have much to do with it. Ibn Sina's father, a liberal-minded physician, taught his son medicine. Ibn Sina learned it easily; he was obviously bright. But a guiding father and a gift for medicine were not enough for the potential of genius to fulfill itself. This required access to libraries of specialized books, the company of accomplished scholars and the free time to take advantage and absorb these treasures. One needed family support, followed by one form or another of patronage. There were no schools or public libraries for science and philosophy. There were contacts and good fortune. Ibn Sina's stroke of fortune was that through his physician father his name became known to the Samanid ruler, Nuh ibn Mansur, who called ibn Sina to come to Bukhara and cure him when he fell ill. Ibn Mansur had a large library, rich in medical, philosophical and scientific works, which ibn Sina was allowed to use. Thanks to ibn Mansur's library, ibn Sina was able to educate himself further in medicine, then proceed to study Aristotle's metaphysics and natural philosophy and Neoplatonic Gnosticism. Rising to become the ruler's personal physician, court astronomer and advisor, he continued to have the library at his disposal, as well as the means to build his own. Success at this level afforded ibn Sina the leisure to study and write his books on science and philosophy and his encyclopedia of medicine, which would give him fame in the scientific traditions of two civilizations.

Royal patronage was essential for doing science, but a scientist could make a living on his own and did not necessarily have to be a physician to do it. Ibn al-Haytham began his education in a madrasa in Basra. His father had an administrative position in the Abbasid government and helped his son pursue his studies. Ibn al-Haytham learned medicine, the one field of Greek-based study that was openly taught. This took him to optics, probably through his study of the eye and eye diseases. From optics he went to astronomy and mathematics. He was able to support himself by tutoring and writing out copies of Ptolemy's *Almagest* and Euclid's *Elements*. Every year he would make a copy of each of the two works and sell each for 150 gold dinars. The 300 dinars a year was enough for him to live and carry on his studies.[7] Hunayn ibn Ishaq and other translators and copyists in 9th century Baghdad received 500 dinars for copying over the period of a month.[8]

Umar Khayyam's father was a tent maker whose profession he followed until he had educated himself in astronomy and mathematics, whereupon his acquired knowledge led to his becoming the royal astronomer in the Seljuq court. How he came to study astronomy and mathematics is not recorded in the sources. He studied in a major madrasa in Nishapur where, according to one report, he befriended two other students. One was Hasan al-Sabah, who went on to become leader of the Ismaili fortress of Alamut. The other found service in the Seljuq administration of Sultan Alp Arslan and rose to the office of chief vizier, becoming famous under the honorific name Nizam al-Mulk. It was through this connection that Umar Khayyam became court astronomer. For ibn Sina, the road to greatness began with medicine. For his fellow Iranian Umar Khayyam, it was astronomy. But in both cases the road was prepared by instruction within the family, family contacts, or family support in some form. From that point on, it was perseverance and the good fortune of finding the right teacher. Physicians, scientists and philosophers were not credited with having attended a certain madrasa or studying in any one place, but with a certain scholar. In a way paralleling the *ijaza* system, it was the mentor who acted as institution and degree-giver to the future scientist, physician or philosopher.

The great scholars in a ruler's court often took on promising young students whose origins were similar to their own. Tutelage was strictly on a personal basis of friendship and never formalized in the court or institutionalized into any kind of royal academy. A cycle of continuity between the courtly and family levels of scientific study fed one into the other. The two levels had parallel lives of their own but were interdependent to the extent that without the private lower-level of study keeping up its liberal family traditions in medicine, astronomy and the natural sciences, the courtly level would have been starved of the young talent it needed to propagate itself. The teacher–student relationship of ibn Baja, ibn Tufayl, al-Bitruji and ibn Rushd in the Muwahhidun court at Cordova is an example of science and philosophy being propagated on the courtly level. However, before reaching the court, each one of the scholars had received medical and philosophical training that could only have been received within the private environment of family, friends and associates. Ibn Rushd's formal training was in Maliki jurisprudence, but within his circle of family and friends, his mind was opened to medicine, astronomy and philosophy. The same could be said for the dynasty of Maragha astronomers, Nasir al-Din, Qutb al-Din and ibn al-Shatir. All three came to the Il-Khanid court knowing enough astronomy to enable them to be taken on as astronomers and to perfect their professional skills in association with the Maragha observatory.

One science did find at least a degree of support in religion. A branch of astronomy called *ilm al-tawqit*, the science of time reckoning, was important in determining time for prayer and the appearance of new moons for religious feasts. The *muwaqqit*, one trained in this rather low-level specialty of astronomical time reckoning, added nothing to the knowledge of observational or mathematical astronomy, but the association of this branch of astronomy with religion sometimes served to sustain the building of observatories and the study of astronomy. In the social mainstream, however, it was generally accepted that science, with the exception of 'ilm al-tawqit, had no bona fide place and so it was neglected, for the most part passively. Hostile attitudes forcefully exerted were rare.[9]

For the exceedingly pious, a society founded on a religious idea embodied in a book that was believed to be the eternal word of God had no need of science. The highest knowledge man could attain was religious: knowledge of the words of the Quran and their meaning; knowledge of Arabic (grammar, rhetoric and exposition) to grasp the meaning of the Book; knowledge of the life of the Prophet and his Sunna, that is, the way of the Prophet, as preserved in the canonical collections of reports that were handed down from person to person and related what the Prophet was heard to have said and seen to have done, collectively known as Hadith; knowledge of the history of the trials, tribulations and triumphs of the Prophet and his companions and helpers in Mecca and Medina; the holy Shari'a law that theoretically came from the Quran and the Hadith; and *fiqh*, the science of jurisprudence. Sufism and scholastic theological dialectics (*kalam*) would be later additions to the religious corpus, both of which were regarded with some suspicion by the rigorously conservative, who held the line of strict traditionalist orthodoxy against all religious and social innovation. For these, *kalam* was often regarded as philosophy dressed up as religion, an abominable plaything of words that in the hands of the ultra-rationalists became poisonous snakes in the garden of scripture.

Unlike in the Latin West, where theology was the queen of the religious sciences and philosophy her handmaid, in Islam the queen was Shari'a law, perceived to have been abstracted from divine sources and needing no logical handmaid other than in drawing analogies from the sacred sources. Formulating the Shari'a through interpretation of the holy sources was considered the most noble of all possible pursuits in relation to exerting intellectual energy, that is, methodical thinking in the service of God and community. And even this mental outlet was blocked when, by consensus of some highly placed ulema, the "Gate of Interpretation" was pronounced closed and the Shari'a completed, by which time the four recognized schools of law had emerged, named after the leading thinkers whose sources of jurisprudence and methods of interpretation differed and whose schools found favor in different regions: Shafi'i in Egypt, Maliki in Spain, Hanifi in Iraq and usually anywhere Turks conquered and ruled and Hanabali scattered everywhere. The Shi'i community developed its own canonical Hadith and school of legal interpretation.

The *shar'i* sciences, that is, the religiously favored sciences having to do with Quran, Hadith and law, were considered the highest category of *'ilm*, science, over and against the Greek sciences, which were at the head of the non-*shar'i* disciplines. Another term commonly used to distinguish the religious sciences from the non-religious was *naql*, meaning transmitted by religious authority, in the sense of individual Hadith being orally transmitted by a supposedly authoritative chain of transmitters going back to a relative or companion of the Prophet. The religious sciences were the *al-'ulum al-naqliyya*. The non-religious category of *'ilm* answering to *naql* was *'aql*, meaning rational. The Greek sciences were the rational sciences, *al-'ulum al-'aqliyya*, also referred to as the "sciences of the ancients," *al-'ulum al-awa'il*. The dichotomy between religious and secular was expressed in several other ways as well, a most common one being the "Arabic Sciences" and the "non-Arabic Sciences," a nomenclature that gives an idea of the importance of Arabic as the language of revelation and religion.

There was never any doubt which category had precedence. Scientists and philosophers accepted the categorization and restated it periodically over the centuries, from Khwarizimi in his *Kitab al-Hisab* (*Book of Computation*) in the 9th century to Kashi's *Miftah al-Hisab* (*Key to Computation*) in the 15th. The social pressure of religious and linguistic bias was so powerful that very rare indeed was it that a Muslim devoted to the exact sciences or philosophy ever went to the trouble of learning the language in which the knowledge had been originally expressed.[10] The Muslims who studied the rational sciences never sought to explore the original sources from which the Eastern Christians had made their Arabic translations. However highly Muslim scientists and philosophers regarded their Greek predecessors, exalted by some to almost religious heights, they nonetheless accepted, with rare exception, something of the religious prejudice of the *jahiliyya*, the age of ignorance and darkness that preceded the revelation of the

Quran expressed in God's language, Arabic, and they did not bother to learn the language of their Greek masters to read their works in the original.

With the exception of those religiously useful branches of astronomy, time reckoning and geodesy, the separation between the religious and rational sciences was deep, their divorce final.

The ulema upholders of the Shari'a and custodians of communal propriety allowed little space in the social mainstream for the open study of the rational sciences. Relegated to the margins of society, science and philosophy remained dependent on the ruler's disposition and patronage, from which came the hospitals and observatories where physicians and astronomers honed their skills and wrote their books. These physicians and astronomers were also natural philosophers, metaphysicians and mathematicians. Khwarizmi, Haytham, Biruni, ibn Sina, ibn Rushd, Umar Khayyam, Nasir al-Din al-Tusi were not patronized for their mathematics, physics, optics or gnostic philosophy but for their astronomy (including astrology) and medicine. Without the ruler's patronage of astronomy and medicine and the observatories and hospitals for them to work and teach in, and thereby pass on their skills, the history of science and philosophy in Muslim civilization would have been a meager volume.

The strict separation between the rational and religious sciences found its social counterpart in the separation between the ruler's court and the sober atmosphere of the mosque and the market place. Members of the royal court enjoyed their own worldly culture of poetry, refined etiquette, studied eloquence, and dress, not to mention the witty repartee and sensual pleasures that went with the wine, banquet table, music, singing and dancing girls. A genre of Abbasid literature called *adab* guided courtiers and secretaries in proper behavior, speech and etiquette. As delineated by its chief exponent ibn Qutayba (d. 889), *adab* culture also included a little geometry and mathematics for practical reasons of surveying, irrigation, construction and accounting, with a bit of practical astronomy as well,[11] but as a body of knowledge and courtly style of behavior, it existed a world apart from the populist culture of the mosque and marketplace.

There was indeed much about courtly life and the caliph's style of rule that disturbed the community of the pious. The caliph had his own political judiciary that circumvented the cumbersome methods of evidence, witnesses and testimony required by the Shari'a courts of law. Particularly disturbing to those who venerated religious law was the muscled executioner standing to the side of the throne, head-severing axe in hand with the leather mat to catch the blood: chilling symbols of the caliph's absolute power over life and death of any believer. The caliph's unbridled power to confiscate the property and possessions of men of wealth and his ministers who fell out of favor, and to put men to death by a mere gesture, was abhorred by the populist ulema who held to Shari'a law and its forms of deciding life and death issues and protecting property.

But what went on in the palace was far away from the ulema and ordinary Muslims, and remembering the strife of revolution and civil war that had sorely tested the community's cohesion since the murder of the third caliph, they grudgingly tolerated these abuses to pious sensitivity that seemed inherent in *siyasi* (political) government, as opposed to *shar'i*. There were those of the ulema who deemed rebellion against unjust government worse than the abuses of tyranny. Extrajudicial taxation, confiscation and execution at the courtly top was tolerable so long as the caliph's government maintained civil peace and order, protected society from heresy and external danger, and enforced the Shari'a as the law of the *umma*. Determining what was heresy was the ulema's task, rooting it out was the caliph's. To maintain good relations with the ulema, or at least keep them quiescent, the caliph was obliged to project an outward appearance of rectitude by acting publicly in accordance to expected form. From the lowly Quran reader in a poor village to the chief judge and leading religious scholar in the capital's largest mosque, the ulema formed a powerful check to the ruler's willful exercise of authority beyond the palace walls.

Most resolutely opposed to the *adab* culture of the courtly elite and the science and philosophy associated with the caliph and his court was Ahmad ibn Hanabal (d. 857) and those after him who followed his example, the Hanbalites. The last of the founding fathers of the schools of legal interpretation, ibn Hanabal was the most extreme in observing the prophetic tradition as the model of prescribed behavior.[12] In his eyes, the caliph's court was a vipers' nest of decadent irreligiosity and heresy. With some notable exceptions, the traditionalist followers of Hanabali populism became formidable opponents of the rational sciences, believing the Prophet's community should be informed only by the sources of truth and guidance that gave birth to it, the Quran and Hadith. Hanbali hostility to Greek science was pervasive. Its influence reached even some who were associated with the royal court. Ibn Qutayba (828–889), the author of *Adab al-Katib*, the prestigious textbook written for the government scribes (the *kuttab*) to master courtly etiquette and elegant prose style in Arabic, abhorred Greek learning and castigated government officials who dabbled in philosophy and astronomy, those poisonous "foreign sciences." However, as a Persian, ibn Qutayba did not feel the same about Sassanian scientific influences: Sassanian learning was domestic, an integral part of the Muslim heritage, whereas Hellenistic learning was foreign.[13] Important as ibn Qutayba was in classical Abbasid thought, this was still only one man's opinion. Many, if not most, of the translators of scientific and philosophical texts came from the *kuttab*, for whom ibn Qutayba wrote and to whom he belonged.

It would take centuries before the negative attitude typified by ibn Qutayba reached a mass critical enough to drain the pool of young men prepared to devote themselves to the arduous work of science. That negativism to the rational sciences – a seemingly imperishable form of fear-fed mental darkness found in all societies – persisted and grew. A century after ibn Qutayba, al-Uqlidisi, a geometer, wrote plaintively that many men hated to be seen with the dust-board used for mathematical calculations and geometry on his hands because of the way people regarded it: a tool of heretics and charlatan good-for-nothings earning their living by astrology in the streets.

Not long after, the geometer al-Sizji complained of the difficulty in finding mathematical works. The reason is obvious: "Indeed, the great mass of people consider the investigation of geometry blasphemous and count ignorance of it a boast. They find it lawful to kill him who believes in its correctness."[14] The relegation of geometry and the rational sciences to astrology and the occult caused such intense social opprobrium for al-Uqlidisi and his profession that he abandoned the dust-board for more expensive pen and paper, the price for passing respectability. Al-Biruni in the 11th century complained of people being proud of their ignorance, liking nothing more than causing harm to those learned in the sciences, whom they call atheistic, while hiding their ignorance behind the question: of what benefit are the sciences? Such people al-Biruni considers "rude and stubborn."

Such were the people who were, unresisted, shaping social attitudes. They would seem to have made much progress by the 12th century, as seen in the autobiographical lament of the mathematician Samaw'al, who could find no one in the whole of Baghdad to tutor him in Euclid's geometry or al-Karaji's algebra.[15]

The religious genesis of the Muslim state endowed the ulema with tremendous social power, far and above anything the Christian clergy enjoyed in the Latin or Greek regions of the Roman empire. From the ulema came the law, the interpretation of scripture, the sanctification of custom, and education. The ulema not only kept logic, science and philosophy sharply apart from religious education but were able to set the social tone in keeping them at arm's length as suspicious subjects of alien provenance that hid in the shadowy corners of society. On occasion they were able to delegitimize them and force rulers to rid themselves of the philosopher–physicians they were patronizing, even to the extreme of having their books burned. But such extreme episodes were most unusual. The conservative ulema were satisfied to keep the "Greek sciences" from contaminating the believers and were for the most part no more than passively hostile to science and philosophy. They could live with it as long as the alien sciences were kept within the confines of the court or in the privacy of the home.

Muslim rulers, as mindful as they were respectful of this mostly quiet social power of the men of religion, lived in peaceful tension with it. Weak rulers who had little popular support and were vulnerable to charges of illegitimacy because of their inordinate behavior, or reasons of political usurpation, collapsing economies, or because of their being challenged by rival princes or generals, were obliged to be unusually amenable to the demands of the ulema in order to win public support. Under such rulers royal patronage was seriously constricted. Those who practiced the "Greek Sciences" would be the first up for exile in a crisis.

Attempts by Abbasid rulers to give the state a voice of religious authority were not altogether successful. Ma'mun's espoused Mu'tazilite scholasticism[16] in hopes it would take root as the unquestioned official state doctrine that Zoroastrianism had in Sassanian Iran, or Greek orthodoxy in Byzantium, was aborted. Caesaropapism would have no place in Islam. The mosque, that indomitable fortress of religion, would remain under the autonomous authority of the ulema, supported by the urban populace of notables, merchants, shopkeepers and craftsmen.

The rejection of Mu'tazilite theology was to have a decisive effect on the place of the rational sciences in Muslim society. Not that Mu'tazilites advocated the harmony of religion with philosophy and science. They were far from being Muslim counterparts of Albertus Magnus and Aquinas. Mu'tazilites generally took a rational approach to religion, applying logical analysis and allegorical interpretation to scripture but without reconciling Aristotle or natural philosophy with scriptural belief. Nonetheless, their logical approach to belief could presumably have opened intellectual passages that otherwise remained closed.

In the broad spectrum of Muslim scholastic dialectics, there were those practitioners of *kalam* who shared philosophical principles and introduced them in their reasoned arguments of scriptural interpretation and doctrinal formulation. These were predominantly, but not only, Mu'tazilites. Some Mu'tazilites in fact engaged in scientific matters, most notably al-Nazzam, a principal dialectician whose interest and activity in science included experimentation.[17] For this reason, modern Muslim reformers regret the demise of the Mu'tazilites. As they see it, the decisive event that in time allowed the West to turn the wheel of fortune against Islamic civilization was the defeat of Mu'tazilite doctrine by what came to be Sunni Orthodoxy. Balanced momentarily on the historical fulcrum of logic, destiny robbed Muslims of the glory and power of their civilization's divine fulfillment. Such a momentous reversal of destiny deserves a moment's look at Mu'tazilite theology.

According to it, God was just. Consequently, people had free will, since a just God would not foreordain them to sin and suffer an eternity in hell. The Quran was not the same as God but was created by God. To hold the Quran as eternal was to equate it to God, which contradicted the logic underlying the belief in the oneness of God. This also meant that the powers attributed to God in the Quran had to be understood as qualities and not essences, as there would be as many gods as essences, since essence is indivisible. Quranic verses that defied reason had to be understood allegorically, such as where God is described in anthropomorphic terms as having hands and face and human emotions, or having a throne and preserved tablets in heaven. The doctrine of divine unity was essential. The logic it demanded could not be compromised. The name these Mu'tazilites gave themselves, believers in God's justice and oneness, proclaimed the pillars upon which their logical interpretation of scripture stood firm.

Their theological positions, along with their logic, went against the populist beliefs of the traditionalist leader, Ahmad ibn Hanibal, for whom free will was a fallacy. For the Hanbalites, God was all-powerful and all-knowing: God had to know and have determined everything; the Quran was eternal, like God; the holy book was the eternal mind and speech of God and as such had to be understood literally, even where the verses expressed anthropomorphism.

The traditionalist Hanbali version was at the time much closer to the ordinary Muslim's beliefs than that of the syllogistic disputation of the Mu'tazilites, many of whom were Iranian, and perhaps for that reason less susceptible than Arabs to believe in the inimitable, miraculous depth, nature and beauty of the Quran's revelations. Traditionalist belief claimed the Quran to be not only the eternal language of God but divinity itself manifested in word. Belief in the Quran's divinity precluded its revelations from being submitted to critical analysis and allegorical interpretation, lest the divine awe and mystery it evoked be diminished and its divinity questioned.

The conflict between traditionalist and Mu'tazilite beliefs became, for some 30 years during the mid-9th century, an expression of the ongoing struggle between the populist defenders of religious autonomy and the caliph's authority.[18]

Theological dispute had always been political in its origins. Going back to the political conflict over the caliphate between Ali and Mu'awiyya, the Kharijites (Seceders) had broken with Ali and formed a revolutionary sect with its own theology and political doctrine. Shi'i Islam began as a political revolt against the Umayyad Caliphate and developed a theology and political doctrine quite opposite to that of the Kharijites. The Umayyad caliphs had supported those religious thinkers who, rejecting free will, advocated belief in God's absolute determinism, a doctrine that conveniently justified the Umayyads, since, as caliphs, they could only be so by the will of God. But the Umayyads did not push their favored school with the same vigor as Ma'mun did Mu'tazilite doctrine. An inquisition (*mihna*) was set up to coerce or drive away the Hanbali-led opposition. Mu'tazilite theologians were only too happy to use imprisonment and torture to convince their opponents of the simple logic that God who created the Quran had to be just, and in order to be, he had to have endowed men with free will. God could also not be attributed having anatomical features or human emotions. God was a bodiless being without dimension, place, or physical sense. Man was not made in his image since God had no image. The wine and virgins of paradise that awaited the believer were allegorical.

The Mu'tazilites were anything but popular. Ahmad ibn Hanibal was among those who suffered their inquisitorial zeal, doing them little good in public relations. Not that it mattered to them. Girded by the power of the caliph, they wielded the inquisitorial sword of their theology as ruthlessly as they did their logical analysis of God and Quran. Driven by power and fervent belief, they exceeded the bounds of piety and tolerance of the great majority of Muslims, losing what little public support they may ever have had. Their cruel inquisitional methods matched the uncompromising logic framing their metaphysics of God, Quran and man. Their excessive application of logic in giving definitive answers to all questions and matters reduced religious belief to an extended syllogism, to the point of, if not absurdity, inanity. One logically unrestrained treatise rigorously defines God in negatives to demonstrate his universality: God is colorless, tasteless, formless and dimensionless, not a liquid, solid or gas, not a point or a geometric figure.

It might be said the Mu'tazilites syllogized themselves to death. Had Mu'tazilite theology restrained itself and the caliphs succeeding Ma'mun, Mu'tasim and Wathiq continued to support them for several generations, quite possibly the school could have become the orthodoxy of Sunni Islam, with a scholastic Aquinas-like modus vivendi holding reason and religion in place as complimentary companions. As it was, the caliphs did not long continue to support them. The orthodoxy that emerged in the course of the 10th and 11th centuries, and that came to define Sunni Islam by the early 12th century, was one of the decisive features that would put Islam and western Christendom on diverging paths of intellectual development. Of course, the seeds of divergence were planted long before the

12th century: They were already in the ground of history back when Constantine saw the cross in the sky and Muhammad received his first revelation.

The decisive difference between Islam and Latin Christendom in their respective responses to pagan Greek legacy goes back to their origins. Muslims created their own society: Muhammad was its founder, its first leader, lawgiver and conqueror, its Moses in Mecca, its David in Medina and its wisdom-filled Solomon in the prophet-embodying Hadith that lived after him. The state was born of a prophet and a holy book, giving credence that God's community needed nothing more. Christ and his small band of followers were, on the other hand, born into an imperial state that had existed for centuries. Laws, traditions and institutions going back to the Roman Republic were well in place when Constantine accepted Christianity. The adopted religion born in provincial Palestine was given its place in the empire. The Greek intellectual heritage and secular Roman law were too deeply rooted and cherished to be replaced by anything the Christians, a small percent of the population at the time, could have conceived. Christ had produced no book of revelation from which his followers could formulate a code of civil, family, commercial and criminal law. For a generation or two after Christ, Jewish law was for many Christians their law, up until Romanized Jews like St. Paul severed the nascent Christian community from its Jewish roots, denying Jewish law without legislating a law of their own, other than the nebulous law of the spirit expressed by Paul as love, hope and charity. Between Jews and Romans, Christians had no room to make law. Their law was of the spirit, Paul told them. Where else could it be without their having a state of their own? The Roman empire's pagan converts to Christianity were far more informed by Greco–Roman civilization than they were by the local customs and traditions of Judaeo–Christian Palestine. Muslims could call the time before the coming of Muhammad and the Quran "The Age of Ignorance," but to most Romans who witnessed Constantine's conversion and saw Christianity becoming the new religion of state, it appeared the age of ignorance had just dawned.

On the Muslim side, there were those religious authorities who had spoken against reason right from the time logic, philosophy and the exact sciences were first being translated and cultivated, but for a long time, the many contending schools with their contradicting political and metaphysical theologies had diffused the energies and attention of religious thinkers, limiting their effect in opposing the rational sciences. Enjoying the royal patronage of strong rulers, science was able to flourish and spread throughout Islam for three centuries, and thereafter for another three in a somewhat patchy fashion with respect to time and place, before fading into a shadow of its former self. Rather than becoming a friend and support of the rational sciences as religious scholarship matured along with the civilization in general over the centuries, as occurred much later in Latin Christendom, religion in Islamdom gained an effective front in opposing natural philosophers and metaphysicians whose portraiture of God, universe and the human condition reflected Plato, Aristotle, Plotinus and Ptolemy rather than Quran, Hadith, Shari'a and the authoritative texts of Quranic interpretation. Those members of the Sunni ulema, who tolerated no competition in the elucidation of truth, gained a defining voice through the works of several dominant theologians between roughly 900 and 1100.

Thereafter, practitioners of the rational sciences, like the Mu'tazilites after they had been deposed and dispersed, were constrained to exist more or less anonymously at the margins of society, except for those fortunate enough to be favored by a ruler's patronage and thereby enshrined in history.

Violent opposition against those who were brave enough to study the rational sciences openly in the bosom of society was rare. Except for a number of notable exceptions, the ulema were in general more passively than aggressively hostile to such religious deviants. During a millennium of history there were only a few occasions when opposition became violent, and even when it did it was books, not scholars, were put to the flames. Social ostracism and exile were the normal punishments meted out to those scholars who incurred the ulema's displeasure. As the ulema embodied the greatest percentage of literates in society, this was effective enough to dampen enthusiasm and keep the rational sciences in their place. In time, the sciences were deprived of the bright young minds needed to sustain a tradition.

The religious leaders of the minority communities often shared the same intellectual prejudices as those of the dominant majority. Compared to the intellectual boundaries permitted by religious orthodoxy in Byzantine and Latin society between 1050 and 1450, the boundary left open by orthodoxy in Muslim society was relatively narrow for anyone wanting to study logic, science or philosophy. In fact, Sunni theology had developed in such a way as to rule out even the possibility of a natural science. The formulation of Sunni orthodoxy, in this regard, is a fascinating study.

The Mu'tazilite logicians lost their place at court when the caliph Mutawakkil (d. 861), searching for religious support against the threat of the Turkish guard that had raised him to the caliphate and now threatened to depose him, turned to the Hanbali traditionalists. Hanbalite popularity among the faithful had been sealed by Ahmad ibn Hanbal's simple piety, courage and devotion to the Prophet, upon whose life he declared to pattern his own, down to the least recorded detail – he is reported to have said that if no credible report of a tradition of Muhammad having done something, he would not do it: a mind-set as extreme in one direction as that of some Mu'tazilite logicians in the other.

Though to traditionalist Muslim thinkers it appeared the Mu'tazilite method of logical analysis had syllogized the heart out of religion and the Quranic mystery, the Mu'tazilite application of logic continued to have some appeal to religious scholars and found its way into orthodox Sunni theology through the scholar al-Ash'ari (d. 923), an Arab who imbibed Mu'tazilite dialectics as a student, only to rebel against the doctrine in mid-career. Like Martin Luther six centuries later, he was at a certain moment assailed by doubt that reason was a valid tool in understanding scripture and a secure scaffolding for religious belief. And like Luther's, his break with the rational scholastic approach to religion resulted in a theological reformulation that still goes by his name. What precisely was behind Ash'ari's strong reaction to Mu'tazilite doctrine is hard to say. His Arab ancestry went right back to the companions of the Prophet, which may have had something to do with his rejection of the doctrine of the created Quran. Quran and Hadith had to be the immutable foundations for a solid dialectic theology. And it is comforting to believe the language of God's mind is the one you happen to speak.

Reason was too slippery a rock on which to found a system of belief, especially in view of the lengths the Mu'tazilites had taken it. Ash'ari broke with them but used the logic he had learned from his Mu'tazilite teachers to build a theology, according to which the Quran was the eternal mind and speech of God. As such, the Quran had to be understood literally, anthropomorphisms and all. Free will was canceled out. All human decision and action had been predetermined by God at the moment of creation. God's greatness could not be constrained to a human concept of justice. Reason with its syllogisms and logical structure of cause and effect was inadmissible as a means of understanding God and the way God controlled nature. Nature did not operate according to causal relationships but by the direct action of God ordering each event, a shadow being caused not by a tree blocking solar rays but by God creating the shadow independent of any object between the sun and the shadow, or hunger being not the effect of not eating but of God's direct act of creating the sensation of hunger. A flaming torch put to a pile of dry wood does not cause the wood to burn: only the command of God does that.

What seems to have triggered Ash'ari's rejection of Mu'tazilite theology, beyond the school's doctrine of the created Quran, was its doctrine of a just God. This has always been a test of belief. Mesopotamian myth reveals the disconnect between humans and their gods when it came to justice. Abraham being commanded to sacrifice his son and the story of Job said it for the Jews. Faith transcended the cruelty of sacrificial murder and the inexplicable suffering of just and good people. Blind obedience triumphed over reason. The abundant injustice in the world with its massive, undeserved death and suffering has not perceptively done injury to the comforting belief in an omnipotent divinity who against all medical and historical evidence is somehow a good and reliable protector. With a mournful sigh that a madman's murder of a dozen school children is all for the best in God's strange way and that their souls are happily resting in heaven, believers are constrained to dissociate God from justice and accept divine inscrutability. Ash'ari is reported to have asked a Mu'tazilite colleague why if God is just would an innocent baby have to die. Because God knew the baby would grow up to sin and be damned to hell, was the reply. Then why did not God save all souls suffering eternal hell fire by having them die as babies? The Mu'tazilite had no answer. Ash'ari's was that the human concept of justice had nothing to do with God's will: Justice, a particular of human construction, could not be imposed on the divine mind, a transcendent universal beyond the ken of human mind. Six more centuries down the corridors of theology, Calvin's answer would be that if God were truly just everyone would go to hell. God was merciful, not just.

Ash'ari applied the scholastic methodology of logical argumentation he had learned from his Mu'tazilite training to Hanbali traditionalism to substantiate the very doctrines the Mu'tazilites had opposed. Stripping away the concept of justice from God presented no problem. God was just, it claims so in the Quran, it was one of God's many qualities, but divine justice was beyond what humans thought justice to be.

God's knowing and preordaining everything introduced the conundrum of the origin of evil. Logically, determinism would have evil coming from God, but this would not do. Here again the Ash'arites brought down the impenetrable mental barrier between human and divine. Evil was present but did not come from God, even though God originated everything. The believer accepted this on faith, *bi-la-kayf*, without asking how, in the famous dictum of Ash'ari. Later Ash'arites would refine this, addressing the doctrine of determinism and the related question on the origin of evil with the theological concept of *kasb*, acquisition, that is, man somehow acquired responsibility for his acts during the course of his life. By acquiring them, he became morally responsible for them, even though the acts themselves were created by God. The introduction of *kasb* provided an escape clause, which saved God from the onus of originating evil. The escape hinged on a subtle distinction between God's foreknowledge and will. What God foreknew was not exactly what he forewilled. Evil crawled out of the crack between God's knowing and willing and was then pinned to human action, to be answered for the coming judgment, followed by an eternity of hellfire, if God so willed. The God of the Quran, if not Ash'ari theology, balanced mercy and forgiveness on the scale of judgment.

With Ash'ari, traditional beliefs were argued into a logical system that, with other refinements, was eventually to become the accepted theology of Sunni Muslims. It was not by any means a triumph of blind tradition over reason, but rather a theological synthesis of Arab tradition and the Aristotelian logic that had been fastened to Eastern religion, first by Nestorian Church fathers and then, in Islam, by Mu'tazilite theologians.

The logical structuring of religious belief did not, however, provide an opening for religion to embrace science and philosophy. Ash'ari's denial of causality destroyed the basis of any possible rational system of natural relationships to explain the workings of the physical universe. With God's direct command of every single natural event annulling causal relationships and religion explaining everything one should know in God's universe, there was nothing else left between heaven and earth to explain or know. As for the appearance of causality in nature, that was exactly what it was, an appearance. It was not real. It appeared to be real, but in reality, causality was no more than an impression created by repetition. What natural philosophers called cause and effect was simply the customary way things in nature worked; but they did not *have* to work in the customary way: God's omnipotence was able to change the customary appearance of order at any instant. In fact, God did change the order every instant. In every instant, in every atom of time, the universe was annihilated and recreated by God's command, so that what appeared to be order, constancy and events occurring in a sequential network of causal relationships was in reality God at work ordering every single effect directly.

Time was atomistic; the intervals of non-time separating the atoms of time were states of nonexistence, instantaneous universal annihilations, which obviated the flow of nature and events by cause and effect relationships. God's atomizing of time in imperceptible instants of annihilation and reintegration made Him the sole cause in all existence, a doctrine called occasionalism, since it was at God's command that each and every occasion and event came into being. Atoms of time and existence imperceptibly separated by existential annihilations replaced cause and effect acting in a time continuum. Quite as al-Ash'ari's fine line between divine knowing and willing provided the escape hatch for evil, his atomism provided a crack in the flow of time into which the accustomed order of nature disappeared; during this interruption anything could happen, whatever God willed. Such instances when he willed what was contrary to what men expected by physical experience were called *kharq al-'adah*, breaking the custom, otherwise known as miracles. Ash'ari's occasionalism thus provided a rational basis for understanding God's miracle-working powers.[19]

Part and parcel of Ash'ari's reversion to traditionalist belief was total acceptance of traditional ways and total rejection of innovation. Traditional ways and practices, called *taqlid* (molding, forming and casting in a mold) was opposed to innovation, *bid'a*, doing things in a way not prescribed by tradition of the Prophet. *Bid'a* was therefore to be avoided in case the particular innovation was religiously reprehensible. This implied avoidance of things foreign or new. The Greek sciences could not help but fall into this category of things to be avoided.

Such was the way it went. Gone were those days of tolerant debates, back when pious Muslims engaged St. John of Damascus in the Caliphal court, and later, at the time of the Abbasid caliph al-Mahdi, when the Syriac patriarch of the Eastern Church, Timothy (780–823), did the same. The debates epitomized by the courtly visits of St. John and Timothy provided the leavening that gave rise to early Muslim theology. It would take generations for it to harden and set itself against the possibility of a natural science, and centuries more for the consequences of that to become apparent.

The question why it was that religious attitudes developed as they did in taking this antithetical stand against the rational sciences has no easy answers. Perhaps it was because science and philosophy came into Islam before an informed religious consensus had been established with the authority to formulate religious ideas into a belief system. The barbed queries of the Eastern Christian theologians may have left the first generations of Arabs in Syria and Mesopotamia feeling insecure and vulnerable. What did they know of Aristotelian or Neoplatonic metaphysical structure, of the relationships of souls, minds, essences, accidents, qualities, substances, particulars and universals? In Damascus, manuals were written in defense of these attacks: if a Christian asks this, the Muslim answers this. The Greek origins of the knowledge made it suspect. What did Muslims, blessed with the last Prophet sent by God, need from Greeks, the early ones being pagans, the contemporary ones Christian infidels, followers of three gods somehow bound up in one? To take anything from them would be an abomination, an admission of Islam's deficiency. The aggressive challenge of Greek thought, coming as it did in Islam's relative infancy, before intellectual defenses had been constructed, may have perhaps scarred the collective consciousness of too many of the ulema.

The religious scholars, judges, jurisconsults and assemblages of men learned in religion who established the contours of Muslim belief and norms of social behavior became a force to be reckoned with, even though they lacked the unity of an overarching organization whose authority stretched across the lands of Islamdom. There was no organization or council empowered to define what was acceptable and what not. Each city and province had its own ulema who represented a religious front unified enough to command the attention of any ruler, regardless of how secure he felt in his position or how lightly he took his religion. The ethereal differences of theology aside, the ulema were in agreement on the basic essentials of down-to-earth belief and practice. A subtle uniformity of belief and behavior came with the mobility of a society in which scholars and preachers moved about from city to city as freely and easily as did its merchants. At the same time, local variations in customary practice across the many regions of Islam, with its different races and cultural traditions, were preserved and Islamized by way of a convenient Hadith or accommodating *fatwa* (legal opinion by a judge learned in the religious law), as long as the custom did not go against the religion's fundamental tenets.

Muslims described Islam as the religion of tolerance (*ibaha*). Nineteenth century apologists would point to the fact that Muslims never burned their heretical philosophers or theologians, the preferred method in the West. Though there are instances enough in Muslim history of theologians being executed: two were executed in late Umayyad times for preaching the doctrine of the created Quran, and Suhrawardi in Ayyubid times for his pantheism.[20] But for all the ulema's liberality as regards Islam's openness to different social customs and its religious protection of Christians, Jews and Hindus (the ravaging of India by Mahmud of Ghazna was a military exploit for plunder beyond control of the ulema), the broad, embracing and easy going spirit of live and let live was not all that much in supply regarding the ulema's tolerance of the rational sciences. However often and profusely philosophers professed their faith and argued the consonance between science and religion, the rational sciences were as often and as profusely condemned within Muslim religious circles. Certain practices and beliefs in Sufism were more heretical than anything Muslim scientists and philosophers did or said they believed, and yet mysticism was provided safe passage into orthodox circles at the same time the rational sciences were finding in religion an unfriendly face, turning on some few occasions into an implacable foe.

There was a tendency among the ulema to consider anyone who pursued science and philosophy a *zindiq*, heretic. Suspicion of involvement or interest in the philosophical sciences could sometimes be enough to get someone accused of heresy. Al-Kindi and others suffered from the charge. The word *falsafa* (logic, philosophy and science) was itself feared to be the very breath of the devil. The pious are reported to have recoiled in horror at the mention of it. A pun was made on the word, *falsafa* becoming *fall as-safa*: "the abyss of stupidity." It did not matter if one of the rational sciences, mathematics for example, was or was not in conformity with religious attitudes or even had anything to do with religion. Mere association with *falsafa* and things from the *Jahiliyya* age of ignorance could, on occasion, be enough for condemnation.

There were degrees of hostility. Attitudes of the ulema varied from region to region, city to city. There was also a difference in outlook among individual members of the ulema in any one region or city. There were even a few of the ulema who expressed enough independence to defend a Muslim's desire to study Greek astronomy, mathematics, logic and medicine, knowledge they considered useful and necessary. Quranic verses were adduced in support and there were many.

Attitudes did not soften in post-Mongol times. The 13th century Hanbali religious reformer ibn Taymiyya spoke for many in Syria, Iraq and Egypt when he said the only true science was that which came from religious sources, such as the "medicine of Muhammad" that was reported in the Hadith or descriptions of natural phenomena described in the Quran. All other so-called science was to be avoided, as it was either false or not useful. A similar attitude was expressed centuries earlier in Andalusia and seems to have been held by practically all the ulema: The only science permissible for a believer was that which was useful to religious practice or in some way benefited religion. All else was useless and would lead the believer astray.[21]

The ulema commonly feared, perhaps with some justification, that anyone studying the rational sciences would end up questioning the truth of orthodox religious belief and its laws. Critical doubt could only weaken belief and the blind acceptance of tradition, leading the believer to heresy. The vocal leaders of the ulema believed that science and philosophy led to contempt of law and religion and those who studied the rational disciplines became so arrogant they made no secret of their disbelief.

With the ulema's unsupportive attitudes prevailing in society, anyone interested in the rational sciences was afraid of being considered a religious deviant if found out. Teachers warned students early in their lives of the dangers of *falsafa* and its affiliates. Those who dared indulge in *falsafa* did it undercover, in the privacy of their homes. The grandson of Nasr ibn Sayyar, the Arab conqueror of North Africa and Spain, outwardly shunned the rational sciences for fear of the ulema, though to confidants he boasted of having mastered all the sciences of the ancients, meaning Greeks. This was the way the rational sciences were perpetuated outside of the protection of royal patronage and the palace walls, clandestinely, in the privacy of one's home, where the scientific traditions were passed on father to son, or among friends, kindred souls meeting in secrecy to read and discuss something of natural philosophy or geometry, like members of a forbidden cult assembling under the cover of night. Those who were found out or were brazenly open in their study of *falsafa* might be shunned socially as irreligious and dangerous.[22]

The socially prominent, those who had much to lose and valued their reputations as pious Muslims and upstanding members of the community, had to indulge their interests in the sciences in secrecy. Merely the knowledge that books related to the *falsafa* tradition were in one's house could be enough to bring disrepute on all members of the household and raise the cry that heretics lived among the community of believers. In early 10th century Baghdad, a high point in the scientific renaissance, the communal pressure of orthodoxy was even then enough to force copyists, outside and inside the royal court, to take an oath to not copy anything from the books of *falsafa*. The famous essayist and humorist al-Jahiz warned that two things must be concealed from society, forbidden drink and suspicious books.

The penalties for being caught with those suspicious books were however not severe, at least not when compared to being burned at the stake. Shame, shunning, finger-pointing, ridicule and exile were the usual penalties. A major event in the tension between religion and *falsafa* occurred in Baghdad around 1200, when a collection of philosophical books and treatises attributed to the Brethren of Purity was publicly burned before a mosque, where a sort of funeral pyre had been constructed for the purpose. As the fire roared, the assembled crowd was ordered to curse the burning books, and along with them, their contents, authors, those who believed what was written in the books, and finally the owner of the books and all his associates and descendants. Because the owner, who was there in the crowd joining in the chorus of curses while watching his books being consumed in flames, was both a member of the Qadiriyya mystical brotherhood, named after the highly venerated Sufi saint abd al-Qadir al-Jilani, and a follower of the Hanbali legal school, both the holy saint abd al-Qadir and the venerated Imam Ahmad ibn Hanbal were unwittingly implied in the list of those being cursed. The event also included the reading of some withering poetry ridiculing the accused heretic, who was forced to listen. In case this was not enough to drive home the lesson in true religion, he was imprisoned for a short period. When released, he renounced his former error and led what was publicly considered a devout life.[23]

As is usually the case in the fever of condemning books, few of the ulema bothered to read what it was they and their colleagues were rejecting. The most vocal of the ulema who declared themselves against logic, philosophy and everything and everybody associated with them knew little about what they were condemning – purposely, for fear of losing their souls. A typical case was that of Shaykh Jamal al-Din ibn al-Abbas (d. 993), who in a treatise refuting philosophy railed against the confusion and harm caused by Greek knowledge. He cursed Socrates, Plato and Aristotle as if they were still alive, calling them ignorant troublemakers and spreaders of heresy, madmen who lived like celibate monks to cease the generation of mankind. The chief of their gang of corrupters and heretics was Plato, "may God curse him." His brothers, Socrates and Aristotle, and someone called Galen, were all heretics and unbelievers. In retribution, God sent a flood against them and they were dispersed.[24]

Some of the ulema whose training was limited strictly to religious subjects described themselves as having felt cold terror upon coming across a page of geometry or astronomy in which the forbidding complex of circles on circles and intersecting lines of planetary models appeared to be diabolical signs designed to conjure the evil jinn and the very devil from the bowels of the earth to seduce the minds of men and destroy religion. The prejudicial pressure of populist religion implored the rulers to curtail their support of the *al-ulum al-awa'il*, those rational sciences of the ancients that had come from outside Islam and threatened to corrupt those whose eyes came into contact with them. The pressure was constant and sometimes effective. Ibn Massara (d. 931), an early Andalusian thinker influenced by the Greek thought that was at the time penetrating Spain from Baghdad, found it wise to abandon Cordova because of the Malikite ulema's hostility to him and "the foreign sciences" in which he showed interest.

Similar examples could be multiplied. Rulers cooperated with the ulema for a variety of reasons, whether it was out of narrow-minded piety or simply the good political sense of identifying the dynasty with populist religion and the ulema's prevailing attitude. The need to gain the ulema's support because of a regime's political weakness was enough to induce the ruler to shift policy. Maneuvering to assert civilian control over the Turkish slave military, the Abbasid caliph Mutawkkil abandoned the Mu'tazilites for populist Hanbali traditionalism in search of public support. The Andalusian usurper ibn abi Amir surrendered support of the rational sciences to curry favor with the ulema in Cordova and sealed the deal by casting to the flames books of logic and philosophy in the caliph's famous library. His is an infamous case in Muslim history and brings into stark relief a weakness that in time allowed the lowly West to take precedence. The palace library, mainly the work of the caliph Hakam II (d. 976), was reputed to have been one of the greatest in Muslim civilization, containing 400,000 manuscripts, a good number of them on science and philosophy.

Hakam II surrounded himself with books. He acquired them from wherever he could and had copies made of scientific manuscripts in Baghdad and Cairo, devoting himself more to intellectual pursuits than to directing the affairs of state. His interest in the sciences seemed to generate a public interest in them, something that would certainly have generated a public interest in them, something that would certainly have not endeared him to the ulema. When Hakam died, his son Hisham, still a child, was declared caliph. Through cunning seduction, intrigue, marriage and murder, political and military power came to be firmly assumed by the vizier, ibn abi Amir, a bright student of law who had begun his courtly career as overseer of the estates of the young caliph's mother, whose bed he is said to have shared.

Ibn abi Amir ran the state efficiently in the name of the caliph, won over the allegiance of the troops and, with himself at their head, delivered a devastating defeat to the Spanish Christians coming down from Leon. With this, he bestowed upon himself the honorific name al-Mansur, the Conqueror, and not long thereafter ordered that his name be mentioned in the Friday sermon and that coins be struck in his name, traditional prerogatives of a Muslim ruler. Usurping caliphal authority in all but name, he won victory after victory over the Spanish Christian armies, pushing them back up against the Pyrenees, and conducting himself in every way as the ruler of the Umayyad realm. In emulation of abd al-Rahman III's Madinat al-Zahra, he had a luxurious palace built for himself outside of Cordova, which he named Madinat al-Zahira, and, like the caliph, wore a gold trimmed silk robe and surrounded himself with all the pomp and ceremony of royalty, even having his son named as his successor. But afraid of going too far in his usurpation of caliphal authority, he never deposed Hisham II or declared the Umayyad caliphate at an end.

He was of a speculative turn of mind, conversant in science and philosophy, but as an illegitimate ruler, ibn abi Amir was obliged to put his appeal for religious support before science and philosophy.[25] In order to win the ulema over, he had the leading astronomer and mathematician of his time, al-Saraqusti, arrested and all the astronomical and philosophical books of the royal library publicly put to the flames. The ashes were gathered and thrown into wells, after which dirt and stones were thrown down to make sure none of the devil's seductions contained in the ashes should rise up from the wells and pollute the air men breathed. The ulema had demanded that all the books of science and logic be burned and ibn abi Amir obeyed, only books on mathematics and medicine having apparently been spared. From that time on, the people of Cordova concealed any interest they might have had in the rational sciences. A little over a century later, the knowledge-starved Latins were in Toledo making their translations and carting off loads of manuscripts on science, medicine, mathematics and philosophy, beginning a movement that, five centuries later, would reverse the positions of dominance that had existed between Islamdom and the West.

The political stability of Muslim states was as fragile as the place of science and philosophy in Muslim society. Without supporting institutions holding the absolutist state together, there was no room for a weak or child ruler. Trouble in the palace meant the beginning of the end of dynasty and state. Ibn abi Amir's usurpation crippled both his own and young Hisham's authority and was not long in bringing to an end the splendid Umayyad caliphate founded by abd al-Rahman III. Ibn abi Amir's son succeeded him upon his death at the end of the 10th century but was unable to establish his legitimacy with the populace of Cordova. The provincial governors abandoned their allegiance to the ruler of the usurped caliphate and, practically overnight, the impressive state of the Spanish Umayyads dissolved. Revolt followed revolt for two decades, as a chaotic series of self-proclaimed caliphs succeeded one another, one from Cordova, others from the provinces, one a Slav military officer, another a Berber, then an Arab, and on it went year after year until finally, before continuing contention over the caliphate destroyed what was left of Islam's political presence in Spain, a council of ulema and nobles met in Cordova and formally declared the caliphate at an end.

With the collapse of central government, Arabs, Berbers and Slavs (the Saqalib) became so occupied fighting each other they failed to defend the Muslim realm, allowing Toledo and the central plateau to fall to the conquistadors in 1085 with little resistance. The rapidity with which the power of the Spanish Umayyads dissipated itself was truly astonishing, outdoing even the Abbasid collapse in the late 9th and early 10th centuries. The rich cultural and intellectual life that had been painstakingly built up in Spain for 2 ½ centuries continued to the end of the 12th century and then went the way of the Andalusian caliphate that had contributed so much to the efflorescence. Strong political rule had initiated and sustained it. Weakness and usurpation had ended it. The lack of a political system girded by institutions fostering stability and legitimacy that could have prevented ibn abi Amir's usurpation or that could have seen the state through the crisis of restoring the dynasty, or of establishing a new one, paralleled the lack of scientific institutions that could have resisted the eclipse of natural philosophy.

The rational sciences were like rare and delicate flowers that could all too easily be uprooted or left to wither. They could grow healthy and flourish only in the rare atmosphere of the ruler's court and only if the conditions there were right, for society outside the palace offered little space. In fact, insofar as Muslim society was a religious society whose attitudes, beliefs and customs were overseen and protected by the ulema, society offered no nourishing soil in which science could take root. Science might blossom in the spring of ideal conditions, but only to shrink away at the first chill blowing into the royal court, as ephemeral as the dynasties that supported it.

Notes

1 Dimitri Gutas, *Greek Thought, Arabic Culture*, Routledge, New York, 1998, pp. 158, 166–175; 191–192. Gutas argues that an established Sunni orthodoxy did not exist until the mid-11th century. Hugh Kennedy puts the opposing self-identifications of Sunni and Shi'i in the late 10th century, *History of the Caliphates*.

2 Gutas, *Greek Thought, Arabic Culture*, p. 190, citing ibn al-Faqih al-Hamadani, *Akhbar al-Buldan*. Mu'tazilites were rationalist theologians; Kharijites were radical egalitarians opposed to Shi'ites and the veneration of Ali.

3 A.I. Sabra, "Situating Arabic Science: Locality vs. Essence," *Isis*, 87, no. 4, 1996, p. 657.

4 The philosophical and mathematical sciences of the Ismaili Brethren found support in the royal patronage of the Fatimid founders of Cairo and its al-Azhar religious institute of advanced Ismaili studies. The caliph al-Hakim founded his *Dar al-Hikma*, or House of Wisdom, for the encouragement of such studies. This was not the science of the royal library and museum of Ptolemaic Alexandria or of the Abbasid House of Wisdom, but what might be viewed today as a brand of New Age cosmology with Man at the center and God throughout, sublimely revealed in the natural sciences and ascending epiphanies.

5 George Makdisi, *The Rise of Colleges: Institutions of Learning in Islam and the West*, Edinburgh University Press, Edinburgh, 1981; and Toby Huff, *Rise of Early Modern Science*, pp. 71–90.

6 Like observatories and hospitals, great libraries were considered by Muslims to be crowns of a proud civilization. Libraries great and small graced the cities of Islamdom, from Cordova and Toledo to Bukhara and Samarqand. In 10th century Shiraz, a library is reported to have had 360 rooms of books. Merv had ten big libraries. Baghdad had thirty separate libraries attached to Madrasas, and Damascus had 150. The palace library in Fatimid Cairo consisted of 40 rooms filled with books, 18,000 of which were on the natural sciences. The royal library of Cordova crowned them all with a putative 400,000 volumes: impressive, even allowing for exaggeration. In comparison, the University of Paris's library possessed no more than 2,000 volumes as late as the 14th century, the Vatican library had 2,257 in the 15th century. Johannes Pedersen, *The Arabic Book*, Princeton University Press, Princeton, 1984, pp. 116–123; Huff, *Rise of Early Modern Science*, p. 73.

7 Hugh Kennedy, "Intellectual Life in the First Four Centuries," in *Intellectual Traditions in Islam*, edited by Farhad Dattari, St. Martin's Press, London, 2001.

8 Saliba, *Islamic Scientific Renaissance*, p. 48; Gutas, *Greek Thought, Arabic Culture*, p. 138.

9 Sabra, "Situating Arabic Science," pp. 668–669, and his "Appropriation and Subsequent Naturalization of Greek Science," pp. 236–242.

10 The exceptional scientist al-Burini was an exception. When accompanying Sultan Mahmud of Ghazna on his Indian campaigns, he learned Sanskrit so he could use Indian sources for his book on India, *Tahqiq ma li'l Hind*.

11 Gutas, *Greek Thought, Arabic Culture*, pp. 111–114.

12 Charles Saint-Prot's *Islam: L'avenir de la Tradition entre revolution et occidentalisation*, Rocher, Paris, 2008, pp. 11–237 provides a well-rounded, detailed analysis of

Ahmad ibn Hanbal's school, Hanbali reformism and the religious tensions in Muslim society to the Mongol period.

13 Scott Montgomery, *Science in Translation*, University of Chicago Press, Chicago, 2000, pp. 100–101.

14 J.L. Berggren, "Islamic Acquisition of the Foreign Sciences: A Cultural Perspective," in *Tradition, Transmission, Transformation*, edited by F.J. Ragel and S.P. Ragep, Brill, Leiden, 1992, p. 263.

15 Berggren, "Islamic Acquisition of the Foreign Sciences," p. 274.

16 For this see D. Gimaret, "Mu'tazila," in *Encyclopedia of Islam*, 2nd ed., vol. VII, Gale Group, Farmington Hills, Michigan, 1992, pp. 783–793.

17 Rudi Paret, "An-Nazzam als Experimentor," *Der Islam*, 25, 1939, p. 228 ff. A more recent study on this is, A. Heinin, "Mutakallimun and Mathematicians," *Der Islam*, 55, 1978, pp. 57–73, argues *kalam* was an active element in the narrative delineating the relations between natural philosophy and religion and that there were in fact two branches of *kalam* theology, a philosophical and a non-philosophical.

18 For a fair account of Ahmad Hanbal's thought and the traditionalist school that took his name see Charles Saint-Prot, *Islam, L'avenir de la Tradition entre revolution et occidentalisation*, pp. 164–242.

19 Majid Fakhry, *Islamic Occasionalism*, George Allen and Unwin, London, 1958, pp. 15–56.

20 Daniel Sahas, *John of Damascus on Islam: The Heresy of the Ishmaelites*, Leiden, Brill, 1972, p. 114, note 4.

21 Ignaz Goldziher, "Die Stellung der alten islamischen Orthodoxie zu der antiken Wissenschaften," in *Abhandlung der Koniglisch Preussischen Akademie der Wissenschaften*, Verlag der Koniglichen Akademie der Wissenschaft, Berlin, 1916, p. 6. (Translated by M.L. Schwartz, *Studies in Islam*, Oxford University Press, Oxford, pp. 185–215).

22 Goldziher, "Stellung," p. 12.

23 Goldziher, "Stellung," p. 14.

24 *Mufid al-Ulum wa Mubid al-Humum*, al-Sharqiyya Press, Cairo (1368 H.), 1940, Chapter 3.

25 Ibn Sa'id al-Andalusi, *Tabaqat al-umam*, edited by C. Chiekho, al-Mashriq, Beirut, 1912, p. 66.

4 Al-Ghazali at the crossroads

With no church organization or papal commander-in-chief to enforce a disciplined order of orthodoxy and pronounce universally on heresy and excommunication, a wide range of individual opinions were allowed among the ulema across the broad lands of Islam. This did at least allow room for exceptions to the ulema's general attitudes of cool dismissal of the rational sciences. The Andalusian theologian and essayist ibn Hazm (d. 1064) appreciated the value of logic and was an early practitioner of it. Ibn Sa'id al-Andalusi, author of the bio-bibliography *Tabaqat al-Umam* referred to in note 25 of the previous chapter, was a religious Maliki school judge (*qadi*) in Toledo in the middle of the 11th century and, at the same time, an ardent devotee of the sciences and practicing astronomer associated with al-Zarqalli and his Toledo Zij, or table of planetary observations, which was to become famous in the West with the 12th century translations made by the Latins in Toledo. Al-Ghazali (d. 1111), living during the height of Seljuq power, wrote extensively on the rational sciences and argued that the application of logic in jurisprudence was similar to its use in philosophy and would be equally useful in theology. By showing how verses in the Quran use various forms of syllogism, he demonstrated the legitimacy of logic.[1] When one of his books, *The Aims of Science* (*Maqasid al-'Ilm*), a compendium of Aristotelian philosophy, was translated into Latin in the 13th century, it so impressed Western readers that in the Latin world Ghazali had the reputation of being an important Muslim philosopher in the tradition of al-Kindi, al-Farabi, ibn Sina and ibn Rushd.

Al-Ghazali was anything but that. He was a spokesman for Sunni orthodoxy in its critical struggle against Ismaili heterodoxy, when the Fatimids were ruling in Egypt and parts of Syria, and the Crusaders from the West were in possession of coastal Syria – not to mention the Ismaili Qarmatians of eastern Arabia who had destroyed the Ka'ba and carried off the holy stone, and the Nizaris, the so-called Assassins, who were terrorizing the Sunni political terrain from Syria to Afghanistan. Al-Ghazali was a young student when the Seljuq Turks drove the Shi'i Buwayhids out of Baghdad in 1055. His thought reflects the crises of his time. Historians consider him a pivotal figure in the intellectual discourse that culminated once and for all in denying the rational sciences a viable place in Muslim society. More accurately, he articulated in relatively moderate tones (moderate, that is, for a religious scholar who saw Sunni Orthodoxy threatened from every corner) a religious position vis-à-vis the Greek sciences. It was a position that had in fact been crystallizing for two centuries. Without reference to al-Ghazali and the outstanding authority his works acquired over the centuries, it would be difficult to understand the place of natural philosophy in Muslim society after the 12th century and how it got there.

As a theologian, al-Ghazali bears comparison to al-Ash'ari in that he and his literary works symbolize the reformulation of Sunni belief and practice in one of Islam's critical periods. Not only did Sunni Orthodoxy appear to be assailed and undermined from every direction, but it was being inundated by the seemingly heretical worship of miracle-working Sufi saints who were seen by conservative ulema to be eating away at the very foundations of Sunni Islam. Ignorance, fear and insecurity had popularized belief in saints, miracles, magic, astrology, talismans, superstition, augury and the grosser forms of alchemy. Even some of the ulema were being seduced by the occult. There was a lively market for it and money to be made. At the other extreme of religious belief, legalistic-minded ulema were starchily rejecting mysticism along with the occult, logic and all of philosophy, including even Greek medicine, mathematics, astronomy and optics, which were considered by many to be beneficial and having nothing to do with religion. In blind fear and ignorance, as al-Ghazali saw it, members of the ulema were themselves undermining the religious and intellectual foundations of the civilization Muslims had built up over the centuries.

As the rector of Baghdad's Nizamiyya Madrasa, Sunni Islam's most prestigious institute of advanced religious learning, established by the Seljuq sultan's Iranian vizier Nizam al-Mulk in answer to Fatimid Cairo's al-Azhar, al-Ghazali enjoyed a position of considerable authority among the Sunni ulema. Like al-Ash'ari, he had been a former Mu'tazilite. He later espoused Ash'ari theology, which was adopted by Nizam al-Mulk, and taught in the Nizamiyya madrasas as an antidote to the seductive Ismaili propaganda being disseminated by religious missionaries (*da'i*) from Fatimid Egypt.

At the height of his career, al-Ghazali suffered a crisis of faith that seems to have resulted in a nervous breakdown, a personal crisis of doubt and mental paralysis that paralleled the crisis facing the Muslim community at large, torn as it was between Sunni and Shi'i, while being inwardly devoured by a popular revivalist fervor called Sufism. In Sufism he came to see something beneficial. Mysticism provided something vital and spontaneous to faith that belief and practice lacked, a deeper and richer experience of personal love and joy. The religion as it was practiced had tended toward a perfunctory and mechanical performance of communal prayer and outwardly oriented obligations, more formal and legalistic than spiritual. The warm beating heart of religion had been buried by a lifeless coda of legal obligation and formulaic prayer and ritual, not to mention the dry scholastic intellectualization of dialectical *kalam* theology. Religion as it was being practiced lacked what the heart required: the joy, love and emotion of the soul's embrace of God. In a word, passion. Orthodoxy had become a shell of communal form and ritual, a meaningless litany of memorized prayer and mechanical movements performed in unison for a transcendent God of law to be feared and obeyed but not to be sung to, embraced or loved. Singing, dancing and chanting opened portals of the soul to God that were deaf to law and ritual. Sufism, al-Ghazali believed, or hoped, would revive religion from the lifeless forms that were smothering it. He strived to be a living example of this.

According to his own account, in the middle of an abstruse lecture on the dialectics of Ash'ari scholasticism, his tongue froze in his mouth and he found himself unable to go on. The words meant nothing. The meaning they were meant to convey evaporated. Ash'arite denial of a connected nature by imposing its atomistic fracturing of time may have seemed as distasteful as did the legalistic and detailed recipes of Muslim religious practice, enough to freeze any critical thinker's tongue in his mouth. In that moment, everything he had been taught fell into doubt. Feeling incapable of continuing his duties as rector of the Nizamiyya, he took a leave of absence, secured his wealth in a non-taxable pious foundation, a *waqf*, the usufruct of which was used for a charity and to support his family, and then set out on a journey to find himself. What he claimed to have found through this search was a rediscovery and deepening of his faith through mystical practice and experience. He described his experience as a spiritual discovery, a journey within himself guided by insight-giving mystical exercises. He thereafter wrote voluminously on mysticism as being at the heart of the Quran and Prophetic mission, but without his being a mystic. The profound mystical experience that dissolves senses and joins the soul to the oceanic consciousness of the oneness of all being, the direct experience of God, escaped him.[2] He admits that he did not experience what the mystics call *hulul* – dissolution, ecstatic union, that state of physical and psychophysiological detachment or self-annihilation, when the physical self is destroyed and the soul is absorbed into the universal soul, the boundary between self and God being annihilated and mystical union achieved. As he puts it:

> The mystical approach to God is the deepest and most meaningful, but it cannot be expressed in words without it sounding mad and heretical, which is the reason so many mystics fall into trouble with religious authorities. The ineffable must remain ineffable. The joyous experience of union with God cannot be captured in words, which is why mystics resort to symbolic poetry to describe it. About such matters silence is best. What was, was, and don't ask about it.

Mysticism had been slowly insinuating itself into mainstream orthodoxy for a century and more, but Ghazili's powerful writings assured him the credit for marrying the two and rendering Sufism a respectable, warm-blooded spouse to the outer forms of religious expression in the sober circles of shopkeepers, merchants, craftsmen, judges, theologians and bureaucrats.

Mystic practice had been regarded by such respectable types with something between suspicion and hostility. These feelings had been general in mainstream orthodoxy since the time of the earliest mystics. This was in part the fault of Sufis themselves. Some had intentionally antagonized the ulema and bourgeoisie by their mad antics in expressing love, passion, joy and divine union. Rabi'a, the woman mystic of Basra, is said to have paraded the streets with a bucket of water in one hand and a flaming torch in the other, meaning she wanted to drown the fires of hell with the one and burn down the gates of paradise with the other so that people would love God unconditionally, neither fearing the pain of hell nor lusting for the sensual pleasures of paradise.

A century later, Hallaj in Baghdad went around shouting like a lunatic that he was the Truth, he was God, that he had God under his cloak, that he was drunk in God, meaning that his soul had dissolved (*hulul*) into God through annihilation (*fana'*) of his physical self and he did not exist except in God. He was in God, God was in him. His gross symbolism was easily misinterpreted in the uptight circles of law, government, bureaucracy and business. Even worse, Hallaj's high-on-God antics mocked established ritual. He built a miniature Ka'bah that he circumambulated in his personalized performance of the Hajj pilgrimage, claiming that true faith was a matter of the heart and love of God and not a timetable of prayer, sacrifice and ritual. Rather than a mystic lost in the ineffable, he was seen by some as a ranting madman claiming to be God and shaking his cloak as if he expected the Almighty to fall out. He was asking to be martyred.[3]

Having long made a public mockery of what passed for mainstream orthodoxy, and having taunted the authorities by demanding that he be executed for blasphemy, Hallaj was at last arrested. He was not immediately executed as a professed heretic. It seems his meaning was understood and appreciated by many common people. Mysticism was popular and Hallaj had a following, including some who were not mystically inclined and saw him as a harmless holy man who was at most a little loose in word and behavior. Some younger Muslims may have admired him for his outrageous challenge to authority and tradition. The last thing the authorities wanted was to make a public martyr of him. However, after years in prison Hallaj finally got his wish. He was flayed, a foot and hand were cut off and his still living body was hung up on a cross where he continued for a day to laugh and cry out his joyous love of God and beseech the Almighty to forgive those who crucified him. A century and a half later al-Ghazali would write a treatise interpreting Muslim mysticism in terms of Christ's love and passion.

Rabi'a and Hallaj were extreme cases. Most Sufis were quiet, respectable Muslims who presented a sober image and followed the prescriptions of orthodoxy, avoiding the wilder expressions of mystical joy. For some exhibitionist Sufis, loss of self in God was a joy that could only be expressed in loss of self-control. In the century and a half between Hallaj's crucifixion and the middle of Ghazali's career as apologist of the mystic way, a large section of the Muslim community had become familiar with, and attracted to, the more restrained style of mystic practice, where public prayer lived congenially with private séances of mystically induced states of rapturous union. In weaving a reasoned interpretation of toned-down Sufi practice and ideas into the fabric of orthodox traditionalist belief, practice and law, Ghazali earned the honorific title of "Reviver of Religion," (or Renewer, *Mujaddid*), an honor bestowed every new century upon the time's most outstanding religious scholar. Ali-Ash'ari had been accepted as one for his having saved the eternality of the Quran from the lesser place given it by the rationalist Mu'tazilites;

and now al-Ghazali was honored for responding to the yearnings of the heart and resuscitating religion from the spiritual doldrums of formalistic ritual and legalistic interpretation. In the succeeding centuries, and until the present day, his books are considered a powerful authority on what is acceptable and what is not in Islamic belief and practice.

Of the many books Ghazali wrote, the largest and most famous is his four-volume *Revival of the Religious Sciences* (*Ihya 'Ulum al-Din*). In it he divides all the sciences into two categories, the blameworthy and the praiseworthy. The religious sciences of course are the latter. The blameworthy sciences have among them those that are beneficial and those that are harmful and wrong. The wrong ones are dangerous. They are the worst of the blameworthy and should not be considered science. Because they are organized and logically structured like a science, they give the deceptive appearance of being one. These harmful ones are those having to do with metaphysics. They must be avoided. As for the blameworthy sciences that are contrary to being harmful, these are the sciences of natural philosophy: optics, mechanics, physics and astronomy. They are beneficial, but also blameworthy because someone who studies them can go too far in his belief that they are the way to truth. Someone weak in understanding could go beyond the limits of scientific truth and unintentionally stumble into heresy. He loses their benefits and gains damnation. Though not in themselves harmful, these sciences, should nonetheless be avoided to be safe. Astronomy in itself is not harmful but could easily lead to perversion of belief, so it is best to avoid it.[4] Logic is another rational discipline that is neither harmful nor beneficial. Religiously neutral, logic is only a structure for rational arguments, but again, one never knows where logic will lead. It is best to be avoided.

Mathematics was like logic, neutral but possibly dangerous. The danger was that believers might fall too much under the influence of logicians, mathematicians and astronomers and be led astray by things they said outside of their fields. Experts could be right about matters in their discipline but wrong when extrapolating or making inferences from it on external matters. Believing that if these experts were right about what they said concerning mathematics or astronomy, simple believers might think them right in whatever they said, even things that touched on religious matters. So in a way these sciences were dangerous, even if not in themselves. But it would have been folly to reject logic, astronomy, mathematics and the other exact sciences because rejection would make Islam, in the minds of intelligent people, look like a religion of ignorance, since what was being rejected was proven knowledge. People who knew that it was proven would leave religion, the result being that Islam would indeed become a religion of ignorance for ignorant people:

> A grievous crime indeed against religion has been committed by a man who imagines that Islam is defended by the denial of the mathematical sciences, seeing that there is nothing in the revealed truth opposed to these sciences by way of either negation or affirmation, and nothing in these sciences opposed to the truth of religion.[5]

Hence, the wisest formula was not to reject natural philosophy but refrain from studying it.

Al-Ghazali's position marks a radical departure from the reasoned arguments of his Turko–Iranian coreligionist of a generation earlier, al-Biruni, whose defense of science can be taken as emblematic of all Muslim natural philosophers. Here astronomy is allied with religion because of its service in a great number of religious and practical matters, such as determining the times of prayer, the direction of the *qibla* to face the right direction for prayer, times of fasting and determining twilight, dawn, new moons, longitudes and latitudes.[6]

Al-Ghazali was fully aware of the critical significance of his position and of its possible dire consequences. But, he believed, in this long hot and cold war of survival that Sunni Islam was waging against Shi'i revolution, the crises of the day required tight intellectual control. It would be the task of the state madrasas to make sure of that.

What was it then that would have to be suppressed in the rational sciences? Above all, metaphysics. Here was the truly dangerous culprit of philosophy. The other disciplines could, if not handled cautiously, lead to serious error, but metaphysics was the pit of error itself, leading nowhere but straight to the trap door of damnation. Ghazali dedicated one of his most important works to debunking it: *The Destruction of the Philosophers* (*Tahafut al-Falasifa*).

Doing what was rare among ulema scholars who attacked the rational sciences, Ghazali took the trouble to study them before warning the believers against them. He did the same with mysticism, but this he argued for, hence the simplistic idea held by some modern historians that his embrace of one and rejection of the other began the decline of scientific study, its place being taken by mysticism. Ghazali's position is at the gravitational center of medieval intellectual discourse. His works are a lens through which can be seen the culminating ascent to social acceptance of one path to the divine, mysticism, and the decisive decline of another, natural philosophy. His was merely a stance that headlined a long-standing attitudinal reality: Mysticism led to God through love; science led to the devil's snare through faulty logic. The balance of respect and fear al-Ghazali expressed for the rational sciences was transmitted through the ulema and society, but among many believers, respect disappeared into the shadow of fear. Mysticism was to be embraced, whereas science and its philosophical correlatives were to be not quite rejected or accepted, but held at arm's length, to be studied only by those with sound religious training. There was little danger that many of the ulema would delve into philosophical and scientific study.

Sufism was a growing movement within society, and it went straight to the heart of religion. Science and philosophy, on the other hand, had little popular appeal and, confined as they were to courtly precincts of royal patronage, or the secluded rooms of private study, presented no immediate challenge to orthodoxy as mysticism did. Al-Ghazali's decisive posture affirmed the ulema's dubious, sometimes unfriendly opinion of science. The ulema, he insinuated, are the shepherds of the flock of believers who must be protected and kept from going astray. It is the shepherd's duty to keep the sheep healthy and away from eating dangerous weeds.

The ulema must act as public censors to keep society pure of any literature that might make believers think in directions that could be dangerous to the individual and therefore to the community.

Given the ulema's coolness to the rational sciences in general, al-Ghazali's qualified declaration not to reject the exact sciences outright could be considered as being on the liberal side of religious argument. Science and philosophy were legitimate fields open to Muslims, but only properly qualified scholars should study them, that is, men trained deeply in religion – the very men who would be the last to come near them. So in effect, his position confirmed that natural philosophy would remain an elite preserve of royal patronage, compressed at the narrow margins of society, where they had been for some time, grudgingly tolerated at best. In some quarters, Ghazali's theology was considered overly liberal. His dalliance with mysticism, logic and science so outraged the Cordovan Maliki ulema under the puritanical Berber Murabitun that his books were publicly burned.

Al-Ghazali died in 1111, by which time the creative period in Muslim science had peaked and was struggling for life support. A generation after the death of al-Ghazali, the algebraist Samau'al, the Jewish convert mentioned earlier, was unable to find anyone in the whole of Baghdad to tutor him in the advanced work of Euclid's *Elements* or the algebra of al-Karaji.[7] Decline in terms of originality and numbers of people involved in the sciences had already set in. Nonetheless, sporadic efflorescences in creativity across the lands of Islam would continue for centuries.

Andalusia, where Ghazali's books were put to the flames a century after ibn abi Amir al-Mansur's public book burning, offers a glimpse of the rise, decline, re-emergence and gradual fading out that characterized the pattern of the rational sciences. The brilliant Umayyad period in Spain ended with ibn abi Amir around 1000. The Berber dynasty of the Murabitun that ruled Andalusia from the end of the 11th century to the middle of the 12th kept a deep chill on *falsafa*. None of it was allowed in the Murabitun palace. But with the succeeding Berber dynasty of Muwahhidun in the last half of the 12th century, a burst of intellectual creativity produced a brief renaissance that included ibn Baja, ibn Tufayl, al-Batruji, ibn Rushd and ibn Maymun (Maimonides), to name but the brightest stars in this late Andalusian galaxy of greats that kept the old flame of science alive, but only for as long as the dynasty was strong. At the height of the dynasty's political power, the ruler's personal physician, ibn Tufayl, produced a philosophical romance in defense of science and philosophy. The book allegorizes the struggle of the rational sciences to survive in a desolate society mentally numbed by sanctification of changeless tradition and memorization of revelation, with innovation regarded as heresy. The device of a philosophical fiction to show the harmony of rational and religious thought had been used a century earlier by ibn Sina. Ibn Tufayl's version is an expansion of essentially the same story and, given a close analysis, should have been considered heretical, and no doubt would have been in less intellectually and politically vigorous times.

A baby washed up onto a deserted island and nursed by a gazelle develops mentally in 7-year stages of ascending levels of consciousness, hence the name

Living, Son of Consciousness: *Hayy ibn Yaqzan*. An alternative version has the baby originating not by procreation but by evolving from an elemental ooze of sun, earth, air and moisture that combine in just the right proportion of elemental matter and heat to create life in a spontaneous reaction. The nature child's mind develops from the primitive imaginative level to the physical level of sense experience and from there, after another 7-year period and another epiphany, to the experimental. This level of consciousness then gives onto the next up the ladder of expanding consciousness, to that of the mathematical and abstract, followed then by the metaphysical level of universals. The final epiphany brings the self-enlightened young man to the ineffable truths of the mystical universe where all is one: consciousness of the unity of existence (*wahdat al-wujud*).

In this culminating illumination, the nature boy discovers that all rational knowledge, from physical to mathematical to metaphysical, is at one with the transcending mystical verity of a divine creator. Having now independently reached the highest level of illumination, the hero meets someone who has come to the island from the mainland in search of spiritual fulfillment through solitude and meditation in natural surroundings. Through a combination of signs, bodily motions and pointing at objects, Hayy learns the visitor's language and they become friends. Hayy communicates what he has learned through outer and inner exploration, while his friend, convinced that Hayy has reached the highest possible level of religious truth, teaches him about the laws and traditions of those who follow the scripturally revealed religion of the civilization on the mainland. They travel to the mainland so Hayy can experience civilization and help the people there reach the same state of blessed wisdom he has reached. It does not go well. The followers of an organized scriptural religion buttressed with hoary traditions and unquestionable books of authority are at first delighted at the wise savage. But then they come to realize the religious implications of self-enlightenment and fear him. His very presence is a proof that other paths to religious truth exist. Feeling their scripture, traditions and authorities threatened by his teachings, they would do away with him. Intuiting this, he returns with his friend to the island. The story ends with Hayy concluding that there are two types of people, those who find a meaningful god through reason, nature and inner discovery, and those who need scripture, prophets, holy law, and learned scholars to guide them as a shepherd guides his flock

The ulema made no demand to have ibn Tufayl reprimanded or his treatise repressed. Either the religious scholars missed the implications of life evolving from natural elements under certain conditions, and of Hayy's reaching God without benefit of scripture, or they dismissed the treatise as being too fanciful to be taken seriously. Ibn Tufayl was fortunate. Another time or place and the author's fate would have likely been similar to that of his student, ibn Rushd, whose philosophical works were far less obviously over the line of heresy, or to that of his contemporary Suhrawardi, whose mystical pantheism led to his execution.

Ibn Rushd (d. 1198) followed ibn Tufayl in his position as chief vizier, physician and astronomer of the Muwahhidun ruler. He was also a learned Maliki judge and a member of the ulema. But his cogently argued resolution of the apparent contradictions between reason and revelation impressed the Maliki establishment as little as did his monumental commentary on Aristotle's philosophy, a work for serious philosophers. For popular consumption he composed a smaller commentary, which adequately explicated the complexities of peripatetic metaphysics. Being more accessible to the non-professional, the smaller work irritated the Cordovan ulema all the more. Matching ibn Rushd's two commentaries are his two treatises on the harmony reigning between religion and reason: the large one, "The Destruction of The Destruction:" *Tahafut al-Tahafut*, which was an answer to Ghazali's "Destruction of the Philosophers:" *Tahafut al-Falasifa* (which was in turn an attack on ibn Sina's Neoplatonic illuminationist philosophy) and a smaller version, "The Decisive Treatise on the Harmony Between Reason and Revelation:" *Fasl al-Maqal*. Together they ingeniously, and exhaustively, demonstrate the logical equivalence between the Quran and Ash'ari theology on the one side and Aristotelian metaphysics on the other. The particular issues are: the eternality of the world that denies divine creation; the dissolution of the individual soul upon death of the body since soul cannot exist without body; God's knowledge of universals but not particulars; denial of God having attributes to preserve the principle that the divine essence is a simple purity; cause and effect; allegorical interpretation to render rational the Quran's irrational verses; denial of judgment day, of resurrection and of anthropomorphism; and some 20 other philosophical objections of lesser import that ibn Rushd showed did not contradict religion.

According to Aristotle, the universe was eternal. There was no time before which it did not exist. Divine creation was a philosophical impossibility, since creating something from nothing was a rational contradiction. Matter had to exist eternally. Body was matter given form by soul. The death of the body meant the death of soul, since one could not exist without the other. They came into being together, and they passed away together. Disembodied soul was an absurdity. Aristotle's God was pure mind, a perfection, and it was this divine mind that transmitted motion to the universe, endowing it with order and harmony. A perfect being, God's only activity was being perfect; the only activity possible for absolute perfection was pure thought: reason. Reason was the activity of God thinking of perfection, that is, of himself, and it was God thinking that put the universe into motion. Pure, perfect and unmoving, the divine mind thought only at the sublimest level of eternal universals, hence the particulars of physical reality and the messy contradictions of the human condition with all their crassness, changes and imperfections were beneath the sublime concerns of the divine mind.

God existed as idea in action; his thinking action sustained universal motion from the sphere of the fixed stars down, flowing from sphere through concentric sphere down to the sphere of the moon, and from there down to the sublunary concentric spheres of fire, air, water and earth, this being the world of matter, change and non-circular motion, in all their forms and actions. God, the unmoved mover and cause of causes, was the mind of eternal principles that governed the cycles of the universe. The principle of cause and effect strung everything in the universe together, from the pure thought of God down to the physical world of change and corruption, life and death. Nothing happened without a cause, of which

God was the first. All this was eternal. Universe, earth and man had no beginning, no end. They were coeternal with God.

The physical world of chance and accident, of wars and natural disasters, did not enter God's mind or come from God's thought or will. Accidental occurrences in the physical world, or in the world of human action, were particulars. God's mind worked only in universals.

With painstaking reason, ibn Rushd reconciled these principles of a transcendent Aristotelian god with the God of the Quran. Not an easy task, marrying an uncaring god of universals to one who is closer to man than his own neck vein, whose attention nothing escapes, a judging, punishing and forgiving god. The closely reasoned arguments ibn Rushd wove in his dialectic were a tour de force in logic and subtlety that drew on every bit of his skill in casuistry as a Maliki judge. Ibn Rushd was the first to characterize philosophy and scripture as milk sisters: two truths nurtured by the same source. Aquinas would modify the relationship, where philosophy became the handmaiden serving theology.

Ibn Rushd's influence was to be far greater in the Christian West than in his own civilization. In his own he suffered exile and oblivion; in the Latin world he was given the honorific title of "The Commentator," the philosopher whose commentaries were the key to Aristotle. This did not save him from being condemned to hell by the Church for his philosophical Averroism, namely, his negation of eternality of the soul and the divine creation of the universe. But the Christian theologians were nevertheless elated to have ibn Rushd's commentaries in Latin translation, so elated that they immortalized his Latinized name, Averroes, in the summas of medieval theology, while casting his corporeal being to the undying flames of hell.

Because ibn Rushd's only support was the Muwahhidun ruler in Cordova, once the ruler found his authority to be declining and in need of the ulema's support, the philosopher–physician and Maliki judge was sent into exile and his books denounced, ignored or destroyed, as had been the philosophical books in the Umayyad Caliph Hakam II's library two centuries before. He was allowed to return to Cordova 10 years later, an old man with few years left. Islam's would-be Aquinas passed his final years in quiet solitude among the gardens and fountains of Cordova.

Both al-Ghazali and ibn Rushd, expiring at opposite ends of the 12th century, were intellectually suited to produce the synthesis between reason and revelation that Albertus Magnus, Thomas Aquinas and Roger Bacon accomplished in the 13th century for Latin Christianity. In Islam this was not to be. Not in liberal Andalusia nor eastwards in the heartland of Islamdom. At the same time ibn Rushd was being humiliated and exiled and his books burned by the strident ulema in Muwahhidun, Andalusia, a 36 year old philosopher, mystic and theologian was suffering worse in Ayyubid Syria.

Shihab al-Din al-Suhrawardi (1154–1191) was an Iranian enthusiast of ibn Sina's illuminationist (*Ishraqi*) theosophy and author of a Gnostic work, *Hikmat al-Ishraq* (*The Wisdom of Illuminationist Philosophy*), a mystically informed commentary on Platonic and Aristotelian philosophy in which existence is seen as a continuum from the lowest level of material being to the highest level of pure consciousness, namely God, symbolized by light. The book's pantheistic implications convinced the Syrian ulema that Suhrawardi was a dangerous heretic whose ideas would poison the minds of the believers. As it had been only a little more than a decade since Salah al-Din had ended the Shi'i Fatimid Caliphate in Egypt and Syria with its Gnostic-espousing theosophy, the Sunni ulema may have been less tolerant than usual. Ostracism, exile and book burning were no longer considered enough to protect the community in the feverish temper of schism and insecurity. The governor of Aleppo, al-Malik al-Zahir, who was Salah al-Din's son and somewhat philosophically minded himself, was as reluctant to execute Suhrawardi as the Abbasid caliph al-Muqtadir had been to execute al-Hallaj. He referred the matter to his father, who was ruling as sultan in Cairo. Salah al-Din was also disinclined to execute Suhrawardi, but the ulema was insistent. Forced to decide between the support of a philosophical mystic and the support of the ulema, he ordered the execution.

Neither ibn Tufayl's philosophical romance nor ibn Rushd's tightly argued reconciliation between Aristotelian metaphysics and Ash'ari theology had any hope of convincing the Cordovan ulema that science and philosophy were windows into the mind of God. Suhrawardi's illuminationism had even less chance with the Syrian and Egyptian ulema. However, the ulema's opposition was not monolithic. Variations in outlook regarding philosophy and science within the ranks of the ulema from Seville to Samarqand certainly existed. Among the ulema who defined the contours of belief through theological argument, *kalam*, were those who rejected some sciences but praised others, particularly astronomy and mathematics for their having religious application. In the words of a leading historian of Muslim science:

> Science was not a direct competitor of kalam in the way that falsafa was, and generally the specialized scientific disciplines were not as such perceived as posing a threat to religion. And yet they had to be reinterpreted in the light of the prevailing kalam metaphysics and given new foundations and new definitions of their scope and value. In addition, the triumphant kalam view of the world and of the observable regularities in it as entirely contingent tended to inculcate certain attitudes to scientific endeavour, if not at first in the minds of the practicing scientists themselves then at least in the minds of the large body of learning-hungry students in the madrasas.[8]

To be sure, there were ulema whose views on the rational sciences were of more liberal hues than those expressed by Ghazali, who took what can be considered a middle position, neither rejecting outright nor condoning. A theologian contemporary to his, abu al-Qasim al-Isfahani (d. 1108), saw in philosophy a tool to sharpen the mind. Philosophy gave definition to things in their general nature as universals, while the Shari'a informed the believer on the level of particulars and details, such as not eating pork, not having sex during menstruation, fasting, praying and not marrying close relatives. He wrote a book on the conviviality of philosophy and religion, which Ghazali is said to have treasured and kept with him.[9] And then there was the well-known theologian Fakhr al-Din al-Razi who in the early 13th century was publishing religious works imbued with Aristotelian logic. There was obviously not enough of his kind to hold the line, and even the liberal ulema who saw nothing wrong, or even saw something positive in *falsafa*, restricted its study to everyone except scholars of advanced religious education. Like Ghazali and ibn Rushd, they feared that in the minds of the untutored the subtleties of *falsafa* would lead to differences, arguments, contradictions, confusion, and finally to heresy and *fitna* – communal turmoil. Believers needed a neat and simple formula to follow, everything spelled out in detail: how and when to pray, how to wash, what not to eat, when not to have sex; a clear, smooth, straight and simple path of life laid out from birth to death. It was the safest way. It saved people from heresy and kept the community safe and free of strife. In this, at least, theologians and philosophers could agree.

By the late 12th century, those who wanted to legitimize philosophical reasoning were marginalized, hanging on as physicians and astronomer–astrologers in the royal courts that offered patronage. Where a powerful dynasty arose and patronage briefly blossomed, as in Mamluk Egypt, Il-khanid Maragha, Timurid Samarqand and Ottoman Istanbul, observatories were built for the continuation of astronomy, but the intellectual basis that gave structure, unity, purpose and spiritual fulfillment to science in general had fallen away. Religious distrust and the narrowing of patronage promised few rewards for any young man contemplating a career in natural philosophy outside of astronomy. The result was disinterest, neglect and near oblivion of the sciences that had for over three brilliant centuries been nurtured and advanced.

Though support for philosophy and natural science after the 12th century had fallen away relative to the 9th and 10th centuries, enough *falsafa* survived to elicit the occasional *fatwa* of condemnation from the conservative side of the broad mainstream. Religious condemnation expressed in a *fatwa* was the only formal weapon the ulema had at its disposal, but it often carried a powerful voice for molding public opinion and of informing and pressuring political authority to do what the ulema could not on its own: ban or burn books, exile philosophers or forbid study of the logical disciplines. A famous *fatwa* issued in 1240, or thereabouts, by a religious judge and scholar, ibn Salah, condemned the use of logic in the most virulent terms, as if to stamp out once and for all the last embers of the brilliant past that had for generations been dimming.

Ibn Salah claimed to have attempted the study of logic but found himself unable to follow the steps of forming a syllogism. Logically challenged, and apparently without understanding what logic was about or how it was structured, ibn Salah issued his *fatwa*, which began by calling *falsafa* "*uss as-safah*," the base of stupidity, a variation on the usual pun popular with the opponents of *falsafa*, "*fall as-safah*," the depths of dumbness. Philosophy, the *fatwa* declares, is destructive and erroneous; it spreads confusion and leads to heretical beliefs. Whoever practices it is blinded to the benefits of the Shari'a and will be ripe for the devil's harvest. As for logic, it is the door to philosophy, and what leads to evil is evil. Logic is denied by religion in general and by the Shari'a in particular. Neither logic nor its terms may be used in relation to the holy law, which has no need of such things:

> Logic and philosophy darken people's belief in the miracles performed by Muhammad that number over a thousand . . . and many times more than that, beyond counting, because in addition to the miracles Muhammad made in his time there are those miracles that are newly made in the centuries after him by the saints of Muhammad's community.[10]

According to ibn Salah, the ruler was obliged to defend the Muslims against those who practiced logic by first forcing them from the schools and then punishing and exiling them. The ruler had to put to the sword anyone who failed to abandon belief in logic or philosophy and follow Islam, since the "only way to root out evil is to root out the source of evil."[11]

The *fatwa* reveals the depth of belief that Muslims had acquired over the centuries in Muhammad's miraculous powers and their worship of miracle-working saints. Earlier considered heretical, miracles and saint worship had become not merely an accepted part of popular belief but an important part, even among the ulema. Logic, philosophy and all things having to do with them had no place in the mind-set of this no doubt large sector of society. Practitioners of logic and philosophy, in their opinion, were heretics deserving of death. Such attitudes from the raucous voice of conservative religiosity would have shrunk the pool of young minds eager to engage the impressive body of natural philosophy built up over the centuries. Lest the believers forgot ibn Salah's message and the fate of those who failed to heed it, the *fatwa* was on occasion reiterated, indicating that people were still around who studied these things. But such ringing denunciations from the conservative wing of the ulema must have had a chilling effect on any inclination to take up logic and its affiliated studies, pushing anyone who did study them more into the shadows.

About the time ibn Salah was composing his *fatwa*, Umar Suhrawardi (not to be confused with his contemporary Shihab al-Din Suhrawardi, [1145–1234], the illuminatist philosopher executed in Ayyubid Syria), a court theologian for one of the last Abbasid caliphs, was declaiming philosophy as nothing more than an infidel conspiracy to destroy Islam. Umar Suhrawardi's uncle had founded a mystical order, the Suhrawardiyya, of which Umar became the leader. The miracles of the Sufi saints, he declared, were dearer to God and more powerful than all the works of philosophers put together. Impressed with the piety and truth of his theologian, the caliph allowed him to wash the ink out from the pages of his personal copy of ibn Sina's philosophical encyclopedia and had other works of philosophy in his library destroyed.[12] A *fatwa* was issued condemning al-Ghazali posthumously because of his liberal position regarding logic.

The courage to think beyond the narrow limits set by those hostile *fatwas* did not die easily, or even completely. Regardless of all the frightening *fatwas* and damning pronouncements against logic and the sciences related to it and the examples made of those who ignored the warnings, the study of logic and the mathematical sciences endured. The sources preserve an account of one brave Cairene who dared study logic and the related sciences at the time of ibn Salah, toward the end of the Ayyubid regime in Egypt and Syria. A well-known and respected theologian and legist who seems not to have concealed the forbidden regions to which his intellectual curiosity took him, he delighted in reading philosophy and natural studies. When discovered, he became the object of fanatic hostility and was attacked by members of the ulema. They drew up a petition declaring him an apostate liable to death. He then fled to Damascus, only to find his reputation had preceded him there. His career over, he lived his life an outcast, scorned and humiliated.

In their efforts to discredit the rational sciences, some of the ulema grouped science and philosophy together with divination and the occult sciences of astrology, alchemy and magic.[13] Evidence indicates the identification endured. In the 1670s, Shaykh Ala' al-Din al-Hasakfi was declaring philosophy a forbidden branch of magic and astrology, since it was used to prophesy the future by the secret art of sand reckoning. Discrediting philosophy by insinuating it into the occult was just another handful of dirt from the grave being dug for it. This was nothing new. Nor was the accusation wholly baseless. The occult sciences and natural philosophy had been holding hands since Hellenistic times. In the golden age of Muslim science, many of the greats, Biruni himself, like almost all astronomers and like Kepler six centuries later, wrote on astrology and cast horoscopes. In fact, Biruni wrote one of the great works on astrology, *Tafhim al-Nujum* (*On Understanding the Stars*), and cast horoscopes for an income, most likely believing that the configuration of the heavenly bodies at the instant of birth had some influence on mood and character, just as certain positions of the heavenly bodies with respect to earth influenced the tides, the weather, caused the seasons and made some animals act in strange ways. Magic, augury and talismans were in a disreputable category of their own.

Astrology and alchemy were rational disciplines structured in method, principle and practice like the other sciences and so were more easily associated with the corpus of natural philosophy. This association was used to slip in the other members of the occult family to reinforce the ulema's case against *falsafa* in the public mind. If magic and augury were the work of charlatans then so too was *falsafa*.

The use of the occult as an effective bogey implies that there was an active market for the occult. One interpretation for the decline of science and philosophy in Muslim society is that unsettled political, economic and religious conditions led to a rise in interest in the occult pseudosciences, a trend fueled by fear and insecurity at the expense of real science. Consequently, an intellectual ossification dried up the creative juices that nourished the exact sciences.[14]

This increased interest in the occult that is reputed to have done such damage to *falsafa* may be a skewed perception of reality. Is the public more addicted to astrology and its occult associates during times of stress? Was the fear that comes from ignorance and insecurity and leads to belief in the supernatural anymore widespread in the 12th century than during the centuries before? Did superstition and belief in magic and the occult become so widespread as to undermine science and philosophy? How does one answer such nebulous questions so many centuries before the art of public polling?

Fear, insecurity, stress and belief in the irrational came with the birth of human consciousness. Every period since the beginning of civilization had its hard times. Muslim civilization was no exception, from the trials of Muhammad in Mecca to the fall of Cordova to the conquistadors and the Mongol destruction of Baghdad. Economic decline resulting from debased coinage, ravaged agriculture, interrupted trade and recurrent deadly plagues: these were not uncommon features of the Muslim historical experience. The Crusades shifted international transport trade from the Muslims to the Venetians, inflationary prices reduced Syrians and Egyptians to eating dogs and donkeys. People dropped in the streets like flies from plagues. Before that, in 968, a low Nile resulted in a devastating famine, with over 600,000 of Egypt's population of several million succumbing. Another low Nile a century later lasted 7 years and repeated nature's reaping of souls. In the early 13th century, plague followed famine, emptying villages and sections of Cairo. The Black Plague that swept across the Middle East and Europe is estimated to have taken up to half of the population.[15] An earthquake shook the Muqattam plateau in Cairo so violently that the terrified people in distant Damascus "thought it was judgment day."[16] The market for charms, amulets, augury, talismans and Fatima's hands to ward off the evil eye no doubt prospered. Yet, for the most part, during these times of travail science prospered. Astronomy and mathematics did for centuries after the classical period of high Islamic civilization.

Superstition and belief in supernatural power, as the occult-causing version of decline goes, were seeping into religious belief and practice, draining the wellsprings of scientific interest and creativity as society, wracked by fear and insecurity in a world perceived to be out of control, turned to the occult and magic. True, the ulema sanctified "the science of letters" (*'ilm al-huruf*), a form of divination based on decoding the alphabetic letters as they occur in Quranic verses, which were believed to be a limitless thesaurus of hidden knowledge. Some of the ulema believed that the Quran contained, in code, not only all the science man has ever learned, but also all the science mankind could or would ever learn. This did no service to the rational exploration of nature.

The extent to which the occult in the 12th and 13th centuries had pervaded public and religious belief beyond the occult accepted by Muslims in the centuries of high scientific creativity would be hard to gauge. Perhaps a more plausible expression of reality is that people always believed in the occult, just as they had in antiquity, in Hellenistic and Roman times, and before that in Pharaonic Egypt and Babylonian Mesopotamia. There is no reason to think that a poll of urban opinion taken in the 9th or 10th centuries would have found a cross section of society statistically any less attracted to the supernatural than in the 13th century or after. Religious belief itself, based as it is on the non-rational and supernatural, is just a larger more organized form of the occult, powerful enough to make its own definitions and make its labels stick. Organized religion makes its own structures and defines its beliefs in the name of God, then relegates everything else supernatural to the occult. Prayer itself is an occult exercise to make contact with a hidden power and direct it.

What society considers the occult, in whatever form and however perceived, is a cultural constant, its social profile a projection of the prejudices of the dominant religion, the empowered cult that excuses itself of being occult. In a religiously oriented society, change in the occult social profile reflects attitudinal change in the dominant cult. In medieval Muslim society, the occult was perceived by populist religious reformers to be a problem only after the challenge of the philosophical sciences had passed. With the threatening dynamism of the philosophical sciences having subsided, the occult sciences became all the more visible as the grave threat, the evil enemy of true religion that the ulema had to combat.

What is known of this imputed inundation of the occult diluting and washing out of the rational sciences? In the late 13th century, a vigorous reform movement arose in Mamluk Egypt and Syria to purify religious practice and belief of all the many ruinous corruptions that were seen to be taking over the Orthodoxy of old, namely the excesses of mysticism, saint worship, miracle workers, magic, astrology, alchemy, augury, talismans and all the superstitions, bogus beliefs and chicanery in the baggage of the occult that played on the hopes and fears of the naïve, greedy and ignorant. The heart of the reform movement was embodied by a popular Hanbali jurist and theologian, the Syrian ibn Taymiyya (d. 1328), who wanted to restore the strict traditional norms established by Ahmad ibn Hanbal. Coming from a family that had for generations been distinguished by its religious scholarship, ibn Taymiyya studied jurisprudence in Damascus where, together with a following of like-minded traditionalists, he led, wrote and preached a program of reform and then suffered the usual fate of the unprotected religious reformer.[17] Unlike the puritanical reformers ibn Yasin and ibn Tumart who had won over Berber tribal armies to carry out the reforms of the Murabitun and Muwahhidun dynasties in Morocco and Andalusia, and centuries later the reformer Muhammad abd al-Wahhab, who had the Saudi tribal leader for protection, ibn Taymiyya had only his pen, his voice, his character, and his popularity. These weapons served him well, at least well enough to make him a threat to the social status quo. Like his spiritual forbearer ibn Hanbal, who was imprisoned and tortured for his opposition to Ma'mun's Mu'tazilite theology, ibn Taymiyya was repeatedly imprisoned by the Mamluk authorities, with encouragement from ulema who putatively feared the challenge he presented to Ash'ari doctrines and certain points of law and the unsettling influence his growing popularity had on the community, seen by his envious rivals in the ulema as being detrimental to established belief. His condemnation of the ulema's profitable trade in selling talismans and occult cures to gullible believers threatened the incomes of many in the religious establishment.

Ibn Taymiyya's conditions in prison were generally lenient. He was allowed visitors, pen, ink and paper, and much of his literary output was written in the prison towers and citadels of Damascus, Alexandria and Cairo. In prison he composed a treatise refuting logic and a number of *fatwas* reaffirming ibn Salah's logic-damning *fatwa* of 1240. When not in prison, he served the Mamluk authorities in a number of ways, one of them using his popularity to unite the populace of Damascus to resist a Mongol incursion. A grievous element of the unsettling times, Mongol attacks on Syria had begun to take on a semblance of normality ever since the destruction of Baghdad and the caliphate, some 40 years before ibn Taymiyya's first arrest. During his final bout of imprisonment, in the Cairo Citadel where he was to die, he was deprived of pen, ink and paper to prevent him from writing his popular *fatwas*.

In accordance with the principles established by Ahmad ibn Hanbal, the only scientific study ibn Taymiyya recognized as religiously legitimate was Prophetic Tradition. Where al-Ghazali two centuries earlier had reformed Muslim belief by reinterpreting Sunni Orthodoxy to accommodate two different if not antithetical planes of religious discourse, Ash'ari theology and Sufism, ibn Taymiyya wanted to reform Islam by joining a moderate form of Sufism to a theology in which Ash'arite dialectics and doctrines were replaced by a simple framework based on conservative Hanbali traditionalism. The great enemies of Sunni belief were seen to be saint worship, miracles and the occult, and ibn Taymiyya and his leading student and follower, ibn Qayyim al-Jawiziyya (d. 1360), were mildly tolerant of the philosophical sciences, for here was an ally that was as inimical to the occult and the irrationalities of miracles and magic as was religion. The enemy of my enemy is my friend, and for the moment, science and philosophy became religion's ally. Common enemies make odd bedfellows, but no more than when religion enlists science to fight the occult.

The traditionalist Syrian reformers allowed a place for logic in understanding the world. In this they were at one with al-Ghazali. Logic was inherent in mathematics and was useful as long as it remained in its proper place and did not enter religion, lest it make of it a rational science. When applied to religion, logic was deceptive, unclear and contradictory, its pitfalls many. Logic could lead as easily to contradiction and falsity as to truth. Since corruption of belief and scriptural misunderstanding were almost inevitable once logic entered, the miniscule truth that the careful application of logic could wring from religious sources was hardly worth the risk. Logic refused to be controlled. It was like a wild horse. It had to be reined in tightly, kept from impregnating religion. Let logic in bed with religion and the bastard offspring would be either a demystified theosophy or a soulless rational system.

What Ismaili propaganda and religious subversion had been to Ghazali, saint worship and the occult were to ibn Taymiyya and ibn Qayyim al-Jawzyyia. In the latter's *Key to the Abode of Happiness* (*Miftah Dar al-Sa'ada*), ibn Qayyim used arguments from logic, astronomy, physics and mathematics to disprove the greater enemy, placing science in the service of religion in battle against the occult.[18] Science and philosophy provided strong arguments, but they contained more that ibn Qayyim rejected than he accepted. Playing both ends against the middle, ibn Qayyim used natural philosophy as far as he could to prove the falsity of occult science but without endorsing the religious legitimacy of the rational sciences.[19]

Yet in comparison to the hostility expressed by centuries of Sunni *fatwas* against logic and natural philosophy, ibn Qayyim's stance was appreciably liberal – in fact not too dissimilar from that of the authorities prevailing in the Roman Church in the 1280s, around the time that conservatives were able to slap the teachings of Thomas Aquinas and his fellow rationalizing scholastics down, as the Mu'tazilite teachings had been four centuries earlier for the same sin of going too far in reducing religion to an exercise in rational thought via Aristotelian logic. Aristotle was struck from the University of Paris syllabus, the works of Aquinas banned and Friar Bacon imprisoned. But while, with respect to Muslim civilization, the issue between philosophy and religion remained for half a millennium pretty much right where ibn Qayyim left it – and even that position was periodically challenged by the ulema who insisted that Muslims have nothing to do with logic or natural philosophy – the scholasticism and natural philosophy of Aquinas and Bacon made a powerful comeback in the Latin West, when enough theologians had caught up to the avant garde of scholastic thinking so that a critical mass was reached that made yesterday's anathema today's orthodoxy.

That indeed may have been the advantage of having a centralized and hierarchical Church organization with a decision-making central committee of theologians concentrated in one place (the University of Paris in the case of Aquinas and scholasticism), to which the bright lights of religious thinkers could gravitate to make their combined intellectual weight felt more effectively. Thanks to the Roman Empire's administrative structure, what the Latins had that Muslims had not was an organized church that provided a center of gravity for consensus, a central committee to make decisions and get on with the job of theology and worldly knowledge. Of course, the Latin success of scholastic incorporation of reason into religion would eventually lead to schism, reformation, freedom of science, enlightenment, secularism and an eventual falling away from belief, the very thing the ulema feared would occur among Muslims if logical analysis and natural philosophy were ever to slip the reins. In terms of the cultural divide that sent Islam one way and Latin Christendom another, the intellectual substance of that watershed is found in the 12th and 13th centuries, in the works of al-Ghazali, ibn Taymiyya and ibn Qayyim on the one side and of Peter Abelard, Peter Lombard, Albertus Magnus, Thomas Aquinas and Roger Bacon on the other.

Mu'tazili scholasticism failed to gain critical mass. The Mu'tazili ulema were dispersed under the hammer blows of Hanbali Traditionalism and Ash'ari dialectics. Centuries later, in Andalusia, ibn Rushd stood by himself, an intellectual loner, an exile. Others like him who supported science and philosophy were scattered across thousands of miles of Islamdom. No central committee, no center of gravity, no rallying point to gather around for advance or defense; no fort to defend and fight their cause. A quarter century separated ibn Rushd's exile and the first papal ban on the advanced version of Averroes-based scholastic reasoning. By demand, the ban was soon revoked and reason continued its advance in informing theology, with Averroes the cutting edge; but the work of ibn Rushd in Muslim civilization never came back from exile and oblivion. Theologians in Latin Christendom went on to become experts and professors of logic, metaphysics and natural philosophy in the universities proliferating across Europe in the 13th, 14th and 15th centuries, but in Islamdom, the marginalization of the rational sciences became a cultural tradition that few ulema dared defy for many centuries, not even when the West, with its manifestly superior weaponry and technical organization, came bursting into the heartlands of Islamdom in the 18th and 19th centuries.

An early note of Muslim recognition of the changing intellectual dynamics that would eventually reverse the political and military relationship of the two civilizations comes in the late 14th century, even while Islamdom's unquestioned world power was represented across the Afro-Eurasian map by a Mamluk empire in Egypt and Syria, a Timurid empire in Iran and central Asia, a Mongol Golden Horde still dominant in Russia and north of the Black Sea and Caspian, and an Ottoman empire rapidly expanding into the Balkans from the east:

> We have heard, of late, that in the land of the Franks, and the northern shores of the Mediterranean, there is a great cultivation of philosophical sciences. They are said to be studied there again and to be taught in numerous classes. Systematic expositions of them are said to be comprehensive, the people who know them numerous and the students of them very many . . . Allah knows better what exists there, but it is clear that the problems of physics are of no importance for us in our religious affairs of our livelihoods. Therefore, we must leave them alone.[20]

Even ibn Khaldun, the greatest Muslim scholar of his time, a historian of science and himself a philosopher in applying logic, sociology and economics to the cyclical patterns of history, can claim that the philosophical sciences have no place in Islam, and without any show of curiosity, can dismiss as irrelevant to Muslims the news about the current intellectual activity going on in the West. It should be no mystery how it was that the astonishing reversal in intellectual precedence between Islamdom and the West came about.

Intellectual activity had indeed been going on in the West for some time. Scholars from England, Scotland, Wales, France, Italy and Germany had, since the early 12th century, been scrambling to Toledo, Salerno and Palermo, old Muslim cultural centers that had fallen to Christendom between 1071 and 1085, in order to avail themselves of the science, philosophy, mathematics and medicine locked up in the treasures of Arabic manuscripts. Ibn Khaldun had heard something of this, and of the rise of universities where the newly acquired knowledge, translated into Latin, was being digested and expanded. His warning for Muslims to stay away from those rational sciences that were taking on new life in the West would a few centuries later begin to exact a terrible price on the pride and dignity of Islam and its loss of place as the leading world civilization.

There was no reason Muslim rulers, merchants and religious thinkers could not have learned what was going on across the way in their Latin neighbor's backyard; only bloated pride. Rather than peer westwards over the wall, more psychological than physical, men of intellect continued to issue dire warnings to stay away from the rational sciences, each warning more strident than the one before. A half century after ibn Khaldun, another well-known historian and leading intellectual light in Egypt, Jalal al-Din al-Suyuti (1445–1505), condemned logic and the philosophical sciences as absolutely forbidden, *haram*:

> Know that, from the time I grew up, I have been inspired with a love of the *sunnah* [exemplary practice of the Prophet] and of *hadith*, and with a hate of *bidaᶜ* [heretical practices] and the sciences of the ancients, such as philosophy and logic. I wrote on the censure of logic when I was eighteen years old, and it was anathema to me. I never heard a problem related to the sciences of the philosophers but I disliked to hear it, nor of a book on any of their disciplines but I avoided reading it . . . As for logic and the philosophical sciences, I do not occupy myself with them because they are *haram* [forbidden] . . . and even if they were permissible I would not prefer them to the other religious sciences.[21]

For al-Suyuti, all the science and medicine that needed to be known were in the Quran. In light of the many extant manuscripts of his book, *al-Hay'a al-Saniyya fi'l Hay'a al-Sunniyya*, which translates roughly as "The Resplendent Astronomy in the Quran and Hadith," al-Suyuti's disdain of the rational sciences can be taken as commonplace, even among those in the highest educated circles.[22] This while the Renaissance was in progress and mathematicians in northern Italy were building the foundations of Galilean physics. Long before the time of Suyuti, this religious conviction that the Quran was the fount of scientific knowledge enjoyed several authoritative adherents, the most renowned being al-Ghazali. He devoted a book to the subject, *Jawahir al-Quran* (*The Gems of the Quran*), the gems being Quranic verses that refer to natural phenomena in showing the power and majesty of the Creator. In place of a natural science based on the study of nature, a natural science based on religion became popular, with the Quran becoming a treasure trove where all sciences could be discovered. Islamic astronomy and prophetic medicine, "sciences" grounded in the Quran and Tradition, vied with the astronomy of al-Battani, ibn Yunus, ibn al-Haytham and al-Biruni and the medicine of Hunayn ibn Ishaq, al-Razi, ibn Sina and Zahawi. Judging from the number of extant manuscripts on the subjects from the 16th and 17th centuries, these became popular in proportion to the steep decline in the genuine study of nature.[23]

At the height of Ottoman power in the mid-16th century, a Turkish historian and religious scholar, Tashkopruzade, issued yet another of the many warnings that "those who practice philosophy are the enemies of God and the Prophets . . . who want to destroy religion . . . and are more harmful than Jews and Christians because they dress as Muslims." Among the accursed enemies of God are listed al-Farabi, ibn Sina and Nasir al-Din al-Tusi. Tashkopruzade apparently had either not heard of or had forgotten ibn Rushd, but he did not forget the usual dispensation for the intellectual elite of the ulema: "Those in whose heart the principles of the Shari'a are deeply planted and whose heart is filled with the greatness of God's Prophet and his law, and whose faith is girded through memorization of the Quran and the Sunna and whose way is strong in knowledge of the religious sciences, they are permitted to pursue the books of philosophy," as long as they did not critically question the Shari'a and did not trespass into areas that contradicted it, except to refute philosophy through religion.[24] This was penned just as Copernicus's *de Revolutionibus* was coming hot off the printing press, scientific revolution riding the wave of technological innovation.

Tashkopruzade's warning came as the last spark of creativity glowed over the long six-century road Muslim scientific genius had traversed. Even the memory of that glorious past in natural philosophy would fade into the dark night of rejection. The rejection could be as literally physical as metaphorical. Recall Sultan Murad III's demolition in 1574 of the one observatory that had been built in Istanbul. Modeled on Ulug Bey's grand observatory in Samarqand, the Ottoman version had hardly been completed when a naval ship approached the towering structure along the coast and, with its artillery, pummeled the masonry to the ground, as the ulema cheered in triumph and pious cries of joy rose up from the crowd assembled to watch God's will blasting from the ships' artillery, as if a great victory at sea had been achieved. The court poet celebrated the victory in verse that claimed the observatory had become useless and was no longer needed since it had achieved its purpose of correcting Ulug Beg's tables. The correction of a set of tables a century and a half old was seen as a monumental achievement.

By way of comparison, while the Istanbul observatory was being triumphantly pulverized, two thousand miles to the northwest in Denmark, Tycho Brahe was building his Uranoborg observatory on the island of Hven. Looking back on the events happening in these two civilizations in the 15th and 16th centuries, it is hard not to picture the educated leaders of Islam digging the foundations of a dark tomb around themselves.

Not every corner of Islamdom succumbed to this destruction in the name of religion that squandered the once great Islamic scientific tradition. As late as 1720, a large observatory built in the Samarqand tradition went up in Mogul Delhi. The mural quadrant was so sturdily constructed and finely calibrated that it was capable of measuring the positions of heavenly bodies with an accuracy of six minutes of arc, as good as Tycho Brahe had achieved.[25] But it did not lead to a comparable result.

Like all fields of intellectual endeavor, the history of science in Muslim civilization abounds with questions that range from unanswered and partially answered to unsatisfyingly and wrongly answered. In this, history is like science, endless: An answer to a question leads to another question, and no question can ever be fully answered to the satisfaction of all. Otherwise historians would put themselves out of work. In relation to the history of Muslim science, two questions have been asked and have received answers that, never satisfying, only lead to more questions: Why did science in Muslim society decline and come to an end and when did it happen? Answers to these questions are like grappling with the ultimate nature of light; the deeper one researches into it the fuzzier the answers become.

One answer that has been given to the conundrum of Muslim science is that there was no decline and end, only a spiritual transformation that preserved

Muslim science from the secular, materialistic, hedonistic, agnostic direction that it took in the West with the Renaissance, Reformation and Scientific Revolution. Another answer is that science in Muslim society never declined: it was only outpaced. Yet another leading historian of Muslim science, who has devoted over half a century to the subject, proposes that decline set in as soon as the inherited Greek science had become naturalized after having been appropriated.[26] The appropriated science was the precious science that Muslim naturalists and philosophers treasured, worked on, organized, reformed and advanced for centuries. The naturalized science was the knowledge deemed useful, such as the medicine, astronomy and mathematics that could preserve health, determine times of prayer, feasts and directions to Mecca and knowledge that could be used to construct buildings, survey land, divide legacies, keep accounts to carry on trade, and so forth. Aristotelian philosophy and its affiliated sciences were rejected. They were appropriated but not naturalized to enter mainstream usage, that is, theoretical inquiry was marginalized.

This was precisely what al-Ghazali prescribed in condemning metaphysics and the theoretical principles that provided methods for an inquiry into nature. Naturalization substituted the philosophical understanding of knowledge with the practical, non-theoretical view of al-Ghazali. According to this interpretation, religion did not oppose science, it accepted science, or some part of it, the part severed from the abstract, theoretical and creative part. This part died, and, no surprise, creative science died with it, while the naturalized useful part lived on, the irony being, according to the author of this interpretation, that science declined not because of religious opposition but religious acceptance, when the philosopher-scientist gave way to the jurist–scientist, al-Ghazali trumping ibn Sina.

Even if it is true that after the death of al-Ghazali more and more men of science were men of religion,[27] religion's successful opposition to the explorative realm of science, inevitably resulting in the decline of science as this process of naturalization progressed, ends up with religion opposing science, no matter how it is viewed. Propping up the lobotomized body to perform mechanical operations left only a shadow of a former self and could not be considered the genuine article.[28]

Accepting that naturalization coincided with decline and was a causative factor, the question remaining is: when did the decline start and when did it finally end the long career of Muslim science? Dr. George Saliba of Columbia University, a diligent longtime researcher in the field who has devoted years to searching through the Arabic manuscript collections of Europe and Asia, persuasively extends the creative life of Muslim science – primarily the astronomy part of it – until the mid-16th century, a full century beyond the Timurid times of the Samarqand astronomers, which for over a generation of specialist historians marked the terminal point of productive science in Muslim civilization. Shams al-Din al-Khafri (d. 1550), an astronomer of first-rate talent as previously mentioned, was carrying on the attempt to perfect theoretical astronomy according to Aristotelian principles in his book, *Hall ma la Yanhall: Solving What Cannot Be Solved*,[29] an apt title for any astronomy based on the Aristotelian system. A generation before al-Khafri, another astronomer was at work reforming the theoretical constructions of the Ptolemaic system.[30] Ghars al-Din al-Halabi (d. 1563), a Syrian astronomer, kept the tradition of perfecting Ptolemaic models going into the second half of the 16th century.

Locating decline

Science in Islamic civilization evolved and expanded in a window of time between the creation of a powerful Abbasid absolutist empire and the formation of a religious orthodoxy whose voice became increasingly more influential as the absolutist empire declined and fragmented. During that window of time, roughly 770 to 1070, or from shortly after the founding of Baghdad to the establishment of the Seljuqid system of religious education that was dedicated to defining, unifying and girding Sunni Orthodoxy in its defense against schism, Muslims had enjoyed a wide arena for speculative thought. The empire of the Abbasid caliphate, and the several autonomous principalities that sprang from it, collectively sustained scientific productivity for centuries. Baghdad, and later the provincial capitals emulating it, had been left relatively free in their liberal courtly life, something that conservative ulema might censor but could not threaten or intimidate, given the doctrinally amorphous state of the ulema as opposed to Abbasid power and prestige that was successfully building on the wealth and affluence of its new capital city situated along the Tigris, in the middle of the great agricultural plain and at the center of world trade. The ulema were forced to accept what they and the pious of the community saw to be the caliph's irreligious courtly life. They disapproved but were nonetheless obliged to tolerate and live with it, even grudgingly accept it, for no other reason than the success the Abbasids achieved in ending the repeated bouts of civil strife that had been tearing the community apart on and off ever since the murder of the third caliph, Uthman, in 656. What was lacking in political purity was gained in social stability.

The many competing religious thinkers with their rival views had acted to diffuse religious authority in a gridlock of contending schools of thought, allowing room for the natural sciences to grow in the shelter of courtly patronage. It took centuries for an established orthodoxy to find its voice and make itself felt, succeeding only after secular civilian government and administration, imbued with the old traditions of Iranian civilization, had given way to military government and administration under Turkish tribal chieftains who had turned themselves into dynastic sultans dedicated to the defense of the Sunni community and its orthodox beliefs.

Prior to the emergence of a dominant orthodoxy girded in Ash'ari theology, much religious energy was consumed in articulating the Shari'a law and all that went into that prodigious enterprise of defining and then interpreting the sources of law. Defining the place of logic, establishing the methodology of making analogical inferences from the Quran, formulating a scientific critique of Hadith and a system of jurisprudence: these were endeavors that defied the control of any overarching authority. And because the endeavors were regional, taking place as they did in different centers of legal thinking – Medina, Basra, Kufa, Baghdad and Syria – they emerged in the form of a multitude of competing schools, of which four became dominant. What might be called a fifth, and has at times been called so by would-be Muslim unifiers, the Shi'ite, was equally a sect with its own political theology, its own collections of Hadith and schools of legal interpretation. The formation of these legal schools ran parallel to the early politico-theological rivalries dividing the Umma: Qadarites (freewillers), Kharijites (radical puritanical egalitarian anarchists representing a reversion to Arab tribal independence and warfare), Jahmites (predestinarians) and Mu'tazilites (rationalists, or upholders of "divine oneness and justice," as they called themselves), to name but some of the most important. It might be considered something of a marvel that a mainstream orthodoxy was able to emerge from what looked to be a hopeless welter of contention, one, in fact, that was concomitant with, and to a great extent the product of, another conflict, this one more violent, namely the internecine wars of the Arab tribal factions that began with the murder of Uthman and the subsequent civil war between Ali and Mu'awiyya over the caliphate. Arab factional strife sporadically erupted right up to the end of the Umayyad Caliphate. Thereafter, as Arab identity politically subsided into the vastly larger community of Muslims that evolved through conversion under the Abbasids, the strife was largely sectarian.

After more than a century into the Abbasid caliphate, a consensus was reached by the ulema of the Sunni community that a holy law, the Shari'a, had been successfully formulated from the holy sources of Quran and Hadith. In fact, a variety of interpretational tools and non-revelational sources of jurisprudence had been brought in to complement the holy sources: personal opinion; analogy based on reasoned textual analysis; consensus of the ulema; benefits to the community; and other sources of lesser importance. When the great jurisprudential effort of interpretation (*ijtihad*) was seen to be getting out of hand with the abounding fabrication of prophetic tradition meant to legitimize one or another sectarian positions, the leading ulema of Baghdad declared the Gate of Interpretation closed. By the beginning of the 10th century, the formulation of law was over and the law deemed complete. The Holy Law that would preside over the Muslim community had been achieved and could not be changed, added to, or diminished. Legislation and legislators became a historical memory of the momentous triumph of the heroes of Sunni jurisprudence.

This decision to close the Gate of Interpretation was a critical juncture in Muslim history, for in a society perceived by itself to be modeled on eternally valid divine commands of obligation and prohibition, down to the smallest detail of personal and communal life, the termination of interpretation of the sources of law imposed a stultifying restriction on intellectual exercise within the circle of the ulema and outside of it. *Ijtihad* had cudgeled the faculties of analytical and analogical reasoning. Closing the gate deprived religious minds of a vital exercise in reasoned thought. Scholars who came after would memorize the authoritative works of their great predecessors. The legal reasoning that went into exploring the law in order to formulate a *fatwa*, a legal decision on a point not clearly delineated in the *Shari'a*, continued the mental exercise that went into thinking out the law but on a vastly reduced scale and, to that degree, the scope of religious reasoning was narrowed.

Closing the interpretive gate of legal formulation was meant to freeze the law in a moment of time. The ulema attempted to freeze time itself by sanctifying tradition and demonizing innovation. The traditional ways of the community, sanctified as preserving the ways and practice of the Prophet Muhammad, the Sunna, were to be the roadmap of a God-pleasing life. This was embodied in the catchword *taqlid*, to mold or solidly encase. Its opposite, *bid'a*, innovation, was morally reprehensible and to be avoided individually and communally. If society or an individual believer faced something not defined by the Shari'a, the ulema were there to pore over the law and come up with a legal opinion, a *fatwa* that resolved the problem according to the *mufti*'s reading of the law. It was a limiting intellectual exercise framed to perpetuate society in a condition that was, in the religious mind, seen to be the highest moment of mankind, when Islam was born, when the Prophet delivered the Quran and against all odds overcame his powerful enemies, the enemies of God. Closing the gate of *ijtihad* in the name of communal unity and conformity closed some of the communal space in which interest in the rational sciences was tolerated. For now, with the formulation of the law and an orthodoxy that held the center, the mainstream ulema were finding a powerful public voice, further isolating the rational sciences at the margins and secret recesses of society and giving would-be liberal-minded dynasts who might have otherwise patronized *falsafa* some second thoughts.

Having legislated themselves out of thinking legislatively, the ulema, armed with their *fatwas* and anathematizing proscriptions of anything that hinted of *bid'a*, could now more effectively legislate scientists, philosophers and religious rationalizers out of the business of interpreting the universe and man's relationship to it. Nature's gates of interpretation were also to be closed and its rational contents emptied. It is not coincidental that in the 10th century, Baghdad, the great center of science and philosophy, began giving way to Bukhara, Khwarizm, Merv, Ray, Nishapur, Shiraz, Tus, Samarqand, Cairo and Cordova, where more effective rulers and economics predominated and the orthodox ulema's doctrine of *taqlid* had not yet become decisive – though it would in time, as instanced by ibn abi Amir's burning of the caliph's books of *falsafa* to the applause of the Cordovan ulema and at the very same time at the other end of Islamdom, in Ghazna, by the severity of Sultan Mahmud's narrow orthodoxy.

Though the gathering voice of Orthodoxy was penetrating courtly circles, the momentum of science and philosophy was too great to be easily or immediately curbed. The orthodox voice may have acted as a reinforcing factor in diminishing the intensity of creativity, among a bundle of other factors, but science would continue advancing for generations, some branches of it for centuries.

Support for the sciences in general faded because of a combination of factors, the most imposing of them being the collapse of Abbasid absolutism, the shrinking of courtly revenues in a declining economy, and the emergence of a formidable orthodoxy. The overall effect was a diminution in the number of natural sciences deemed worthy of study and support. Astronomy and its helpmate mathematics were deemed religiously and practically worthy, for obvious reasons. For almost half a millennium after the high period of scientific creativity, Muslim astronomers and mathematicians would hold their own as the world's leaders.

Between 850 and 950, the caliph's fortunes went from bad to worse. The economics and material wealth that sustained Abbasid absolutism rapidly dissipated when the caliph's authority was overridden by his Turkish slave military, whose contending and unruly chieftains made for political instability, leading to civil turmoil, administrative paralysis, fiscal chaos and infrastructural neglect. The system of agriculture between the Tigris and Euphrates that had been carefully built and maintained as a source of state revenue to support the palace and the civil and military administrations started to break down. Productivity declined, and government revenues fell short of expenses. Annual government income from the Sawad area in lower Iraq fell from around 100 million gold dinars to slightly over 30 million. To meet the financial crisis, the government issued land grants to officers in lieu of pay. The land grant was meant as a temporary measure, but once employed, the quick fix took on a life of its own and in time became customary. The landholding grant (*iqta'*) was issued for a stipulated period of time to an officer whose salary, and those of the troops under him, were paid out of the taxes collected from the land. This gradually replaced the civil administration by a sort of military feudalism, as it were. The effect was to diminish the central government's control of the land and starve the treasury, which at one time had afforded caliphs the means to maintain a splendid court, build observatories and hospitals and lavishly support the scholars of their choice.

Ma'mun's House of Wisdom faded into history not long after the Turkish slave guard's murder of the caliph Mutawakkil and the ensuing decade of guard rule with its administrative and financial mismanagement and concomitant civil disturbances. The caliph Mu'tasim's (833–842) relocation of the capital from Baghdad to the new city of Samarra, built on an unpromising and infertile stretch of Mesopotamia, where the capital would remain some 60 years before returning to Baghdad, added to the political and economic dislocation. In 935, the position of the caliph's Turkish guard was officially established. The caliph was henceforth to be called the "Commander of the Believers," while the Turkish chieftain became the "Commander of the Commanders," meaning that the Turkish military commander-in-chief was now the "protector" of the caliph. The new titles said it all. The collapse of the caliph's authority was accompanied by a shift in the relations between religious and political power in favor of the guardians of orthodoxy. The royal patronage of old that had sustained science and philosophy was now, in addition to being diminished for economic reasons, effectively challenged by the ulema. Although Mu'tazilite rationality had survived in Ash'ari's theological reformulation of orthodoxy, the ulema's hostility to it, to *kalam* in general, to natural philosophy in particular, and above all to metaphysics, continued to be adamantine. The contention among rival sects and theological schools that had consumed much of the collective ulema's energies and allowed for a relatively unimpeded and impressive philosophical and scientific efflorescence had come to an end.

By 1100, with Ghazali's unique synthesis of Ash'arite theology and mysticism, and the Seljuq vizier Nizam al-Mulk's madrasa system of public religious education designed to propagate the reformulated orthodoxy, the Sunni ulema was organized into a social force more potent than at any time previous. Supported by the Seljuq government, the ulema could now speak with one voice to a public as eager as ever to accept solace in miracles, promises of paradise and a literalist interpretation of scripture: palliatives that excited the imagination without straining the mind. As for the office of caliph, in losing its political authority, it gained religious prestige. A weak caliphate that was no more than a ghost of its former self, but still carried the dignity and prestige of the Abbasid family with its aura of past glory, could be a political asset of the state in its support of Sunni Orthodoxy against heresy and the revolutionary Shi'i movements.

The office of the caliph became essentially a ministry of religious affairs and the person of the caliph a symbol of the orthodox establishment's unity and leadership. The transformation provided the politically enfeebled caliph with a palace and a place of continued prestige and dignity in the community of the believers and preserved the fiction that the caliphate of old lived on. However, in such altered circumstances the caliph's court, surviving on a yearly stipend provided by the military government, could no longer provide the patronage of old, even if it had wanted to. The ulema, on the other hand, found staunch supporters in the Ghaznavid and Seljuq sultans, who paid considerable attention to religion and prided themselves as guardians of Sunni Islam against the Shi'i challenge. The symbiosis of Caliphate and Sultanate put a further squeeze on science as a rewarding career.

The Seljuq defense of Sunni Orthodoxy was transmitted to Egypt when the Ayyubid dynasty founded by Salah al-Din replaced the more liberal theosophical-minded establishment of the Fatimid Caliphate. As the Seljuqs had ended Shi'i Buwayhid rule and imposed a strict orthodoxy in the east, Salah al-Din ended Fatimid Shi'i rule and imposed a similar orthodoxy in Egypt and Syria. The political temper of the times allowed little tolerance for perceived heresy from whatever quarter, as witnessed by the Ayyubid ruler's execution of the Gnostic philosopher Suhrawardi. Salah al-Din converted Fatimid al-Azhar with its corpus of Shi'i studies and Neoplatonist natural philosophy into an institute of Sunni learning patterned on the Seljuq Nizamiyya. As the Seljuqs had done in Iraq and Iran, Salah al-Din built state-funded madrasas all over Ayyubid Egypt and Syria. With their salaried teachers drawn exclusively from the ulema, these schools, the only schools available, inculcated the principles of Orthodoxy in the minds of young students, effectively preserving them from the dangers of intellectual curiosity about the Greek sciences.[31] The Mamluk Sultanate, which grew out of Salah al-Din's slave military system, carried on this praetorian protection of Orthodoxy.

In sum, centuries of unstable political rule and economic vicissitude, and the concomitant coalescence of an autonomous Orthodoxy whose cause was championed by a succession of dynasties, Ghaznavids, Seljuqs, Ayyubids and Mamluks, resulted in a situation in which the ulema was able, *fatwa* by *fatwa*, generation by generation, to create an intimidating atmosphere that encouraged neglect, if not fear and suspicion, of the rational sciences, even in places where the study of them had been formerly tolerated. The atmosphere that had once given light and sustenance to the scientific tradition was drained away. Interest in natural philosophy became a pale shadow of its former self. With the exception of the professional *muwaqqit*, careers in natural philosophy vanished, until by the end of the 16th century, all that remained of the great scientific enterprise was the memory, and even that had grown vague.

Notes

1 This is found in his *Qistas al-Mustaqim*, (*Just Balance*) and even reconciled al-Ash'ari's occasionalist theology, with its rejection of natural causation, to ibn Sina's logic, with its Aristotelian cause and effect and uniformity of nature. [Michael E. Marmura, "Ghazali's Attitude to the Secular Sciences and Logic," in *Essays on Islamic Philosophy and Science*, edited by George Hourani, New York University Press, New York, 1975, pp. 100–111.

2 In his short autobiography, *Deliverance From Error* (*Munqidh min al-Dalal*) [Richard J. McCarthy, S.J., *Freedom and Fulfillment: An Annotated Translation of al-Ghazali's al-Munqidh min al-Dalal and Other Relevant Works of al-Ghazali*, Twayne Publishers, Boston, 1980; also translated by Montgomery Watt as *The Faith and Practice of al-Ghazali*, One World Publications, London, 2000.

3 For Hallaj and his passion for martyrdom see the classic work on the subject by Louis Massignon, *The Passion of al-Hallaj: Mystic and Martyr*, Princeton Press, Princeton, NJ, 1986.

4 *Ihya 'Ulum al-Din*, vol. I, edited by Badawi Tabana, Issa al-Babi al-Halabi, Cairo, n.d., p. 31.

5 *Deliverance From Error*.

6 Aydin Sayili, *The Observatory in Islam*, Turk Tarih Kurumu Basimevi, Ankara, 1988, p. 155.

7 J.L. Berggren, "Islamic Acquisition of Foreign Sciences: A Cultural Perspective," in *Tradition, Transmission, Transformation*, edited by F.J. Ragel and S.P. Ragep, Brill, Leiden, 1992, p. 274.

8 A.I. Sabra, "Science and Philosophy in Medieval Islamic Theology," *Zeitschrift fur Geschichte der Arabischen Wissenschaften*, 9, 1994, p. 41.

9 Mustafa abd al-Raziq, *Tamhid li Tarikh al-Falsafa*, Cairo, 1966, p. 82.

10 Ignaz Goldziher, "Stellung," p. 35.

11 Abd al-Raziq, *Tamhid*, p. 86.

12 Gustave von Grunebaum, *Classical Islam*, Routledge, London, 1970, pp. 198–199.

13 abd al-Raziq, *Tamhid*, p. 89.

14 *Classicisme et decline culturel dans l'histoire de L'Islam*, edited by R. Brunschvig and G. von Grunebaum, *Actes du Symposium international d'histoire et de la civilization musulmane*, Paris, 1956.

15 For a review of the ecological and natural disasters, the invasions, and demographic and economic factors contributing to the diminishing resources of the central lands of Islam between the 10th and 15th centuries, see A.Y. al-Hassan, "Factors Behind the Decline of Islamic Science After the Sixteenth century," in *Islam and the Challenge of Modernity*, edited by S.S. al-Attas, International Institute of Islamic Thought And Civilization, Kuala Lumpur, 1996, pp. 351–389.

16 Ibn Iyas, *Bada'i al-Zuhur fi Waqa'i al-Duhur*, Cairo, 1960, vol. 1, p. 123.

17 Charles St. Prot, *Islam, L'avenir de la Tradition entre revolution et ocientalisationc*, Rocher, Paris, 2008, pp. 211–231 for an informed and critical interpretation of ibn Taymiyya's life and thought.

18 J. Livingston, "Ibn Qayyim al-Jawziyyah: A 14th century Refutation of Occult Science," *Journal of The American Oriental Society*, 112, 1992, p. 598 ff.

19 Namely, he rejected as heretical that God knew universals but not particulars; that God could not reverse the customary order of nature; that the universe was eternal; that devils, jinns and angels did not exist; and that belief in resurrection and judgment day were fantasies of the frightened and ignorant. Mysteries or images that went beyond reason in the Quran could not be explained away by allegorical interpretation.

20 Ibn Khaldun, *Muqaddima*, translated by Franz Rosenthal, Princeton University Press, Princeton, 1967, vol. 3, pp. 251–252. For more on ibn Khaldun's refutation of philosophy see Sabra, "Appropriation and Subsequent Naturalization of Greek Science",. p. 239.

21 Antun Heinen, *Islamic Cosmology: A Study of as-Suyuti's al-Hay'a as-Saniya fi'l Hay'a as-Sunniya* Franz Steiner Verlag, Beirut, 1982, p. 13.

22 Heinen, *Islamic Cosmology*, p. 25, p. 33.

23 But to be fair to the Ottomans, who are even now still being blamed for the Muslim decline in learning, it must be noted that Sultan Mehmet II, conqueror of Constantinople, had St. Thomas Aquinas' *Summa Contra Gentiles* translated into Turkish from the Greek translation, along with other Greek works that came into the possession of the sultan with the fall of the Byzantine capital. Mehmet II also apparently requested the Persian poet Jami to compose a treatise on the respective values of theology, mysticism and philosophy. See Gutas, *Greek Thought, Arabic Culture*, Routledge, New York, 1998, pp. 174–175.

24 Mustafa abd al-Raziq, *Tamhid al-Falsafat al-Islamiyya*, Cairo, 1966., p. 87.

25 M. Abdus Salam, *Renaissance of Sciences in Islamic Countries*, World Scientific Press, Singapore, 1994, p. 18.

26 Sabra, "Appropriation and Subsequent Naturalization of Greek Science," pp. 239–242.

27 Saliba, *Islamic Science and the Making of the European Renaissance*, pp. 186, 189, 243, where the men of religion who were men of science were mainly trained in practical astronomy. This confluence of religion and religious uses of astronomy, embodied by the *muwaqqit*, practitioners of '*ilm al-tawqit*, is seen as blunting the idea of an overall opposition to science.

28 On this see A.I. Sabra, "Situating Arabic Science: Locality vs. Essence," *Isis*, 87, no. 4, 1996, pp. 668–669.

29 Saliba, *Islamic Science and the Making of the European Renaissance*, p. 242.

30 Muhiy al-Din ibn Qasim, known as al-Akhawayn, who authored *al-Islahat fi 'ilm al-hay'at*: Reforms in Theoretical Astronomy. Saliba, *Islamic Science and the Renaissance*, pp. 111–112.

31 H.A.R. Gibb, *Mohammedanism*, Oxford University Press, Oxford, 1962, p. 145.

Part II

The Latin connection

From Greco–Arab classical
to European modern

5 The Latin connections
Translation and transmission

A mental wall separated the worlds of Islamdom and Latin Christendom, a barrier as formidable as the Pyrenees that for centuries separated the two civilizations geographically. Perceived as a land of primitives, lower forms of the human race whose minds and culture befitted their natural conditions in the cold, wet, inhospitable forests of their northern clime, Muslims had little interest in what lay beyond the wall. Whatever was in the lands of Christians that Muslims might want could be brought by Frankish merchants. Slaves were one commodity in demand. Where the Abbasids found theirs in the Turks from the east, the Spanish Umayyads found theirs in the Slavic peoples they bought from Christian slave merchants. Christian merchants from the West traveled to Spain, North Africa, Egypt and Syria, Iran and beyond, but not long distance; Muslim traders seldom ventured into the lands of Latin Christendom. Except for a handful of professional geographers and travel writers, anyone going to France, Germany, England or Scandinavia to satisfy intellectual curiosity concerning the ways of the brutish Franks of those frozen regions would have aroused concern over the person's mental state. Muslim movement was toward the East, to India, Indonesia, Malay, Maldives, Philippines, China, lands of warmth, riches, spices and beautiful things of silk, jade and porcelain. The West had little to offer other than slaves.

Attitudes were different on the other side of the wall. Latins imagined Cordova and Muslim Spain to be lands of magnificence, opulence and knowledge; garden paradises of art, palaces, tiled fountains and libraries, where orchards of exotic fruits and exquisite delicacies abounded. And the Latins knew that somewhere farther to the East lay the land of spices that would bring great profits to those who brought them to the West. The draw was eastwards.

As early as the late 10th century, a Venetian trading colony was established in Fatimid Egypt. This was to be followed by colonies from other Italian cities planted in North Africa and Syria in the 11th and 12th centuries. In the 13th century Marco Polo, an emissary of the Venetian merchant state, journeyed to the Mongol capital to establish trade relations. The East attracted Latins in other ways. The Holy Land was in the East. Latin monks continually journeyed to Jerusalem to see the birthplace of their savior. They were followed at the end of the 11th century by Frankish armies conquering and plundering in the name of Christ. Political, economic and intellectual expansion marched together. By Marco Polo's time, a great fund of Muslim learning had been imported into the Latin West. Between 1100 and 1300, the West crawled out of its economic and intellectual impoverishment. It was a long crawl out of a deep pit. Almost everything of what little Greek learning the Romans had brought to their western provinces during the days of empire had disappeared with the withdrawal of Roman garrisons under the impact of imperial bankruptcy and the Germanic invasions. Precious little medicine, natural philosophy and mathematics was known in the West before 1100. The extent of geometrical knowledge was limited to a cursory summary of it in Boethius' *Quadrivium*. Of Plato's works, only the mythos-ridden *Timaeus* was available in Latin.

However familiar the names Pythagoras, Socrates, Plato, Aristotle, Euclid, Archimedes, Hippocrates, Galen and Ptolemy may have been to the educated elite in the Latin Church, the ideas that went with the names were little known. Being innocent of the knowledge embodied in those classical thinkers, Christian theologians did not consider them beyond the pale. St. Augustine had considered Plato a pre-Christian Christian, equating the philosopher's idea of love of the Good to the Christian's love of God. Boethius allegorically interpreted Greek myth into Christian beliefs and values in his *Consolation of Philosophy*; his categorization of knowledge into the Seven Liberal Arts, regardless how little was remembered of the seven at the time, infused a thin strain of the rational classical tradition into the bloodstream of Christianity so that centuries later, when science and philosophy poured in, an immune system was there to help fight off pious rejection of what might otherwise have been regarded as alien transplants. Nevertheless, for half a millennium, roughly 500 to 1000, the substance of the liberal arts was indeed quite alien to the Latin West. When the names began acquiring substance, philosophy and the sciences were accepted into religion and society, with relative ease when compared to Muslim experience.

Christianity was historically well prepared to accept reason and the rationalization of religion that went with science and philosophy. This had to do with adaptability. All religions need to adapt to intellectual changes if they are to have continuing meaning and relevance in society. A religion that has meaning and relevance also has influence in shaping values and attitudes. Adaptation is an ongoing process. The process is never easy and, in the long run, can result in religion adapting itself into a marginal existence, its place usurped by a secularized form of the old religion. Hence the conservatism found in most religions. The polar tension between adaptation and holding the line, that is, the dynamics determining which of the two poles will give in to the other, depends on the balance of forces within society. Since religious organizations are not at all independent of the social body they seek to instruct and guide, dependent as they are on that body's support and on its members volunteering to assume the duties required for the religious organization to carry on, the same tensions and resultant balance will be at work within the clergy, pulling its formulations this way and that. Islamic and Christian societies each had their own unique system of dynamics determining the direction of balance.

Christianity had a much easier time than Islam adapting itself to the Greek heritage because of the wrenching interpretive transformation forced upon leading

Christian thinkers in the 4th and 5th centuries, after Christianity had been adopted as the state religion of the empire. Over half a millennium later, as the Latin West began transforming itself economically and intellectually, Christian theologians were again obliged to reinterpret their religion: through logic and philosophy in one area, and commerce and taking interest on loans in another. In the process following a period of internal debate between conservators of tradition on one side and theologians who were prone to recognize and accept economic and intellectual reality on the other, the Church became a strong supporter of trade, banking and science – and in the case of science, not merely a supporter, but a rich source of leading scientists. If science is today one of the West's leading endeavors and indicators of intellectual creativity and thriving civilization, it is because theologians studied science and became scientists in the 12th, 13th and 14th centuries. Even into the period of the Reformation's religious wars, the Latin Church was at the forefront of scientific thought.

The completely different conditions in which Islam and Christianity originated, and the conditions attending their respective developments, were powerful determinants of the different paths the two religions took in their regard of science and philosophy. Having begun as a pacifist Jewish sect in a vast empire, and then having doggedly advanced to a persecuted religious minority, and then to a state religion, Christianity had been obliged to come to terms with the imperial values and laws of late Roman society, and in doing so, was refashioned into a triumphant and muscular faith. Islam, on the other hand, was immediately triumphant and created its own values and law; or, it could be said, the values and laws were inherited from Arab tribal culture, softened by the Quranic revelation, and given Islamic birth certificates. Compared to theologians in the Latin Church, the Muslim ulema was much more advantageously placed to stick toxic labels on values and intellectual traditions that the religion's accepted protectors took to be alien and contradictory to its origins, and make them stick. Islam's genesis made the religion less vulnerable to the exertions of those who would call for change, innovation and revaluation.

The radical transformations Christianity had to make in rising from its humble origins to become the state religion of an empire were a leaven that would later make possible the cerebral expansion of its theology to embrace reason and natural philosophy. Love, forgiveness, the meek inheriting the earth, the rich never entering paradise, turning the other cheek, denying world and wealth, surrendering family for God, plucking out the sinful eye, all this humility and love became pious fictions thinly veiling a reality of aggressive war, wealth, power and position. In values, attitudes and structure, the Latin Church mirrored the society in which it had grown. The transformation had been made long before the Crusades. Born in a world that was Roman, Christianity became Roman, and the universal church was known as such: Roman Catholicism. The state was Roman, the government was Roman, and the civil law was a creation of human rational thought, openly professed. Law was not God-given. It could be created, changed, rejected and replaced with new law.

Christian converts in Roman society found this to be the natural order of things. The idea was expressed in the reported words of Christ: "Render unto Caesar what is Caesar's." 'Isa, or Jesus, was a subject of the Roman Emperor and his preaching that became holy scripture accepted the separation of man's relationship to the state and its law from his relationship to God and God's law. The Christian emperor Justinian was called "the Great" because, among other achievements, he codified law. Christians had no pretensions that the law came from God. Once the early Christians had worked themselves out of the fundamentalist hysteria of self-punishment and rejection of the world, Christianity matured as a religion, fortified by the deeper interpretations of Jerome, Ambrose, Augustine and Boethius and by the lifestyle of its imperial adherents. The medieval theologians who cursed logic and the sciences as footpaths to hell paved by the devil did not have a chance in the long run. In the never quite exact game of historical analogs, Latin Christendom's theological mainstream was analogous to Islam's Mu'tazilites who had been swept from the field.

Around the middle of the 11th century, Latin Christian society began its tremendous transformation, having suffered six centuries of bare survival at the most primitive level of agriculture, worsened by infelicitous weather, uninterrupted internal feudal warfare that bordered on political chaos and repeated invasions by Arabs, Vikings and Magyars, all of which had combined to make life short, brutal and hungry. When the worst of the West's medieval travails had at last ended, the hardened westerners were able to gain balance, take a breath of relief and look out at the world. With popes elbowing their way to the ranks of kingly power as they avidly pursued the policy of Papal Monarchy, Latin society surveyed the surrounding horizons and espied possibilities that had only shortly before been beyond reach or contemplation. One way to express this new internal strength was to join the conquistadors in crushing Muslim rule in Spain. The sequel was the Latin invasion of the Muslim heartland, essentially a land grab by the French nobility dressed up as a holy crusade. Another expression was the exodus of a legion of Latin scholars from their home countries to go in search of Muslim knowledge of medicine, astronomy, mathematics, logic, natural philosophy and metaphysics, but particularly in search of the works of Aristotle, upon which could be built a rational system of belief, a quest that had first come alive when young Abelard started lecturing just at the moment the Crusades were getting under way.

Both conquests, by sword and by mind, would take the West into the world of Islamdom and become enduring features of Western Christianity and society, revolutionizing Christ's "love thine enemy" and Paul's "God's foolishness is wiser than man's philosophy" by bringing them down to the level of human nature, that is, the materialist instinct to take and possess and the religious one to understand the universe and one's place in it.

The Latin Christian search into the founts of Arabic–Muslim learning began even before the rise of the West. Westerners had come into occasional contact with Muslim learning during the middle of their intellectual eclipse and brought some little bit of it back with them. Diplomatic exchanges between the Spanish Umayyads and Frankish Carolingians in the 9th century reached a high point in the 10th when Otto the Great (936–973) sent the monk John, Bishop of Gorza, on a mission to the Caliph abd al-Rahman III in 953. Bishop John of Gorza, who had a keen interest in astronomy and mathematics, was in Cordova for three years at the height of the Spanish Umayyad caliphate's power and the early period of its scientific renaissance. He may have learned Arabic from a Spanish Jew who assisted him during his residency there. Inferential evidence has Bishop John returning to Lorraine with Arabic scientific manuscripts. This is not recorded in the sources reporting on his mission, but it would explain why the schools of Lorraine in the later 11th century became the first centers for study of the abacus, a computing device composed of columns of moveable bead-like counters arranged in rows of ten, nine rows for the integers based on the Hindi–Arabic decimal system and one for zero. It would also explain why Lorraine produced the earliest eminent mathematicians in the West.[1]

Some 15 years after John of Gorza's mission, another religious scholar crossed the Pyrenees. Gerbert (945–1003), a scholar of humble origins who would later become Pope Sylvester II, was the most important of the early initiators of intellectual appropriation of Muslim learning. His was a personal quest for knowledge. As a young monk he went to Spain and studied mathematics and music for several years under a bishop in the Catelonian town of Vic in the north of Spain. The future Pope may have learned Arabic along with some science and philosophy.[2] Gerbert returned from Spain with the concepts of the abacus, which introduced the decimal system to a West that was not quite ready for it. That a bright man so ahead of his time in his knowledge of mathematics and the rational sciences could be elected pope says a great deal about the Church and Latin civilization at the turn of the millennium.

Equally extraordinary was Constantine the African (1020–1087), another early pioneer who anticipated, by almost a century, the Latin world's intellectual dawning. Little is known of Constantine's origins.[3] Even his name is a mystery. Carthage is said to have been his birthplace, hence his identification as African. He knew Greek and the name Constantine suggests Greek lineage, though he was a Roman Catholic who spoke and wrote Latin. He may have been descended from a Greek family going back to the time when North Africa was part of the Byzantine Empire. How he became Latin is an open question. He may simply have gone from one sect to the other. He also knew Arabic, which would have been natural if he had in fact come from Carthage. His knowledge of Arabic facilitated his journeys to Syria, Iran and India, where he went in search of medical texts, several of which he translated into Latin. Here was an Arabic-speaking Latin Christian searching the lands of Islam for medical knowledge at the pinnacle of Muslim efflorescence. After several years of travel, study and translation, he settled in Sicily. The island had by then been ruled by the Arabs for almost three centuries, and its leading city, Palermo, was a center of Islamic civilization rivaling Cordova. Constantine collected more medical manuscripts there, and from Palermo he then went to another center of Arab civilization, Salerno in the south of the Italian mainland. The Arabs had ruled it from Sicily, but by the time of Constantine's arrival, Salerno had fallen to Christian rule under the Normans. Sicily was soon to follow.

The religion of the rulers of Sicily changed but not the civilization, as the Norman kings of Sicily and southern Italy emulated the civilized traditions of the

Muslim rulers they replaced, with patronage of scholarship continuing to be a prominent feature of the royal court. Under Norman rule a medical school was founded in Salerno, based on Muslim medicine. Religious tolerance was another tradition the Norman rulers inherited. The Normans patronized Muslim scholars as well as Christian. One of the greatest Arab geographers, the Moroccan al-Idrisi, was supported at the court of the Norman king Roger of Sicily, to whom Idrisi dedicated his famous world geography, *Kitab al-Raja*, or *Book of Roger*. Owing to Norman patronage, Salerno became Latin Christendom's first reputable medical center and one of the important sources from which the Latins gained knowledge of Muslim medicine. It was in Salerno that Constantine the African became a monk in the Benedictine monastery of Monte Cassino.

As a physician and translator at the Salerno medical school, Constantine translated 87 treatises from Arabic to Latin, two of them major encyclopedias, giving Salerno uncontested primacy as Latin Christendom's principal source of Muslim medical knowledge. Among his translations was Hunayn ibn Ishaq's *Medical Questions*, which may have been the first Arabic medical text put into Latin. It will be recalled that it was the Nestorian Christian Hunayn ibn Ishaq who was the greatest translator of Galen and Greek medicine into Arabic in Baghdad two centuries earlier. With Constantine, a new round of medical translation begins, from Arabic to Latin, forming a chain of transmission that reached from Salerno back to Hellenistic Alexandria, with Edessa, Nisibis, Jundi Shahpur and Baghdad providing the bridges.[4]

Constantine's translations were a prelude to the movement that was to begin half a century after the fall of Toledo to the conquistadors (1085), when Latin scholars started arriving there to translate Greco–Arab learning to Latin and to bring back Arabic manuscripts, paralleling the translation movement in Baghdad three centuries earlier. Though the translations of the 12th century improved on Constantine's work, his translations were nonetheless read in Europe's medical schools into the middle of the 16th century.

The rapid spread of copies of Constantine's translations indicates how eager Latin physicians were in the late 11th century to end their deprivation of medical knowledge, of which they were becoming painfully aware.[5]

Gorza, Gerbert and Constantine were individuals who ventured to the lands of Islam and gathered there what learning they could for themselves and their civilization at the moment its intellectual awakening was dawning. Within a generation after the death of Constantine, those two great movements relating the Latin West to the Muslim world were underway, conquest in the Holy Land and translation in Spain and Sicily: one the work of soldiers, the other of scholars. One of the few exceptions on the Crusader side of this Latin incursion was Stephen, a scholar among warring soldiers and princes. Hailing from Pisa, Stephen arrived in the Holy Land in the mid-1120s, settled in Antioch, learned Arabic, translated medical texts and became known to history as Stephen of Antioch. Having come to learn rather than plunder, Stephen must have been a pleasant surprise to the local Muslims and Christians, accustomed as they were to the crude, swaggering, unwashed and unlettered soldiers, whose art of healing was as bloody as their warlike demeanor. One of the books Stephen translated was the popular medical compendium of Ali ibn al-Abbas, the same work that Constantine had translated in Salerno a half century earlier. Stephen's translation contained a helpful glossary of Greek and Arabic technical terms with Latin equivalents. The translation had a long life in the Latin medical tradition. Printed first in Venice in 1492 and again in 1539 in Basle, Stephen's rendering of Ali ibn al-Abbas was being used as a medical school textbook almost up to the time of William Harvey in the late 16th century.

A generation after the first Crusaders had set off eastwards with swords and siege engines to take possession of the Holy Land, their scholarly counterparts had, with ink pots and rolls of vellum, begun trooping southwards over the Pyrenees to take possession of Muslim knowledge. While the crusaders eventually withered away in neglect, ignominy and defeat, the scholars came over the Pyrenees and returned with their translations and mule-loads of Arabic manuscripts, the raw material that prepared the way for the scholastic reinterpretation of Christianity in the light of science, philosophy and logical argument. Just as the scholastic movement was reaching its height at the close of the 13th century, the last Crusader fortress fell back to Muslim rule, ending the West's first attempts to colonize Muslim territories.[6]

That the translation movement and the active period of the Crusades began and ended at about the same times is not accidental. The two movements were manifestations of the social energy of Latin Christendom that was now surging. In a couple of centuries the transformed West would be leaving its neighboring civilizations of Byzantium and Islam on the less advantaged side of the balance of social power. On the intellectual side of social potential, those Latin cathedral schools that had been established in the 11th and 12th centuries were in the process of becoming universities in the 12th, 13th and 14th centuries, creating a demand for school texts in medicine, astronomy, mathematics, natural philosophy and metaphysics. Toledo offered a banquet of learning to the eager Latins arriving at the dawn of their intellectual awakening. As Abelard was growing famous in Paris with his critical lectures on logic and scripture and delighting his students by embarrassing the Church authorities with his list of scriptural contradictions, his fellow scholars were crossing the Pyrenees and descending on Toledo in increasing numbers. It would be their work of translation that provided the Aristotelian logic that Abelard's scholastic successors required in order to resolve, or hide in syllogistic circumlocution, the scriptural contradictions he had listed.

The earliest translations in Spain left much to be desired. This was rectified by the archbishop of Toledo, Raymond II (1126–1151), who played a key role in establishing a school of translation in order to give some organization and methodology to the work of the translators. The school attracted most of the great figures of the translation movement and contributed to Toledo's position as the primary city of Arabic–Latin translation.[7]

The translations were undertaken to supply texts for courses in the universities that were just beginning to rise. In the largest sense, translation was an enterprise in the interest of faith through the deepening of knowledge that could lead to a rational understanding of religion. With that came understanding the universe. This was the intellectual connection with the earlier translation movement of Greek to Arabic: the conviction that reason as applied to discovering and organizing knowledge of heaven and earth was a valid means of complementing scripture in regard to man's innate urge to know himself, the universe around him and the relation between the two. For naturalists and metaphysicians of both religious civilizations, the rational pursuit of knowledge was as much a need for the intellectual satisfaction of answering curiosity as it was a spiritual quest to reach for what transcended reason.

The two translation movements also shared a pragmatic motivation. Astronomy was fundamental in determining a calendar regulating the economic and religious life of civilization with respect to solar and lunar positions through the days, months and years of a cycle. The calendar, the sundial and the clocks of medieval life were products of astronomy. And then there was medical knowledge: medicine for the body, metaphysics in harmony with scripture for the soul and science for the mind. Reason married to religion as a source of spiritual wisdom was a logo the scholars of both movements held to be unquestionable.[8]

Latin Christendom had its political equivalents to the classical caliphs of high patronage. Fredrick II in Sicily and Alfonso the Wise in Castile and Leon patronized translations and encouraged the study of science as energetically as had Mansur, Harun al-Rashid and Ma'mun in Baghdad. In other ways the two translation movements widely diverged. For instance, the Syriac Christian and pagan Harranian translators who conveyed Greek science into Arabic were creative scientists, professionally involved in most or all the sciences and paid by the caliphs to be their physicians and astrologers. In the Muslim experience, the rulers funded the translations and provided a congenial milieu for science. The Latin translators, on the other hand, were centered in the mainstream of the secular and religious currents that formed their society and were at most rudimentary scientists, philosophers and physicians, neither creative nor accomplished; nor were they in the service of political rulers who held the translation enterprise in their hands. They were generally religious scholars educated in cathedral schools and working within the church as translators and logicians devoted to scholastic enterprise. Some of those who made translations into Latin had come to Toledo, Salerno and Palermo in order to advance their studies in becoming physicians, astronomers or mathematicians. Taking knowledge from infidels or learning Arabic to translate texts gave no religious or psychological problems to the Latins, whether they were men of the Church or not. Latin was the language of religion and learning but it was not sacred; it was only liturgical. The son of God spoke Aramaic. There was no talk of God thinking in that language. Scripture came in Hebrew and Greek. Translating Arabic science to Latin in the 12th century came as naturally as had Jerome's Latin translation of Hebrew and Greek scripture in the 4th century.

Described as wanderers and homeless intellectuals, the translators came to Spain from all parts of Latin Christendom. They came as individuals, paying their own expenses in pursuit of their own goals, in search of manuscripts corresponding to their own personal desires, and not as envoys of a sponsored church undertaking.

What in the world did these eager Latins think of the Muslim culture they entered, so different from their own? They came from an agrarian, feudal, monastic and hierarchical society to a society of cities with maze-like covered markets, supplied by a network of long distance trade routes and resplendent with the rich aromas and colors of spices, fruits, delicacies, fabrics and dyes from all over the African and Asian world, with stall after stall of leather goods, gold and silver jewelry, porcelain and ivory. The streets were filled with freely mingling people of all colors, and the surroundings were lush with gardens, fruit trees, fountains, parks, villas and palaces. To the Latins, coming from a celibacy-obsessed sacerdotal society, life in the Arab cities of Spain must have been unsettling in its free, secular, urbane and sinfully indulgent licentiousness.

The Latins appear not to have wanted anything from the Muslims but their knowledge, which they took stripped of the culture that bore it. What they took away was an uncontaminated, culturally neutral body of knowledge that was in keeping with the West's negative views of Islam,[9] a sentiment perhaps somewhat reminiscent of present day cultural attitudes harbored by many Muslims who, while hostile to Western policies, styles of life and values, find themselves obliged to take from a West they regard as morally obscene and depraved.

The Latin translators admired Muslim learning; they admired the science, medicine and philosophy of Muslims, but they disdained the culture of Arab Islam, as Muslims had admired the knowledge of the Greeks while shunning the culture that bore it. Some few Latins may have freed themselves of their religious prejudgments to see something fine in the host culture, but nothing of this sentiment is evident in what they wrote. They came on a mission to translate and collect manuscripts and no more. Making cultural contact or learning anything more than what was in the select category of literature they were after had no place in their mission. No bridges were intended and none were built. They came, they did their work, and left. Going to Spain was a lunar mission, manuscripts in place of rocks.

Who were these adventurous frontiersmen of the Western Awakening braving the Pyrenees? One of the earliest was a Welsh philosopher, mathematician and scientist, Adelard of Bath (d. 1142). Setting up at Raymond's Toledo School, he translated the tables of the Andalusian astronomer Majriti, Euclid's *Elements*, and most of Khwarizmi's *Algebra*. Adelard's mission went beyond mathematical scholarship. Collaborating with Herman of Dalmatia (d. 1143), who was in Toledo translating Ptolemy's *Almagest* (not to be confused with Herman the German who was translating in Toledo during the middle of the next century), he translated the Quran, the first translation of it in any language.[10]

Among Herman's major translations was al-Majriti's critical revision of Ptolemy's *Planisphere*. This introduced the Latins to the technique of stereographic projection, that is, projecting onto the two dimensions of a plane the arcs on the three dimensions of a sphere, a skill basic to making the plates of an astrolabe. The Latins attained mastery of stereographic projection by the end of the 1140s, thanks also to translations made by Plato of Tivoli (fl. 1134–1145).[11] Adelard of Bath and Herman of Dalmatia were shortly followed by another of the great translators, the

Englishman Robert of Chester, who arrived in Toledo in 1145 and made the first complete translation of Khwarizmi's *Algebra*.

The greatest of the Toledo translators, by any measure, was the Italian polymath Gerard of Cremona. He had come to Toledo in 1167 in search of a copy of Ptolemy's *Almagest*. Unable to find one in Latin Europe, where even at the end of the 12th century Ptolemy was little more than a name, Gerard made Ptolemy his mission. He learned Arabic in Archbishop Raymond's school and remained in Toledo almost 40 years, during which time he translated close to a hundred books and treatises, including the *Almagest* for which he had come. Working from the Arabic translations of it from the Greek by al-Hajjaj and Hunayn ibn Ishaq as revised by Thabit ibn Qurra, Gerard spent 10 years putting the *Almagest* into Latin. Apparently without knowing it, he used Hunayn's method: collecting as many copies of the manuscript as he could in order to come up with what he hoped would be a complete Arabic text for translation.[12] He followed the same method in translating the medical equivalent of the *Almagest*, ibn Sina's monumental encyclopedia, *The Canon* (*al-Qanun*). It is for having brought the *Almagest* and *Canon* into Latin that Gerard gained fame.[13]

In addition to the *Almagest* and *Canon*, Gerard made another translation of Euclid's *Elements*. He also translated many works of Aristotle, Galen and Hippocrates. Among these were 17 books on mathematics and optics, 14 on natural philosophy and 24 on medicine, encompassing the chief works of al-Kindi, Thabit ibn Qurra and ibn al-Haytham. Enhancing the value of his translations was his transliterations of Arabic terms for which Latin had no equivalent. Hence the many Arabic terms in mathematics, astronomy, horticulture and chemistry that are found in the modern languages of Europe. In terms of the quantity and refinement of his translations and his creation of loan words, Gerard is reminiscent of his great predecessor, Hunayn ibn Ishaq, both of whom were work horses in the transmission of an intellectual heritage that went back to antiquity.

One of the last of the important translators to work in Toledo was a Scotsman, Michael (d. 1235), whose mission was to translate Aristotle's works into Latin. Like his predecessors, he first had to learn Arabic. This he did within three years of his arrival in 1217. Working intensively, he translated the Arabic version of Aristotle's *On Heaven and Earth* (*de Caelo et Mundo*) and ibn Rushd's commentary on it. He also translated al-Bitruji's non-Ptolemaic *Astronomy*, this with the help of a Jewish scholar. He continued his work of translation in Salerno where he was Fredrick II's court astrologer.

Spanish Jews and Christians assisted the Latin scholars in their quests, particularly those Christians who were Arabic in language and culture but remained true to their ancestral religion during the centuries of Muslim rule. These were the so-called mozarabs, from Arabic Musta'rib: those who would be Arab. Among the most notable of them were John of Seville (d. 1142), a Jew until converting to Christianity midway through his scholarly career; Hugh of Santalla (d. 1145); and Mark of Toledo (d. 1216). Spanish Jews literate in Hebrew, Arabic and Spanish played a role in Arabic–Latin translations just as the Eastern Christians, literate in Greek, Syriac and Arabic, had translated Greek and Syriac manuscripts into

Arabic. One of the most famous of the Jewish translators was Abraham ben Ezra. He made a Latin version of al-Biruni's commentary on Khwarizmi's astronomical tables and also translated several astrological treatises of the Baghdadi Jew Ma sha' Allah, Masahalla in the Latin tradition. Another Spanish Jew, Moses ibn Tibbon, translated many Arabic works, most notably ibn Rushd's commentaries on Aristotle that were in great demand by the scholastics of the philosophy departments and theological colleges of the emerging Latin universities.

As in the Muslim experience, the Latin quest of science transcended religious differences. Jewish translators sometimes teamed up with Christian scholars to make translations, the Jew working from Arabic to vernacular Spanish, the Christian taking the text from Spanish to Latin. Two well-known teams were at work in the middle of the 12th century in Toledo. One was Dominicus Gundisalvus working with ibn Da'ud (Aven Death in the Latin tradition), a Spanish Jew who converted to Christianity taking the name John and becoming known as John of Spain.[14] Ibn Da'ud's translations from Arabic to Spanish were then put into Latin by Gundisalvus. Together they turned out translations of works on astronomy, astrology, mathematics, medicine, geography, and philosophical treatises by al-Kindi, al-Farabi, ibn Sina, and al-Ghazali (who, as already mentioned, so thoroughly explicated the principles of philosophy that he was mistakenly assumed by the Latins to be a great Muslim philosopher). Gerard of Cremona and a Mozarab named Ghalib formed a team operating in the same way as Dominicus and John. These two teams were responsible for most of the philosophy translated into Latin in Spain during the 12th century.

The usual route taken by Latin translations of works written originally in Hebrew, or translated to Hebrew from Arabic, is that of the *Book of Measurements* by Abraham Bar Hiyya (d. 1136), a Sephardic mathematician whose work, based on Arabic sources, exemplifies the joint labors of Jews and Muslims in scientific production. Bar Hiyya's book was translated by his contemporary, Plato of Tivoli, an Italian who worked in collaboration with an anonymous Spanish Jew, possibly the author Bar Hiyya himself. The Jewish collaborator, whoever he was, made an Arabic version of the original Hebrew text; Plato of Tivoli then put the Arabic into Latin. His knowledge of Arabic enabled him on his own to translate the works of Thabit ibn Qurra, al-Khwarizmi and ibn al-Haytham, ranking him among the most important of the 12th century translators.

Plato of Tivoli and Gerard of Cremona were typical of these scholars. They came over the Pyrenees and took the trouble to learn Arabic, if they did not know it already, so they could translate from Arabic to Latin unassisted. Translations from Hebrew to Latin usually required the collaboration of Jewish and Christian scholars, with a vernacular Spanish version bridging the original language and the final product. The work of translation into Latin, like the earlier undertaking of translation, into Arabic, was, like the work of science itself, a unifying and high-minded endeavor that brought together people from different regions, religions, languages and cultural traditions.

The Toledo school of Archbishop Raymond II, which played so powerful a role in the westward transmission of Greco–Muslim science, mathematics, medicine and philosophy, lasted a century and a half, 1135–1284. Before closing its doors, the school can fairly be said to have overseen the translation movement from beginning to end. Two years after it closed, an institute of Islamic studies was founded near Toledo by the Order of Preachers, an order that had been drawn there by the scholarly prestige of Archbishop Raymond's school. The Order of Preachers, however, did not go there for purely scholarly motives. Its purpose was to educate missionaries in Arabic and Islam in preparation for their mission to convert Jews and Muslims of the Middle East to Christianity. Peter the Venerable's sponsorship of the Quran's translation into Latin was part of this proselytizing assault that complemented the military assault of the Crusades, which by 1250 were coming to an inglorious end. On a more enlightened level in the contention between the Islamic world and the West, Raymond's Toledo school was reopened in the early 1990s as a language and research institute for comparative studies in Islamic and Western civilizations. The institute presently occupies the restored building of Raymond's school.

While the Toledo school was functioning in the 13th century, several other translation and study centers opened in emulation of it, one in Seville founded by Alfonso X (d. 1284), king of Castile and Leon and another in Palermo founded by Fredrick II, who knew the Arabic language and favored Arab civilization. The Norman rulers of Sicily, especially Roger II and his grandson Fredrick II, were enthusiastic enough in their preference and patronage of Arab civilization and learning to be labeled "baptized sultans" by the disapproving pope. It is worth noting in this context that Naples, owing to Fredrick II's founding of the university there in the early 13th century, also became an "acculturating point" of Arab/Muslim learning in the south of Italy.[15] These Christian sultans also patronized classical learning that had been sustained by the Greek community in Sicily, where it had collected, copied and preserved Greek manuscripts. When Michael the Scot arrived in Salerno in the 1220s to translate scientific manuscripts, Fredrick handsomely supported his work and appointed him court astrologer for the 10 years Michael was working there.

While performing his duties as court astrologer, Michael made translations. He translated medical treatises of al-Razi and ibn Sina for a Danish physician who was at the time working in the medical school at Salerno while writing a Latin text on bloodletting and surgery based on Arabic sources. Among the many other things the Scotsman Michael translated was ibn Sina's version of Aristotle's *Zoology*, a subject dear to his royal patron's heart. Fredrick wrote a book on falconry, based on Arabic sources, as almost everything was during the first three or so centuries of the Western intellectual ascent. Indeed, a 13th century king writing a scientific book was for its time as remarkable an event as a 10th century future pope going to Spain to study mathematics and science. If a civilization's strength has anything to do with curiosity about nature and the intellectual richness of other civilizations, a visitor from outer space would have observed that the Latin West of the 12th and 13th centuries had much to anticipate. The Crusades were being lost, but a greater conquest was being won. Translating Arabic–Muslim knowledge was one front in this conquest. Translating the original Greek corpus was another.

Sicily had been ruled by the Byzantines for centuries before the Arab conquest. The Greek community that had endured on the island over the three centuries of Arab rule preserved Greek learning and possessed troves of Greek manuscripts. This was something Spain did not have: a Greek community maintaining a continuity of contact with science and philosophy as knowledge expanded and moved from Athens and Alexandria to Baghdad, Bukhara, Cairo, Cordova, Toledo, Salerno and Palermo. In the words of a distinguished authority:

> Easy of access, the Sicilian capital stood at the center of Mediterranean Civilization, and while the student of Arabic science and philosophy could in many respects find more for his purpose in the schools of Toledo, Palermo had the advantage of direct relations with the Greek East and direct knowledge of works of Greek science and philosophy which were known in Spain only through Arabic translations or compends.[16]

The Belgian Dominican William of Moerbeke (d. 1286) was one of the first Latins to exploit the Greek riches of Sicily. Having versed himself in classical Greek before his arrival in Palermo in search of manuscripts, he began the work of realizing his life's passion, delivering Aristotle to Latin direct from the Greek original, or as close to the original as he could come. He translated Aristotle's *Politics*, *Zoology*, *On the Heavens* and other works by Aristotle, in addition to Greek treatises of Archimedes he found in Palermo.

Jews also played an important part in the translations made in Palermo, where a small Jewish community lived alongside the Greeks. As "People of the Book," like Christians and Sabians, the Jews in Islamdom had a protected status and they continued to enjoy it in Sicily even after the island had fallen from Arab Muslim to Norman Christian rule. Normally Christian rulers were much less tolerant of Jews than Muslim rulers were, but the Sicilian Normans were an exception. The Norman kings of Sicily provided a congenial environment for Greeks, Arabs, Jews and visiting Latin scholars to interact and translate texts. It was not unusual for scholars of the Jewish community to know Arabic, Hebrew and Latin, and one of them, Faruj ben Salim, performed a great service to the medical future of Latin Christendom in 1282 by translating a copy of al-Razi's encyclopedic *al-Hawi* (*Continens*) into Latin.

Earlier translations of Razi's medical encyclopedia had been made in Toledo, but just as in the Greek to Arabic translation experience, texts of major importance (such as those of Aristotle, Ptolemy, Euclid, Galen, Khwarizmi, al-Razi, Ali ibn Abbas, ibn Sina, ibn al-Haytham) received several translations, new ones being intended to improve on the previous ones. In the earlier movement, Euclid's *Elements* had received three translations, each new one being better than the one before it, with the final and best being that of Baghdad's master translator, Hunayn ibn Ishaq. Paralleling that, Euclid received the same number of Latin translations, one by Adelard of Bath, a second by Herman the German, the third and best by the Latin world's grand master, Gerard of Cremona.

The earlier translations were more likely to be incomplete or flawed, but this was not always the fault of the translator as much as it was defects in the copy of the text being translated. When better and more complete copies became available, their translations came closer in meaning to the original. Also at work was an abiding rivalry among the Latin translators. A better translation of a major work could bring scholarly fame and prestige. Clarity and simplicity were important factors in judging the excellence of a translation.[17] This explains the multiple versions of ibn Sina's *Canon*, Ptolemy's *Almagest*, al-Khwarizmi's *Algebra*, ibn al-Haytham's *Optics*, Razi's *Continens*, ibn Rushd's *Colliget*, Ali ibn al-Abbas's *Complete Art of Medicine* and al-Zahrawi's *Tasrif.* Because Greek texts were valued as direct connections to the golden stream of philosophy, science, mathematics and medicine, Greek copies of the works of the great philosophers and Hellenistic scientists were more likely to receive multiple Latin translations, just as they had in Baghdad. The Greek manuscripts preserved by Palermo's Greek community brought some Greek knowledge directly into Latin Christendom, giving Palermo a special place in the early transmission of science. But even with Greek copies being available, their Arabic versions were still highly prized, since they significantly expanded on the Greek originals by way of their critical commentaries.

In addition to Toledo, Palermo and Salerno, other Latin centers contributed to the translation and transmission of Arabo-Islamic science. The aforementioned Alfonso X, called el Sabio, The Wise (d. 1284), ruler of Leon and Castille, provided a center in his court. Alfonso ardently promoted astronomy, being himself an amateur astronomer. One of his greatest contributions to the science was his support of the translation of a great number of Arabic astronomical works, the most famous being the tables of the Cordovan astronomer al-Zarqali (d. 1087), which in Latin became known as the Toledan Tables. Alfonso also had translations made from Arabic into Spanish, one of the earliest examples of translation into a vulgate language. Another important contribution he made to the rise of science in the West was his sponsorship of the translation and publication in a single volume of a collection of 15 Arabic astronomical treatises, collectively known as *Libros del Saber de Astronomia*. Composed of star catalogs and descriptions of observational equipment such as astrolabes, quadrants and gnomons, the *Libros del Saber* became popular among the widening circle of science enthusiasts in the Latin world during the 13th century.

Muslim science was by then coming from many diverse sources, mainly of course from regions that had fallen to Christendom – Toledo, Seville, Saragossa, Palermo, Salerno, and to a much lesser extent Syrian/Crusader Antioch. And then there were those scholars who ventured to lands firmly under Muslim rule. One was the intrepid mathematician Leonardo Fibonacci of Pisa (d. 1240), the West's first great mathematician and the first Latin to contribute something original to his field of study.

Leonardo's father was a merchant and consul of the Pisan trading colony established in the Muslim coastal city of Bugia in present day Algeria. The energetic pursuit of wealth and social place through peaceful commercial enterprise had, by Fibonacci's day, replaced the cruder methods of the Crusades. The upsurge in

European commerce was led by Italian city-states, like Venice, Genoa, Pisa, Bari and Livorno. Some Italian cities had established merchant colonies in Egypt and North Africa as early as the late 11th century. Commercial enterprise was the springboard of Leonardo's venture in search of the treasures of Arabic knowledge. His merchant father entrusted his education to a North African shaykh who taught him Arabic and some mathematics. His father's concern was not completely altruistic. The business of international trade was in need of people trained in accounting, bookkeeping, computing compound interest, cost and profit sharing, cargo and insurance rates and volume and surface measurements of irregular shapes. Even computing the volume of a sphere, cube or parallelepiped would have been a challenge to the vast majority of maritime traders at the time. A mathematician in the family would be a valuable business asset. The rise of mathematics in the West had the same pragmatic basis as it had had in Muslim civilization.

Leonardo was confidently traveling on his own through north Africa, Egypt and Syria as a young man, all the while applying himself to the study of mathematics through the works of Khwarizmi, abu Kamal al-Shuja', al-Karaji, ibn al-Haytham and Umar al-Khayyam. Leonardo could read these in the original Arabic or, if the Arabic manuscript was not available, in the Latin translations of Gerard of Cremona and Adelard of Bath. Ibn al-Haytham's work on cubic equations and his method of extracting cube roots particularly fascinated young Leonardo. He mastered the abacus and in 1202 published a book on it, *Liber Abaci*. The book offered more than the mechanics of the abacus. It employed Hindi–Arabic numerals, at the time known to only a handful of Latin scholars, since the Latin translation of Khwarizmi's book had not been available long enough for its contents to be digested.

An early vehicle conveying Arab numerals and the decimal system to the Latin world, Leonardo's book was quick to become famous. It reviewed the numerical place-value system of units, tens, hundreds and thousands and explained the decimal system's arithmetical operations related to practical applications in solutions of business problems. Most of the book was devoted to pure mathematics: proportions, extraction of roots, geometry and algebra.

Leonardo's fame reached Fredrick II. The emperor invited the mathematician to come to Palermo for a visit. Leonardo accepted. It was a fruitful visit in several ways. Bonding by a shared fascination in mathematics, Fredrick and Leonardo came to think highly of each other and carried on an enduring friendship, even after Leonardo left Sicily. One of the fruits of the relationship was their long correspondence. Fredrick and the mathematicians in his court would send Leonardo problems and Leonardo would send the solutions back. Leonardo dedicated to the king his *Liber Quadratorum* (*Book on Squares*), an analysis of quadratic equations which contained some original methods of solution. Considered his greatest work, it elevated Leonardo to the rank of a pioneer mathematician breaking new ground.

In his analysis of quadratics, Leonardo gives a solution for working out a number that added to or subtracted from a square gives another square ($x^2 = y^2 + a$). One of his contributions to number theory is known as the Fibonacci sequence. A pair of rabbits, one male, one female, are put in a pen. How many pairs can be produced from the original pair in a year if each pair reproduces itself every month? The solution produces a recursive sequence, each term being the sum of the two preceding terms, 1, 1, 2, 3, 5, 8, 13, 21, 34. . . Leonardo's original solutions of binomial and trinomial equations based on the mathematics of Khwarzmi and abu Kamil show that Latin science was at a stage where translation, assimilation and an urge for creativity were simultaneously at work, the same overlapping stages that were apparent in the Muslim experience toward the end of the 9th century. In both the Muslim and Latin instances, it took about a century to go from the beginning of translation to the earliest sign of creativity.

The Fibonaccis, father and son, each in his own way, bring to light early signs of the parting of the ways between Islamic and Western civilization. First, Leonardo's father was consul of a merchant colony in a Muslim state. Led by the Italian commercial city-states, Genoa, Venice, and Pisa, Europe was expanding commercially throughout Islamdom. Later in the same century Marco Polo would be heading for China to make a trade deal with the Mongol Khan on behalf of the Venetian Republic. Muslim merchants did not set up colonies to trade with the West. A merchant might enter Christendom. A Persian came to Byzantium in the 14th century, but no Muslim colonies were established there. Settling in the West to promote trade was as alien to Muslim merchants as the task of learning Greek was to Muslim scientists, physicians and philosophers. Commercial enterprise took Muslims to Indonesia, Malaysia, Maldives, the Philippines, where they settled and spread their religion, but the Christian West was inhospitable territory. The lands of Greek and Latin Christendom were on the frontier of the *dar al-harb* (abode of war) in the world of war and peace, as formulated by religio-political theorists in Abbasid times. The Buddhist east, on the other hand, was fertile soil for trade and proselytizing. In any case, Muslim merchants saw little in the Latin West to attract them. A Sufi missionary heading westward to spread Islam among the Latins would have been thought to be out of his mind. The Christian conquest of Spain followed by the Crusades combined to give the lands of Christendom a renewed aura of war, not a place in which to settle and engage in peaceful commerce and preaching.

There was yet another Latin channel to Muslim learning that warrants comment: Byzantium. In 1295 an Orthodox bishop residing in Tabriz in northern Iran made a Greek translation of the *Zij-i Sanjari* and *Zij-i Ala'i*, astronomical tables that were made in late Seljuk times. The translations contributed to an ephemeral Muslim-inspired Greek revival of astronomy that would feed into the rise of science in the West. Here it was war operating, not commerce. When it became apparent that the walls of Constantinople were eventually going to give way to Ottoman artillery, Byzantine scholars began arriving in Italy with Greek and Arabic manuscripts. Among the Arabic manuscripts were two copies of Euclid's *Elements*. They were quickly translated into Latin and published in Rome. Western scientists were then able to gain access to Greek originals, some of which had never been translated into Arabic or had been translated and lost before the start of the Latin translation movement. But with their critical commentaries and contributions of substance, the Arabic versions of the Greek originals were the richer sources. This would be a fact ignored or denied by the time of the Renaissance, when adulation of classical antiquity, a projection of self-adulation, required that Europeans forget their debt to Muslim civilization.

Sometime around the middle of the 13th century the Arabic to Latin translations came to an end, by which time the new knowledge was being assimilated in the rising universities. Astronomy, through Gerard of Cremona's and John of Seville's translations of the *Almagest*, led the way in the quadrivium (astronomy, arithmetic, geometry and music); Aristotle, through translations of ibn Rushd's *Commentaries*, in the trivium (rhetoric, grammar and logic). The queen of the sciences was astronomy, to whom the other three mathematical sciences made a reverential bow. The subject was divided into two parts, introductory astronomy and advanced, the first being taught as part of the quadrivium in the college of arts and sciences. This was a basic course whose textual materials were simplified versions of the *Almagest*. The common text students used to enter the complexities of Ptolemaic astronomy was John Sacrobosco's *Tractatus de Sphaera* (published in 1220), supplemented by various other texts going by the name of *Theorica Planetarum*. Both the *Tractatus* and *Theorica* were popularized handbooks that avoided the more challenging problems. The *Almagest* itself was what serious students of astronomy studied in the advanced course at the graduate level.

As assimilation progressed through the last half of the 13th century, a certain chauvinism came to the fore in that the Latins progressively diminished the importance of their Arab/Muslim sources by not referring to them, while emphasizing the "true" sources of scientific and philosophical knowledge, the Greeks. The earlier adulation of the Muslim masters was being erased from memory. This late medieval/early Renaissance ennoblement of the Greek sources was an attitude that just as well could have been borrowed from the Arabic sources, which also ennobled, and at the same time criticized, their Greek masters. Roger Bacon, a Latin rebel with many causes, was a lone voice calling for Arabic to be taught in the universities and for scholars to learn the language in order to read the sources themselves rather than translations of them. Not only was his call rejected, but as Greek manuscripts were found and translated, Arabic loan words were replaced by Greek equivalents or Latin neologisms, making it appear that the knowledge had come from the Greeks. The adulation that had prevailed among the Latins for Muslim learning between 1100 and 1270 was being methodically reversed, to be lost in a memory hole. Arabic was not necessary, was the answer to Bacon: science came from the Greeks; the Arabs were mere inheritors and transmitters of the Greek legacy. Or as the mantra goes today, Arabic science was nothing but Greek preserved in cold storage for six centuries. Accordingly, Aristotle was given full stage and ibn Rushd downplayed. Khwarizmi's mathematics, Haytham's optics, ibn Sina's medicine, ibn Rushd's expansive commentaries on Aristotle, and all the rest that carried Arabic–Muslim science beyond its Greek legacy, were washed out of the record in what has been called "The progressive deleting of the Arabs from the legacy . . . an untold story of medieval European history."[18] By 1350, the Arab/Muslim interlude could be overlooked: knowledge had come from the Greeks. By the late 1300s, any overt recognition of the vast Arabic contributions to Latin science had been largely distilled down to a small handful of names and texts:

> In the end, the things late medieval Europe found of greatest interest and use – astronomy, mathematics, mechanics, cosmology, philosophy in particular – were eventually attributed to the Greeks. Broadly speaking, only in astrology, alchemy, and, to a significant degree medicine, did the contribution of the Arabs retain the highest level of recognition, as fully original thinkers, down to the High Renaissance . . . In most areas an entire cosmos of Arabic sources was recast, century by century, until it had been reduced to a few visible orbits around the gleaming suns of Aristotle, Euclid, Ptolemy, Plato, Galen, and Archimedes.[19]

But traces of the earlier admiration for Muslim learning and moral character are still there to be read in Dante's *Purgatorio*.

A century and more before the fall of Byzantium, the scientific impetus in the West had reached a stage of maturity whereby scholars had become critical enough of the early Latin translations and their flaws to demand better translations and better originals from which to make them. At least this was the Latin argument for now turning from the accomplishments of infidel Muslims to these of the pagan Greeks – including living Greeks: Byzantines, the living descendants of Aristotle, Plato, Euclid, Archimedes, Appolonius, Galen and Ptolemy, as Western scholars fancifully preferred to think. While Greek manuscripts were being purchased in Constantinople, Greek scholars were invited to Oxford and Paris to teach classical Greek to Latin scholars so they could read those Greek sources.

The elevated position that the ancient Greek masters were given in late Latin civilization may have sold their Muslim teachers short, but at the same time, it showed the growing sophistication of religious and secular scholars in the 13th and 14th centuries. Though theologically and historically pagans, Aristotle, Pythagoras, Euclid, Ptolemy, Galen, Hippocrates and Archimedes were awarded by the Latin Church the status of pre-Christian patron saints of the sciences or what amounted to an honorary doctorate for venerated pagans of antiquity. A select group of them were held to have perfectly mastered one of the seven arts and sciences: Aristotle for logic, Pythagoras for music, Euclid for geometry, Ptolemy for astronomy and Galen for medicine. For grammar and rhetoric, of course it had to be Roman masters.

Reflecting on the Islamic experience and the ulema's antipathy toward logic and natural philosophy, the liberality of the medieval Church in dispensing a venerable form of honorary sainthood on pagan scientists and philosophers was a revolutionary reinterpretation of religious doctrine. By dying before the coming of Christ, virtuous pagans of reason had missed their chance of being baptized and so could not climb Mount Purgatory to be purged of sin and enter paradise. But with the Church now turned from crusading in foreign lands and busily engaged in the work of building high civilization, the scholastic theologians had no mind to deliver these geniuses to the fires of hell, and so were obliged, guided by logic of course, to create a Limbo where virtuous pagans could sit and think for eternity. Even a few virtuous Muslims were put there. Ibn Sina (Avicenna) was in this way honored; so too was Salah al-Din (Saladin), the Kurdish ruler from northern Iraq who defeated the Crusaders and restored Jerusalem to Muslim rule. This was true religious generosity. What nobler reward could the Church give in recognition of a Muslim physician's great contribution and a Muslim warrior's humane and magnanimous treatment of Crusader prisoners and faithfulness to treaties made with Christian kings?

Christian Limbo was offered to pagans and Muslims with a certain prejudicial selectivity. Aristotle got limbo but Averroes (ibn Rushd) got hellfire for providing the Latin translators and scholastics with his monumental grand commentary on Aristotle. That was ingratitude, especially since Aquinas won sainthood for using the Andalusian's commentary to rationalize scripture and give Christian theology an Aristotelian foundation. It was easier for the Church to swallow the loss of Jerusalem than to accept a philosophy that had the individual soul dissolve with the death of the body, a principle just as Aristotelian as Averroist. And yet Aristotle got to go to Limbo. But even here a gesture of Church magnanimity shows through, for though Averroes was sent to hellfire everlasting, he was nonetheless honored with the title of "The Commentator," and as such was theologically chained between Aristotle and Aquinas, assuring him the fame in the Christian West that was denied him in the Muslim East.

The glittering cast of pagan and infidel superstars waiting out resurrection and judgment in Limbo is as powerful an expression of the universalistic and integrative spirit of this period in Latin civilization as those great contributions born of that spirit: the scholastic synthesis, Dante's *Divine Comedy*, Gothic cathedrals, and the lively fermentation of scientific knowledge beginning to bubble up at Oxford, Paris, Bologna and all the universities sprouting up across the Latin world in the 13th and 14th centuries.

The wellsprings of this confidence and optimism that Church and university had in reason and science went many centuries back, to St. Augustine who embraced Plato's love of the good on the wings of Reason, to Boethius who philosophized Christianity by interpreting classical mythology in terms of Christian belief, and Cassiodorus who defined the Quadrivium and Trivium. The Church's quest for Aristotle had brought with it Archimedes, Euclid, Galen, Ptolemy and the rest of the honored pagans. That the men of the Renaissance could regard the centuries before them as sterile and unimaginative is a measure of their unbounded narcissism. Perhaps only by surveying the Islamic experience in regard to science and metaphysics can the liberal stance of the Latin Church be appreciated for what it was.

The part the Church played in the vigorous rise of science in the West evokes a speculative wonder at what the course of science might have been in Muslim civilization if the caliph Ma'mun, patron of both Mu'tazilite theology and scientific translation, had succeeded in creating a hierarchical structure of the ulema with the caliph presiding over an official orthodoxy of Mu'tazilite theology, thus precluding the Ash'arite school that came to predominate. Would ibn

Rushd have been the Thomas Aquinas of Islam? Would development of the two civilizations have been parallel? The same years al-Ghazali was cautioning against the pitfalls of logic, philosophy and mathematics, Anselm of Canterbury (1033–1109) was applying reason to scripture and theology, while Peter Abelard (1079–1142) was initiating a logical analysis of scripture. From that point on, divergence started to become more pronounced: Abelard's student, Peter Lombard, carried on his teacher's work of reconciling reason and scripture; Lombard's contemporary, Thierry of Chartres (c. 1156), rendered the biblical account of the six-day creation in terms of natural philosophy, mixing Platonic cosmology with Stoic and Aristotelian ideas; Adelard of Bath (d. 1142) claimed that everything in religion that could be explained rationally through nature should be. In his *Philosophy of the World*, William of Conches (d. 1154), tutor of the future king of England, Henry II, composed a fully worked out earthly and heavenly cosmology based on Platonic principles. Answering his Christian critics who were as opposed to this mixing of religion and philosophy as were the Ash'arites on the other side of the divide, William framed the defensive posture that the naturalist theologians would take against those who supported God as the direct and immediate cause of everything in the universe: that those who opposed science did so out of ignorance. Unwilling themselves to study nature, they wanted no one to. They desired everyone to be like them, as ignorant and uninquisitive as peasants about the causes of nature, calling heretics anyone who sought the causes of things.[20]

By the time of Albertus Magnus (d. 1280), it was considered a breach of scientific integrity to resort to divine causation in explaining nature. God may have designed and created the physical universe, but divine cause was no longer needed to comprehend its operations. The creator had been reduced to an architect who measured the world and built it as a magnificently ordered organism imbued with soul and intelligence, with man, the masterwork of creation, a microcosm of the whole.

It was not all smooth sailing for the rationalists. In the view of some important theologians in early 13th century Paris, the heavy doses of Aristotelian and Platonic philosophy being pumped into the sinews of Christianity were threatening to tear the cherished mysteries out of the heart of the faith, reducing religion to a pantheistic theosophy. The conservatives gathered enough momentum to halt temporarily the further philosophization of Christianity by having a provincial council in Paris condemn Aristotle's writings and proscribe them from the academic curriculum at the University of Paris, heart and soul of the scholastic movement. A ban was placed on teaching natural philosophy and Aristotle's metaphysics in 1215. The ban also depressed the study of the Latin versions of al-Kindi, al-Farabi and ibn Sina – and al-Ghazali, considered by Latins to have been a disciple of ibn Sina!

The scholastic movement was too pumped up to be held down for long. By 1230, protests and strikes by undergraduate and master's degree students had succeeded in restoring Aristotle to the curriculum. From then on, Aristotelian philosophy, with its Muslim commentators, was an essential ingredient in the university's intellectual diet. In the 1240s, the church ban was officially lifted. In 1255, university lectures on Aristotle were mandated as obligatory by the Church.

Supplied with Latin translations of ibn Rushd's commentaries on Aristotle and ibn Sina's Neoplatonism, Albertus Magnus and Thomas Aquinas planted the seeds of reason deep in the soil of theology. The immutable chain of reasoning that revealed nature's laws did not contradict God's omnipotent will: They resided in harmony as two concentric spheres turning in unison, the highest being God the creator, whose will directed the movement of the universe, the other being the natural laws that God created and endowed with autonomy of action. To human nature the Creator imparted a freedom of will that was free to act rationally or otherwise. Within this general schema, scholastic theology legitimized science and philosophy in the service of religion. Philosophy became the handmaiden of theology, similar to ibn Rushd's metaphor of philosophy and scripture being breasts of the same wet nurse, both of which nurtured the human soul.

To save the mystery of faith and scripture, Aquinas declared certain religious beliefs to be beyond the reach of reason. Creation, transubstantiation, the Trinity, God having a son, crucifixion of the son of God, judgment and resurrection, these mysteries existed in a supra-rational realm that had to be accepted on simple faith, and anyone who too zealously pursued reason to the point of questioning these things, or who criticized or questioned the Church's authority to establish what was beyond question and reasoning, would be in for trouble. The theologians who, Ghazali-like, suspected philosophy and reason to be too uncontrollable and ambitious to accept to remain long in the subservient role of handmaiden, rejected the limits set by Aquinas as being overly generous. Once again, the conservative wing charged the rationalists of having overstepped the bounds by their locking religion in the lion cage of reason. Reason would devour divine truth, rationalize it out of existence.

The charge was essentially the one leveled by Muslim conservatives against the Mu'tazilite scholastics four centuries earlier. In both instances there was cause for the charges. In the Latin case, Aristotelianization of religion was indeed breaking out of its restraints. Some younger extremists at the University of Paris were attempting to reduce the whole of religion to reason. Nothing was sacred. One of them was Siger of Brabant (d. 1284), a young and audacious scholar teaching in the University's faculty of liberal arts. Arrogant, brilliant, argumentative, radically rationalist, Siger was a kind of late 13th century reincarnation of Abelard returned to Paris to excite and enthrall once again the crowds of university students who flocked to his lectures. Siger called for a maximalist scholasticism that would have overrun the sacred ground of mystery by pushing as far as it would go the metaphysics of Aristotle, or rather of ibn Rushd, fount of Averroism. Nothing was to be spared by Siger and his followers, not the Trinity, the miracle of the host, the existence of heaven and hell, the divine act of creation from nothing, immortality of the soul, nothing. Everything was fair game in the crucible of reason. Anything that gave reason a problem was to be rationally reinterpreted. If the mysteries were indeed truths dressed in mystery, they would have no trouble withstanding the critical acids of logical inquiry and analysis. As it appeared to conservatives and moderates, Siger and his fanatic band of scholastic radicals were subverting religion. They were sacrificing God's power to a full blown philosophical naturalism, undoing the marvelous synthesis that had been painstakingly worked out during the century and a half between Abelard and Aquinas.[21]

In 1270, and again in 1277, conservative reactions led by theologians at the University of Paris resulted in the condemnation not only of the scholastic radicals, but of Albertus Magnus and Aquinas as well. The new doctrine dictated that Christians could no longer believe that nature was an autonomous system of secondary causes comprehensible in itself. Everything was to be understood as being directed by God as universal cause. Other matters were also forbidden to reason by the 1277 condemnation: creation, eternity, denial of personal immortality of the soul, free will and determinism. Unlike the anti-Mu'tazilite determinists in Baghdad, the Latin Christian conservatives espoused the doctrine of free will. Man was free to choose his acts just as God was free to alter nature. The symmetry was not contradictory. In conservative logic, God was the direct cause of everything, but free will prevailed. God's freedom to play with nature, for the sake of miracles, translated to man's freedom to play in the field of choice, for the sake of divine justice. Muslim orthodoxy had it the other way. God's freedom bound humankind in iron chains of determinism; free will, like cause and effect, was a phantasmagoria.

In their condemnation of 1277, the conservatives listed 219 items that were to be grounds for excommunication. One or more of these items was used to bring down the Franciscan philosopher Roger Bacon, whose innovative ideas in natural philosophy had for many years been arousing the hostility of the head of the Franciscans at the University of Paris, Giovanni di Fidanza (Bonaventure, d. 1274), moderate scholastic and future saint. Bonaventure feared Bacon's naturalist theology, but it was Bacon's stinging criticism of Church corruption, its puerile teaching methods and uncritical adherence to old authorities, and above all, Bacon's unbounded praise for the ethical values of Islam as exhibited by al-Farabi, ibn Sina and al-Ghazali, that drove Bonaventure so against Friar Bacon. That a fellow Franciscan could dare hold these Muslim thinkers up as exhibiting the spirit of true Christianity was unconscionable. Roger would have to be made to see the light.[22] Bonaventure waited for his chance to shine it in his eyes but died three years before the chance came, in the form of the Condemnation of 1277. The following year, the theologian who admired Muslims was charged with heresy by his fellow Franciscans and imprisoned. He died in prison 14 years later.[23]

As for the other firebrand at the University of Paris, Siger of Brabant, he lived by the pen and died by it, literally. Six years after Bacon's imprisonment, Siger's tempestuous life was cut short by his secretary, who in a fit of anger plunged the quill of his pen deep into the theologian's breast, penetrating his heart and killing him on the spot. With the wings of the wild bird of philosophy clipped, the radicals in Paris kept their profile low, their mouths shut and what they read secret. After a half century underground, scholasticism burst to life again with a new generation of theologians grounded in the teachings of Aristotle, Averroes, Albertus Magnus, Aquinas and Bacon. The condemnation of 1277 was reversed as it related to the writings of Aquinas. His summas were read openly.

Aquinas was canonized shortly after his death. His sainthood, in tribute to his crowning contribution to scholasticism's synthesis of scripture, metaphysics and natural philosophy, completed the Church's eschatological spectrum of Aristotelianism: Aristotle in limbo, ibn Rushd roasting in hell and Aquinas sublimely smiling down from paradise. Thanks to the 12th and 13th century scholastics, the University of Paris where they flourished and the radicals who taught there, Siger and Bacon in particular, both of whose faith in the human mind and courage to question authority kept expanding the envelope of intellectual freedom, critical inquiry became a permanent feature of the Western intellectual tradition.

Though it turned out to be a hiccup in the rational formulation of theology, the Condemnation of 1277 had struck the liberal community of scholars at the University of Paris like a bolt of lightning. It destroyed some of the extreme rationalists and scared the moderates. The memory of it remained as a warning. When concluding a lecture or treatise on natural philosophy and religion, rationalist theologians, always fearing they may have unwittingly trod into the forbidden garden of mystery, used formulas such as "But this must be wrong since it goes against what the Holy Church has prescribed," as Muslims protected themselves by quoting a verse from the Quran to preserve the appearance of faithfulness to orthodoxy.

There is perhaps another side to the 1277 Condemnation: its proclamation that several worlds could exist, and that in one of them it was possible for the heavenly spheres to follow rectilinear motion rather than circular. This was intended to restore God's power of miracles. But according to some scholars what it did was separate astronomy, and hence natural science, from the ancient principles of Aristotle and Ptolemy. This in time allowed for the embryo of modern science to develop.[24]

During the 13th century, natural philosophy had more than come into its own. At the University of Paris its principles permeated the sacred lecture rooms of theology. This was not because of the intellectual belligerence of philosophers in the College of Arts and Sciences. The philosophy professors knew from experience to steer their reasoned arguments clear of the lethal minefields of theological issues, where the mind-bending fundaments of eternity, material creation out of nothing, the miracle of transubstantiation, the trinity, or the possibility of God being able to create a vacuum, were declared by theologians to defy reason and were jealously guarded by the College of Theology as off-limits to all but its own professionals. There indeed existed a lively rivalry between the College of Arts and Sciences and the College of Theology, but it was the theologians themselves who brought natural philosophy into religion, not the philosophers.[25] Like the ulema in Islam, the Latin theologians wanted to lay claim to all knowledge and were aggressive in doing it.

In the early 14th century a modus vivendi of sorts was reached between faith and reason. This was through the efforts of two pragmatic Franciscan theologians, the Scotsman Duns Scotus (d. 1308) and the Englishman William of Ockham (d. 1349). Theology and philosophy were perceived to exist in their own separate domains of inquiry and analysis, but not altogether separate. An overlapping intermediate zone allowed philosophy and theology to merge in a pursuit of knowledge that was considered neutral, neither sacred nor profane. Duns Scotus and William of Ockman produced epistemologies that cut natural philosophy loose from theology. Science was seen to be independent of, but in harmony with, religion. Natural philosophy's divorce from the divine was a vital step toward the secularization of science. It was also a vital step toward freedom of thought and all other freedoms associated with Western civilization.

Such were the long-term consequences of the translations. The translation of Muslim science, and the Greek heritage it embodied in Arabic, like a palimpsest of Arabic over Greek, now written over again, this time in Latin, established the focus and framework and defined the problems of scientific study and debate in the West right up to the 17th century and the Scientific Revolution.

Notes

1 J.W. Thompson, "The Introduction of Arabic Science Into Lorraine in the Tenth century," *Isis*, 12, no. 2, 1929, pp. 184–193.
2 Joseph Puig, "The Transmission and Reception of Arabic Philosophy in Christian Spain," in *The Introduction of Arab Philosophy into Europe*, edited by C.E. Butterworth and B.A. Kessel, Brill, Leiden, the Netherlands. 1994.
3 According to an account of his life by the 12th century physician "Magister Mattheus F." of Salerno, Constantine was a Muslim merchant in Carthage who came to Salerno on a business matter. He then returned to north Africa and translated medical texts from Arabic to Latin. He converted to Christianity and became a monk in the Benedictine Abbey of Monte Cassino, where he continued making medical translations. John Freely, *Aladdin's Lamp: How Greek Science Came to Europe Through the Islamic World*, Vintage Press, New York, 2009, p. 122.
4 In addition to Arabic versions of Hippocratic and Galenic treatises, Constantine translated the encyclopedic compendium of Ali ibn Abbas (Haly Abbas). The title Constantine gave to his translation of ibn Abbas's *Kitab Kamal al-Sina' at al-Tibbiyya* (*The Complete Art of Medicine*) was *Pantechna*, a Greek rather than Latin rendering of the Arabic title. Why he chose a Greek title for a Latin translation of an Arabic work is another mystery, considering that he gave Latin titles to the Greek medical texts he found and translated in Salerno.

  Constantine's work in the Salerno school added greatly to its reputation. Medical students from all over the Latin West are said to have flocked to it. A century and a half after Constantine's death, the Holy Roman emperor Fredrick II ordered that any physician practicing in the empire had to stand for the Salerno medical exam.
5 Arab accounts of the practices of Frankish physicians who accompanied the Crusades make Latin medicine appear practically non-existent and its methods as brutal as they were irrational. Their cure for a broken leg? Hack it off with an axe above the fracture. Treatment for headache? Cut a hole in the scalp to let out the evil demon. Few patients survived the surgeon's operating table. For a humorous, and horrific, description of Frankish surgery from a Muslim point of view see Philip Hitti, *Chronicle of an Arab-Syrian Gentleman, and Warrior in the Period of the Crusades: Memoirs of Usamah ibn Munqidh*, Columbia University Press, New York, 1929, pp. 161–170.
6 Acre, 1291, to the Mamluk Sultanate ruling in Egypt and Syria, Sultan Qala'un finishing what the Seljuq atabey Imad al-Din al-Zangi of Mosul had begun in 1144.
7 Though the importance of Raymond's institute has been questioned. Montgomery Scott, *Science in Translation*, University of Chicago Press, Chicago, 2000, p. 143. Toledo was

not the only center of translation in Spain, however. There were several others, Saragossa, Barcelona, Pamplona, Segovia and Leon.

8 The difference was that in Islam, with very few exceptions, the ulema did not participate in natural philosophy or share in the view that it was a source of wisdom. Archbishop Raymond's Church-funded Toledo school of translation was the closest thing in Medieval Latin Christendom to Baghdad's House of Wisdom, which was funded by a political ruler and staffed by eastern Christians. The ulema had little to do with it. The institution was purely a product of royal patronage.

9 Scott, *Science in Translation*, pp. 166–167.

10 Adelard and Herman made their Latinized Quran at the request of Peter the Venerable, Abbe of Cluny. Peter was a singularity in that he engaged himself in both great movements of the time, translation and crusades, his Latinized Quran being intended to refute Islam on the logical, literary and propagandistic front in Christendom's conflict with its monotheistic rival.

11 Henri Hugonnard-Roche, "The Influence of Arabic Astronomy in the Medieval West," *Encyclopedia of the History of Arabic Science*, vol. 1, edited by Roshdi Rashed, London and New York, Routledge, p. 285.

12 Scott, *Science in Translation*, p. 156.

13 His *Canon* especially was honored as a model of translating excellence and became one of the most esteemed texts in Latin medicine. No less than 60 editions of it were made between 1500 and 1674. Printed and reprinted as a basic medical text in the universities well into the 18th century, the text was in fact being read at the University of Padua medical school up until 1767, showing not only the respect European physicians had for ibn Sina and Gerard's translation, but also the lack of medical progress made in Europe relative to the other sciences.

14 Ibn Sina's *Kitab al-Shifa* was first translated by ibn Da'ud to Castillon and then translated from that to Latin by Gundisalvus. For its importance as the introduction of Muslim learning to the Latins, see Alain de Libera, *Penser au Moyen Age*, Seuil, Paris, 1991, pp. 111–112. The Jews in Spain, collectively known as Sephardim, were an intellectually creative community that had contributed substantial works in science and philosophy in both Hebrew and Arabic. One of the most renowned of the Sephardim was ibn Maymun (Maimonides), a contemporary of his fellow Cordovan, ibn Rushd, both of them learned legal scholars, philosophers and physicians. Ibn Maymun's *Guide to the Perplexed*, which harmonized Judaism with science and philosophy, paralleling ibn Rushd's reconciliation of Islam and Aristotle, was written in Arabic, translated to Hebrew and later to Latin.

15 A zealot in the cause of scientific knowledge, Fredrick II was reported to have conducted some rather brutal experiments, such as sealing a man up until death in order to detect the presence of his soul; cutting men open after their having eaten to observe the action of digestion; isolating a newborn baby to find what language it would eventually speak. Libera, *Penser au Moyen Age*, pp. 169–177.

16 Charles Haskins, *Studies in the History of Medieval Science*, Harvard Press, Cambridge, 1927, 156.

17 The Catalan physician Arnarld of Villanova, for example, relied on ibn Sina's *Canon* in his practice of medicine but said that he in fact preferred Razi's *Continens* because its Latin translation explained Galen more clearly.

18 Scott, *Science in Translation*, p. 171.

19 Scott, *Science in Translation*, pp. 178–179.

20 Scott, *Science in Translation*, 146.

21 For the life and thought of Siger see Tony Dodd, *The Life and Thought of Siger of Brabant, Thirteenth century Parisian Philosopher: An Examination of His Views on the Relationship of Philosophy and Theology*, E. Mellen Press, Lewiston, New York, 1998.

22 Roger Bacon's impassioned castigating of the Church for its moral and intellectual corruption, while praising Muslim virtue, scandalized even the liberal faction of the

Franciscans. Muslims who were better Christians than Christians! Six centuries later, reforming Muslims would be claiming the makers of modern Western civilization to be more faithful to Islamic principles and truer Muslims in spirit than those born into the religion. In this way, Islamic virtue was credited for the success of the modern West.

23 *The Opus Majus of Roger Bacon*, edited by J.H. Bridges, Clarendon Press, London and Frankfurt, 1964, vol. I, xxxi.

24 Arun Bala, *The Dialogue of Civilization in the Birth of Modern Science*, Palgrave Macmillan, Basingstoke, 2006, p. 121. The author traces the theological parentage of the proclamation to the translations from Arabic, particularly of al-Ghazali's works. The Condemnation as a whole is certainly a delayed result of the translations and the rational knowledge they sent streaming into Latin academic circles. As for al-Ghazali, it was, according to Bala (p. 122), his belief in God's power to change the order of the world as God willed that influenced the thinking in the College of Theology at the University of Paris in the 1270's, moving the conservative Latin theologians to an Ash'ari position.

25 Edward Grant, *Science and Religion, From Aristotle to Copernicus*, Johns Hopkins University Press, Baltimore, 2004, pp. 176–190.

6 Latin assimilation and ascendancy

Assimilation

The Toledo School functioned for a century and a half. By the time it closed its doors in 1284, the heart of the Arabic corpus of mathematics, philosophy, science and medicine had been translated. The heart, but not the whole body. Astronomy and mathematics continued to be pursued creatively in a few regions of Islamdom into the 16th century. Much of this achievement was not translated until later and would not be fully assimilated in the West, not even by the time of Copernicus. There were some features of Maraghan astronomy the Polish astronomer seems not to have figured out.

The heady stream of Muslim knowledge introduced into the Latin world took time to be digested. It contained much that challenged the Latin scholars, such as, for example, the trigonometric algorithm to determine parallax constructed by the mathematician–astronomer Habash al-Hasib. But just as young ibn Sina was at last able to master Aristotle's metaphysics by repeated readings, the Latins mastered Muslim mathematics and with it Habash al-Hasib's trigonometric algorithm for parallax, which in later centuries was renamed "Kepler's Equation." Copies of Gerard of Cremona's translation of al-Zarqalli's Toledan Tables were diffused widely through the Latin scientific community, as were the tables of al-Khwarizmi, introducing Latins to Ptolemaic astronomy and observational techniques. The Muslim craft of making astronomical equipment, such as quadrants, sextants, armillary spheres and astrolabes and their use in stereographic projection was mastered during the 13th century.[1]

The translation of star catalogs resulted in Latinized Arabic names being adopted for many stars, though many were later replaced by Greek and Latin names. The sine relations for astronomical computations on a sphere using the triangulation of great circles, developed by 10th century Muslim astronomers in Iran, were learned and gradually adopted through the translation of the 12th century Andalusian astronomer Jabir ibn al-Aflah's *Islah al-Majisti* (*Reform of the Almagest*); though it was not until centuries later when Austrian astronomer–mathematician Regiomontanus wrote his popular *de Triangulis*, based on the translation of al-Aflah's book, that the relationship would come into general use. The theorem in Latin was named after the author of the source from which it was learned, Jabir becoming Geber. Translation and assimilation of this new knowledge was gradually making Latin the third major language of science after Greek and Arabic.

In a manner approximating the Muslim experience, the work of translation, assimilation and early originality overlapped, as seen through the scholarship of Leonardo of Pisa, Sacrobosco, Robert Grosseteste, Albertus Magnus, Thomas Aquinas, Roger Bacon and Duns Scotus, whose lives spanned the last half of the 13th century. While Leonardo was learning Muslim mathematics, working out his Fibonacci sequence and making original contributions in theoretical geometry and algebra, the Augustinian monk John of Holywood (known by the Latinized version of his name, Sacrobosco) was at the University of Paris writing student texts on elementary algebra, geometry and astronomy based on the Latin translations of Muslim texts in these fields by Gerard of Cremona, Robert of Chester, Adelard of Bath and Michael the Scotsman. For having assimilated the rudiments of Muslim science, Sacrobosco was at the time considered the leading professor of mathematics at the University of Paris and the West's most famous astronomer. Yet the Latins still had a long way to go before attaining the mathematical and observational expertise reached by the best Muslim scientists.

Across the channel, Sacrobosco was rivaled in mathematics by his fellow countryman and contemporary, Oxford's Robert Grosseteste. Sacrobosco had studied at Oxford, which, like the University of Paris, had grown within the span of the translation period from a divinity school to a university that would rival Paris in the 14th century. Theologian and professor of philosophy and mathematics, Grosseteste wrote commentaries on the Bible, on Aristotle's Physics and Posterior Analytics, studied optics, and prepared the way for the scholastic synthesis that would be completed in Paris by the Dominicans and Franciscans. While assimilating the new knowledge coming in with the translations, he engaged in experimental science that provided the groundwork for what would become his country's empirical tradition. Based on his study of the translations of ibn al-Haytham's work, his treatise on optics began the Latin tradition of the geometric analysis of light, with a hint of originality. Reporting on his study of lenses he envisioned the possibility of the telescope and the microscope:

> This part of optics, when well understood, shows us how we may make things a very long distance off appear as if placed very close . . . and how we may make small things placed at a distance appear any size we want, so that it may be possible for us to read the smallest letters at incredible distances or to count grains of sand, or seeds, or any sort of minute objects.[2]

Light was for Grosseteste more than a natural phenomenon. It was spiritual. It illuminated a reality beyond the natural realm. His theology, heavily Neoplatonic, was akin to the illuminationism of ibn Sina and Nasir al-Din al-Tusi. Nasir al-Din was in fact Grosseteste's contemporary: two philosopher–scientists, one in Iran, the other in England, both simultaneously propounding an epistemology of light. According to Grosseteste, light was the first thing to have emanated from unformed primal matter. Beginning as a dimensionless point and spherically expanding to the geometric limits of the universe, that is, to the sphere of the fixed stars, light was God's instrument in universal creation. Light mediated communication between body and soul and energized the senses that gave knowledge of the physical world, just as the soul gave knowledge of the divine world. Being fundamental to physical and divine knowledge, the study of light for Grosseteste was as much a religious as a scientific exercise, oriented as much to theology and philosophy as to geometry. Nonetheless, for all his mathematical ability and for all his work on ibn al-Haytham's optical studies, credit for founding a purely mathematical optics in the Latin West is not generally attributed to Grosseteste, but to a Polish scientist a generation later, Witelo (d. 1281).

Another one of those pioneering professors of science at the prestigious University of Paris, Witelo studied light without imputing to it any spiritual or philosophical attributes. His work shows he had assimilated the optics of ibn al-Haytham.[3] Shortly after publication of Witelo's book, a third important work on optics appeared: John Pecham's *Perspectiva Communis*. This was little more than a summary of Witelo's book, itself basically a recapitulation of ibn al-Haytham's *Manazira*. Haytham's and Witelo's work, and to a lesser extent Pecham's, continued for centuries to be the authorities for the study of optics in the Latin universities. Not until the 17th century can the study of optics in the West be considered to have definitely advanced beyond its inherited Islamo–Hellenistic legacy.

Assimilation of the Aristotelian inheritance and its ultimate achievement in a scholastic synthesis came much sooner. Aristotle had been the coveted gem at the heart of the translation movement. The application of Aristotelian logic and natural philosophy to religion was what scholastic theology was all about and was intensely pursued by Latin Christendom's best minds. But it brought more than that. Without the search for Aristotle in Toledo and the scholastic application of Aristotelian philosophy to theology in Paris and Oxford, there would have been no optics, astronomy, mathematics, physics and medicine. Having accompanied the sought-after works of the master philosopher, the sciences were used to complete the cosmic portrait. Astronomy revealed how God designed and moved the heavenly bodies; medicine provided health for the body; and theology, queen of the academic disciplines, provided health for the soul and, above all, explained how God worked. Aristotelian philosophy, theology's bride, delivered natural philosophy in the aura of religion. Scholasticism and science arrived hand in hand.

Aristotle in Latin Christendom was institutionalized by the German Dominican Albertus Magnus (d. 1280). For this, Albertus was called "The Great," the only theologian and scholar to be called so. He enjoyed the title even while he was alive, though his soul had to wait until 1931 before the Church pronounced it to be saintly. Before becoming a Dominican, Albertus had studied the liberal arts in Padua. This prepared him for the theological and scientific work he would accomplish in midlife. As a professor of theology at that bastion of liberal thought, the University of Paris, he imbibed Aristotle from the recently made translations, in particular the translations of the commentaries of ibn Rushd, whose death preceded Albertus' birth by only two years, so close on the heels of Muslim accomplishment were the Latins rapidly following. With Albertus, the fruits of the century-long labors of translating Aristotle from Arabic were being harvested.

Albertus's devotion to "the Philosopher" was not without critical limits. The knowledge of science he gained at Padua endowed him with the sense to reject Aristotle where ibn Rushd had not: in the latter's acceptance of Aristotle's astronomy of concentric spheres over Ptolemy's eccentrics and epicycles.[4] Beginning in the 1240s, by which time he had mastered the knowledge available in the translations, he devoted himself to writing commentaries and paraphrases on Aristotle's natural philosophy and metaphysics, a labor of love that lasted over 20 years and 8,000 pages. His interpretation of Christian doctrine and scripture in the light of Aristotle, with dashes of soul-saving Neoplatonism where needed, practically concluded the task scholastics had set for themselves. In his commentary on Aristotle's *Physics*, he wrote that his purpose was to provide the brethren of his order with a book that would help them understand the science of nature and all the philosophy of Aristotle – logic, rhetoric, poetics, politics, economics and metaphysics. The explicit assumption was that nothing in natural philosophy, correctly understood, could be contrary to religious belief; its metaphysical corollary was that a Christian theology was impossible without a sound understanding of Aristotle. Revelation and faith led to the same truth as philosophy and science. Wherever Aristotle's principles gave trouble to Christian sensitivities, such as belief in the eternity of the universe and death of the individual soul along with the a physical demise of the body, Albertus simply substituted the appropriate parts of ibn Sina's Neoplatonic philosophy. As smoothly as it had served Muslim thinkers who philosophized their scriptural religion, Neoplatonism served their Christian counterparts in structuring a rationalized theology.

At the heart of Albertus' cosmology was the omnipotent first cause running the natural world through a system of secondary causes. Following his rationalist predecessors of the previous century, particularly Adelard of Bath, William of Conches and Thierry of Chartres, Albertus posited God as the author of creation and everything that occurred in it. Between God and nature, a set of natural causes accounted for everything under the sun, including the Biblical miracles: the flood of Genesis, the Red Sea parting of Exodus, the sun stopping for Joshua, and so forth. Nature operated in accordance with secondary causes. Their study and explanation constituted the work of philosophers. Theologians devoted themselves to the first cause and God's will. The boundary of causes between primary and secondary should never be crossed. Natural science was not to be confused with theology and its metaphysics. They were two entirely different things. Introducing divine causality to explain natural philosophy was to Albertus Magnus a violation comparable to, in today's world of science, introducing "Intelligent Design" to explain evolution. God certainly designed the universe intelligently, but secondary causes were to be understood on their own and not referenced to any agent operating outside the box. His interest in the branches of philosophy below the metaphysical level was for the service they could render to theology, but the formulation of theology should have nothing to do with secondary causes. Here was the seed of that divide between religion and secular science that a few centuries hence would become decisive and final with Galileo.

As a student of natural philosophy at Padua in his early years, before going theological, Albertus had valued science as something in itself, and this appreciation, perhaps inspired by his study of Aristotle who combined metaphysician and naturalist, was never quite lost. Science, he believed, could by analogy address theological questions. Just as Aristotle had observed the evolution of the chick from its embryonic form through a window he had made in the shell, Albertus observed animal embryos in a quest to find the moment the soul entered a human embryo. His study was itself an embryo in its introduction to the Latins of experimentation, observation, and the description and classification of results.

At the tail end of this second remarkable century of Latin Renaissance, with its assimilation of Greco–Muslim thought and progressive rationalization of cosmological theology, came the singular Franciscan friar Roger Bacon, scientist, theologian, philosopher and visionary. Combining the intellectual insights of both his great predecessors, Grosseteste and Albertus, Bacon added to them a prophetic vision of technological wonder. His advocacy of experimentation in investigating nature contributed to the work of Albertus in laying the basis of the modern scientific method, while his futuristic ideas in applied science – reading glasses, submarine, gunpowder weapons, automobile, airplane, power driven ship and weight lifting machine – rivaled those visions captured in the sketch books of Leonardo da Vinci three centuries later.

Roger Bacon was a fitting end to an intellectually dynamic century. Educated at Oxford, where he studied under Robert Grosseteste, and then Paris, where he lectured in the Faculty of Arts on Aristotle and was a colleague of Albertus Magnus, Bonaventure and Aquinas, he became a star among the brightest scholastics, and a polymath of sorts: prolific philosopher, competent mathematician, less of an astronomer, but thanks to his student days studying optics under Grosseteste at Oxford, an experience that endured the whole of his life, he became a leading theorizer in it. Bacon's optical works appealed to a public beyond the academic community and circulated widely among the steadily expanding literate population, spreading the name and knowledge of al-Kindi and ibn al-Haytham. His own theory on optics was a syncretism of the extramission theory of Plato, Euclid, Ptolemy and al-Kindi on the one hand and of al-Haytham's intromission theory on the other. Bacon's quintessential scholastic mind would not accept what to the ordinary observer would appear to be simply a contradiction in theories, one of which would have to be wrong. Both theories had to be right. And so like a true scholastic trained in logically synthesizing what might to others look to be contradictions, he took each theory to describe an aspect of the physical nature of light, and resolved the dualism in a unifying geometry of extramission and intromission – a medieval version of today's experimental conundrum of light as particle and wave.

In the course of his optical studies, he explained the geometrical optics of focal point, center of curvature, refraction of light passing through convex lenses, projection of rays and magnification – seeing distant objects as if they were near. Yet for all his imaginative descriptions of future machines, he did not apply his knowledge of optics to craft something useful, like a telescope, or even simpler, reading glasses. He may have lacked the skill of hand of a Galileo to assemble the lenses – or even to grind them. The combination of optical science and craftsmanship was centuries away, although reading glasses, first appearing in Vienna, became a reality toward the end of Bacon's lifetime.

As much as he was a sublime theorist, he did nonetheless pursue experimental alchemy. And he believed in astrology. Searching to unify the physics of heaven and earth as he thought he did optical theory, he anticipated William Gilbert some three centuries by likening planetary influences to a giant magnet suspended over an earthly field of iron. Certain configurations of planetary positions could, like the attraction and repulsion of a magnet, produce influencing forces. His epistemology was no less innovative than his theories. It was the critical nature of his epistemology that offended his Franciscan superiors at the University of Paris.

Bacon was continually under investigation. Bonaventure resented that he had his books published without first submitting them to him for the Franciscan Order's approval. Bonaventure also suspected that Bacon's philosophy exceeded the limits established by Albertus and Aquinas. But above all, what Bonaventure and other Franciscan scholars in Paris most resented in Bacon was his biting criticisms of the archaic and ineffectual methods of education practiced by the Franciscans and Dominicans. His ardent advocacy of educational reform rankled establishment scholars and alienated his Franciscan superiors by calling their pedagogy, in so many words, stagnant if not regressive. He openly criticized his fellow scholastics for believing certain books to be authoritative when they were nothing of the sort. He pointed out their lack of critical questioning that allowed ignorance dressed in eloquence to pass as wisdom, nor did he hide his impatience with their fear that kept the true understanding of scripture from being reached through reason.

Bacon's theory of knowledge departed radically from the epistemology found in the scholasticism of Albertus and Aquinas, even from Siger of Brabant, that other firebrand of extreme rationalism at the University of Paris who advocated dipping even the Trinity and transubstantiation into the acids of analytical reason. Knowledge according to Bacon came from reason working through experience. Experience was of two kinds. One was that of external sense perception that could be enhanced by instruments that measured physical nature, expressed in the precise language of mathematics. The other was that of internal perception which mystically illuminated the soul and lifted it to higher levels of consciousness. Natural science, the product of the former, led to knowledge of the creator through knowledge of creation, at which point both sources of knowledge, the external and internal, became one.

Mathematics, optics and experimental science were all necessary elements in reaching natural truth. Also necessary were Hebrew, Greek and Arabic languages: how else could one study the original sources of scripture, science and mathematics? He called for these languages to be required study not only to reach science and scripture but also to understand Judaism and Islam. Since Christianity encompassed only a small fraction of humanity, common ground between Christianity, Judaism and Islam had to be identified to provide a bridge for the peaceful conversion of Jews and Muslims, using rational argument based on a common core of religious identity instead of the senseless violence of crusading wars and slaughter. The word, not the sword, would triumph. Winning converts was important to Friar Bacon, but winning them through reason, that universal language of mankind, not by brute force, was the only true way to win a soul. His recognition of a core of belief held mutually by Christians, Jews and Muslims was for his time as remarkable as his scientific theories and futuristic vision. A common core of religious identity in Judaism, Islam and Christianity? The Jews of Judaism who killed Christ and the Muslims of Islam who seized from Christendom Anatolia, Syria, the Holy Land, north Africa and Spain? Little wonder his Franciscan superiors wanted to shut him up.

Bacon's work indicates that toward the end of the 13th century Latin knowledge in many areas of science and philosophy was upwardly converging with the Muslim standard. Muslim mathematics and astronomy would continue advancing for another two centuries, but overall the Latins had reached what might be considered a certain commonality with Muslim learning. Bacon's study and theory of optics, and the illuminationist thought related to his optics, bare striking similarity to that of his Muslim scientific contemporaries, who were at the time philosophizing on light while reforming Ptolemaic astronomy in Iranian Maragha. Nasir al-Din al-Tusi and Qutb al-Din al-Shirazi were, like Albertus Magnus and Roger Bacon, mystically inclined scientists for whom light was the source of spiritual wisdom and physical knowledge. Light symbolized the truth of both worlds. The inner light of the soul illuminated by faith and self-reflection shone mystically on truths that transcended the physical world of sense perception, the world in which it was optics that illuminated reality. Following his teacher, Grosseteste, Bacon theorized that the geometric expansion of light underlied all physical reality, while Qutb al-Din, in his *Sharh Hikmat al-Ishraq*, essentially an explication of illuminationist theosophy, described light as the source of all physical and celestial motion.

In regard to the purely scientific study of light, while ibn al-Haytham's experimentalism was being revived in the Muslim east by Kemal al-Din al-Farisi's recension of his *Book on Optics*, Grosseteste, Bacon, Witelo and Theodore of Freiburg, were well on their way to mastering ibn al-Haytham's optical studies. Though the celestial theosophy of light in Ishraqi illuminationism endured as a religious path to higher consciousness of divine wisdom (*hikma*), al-Farisi's experimental studies would find few followers to keep optics alive and advancing in the East. In the West also, the level of optical science would for centuries remain where the 13th century scholars working on the Latin translations of Muslim optics had brought it.

Around 1300, Islam and the West had much in common intellectually, though there was still a good deal of Muslim intellectual property that the Latins had not yet appropriated. It would be a couple of centuries more before a point of convergence was reached where science in the West was roughly at the level Muslims had achieved. But when that point of convergence was reached and the West had nothing left to take from Muslim science, the trajectory of scientific production in the West would be headed steadily upwards, while in the world of Islam it would slope downward: alive but not advancing at the rate of earlier centuries.

Given the commonality of philosophic, scientific and religious background shared by their respective civilizations, the similarity of thought shared by the contemporaries Roger Bacon and Qutb al-Din, or a generation later, by the contemporaries Kamal al-Din al-Farisi and Theodore of Freiburg, is not all that surprising. The two civilizations had reached an approximate intellectual parity. This would last at least until the late 15th century, after which the scales of scientific knowledge (excluding astronomy and mathematics) would begin to tilt, convergence turning to divergence during the first blush of the Florentine Renaissance. The West would not reach the level of Muslim astronomy until the 16th century, during the lifetime of Copernicus, but thereafter the divergence would steadily increase.

Bacon spoke presciently, prophetically in fact, of the practical application of science. Practicality would join hands with experimentalism, that is, with the hands-on experience gained from craftsmanship that could be applied to simulating, manipulating and mechanically squeezing nature into the chokehold of experiment to cough up answers. In this the West would diverge sharply from its Muslim rival, as sharply as the difference between them in regard to religious and public investment in science education. The earliest example of this aspect of divergence can be seen in the optical studies of the 13th century and in the speculation on possible applications. By experimentally studying convex lenses and observing the gathering of light rays concentrated on an object placed at the focal point of the lens, Grosseteste and Bacon were able to consider the use to which the magnifying power of a lens could be put as a reading aid. Practicality and usefulness were aspects of what Bacon was referring to when he wrote that civilization grows strong with science. His knowledge of mathematics and astronomy was nowhere near that of his Maraghan contemporaries, but his technical imagination was of another order, leading him to speculate on the production of explosive powder, perpetual lamps, and the other machines previously mentioned. He believed science was an ongoing process of one discovery leading to the next. He believed its practical applications would be revolutionary.

Ascendancy

The impressive rise of Latin science was based on a solid foundation of educational, technological and economic expansion. The expansion began during the last half of the 11th century and continued uninterruptedly, in spite of the 14th and 15th century setbacks of war, bubonic plague, bad weather, famine and peasant rebellion. Technology fed into educational advance. Paper manufacturing, learned from Muslims in Spain in the 12th century, made books several times less expensive. The printing press, invented in the middle of the 15th century, did the same. These inventions made learning more widespread and, hence, religious and political authority less critically impervious to questioning. The compass and artillery, other late medieval innovations that, like paper and the rudiments of printing also originating in China, made overseas navigation less hazardous and gunpowder ballistics, respectively, a science in demand, along with, by extension, the study of bodies in motion.

Economically, the West had been gaining on Islamdom since the middle of the 11th century, when windmills and watermills began increasing production by replacing animal and manual labor in crushing grain, pressing olives, sawing wood, and bellowing iron forges. Productivity in paper manufacturing was multiplied by the introduction of mills to crush the pulp, replacing the traditional method of animal power that continued for centuries in the East. The invention of the heavy moldboard plow which rendered the soil of northern Europe more amenable to cultivation increased agricultural productivity. The 3-field system of crop rotation was introduced, as was the cultivation of new crops such as barley, oats and legumes, adding richly to the heretofore meager European diet. Other new crops such as clover and alfalfa, fixed the soil's nitrogen so that fields no longer needed to lie fallow. The invention of the horse collar and tandem harnessing increased the amount of work a peasant could accomplish in a day. Year by year over the decades between 1050 and 1300, cultivable land was expanded with the clearing of forests and draining of swamps. Combined with the technological innovations, new crops and methods of farming, the expansion culminated in what amounted to an agricultural revolution.

On one hand, this led to an improvement in diet, health and longevity, and on the other a weakening of the feudal system by the emergence of local market economies. The continuing pressure of growing population, agricultural output, urbanization and regional specialization in food products and manufactures expanded the markets from local to provincial, then to national and international. When agricultural production exceeded local consumption, the surplus served to free the serfs from the land, some buying their freedom with their surplus and going off to explore new possibilities of life in the growing cities, others becoming free farmers by buying or renting land from the lords of the manor, who were finding that selling or renting was more profitable than holding their serfs to vassalage. Some serfs ran away to the city. If their lords did not find them within a year and a day they were legally free.

Cities increasingly became the focus of economic activity as the population increased and markets and trade expanded on an international scale, with a moneyed economy of commerce and manufacturing steadily gaining on the old manorial economy based on land. As the manorial system declined, its companion institution of feudal relationships that contractually bound vassal and lord in a hierarchical political structure was squeezed almost out of existence over the centuries. Agricultural surplus, commercial expansion and increasing manufacturing supported a burgeoning urban life with a population that doubled and trebled in the 12th and 13th centuries. By the middle of the 13th century, Italian trading cities such as Genoa, Naples, Livorno and the maritime giant, Venice, were establishing commercial relationships with North Africa and Asia. Leonardo of Pisa's father was a merchant established in modern day Tunisia. Marco Polo's journey to Mongol China was only one of many commercial and diplomatic missions opening trade between Venice and the East and reviving the old silk route. To expedite and

expand trade and manufacturing new business devices and institutions were created in the 14th century: letters of credit, clearing houses, double entry accounting,
insurance companies and family-owned banks with branches in major cities.

An early sign of Latin vigor had been the Crusades. Economic opportunity
within the borders of Latin Christendom redirected the social vigor of young men
from overseas military adventures. By the 13th century, crusading had become
an anachronism. European energies were so consumed by internal development
that the Crusader states were left to waste away in neglect. As the energetic and
enterprising bourgeoisie asserted its place in the economic and political life of the
West, successful merchants became bankers to rulers and gained titles of nobility
through wealth and service to royalty. In the independent trading city-states of
Italy and the Hanseatic League on the North Sea coast, wealth, not lineage, made
for nobility. Even the withering spirit of crusading was put to commercial enterprise. A Crusade led by Italian city-states at the beginning of the 13th century
targeted Constantinople in a successful grab to take over Byzantine trade. The
Crusade had been originally meant for Egypt, but the trade of Italian cities with
that country under Salah al-Din's Ayyubid dynasty was too lucrative to be wasted
on a religious war.

A standard coinage was coming into use for trade, analogous to the standard
gold dirham and silver dinar that stood as the recognized international currency of
high Abbasid times. European society was being transformed from a landed feudal
to a moneyed commercial economy. Islamic society had been earlier devolving
into a military landholding economy, reminiscent of what Latin society was now
leaving behind.

The changes that so oppositely transformed Muslim and Latin societies from one
another was a curious turnabout. Islam was founded by a merchant prophet in a
merchant society. The Quran contains many references to business. Worldly, business oriented, forged in war and triumph, Islam, from the point of view of its scripture and the career of its founding Prophet, would of all religious communities prior
to Calvinists, Quakers and Mormons be the one considered most congenial to material success and enjoyment of the physical pleasures that success provides. Commercially oriented, commanding communal and individual responsibility and duty
in protecting and caring for the community, there was in Islam's genesis no room for
asceticism or isolated life in personal meditation apart from the community of believers. The Prophet was at first a merchant, then a warrior, a statesman and legislator.
The Umayyad caliphate stemmed from a wealthy family of merchants; the Abbasids
who followed were rulers of an empire whose merchants carried trade from Baghdad
to China, the Maldives and islands of Southeast Asia in one direction; to the western
Mediterranean in another; to Scandinavia in the north; and Central Africa in the
south. War, conquest, material wealth and trade joined comfortably in the Islamic
ethos. The merchant and warrior were equally honored figures in Muslim society.
Islam's dominant position in the central lands from Spain to China would seem to
have destined global commerce to become a Muslim monopoly.

Early Christianity on the other hand rejected riches and war. Wealth was a burden that turned one from what was essential – salvation. The believer was told to

forgive, love and suffer in a world that was about to pass away. The physical world and its pitfalls were to be avoided, material success rejected. Taken at the face value of its professed principles, Christianity looked to be a poor religion for an empire. Once becoming the Roman religion of state, however, Christianity accommodated itself to the exigencies of war and enjoyment of the riches war brought to those in position to reap them.

As the West began its ascent, the Seljuqs in Iran, Iraq and Syria, were expanding the system of military land grants; the Fatimids in Egypt, after having dominated Mediterranean trade, were in decline. The commerce-oriented caliphs in Baghdad, Cairo and Cordova of centuries earlier, with their civil governments ruling collectively over an Afro–Asian empire girded by international trade and a highly regarded currency, was by the 12th century a nostalgic memory of grander days. The transformations are poignant. While al-Ghazali was speaking for an otherworldly Sufism of inward withdrawal, meditation and universal love that had been transforming mainstream Islam for over a century, the restless young men of the West were sharpening their swords and donning their Crusader robes to kill and conquer in the name of Christ. Instead of deploring these invading Latin knights seizing Jerusalem and coastal Syria and Palestine, the venerated theologian of Islam wrote a treatise identifying Sufism with Christian otherworldly love.

Islamdom's transformation, from a moneyed, mercantile and urban society united in a vast empire to a fragmented military landholding society, opened the way for the penetration of Western trade. As early as the Fatimids, Muslim potentates were issuing certificates of trade privileges to the merchants of Venice and other Italian maritime cities. In addition to a favored rate of customs duties, the merchants would later be privileged with the extraterritorial protection of their consuls general. Salah al-Din, who supplanted the Fatimids but failed to maintain their commercial fleet and their focus on international trade, boasted of using the profits he made on Frankish merchants to buy weapons to be used against their crusaders. Had Muslim merchants set up trade outposts in the West as Europeans were doing in their world, they might have become aware of, and shared in, the great changes that were transforming their Western rival. Who knew, but centuries later the economic and technological elements that informed that transformation would have come to be considered not purely "Western" but just that much more common property in the already considerable heritage that Muslim civilization and the West shared in their history of cultural and intellectual transfer.

Emblematic of the rise of the West is the advent of its universities. The university embodied the many concomitant changes previously mentioned that had been transforming Latin society for several centuries preceding the Renaissance. Its scholars organized and framed the new Greco–Muslim learning into a core of arts and sciences, while interpreting the learning as the basis of a Christian theology that turned science and reason into supports of religion. Jointly buttressed by Church and State, even when the two were bitter rivals, universities proliferated throughout Europe, from Spain, France and England to Scandinavia, Poland and Hungary.

The origin of the university was in the primary schools that had been established in bishoprics and monasteries as early as Charlemagne's reign in the late

8th century. For reasons of state, the emperor supported education and had no choice but to use the Church in this endeavor, as it was the only viable institution of education to be had. The main discipline in these schools was Latin grammar, hence "grammar schools." Around 1100, cathedral schools began to grow in number and size in step with urban, economic and demographic growth. In 1179, the pope ordered that every cathedral should provide one teacher to instruct free of charge anyone who wished to be taught, young or old. The intent was to train young boys to become priests. Non-religious courses of instruction were added to the curriculum to train young men to become government officials.

During the 13th century, secular education became increasingly important, giving rise to alternate schools unaffiliated with cathedral or church. These new schools taught law, Cicero, Virgil, and computational skills to sons of merchants. Except for the Latin classics, instruction was in the local vernacular. With Dante, Petrarch, Boccaccio and Chaucer, a new set of classics in the popular language was growing. In both the older cathedral schools and these secular alternate schools, the earliest universities found their roots. Even as the grand synthesis of Grosseteste, Albertus Magnus, Aquinas and Bonaventure was being formulated, education was breaking free of church monopoly, owing in great part to the rising commerce that was becoming a prominent feature of life and means of livelihood.

Each university had its own particular strength. The university at Bologna, a secular institution, was the earliest university in the south and specialized in law. It would be the model upon which universities south of the Alps patterned themselves. To the north, the model was, of course, the University of Paris. Originally a cathedral school, it grew to become a university at the beginning of the 13th century, specializing in theology, philosophy and the liberal arts. Though it had been the center of philosophical and theological study in the 12th century, the University of Paris did not officially become a university until 1200, when several cathedral schools were unified as one. This product of a cooperative agreement between two giant personages of the time, King Philip Augustus of France and Pope Innocent III, the foundation of the University of Paris, was one of the happier events that involved a king and a pope.

Just as primary schools had proliferated in the 12th century, universities did so in the next. Oxford, Cambridge, Montpellier, Bologna, Naples, Salamanca: they sprouted like mushrooms after a spring rain. From Italy, France and Spain, universities spread in the next century to Germany and further east: Heidelberg, Cologne, Erfurt, Vienna, Prague, Cracow, Budapest; and in the 15th century, Leipzig, Nuremburg, Ingolstadt, Frankfurt, Basel, Politiers, Bordeaux, Toledo, Valencia. A city was not a full city without a university. As many as 60 had been established by 1500, with a total student body of 60,000 to 70,000, many of them studying science, and all crowded into that small nub of territory at the far extremity of the Afro-Eurasian world, an intellectual time bomb that was soon to go off.

Like the medieval cathedral with its four spires, a full university would have four colleges: liberal arts, theology, law and medicine. Few had all four, but they all had to have a college of liberal arts. This was the core curriculum of any university, the seven liberal arts, grouped in the trivium and quadrivium, going back to Boethius. The university awarded a bachelor of arts degree after four years of successful study in the liberal arts and a master of arts in law, medicine or theology after an additional three or four years of study. A doctorate in theology required a total of 16 years of intensive study following the bachelor degree, meaning a total university training of some 20 years. Early universities were composed of corporate guilds of either students or teachers, depending on the university. The guilds set the curriculum and fees, hired and fired and generally ran the institution. Owning no property, the universities rented buildings and were free to move if the city fathers made university life unpleasant. Students often made life unpleasant for the townspeople. A boisterous, unruly and irreverent lot, given to drinking, singing, brawling, dueling and carousing through the streets at all hours of the night, students at the medieval universities were not your typical non-alcoholic Nizamiyya Madrasa or al-Azhar types memorizing the Quran and praying five times a day; Muslim students would have been as aghast at the behavior of their Latin counterparts across the sea as they would have been at the subjects they were studying – logic, metaphysics, mathematics, music, geometry, philosophy – trapdoors to hell's fires, snares of the devil that the ulema's *fatwas* had been railing against for centuries.

The student population multiplied in those universities being founded between the Atlantic coast and the Vistula. During the 13th century, as many as 7,000 students attended the University of Paris; 2,000 were at Oxford. The increasing number of students made for a corresponding increase in the number of teachers. Literacy expanded. It is estimated that less than 1 percent of the European population was literate in 1050, primarily the clergy. By the end of the 15th century, the figure had risen to about 40 percent. Generation by generation, the mounting intellectual current that would carry the West to its moment of world dominance was drawing interest from the capital that centuries earlier had been invested in its cathedral schools and universities. In Islam as we have seen, science had been hatched and nurtured in the nest of royal patronage. The nest could often be a cage. Science had grown its own wings, but there was limited social space for it to fly. If the royal master went down, the caged bird went with him. In Latin Christendom the university was the nest, diverse in origin and not wholly dependent on any single ruler or institution. Some universities, for example Oxford, originally a Franciscan theological seminary, were under the wing of the Church. The University of Paris, as mentioned, was a joint venture of king and pope, with the church at the center. This was not the case with the secular auxiliary schools. Some universities were autonomously self-governed by their corporate guilds of students or teachers who decided on curriculum, tuition fees, teaching staff and salaries. Such were the universities of Bologna, Paris and Heidelberg. The university at Naples was founded by the reigning Norman king (see Plate 4).

Since the natural sciences were included in the core curriculum of the quadrivium in the college of arts and sciences, science was far more institutionally rooted in society and far less immediately vulnerable to political and religious vicissitudes than in Muslim realms, where patronage proved to be a brittle foundation. Western science grew its own wings in the universities of the high medieval period and was not put into question by contending epistemologies, no matter how conservative a Dominican or Franciscan theologian may have been. Both the Platonic Realists who maintained that knowledge of the material world came through cognition of ideas and forms in the realm of the mind and their epistemological rivals, the Aristotelian Nominalists who held that it came through sense perception, invested much thought in figuring out the world of nature, its compositions and constructions, its operations and how it was to be investigated. That the study of it was a threat to religion and society was never pressed for long, to any consequence, by any important theologian.

A 14th century theological controversy on the place of science, not its validity, proved to be a boon to science. The Franciscan William of Ockham, a Nominalist theologian at the University of Paris whose word carried much weight, convincingly argued that natural philosophy was a realm of knowledge apart from theology. This was giving science its own wings and opening the cage. A principle of scientific investigation emerged from Ockham's nominalist thought, that nature did nothing deviously or in excess but acted in the most direct and simplest way in going from one state or condition to the next. Nature never did in three steps what could be done in two, a dictum going back to Aristotle but credited to the Englishman as "Ockham's Razor."

Concomitant to that epistemological shift, a new approach to faith and theology was gaining favor among the believers in the late medieval period. This was the mystical dimension. As had been the case with Sufism in the Islamic experience, mysticism's rising popularity was in reaction to the uninspiring dryness of logical disputation, but perhaps it had more to do with the believer's revulsion with the Church's growing worldliness that came with the policy of "Papal monarchy" which cast the pope as Christendom's king of kings. The Crusades had been an early example of the policy. Papal Monarchy was seen by many Christians of conscience as a corruptive policy that turned the Church from its spiritual mission of moral guidance into an institution of greed, profit, power and wealth, in which moral laxity, scandal, simony, nepotism, selling of indulgences and venality of every sort were orders of the day. In the words of Dante:

> Of old, when Rome reformed the world, she showed
> Two suns to lighten the twin ways that went
> One with the other: world's road and God's road;
> But one has quenched the other; the sword's blent
> Now with the crook; when one and the other meet
> Their fusion must produce bad government.[5]

Famine, war, the Black Death, Church corruption, and social upheavals: miseries abounded to turn believers toward the spiritual refuge of mysticism. With natural philosophy so imbedded in theology and intellectual life, the mystic current could not but impinge upon it. But rather than denying or denigrating the study of nature, as Sufism sometimes had a tendency to do in the Islamic experience, Latin mystical thought, for the most part, flowed into natural philosophy, as seen in the religious thought of Nicolas of Cusa (d. 1464).

A Paduan-educated German cardinal of the Church, Nicolas was a mystic as well as one of the 15th century's leading mathematicians and natural philosophers. The rationality of mathematics and Gnosticism, combined with the ecstasy of the mystic's embrace of the boundless cosmos, made for a pantheistic theology: the universe mirroring divine reality. As a mystic, he found logical constructions as a path to understanding God insufficient. The application of logic was limited to the finite world. Being infinite, God could only be intellectually comprehended through mathematical symbolism that transcended logic. Nicolas compared a line to a circle to show what he meant. A line is finite. It has a beginning and an end. A circle does not. But the circumference of a circle of infinite diameter becomes a line. Logical opposites found their resolution in infinity, in God. If the universe mirrored its infinite creator, the reflection must also be infinite, though only relatively with respect to God's absoluteness, which was absolutely infinite. Being relatively infinite, the universe had no center, and so the earth, displaced from its centrality of position, became just another planet. It was not at the center of anything. Nor was it stationary. It moved about in infinite space along with the other planets.

The mystical cardinal expanded scholastic discourse to include mathematics and the number theory of Pythagorean and Neoplatonic cosmology, uniting God, angel, soul and body in a geometrical field that extended from one to infinity, where all contradictions were reconciled. In an infinite universe, Nicolas reasoned, the earth would not be the only planet to sustain life. This was getting into dangerous territory. The Church could not accept that Christ went infinitely from planet to planet getting crucified and saving mankind. Nicolas of Cusa was charged with having beliefs contrary to the Christian doctrine of the uniqueness of creation and singularity of Christ as savior and was called before the pope and a council of cardinals to defend himself. Nicolas obeyed, and so brilliantly did he argue his case against pantheism that the pope elevated him to the highest level in the College of Cardinals. A century later, in the storm of Protestant revolt, the Church would not be so liberal. Brought to Rome on similar charges of philosophical radicalism, Giordano Bruno would be burned at the stake.

The scientific accomplishments of Cardinal Nicolas are no less impressive than his mystical–philosophical speculations. He authored a number of treatises on mathematics, geometry and the physical sciences, some of which were of a practical nature, expressing the growing importance naturalists were giving to measurement in the 15th century. One was the simple but insightful method he devised to determine atmospheric humidity. This involved measuring the variation in weight of a standardized quantity of wool. It was, in fact, this very idea of measurement and the quest for exactitude it implied that undid the logical fabric so meticulously woven by the 13th century scholastics. The assault would come on two fronts: the study of objects in motion on earth and of the orbiting planets in the heavens. In regard to earthly motion, Church scholars devoted to natural philosophy at the universities of Paris and Oxford in the 14th and 15th centuries unwittingly undermined the scholastic synthesis of their theological predecessors by continuing the Muslim criticism of Aristotle's theory of motion. It will be recalled that ibn Sina, ibn Baja and Barakat al-Baghdadi had followed the lead of the 6th century Byzantine natural philosopher, John Philoponus. In the Arabic critique, that which impelled an object in its motion was *mayl*, that is, the tendency to move. *Mayl* was imparted to an object by an act prior to the object's motion: an arm throwing a stone, an arrow sent forth by the tension in a bow, and so forth. In Latin, *mayl* became *inclinatio*. According to theory, once this tendency or inclination to move was used up, the object would fall abruptly to earth. This approach to the analysis of motion was advanced concomitantly by scientists on either side of the English Channel during the early decades of the Hundred Years War, primarily by Thomas Bradwardine (d. 1349) at Merton College, Oxford, and Jean Buridan (d. 1358) at the University of Paris.

Bradwardine, a mathematician but like so many natural philosophers also a theologian – he was archbishop of Canterbury – was the first to analyze motion in terms of velocity. In his 1328 *Tractatus de Proportionibus Velocitatum*, velocity was given to be proportional to the ratio of force to resistance ($F = V/R$). Because the men who continued Bradwardine's work were also scholars at Merton, their analysis of motion and methods of calculating average velocity became known as the Mertonian method. Buridan in Paris was, meanwhile, working out his theory of impetus. A spinoff of *mayl* and *inclinatio*, Buridan's impetus was a function of both velocity and the quantity of matter in the body in motion, which corresponds conceptually to the modern concept of momentum, the product of velocity and mass. Impetus acted to perpetuate the motion of a body while being acted against by two independent forces: the pull of gravity and air resistance. Anticipating Galileo, he analyzed the acceleration of freely falling bodies and theorized that increments of impetus were added onto the body, whose rate of fall increased under the influence of the force of gravity.

Where Bradwardine remained the major figure in the Mertonian school of motion studies, Buridan was surpassed in Paris by his brilliant student Nicolas Oresme (d. 1382), bishop and theologian. Like Bradwardine and Nicolas of Cusa, Oresme is another of the many examples of high level churchmen at the cutting edge of scientific creativity. His geometric analysis relating uniform and non-uniform velocity to distance traversed was groundbreaking. Oresme geometrically represented the distance an object would travel moving at constant velocity. One side of a rectangle represented velocity, with its adjacent side, time, and their product, the area of the rectangle, being the distance. For an accelerating object, the distance traveled was the product of time and mean velocity, the mean velocity determined simply by dividing the maximum velocity by two, which in geometric terms was half the area of the velocity-time rectangle, or right triangle. Oresme's method became known as the "mean-speed theorem." At Oxford, the theorem was known as the "Merton rule."

By relating Buridan's theory of impetus to heavenly motion, Oresme took an important step in reuniting what Aristotle and his followers had severed. According to Oresme, God had imparted impetus to the celestial bodies, the same impetus that moved physical bodies on earth; but whereas the impetus of bodies in the physical realm diminished and became exhausted over time, that of the heavenly realm remained constant. Undiminishing heavenly impetus kept the celestial bodies in their clocklike motion; as it was in heaven, so it was on earth: impetus was impetus. Oresme did not go so far as to state it as such, but a giant step toward uniting the causal force of heavenly and earthly motion had been taken. By implication, the theory of impetus did away with those angelic intelligences thought to provide the spheres with their constant motion. Buridan and Oresme also discussed the possibility of the existence of other worlds as a product of God's power, but, like the medieval scholastics who speculated on questions beyond what papal councils declared permissible, they were obliged to reject the idea of God having done that. In a similar vein of scientific speculation, they considered the possibility of a moving earth, and rejected it because of religious orthodoxy, homocentrism and common sense. Whatever they believed in their heart of hearts, the lively consideration of heretical possibilities shows a wide stretch of the religious mind, analogous to Muslim speculations on a sun-centered astronomy.

The study of motion by Bradwardine, Buridan and Oresme was purely an exercise in mental geometry. Acceleration of freely falling bodies was simply theorized as the geometric accumulations of increments of impetus. No one in the 14th century thought of dropping stones from towers, tracking the trajectory of an arrow or timing earthly objects in motion with a length of rope and water clock. Nonetheless, this modest chipping away at a corner of the Aristotelian system prepared the way for the ultimate collapse of the whole edifice two centuries hence, taking down with it the Ptolemaic structure of the heavens.

The theory of motion was not the only fundament of the medieval world view being chipped away. While Nicolas of Cusa was abstractly thinking of infinity outside the spheres of Aristotelian cosmology, the Portuguese Prince Henry, known as the Navigator, was setting up a center for studies in astronomy and navigation. This was not an effort to find a better system of the world but a response to the challenge of finding a way to the spices of India. This was serious science. There was money to be made. At the end of the century a world was discovered that had never been mentioned by Aristotle, whose structure of the cosmos and natural philosophy was about to crumble like dry clay.

The year 1492 was significant. In addition to marking the end of all Muslim political presence in Spain, it marked the beginning of the West's grip over the globe, and with it, control of the maritime trade in the south seas of Asia that had for centuries been carried by Muslims. Not long after the voyages of Columbus, the globe became laced with trade routes dominated, for the moment, by Portugal and Spain. Owing to Iberian profligacy, state policies of strict mercantilism, and the noble airs of an aristocratic warrior mentality inculcated during the long centuries of fighting the Muslims, the weighty mantle of world dominance fell from Iberia to more commercially minded, proto-capitalist countries: Holland, England and France. The European venture of global domination was exploitive, cruel and rapacious, and the Dutch, English and French were no less ferocious than their Iberian predecessors. The new institutions of slavery and plantation colonization exceeded even the Crusades in cruelty, but unlike the Crusades, the profits extracted from this new burst of expansion lifted Western Europe to a level of wealth no civilization before had reached.

The other significant year of the 15th century had come 40 years earlier. In 1453 Constantinople, the fabled city that had eluded the early Arab conquerors fell, impoverished and depopulated, to the Ottoman Turks. The fall was in some measure an intellectual boon to Western Christendom because of the Greek scholars and manuscripts that found their way to the West, a transmission that had been going on for some time but to a lesser degree. Some of the Greek manuscripts that arrived contained material that had not been translated into Arabic and so had not made it into Latin, though there was little in these manuscripts that could add to the science that had been gained from the Arabic sources. The rush for Greek originals was an impulse of Eurocentric chauvinism meant to bury the Arabic–Islamic legacy. In the same year, 1453, Bordeaux fell to the French, ending their Hundred Years War against the English, the war that had begun with arrows and catapults and ended with artillery. Both Constantinople and Bordeaux fell to the recently developed weapons of mass destruction that blew open their defensive walls, powder-driven lead pummeling stone and mortar. With England's dynastic claim on French territories crushed once and for all, both monarchies were free to concentrate their energies on internal affairs and the building of strong, centralized nation states to replace the debilitated structures of feudalism. There was an upsurge in demographic growth, urbanization, commerce, investment banking, entrepreneurship, technological advance and, thanks to the application of the printing press, literacy and education. By 1453, at the early stage of its Renaissance, the population of Florence was 75 percent literate.

As this was going on, Ottoman armies were advancing along the eastern flank of Europe to the triumphant blasts of cannon and drum, while Portugal and Spain were more quietly venturing out in ships, westward to the Azores and southward along the Atlantic seaboard of Africa, in search of gold, slaves and India. In the overall balance of power, Vasco da Gama's rounding the Cape of Good Hope, Magellan's circumnavigation of the globe and Albuquerque's fortification of the Straits of Hormuz in the Persian Gulf more than answered to the Ottoman conquest of the Balkans and Hungary and almost Austria. Each trading and naval station established along the shores of the Persian Gulf, the coast of India, Malay and the Philippines was a step in the monopolization of maritime trade by the West, and a loop in the hangman's noose that would strangle Muslim economic autonomy, making Ottoman conquests Islamdom's last hurrah.

It took some time for the deeper transformations of society resulting from the changing balance of economic power that had been taking place in the 15th and 16th centuries to manifest themselves on the field of battle. Even as Ottoman armies were advancing victoriously, and as Safavids and Moguls were enjoying their moment of Muslim pomp, glory and renaissance, the economic rug was being pulled from under Islamdom's grand empires. It took until the end of the 17th century before Western economic growth and technical expertise was translated into decisive and irreversible military success on the battlefield, with Spain and Portugal joining the Ottomans in the downhill slide of empire at the hands of Holland, France and Britain. The gold and silver that the Iberians plundered and converted into currency was greatly wasted as far as their economic growth was concerned, for they used their fortunes to import manufacturers from England, France and Holland, unwittingly expanding the volume of trade and industrial production of their rivals. Some fraction of the wealth generated in western Europe through colonization and the trade in goods and human beings found its way to education and the patronage of science, to a degree at least beyond anything the medieval period had known. The West's intellectual investment steadily increased with the rising economy and was soon to draw dividends. Existing universities continued to grow and new ones continued to be founded in cities to the north and east: Glasgow, St. Andrews, Aberdeen, Leiden, Rostock, Copenhagen, Uppsala and Pressburg, bringing more and more young minds into the academic arena and science:

> And so it is with all the material, moral and intellectual factors involved in the changes of the sixteenth century – somewhat suddenly they passed the critical stage. Growing wealth increased knowledge, and new knowledge in its turn created wealth. The whole process became cumulative, and advanced with accelerating speed in the irresistible torrent of the Renaissance.[6]

Universities in the German-speaking region of Europe were relatively late in making their mark in the burgeoning interest in science and its assimilation. Long overshadowed by Paris and Oxford in the north of Europe and Bologna and Padua in the south, which collectively dominated scientific and medical studies in the 13th and 14th centuries, German universities did not begin producing scholars rivaling those in France, England and Italy until the middle of the 15th century. Two scientists in particular marked Germany's arrival, the astronomer–mathematician George Peurbach (d. 1461) and his student assistant Johann Muller (d. 1476), more commonly known as Regiomontanus, the Latinized version of his Austrian hometown, Konigsberg. Both were graduates and teachers at the University of Vienna, which was coming into prominence. Regiomontanus, something of a child prodigy, enrolled in the university at 14, had his bachelor's degree at 16, his master's a few years after that, and was then collaborating with his teacher Peurbach on a series of astronomical endeavors: calculating ephemerides, observing comets and lunar eclipses, and correcting the Alfonsine Tables – so-named because Alfonso X had in 1250 commissioned a Latin emendation of tables made centuries earlier by Muslim astronomers in Spain. The collaborative observations of Peurbach and Regiomontanus signal that Western astronomy had reached maturity. In 1456, they observed a comet that two centuries later would be given the name Halley's Comet. Their calculations of ephemerides and eclipse tables, and their measurement of the obliquity of the ecliptic, represent the highest achievement Western observational astronomy would reach until the time of Tycho Brahe a century later.[7]

With the exception of the refinements made by the Maragha astronomers, namely Nasir al-Din al-Tusi's double epicycle system and its modifications by ibn al-Shatir, whose influences would not be evident in European astronomy until Copernicus, the teamwork of Peurbach and Regiomontanus shows that the West had at last absorbed, mastered and refined the Greco–Muslim astronomical heritage. Regiomontanus' work in trigonometry and algebra makes it evident that Western mathematics had come close to the level prevailing among leading Muslim mathematicians of the 15th century, principally those in Timurid Samarqand. In fact, the sudden rise in the level of Western astronomy, as seen in the work of Peurbach and Regiomontanus, has been ascribed to influences coming from Samarqand. The connection is not far-fetched. The astronomy of al-Kashi at Samarqand, with its Maraghan influences, was transmitted to the Ottomans whose empire bordered Austria. Similarities of observational equipment, of parametric values and the specific problems astronomers focused on, and the identity of methods used to determine solar and planetary parallax, suggest that Peurbach and Regiomontanus were somehow aware of the latest advances made by Muslim astronomers. This could have been by way of an Austrian astronomer who had visited Istanbul, but more likely a Byzantine astronomer who had fled the city after its fall to the Ottomans.[8] Possibly it was someone residing in post-Byzantine Constantinople who was knowledgeable in astronomy and came to the West, someone such as Bessarion, whose name in the annals of science is intimately associated with Peurbach and Regiomontanus.

Born in Trebizond, Bessarion, a Greek Christian of the Latin Church, became, in turn, Bishop of Nicea, Cardinal, and then, after the city had fallen to the Ottomans, Papal Patriarch of Constantinople.[9] He had earned a name for himself in science through his Latin translations of a number of Greek scientific manuscripts from the collection he brought with him to Italy a few year after Constantinople fell. He had come to Vienna in 1460 on a papal mission to seek Hapsburg help in forming a latter-day crusade to restore the fallen Byzantine capital to Christendom, perhaps wishing to forget the last time the West had organized a crusade to assist the Byzantines against Turks, specifically the Seljuqs. In that crusade, the Italian organizers ended up seizing and ruling Constantinople to collect commercial debts the Byzantines owed them, and thereafter for 60 years keeping the city for their own profit, they ripened it for the Ottoman plucking. Cardinal Bessarion's intended crusade came to nothing, but while in Vienna he sought out Peurbach and Regiomontanus, who by then were known as leading astronomers. During one of their conversations Bessarion happened to mention that among the collection of Greek manuscripts he had brought to Italy from Constantinople was a good copy of the *Almagest*.[10] He was eager to turn it into a Latin translation that would surpass the 12th century one made by Gerard of Cremona, as well as the one his compatriot George of Trebizond had made from a Greek manuscript in 1451, just before the fall of Constantinople. He was sure his Greek copy could be the basis of a superior Latin text if translated by experts.

Bessarion also wanted to compose a handy abridgment of the *Almagest* to be used as an undergraduate textbook for astronomy students, but his duties as

Cardinal, Bishop and Papal Legate left him little time and he suggested to Peurbach and Regiomontanus that perhaps they might consider taking leave from their teaching duties in Vienna to return with him to Italy where they could join their efforts in realizing these projects. The suggestion was no sooner made than accepted. The allure of working in the salubrious climate of northern Italy must have added to the excitement of producing a first class translation of the *Almagest* from its original language, while the newly invented printing press offered the chance of making their work readily available to all the astronomers and universities of Europe.

The three men left for Italy that same year. Bessarion taught Regiomontanus Greek, and together they worked on the Latin translation, while Peurbach, foregoing the joy of learning Greek, worked on condensing their translation to the *Epitome*. It was a harmonious working relationship, for as long as it lasted. Within a year of their arrival to Italy, Peurbach was dead, his productive career cut short at 38. Regiomontanus and Bessarion carried on. When they had finished the translation of the *Almagest*, Regiomontanus completed what Peurbach had left unfinished of the *Epitome*, faithful to the vow he had made in response to his beloved teacher's deathbed request. The *Epitome* was published under both their names in 1461 and dedicated to Bessarion, who had made it all possible. Regiomontanus died 15 years later, having just turned 40. Much of the fame achieved by the two astronomers came from their clearly written, elegantly printed and widely distributed translation of the *Almagest*, and the condensed *Epitome* they made of it, the latter becoming the standard textbook for astronomy through the Renaissance up to the time of Copernicus.

In effect, Peurbach, Regiomontanus and Bessarion, aided by the printing press, accomplished in Western astronomy what the Harranian al-Battani did in Muslim astronomy six centuries earlier, that is, established astronomy on a sound Ptolemaic basis as a point of departure in going from the assimilative to the creative phase. The difference, of course, was that in addition to Ptolemy, Western astronomy had inherited, in Latin translation, the advanced mathematical methods and parametrical refinements that astronomers in Islamdom had given the system.

So then why the joyous rush to have a Latin Ptolemy from the original Greek that lacked the very significant parametrical, mathematical and methodological refinements Arab/Muslim astronomy had applied to Ptolemy and that had long been translated into Latin? One reason was certainly a rational one: the purity of scholarship required a translation of a text from the original language, regardless of its practical value; its being several centuries out of date would have been irrelevant to the scholars. Another reason was more psychocentric than rational. Renaissance humanism had abandoned reason and history in its adulation of Greco–Roman antiquity.[11] Covering the traces of the West's debt to Arabic sources justified and purified that adulation. The Islamo–Arabic intermediary between the Greeks and the Latins was erased by its being omitted. Everything came from the Greeks. Making the suppression of memory regarding the Arabic sources all the easier for Europeans was the threat that the Ottomans posed to Christian Europe at the time. While the Ottomans pressed in, from the battle of Kosova in 1389 to the fall of Constantinople in 1453, and the Renaissance flourished, fear of the former and veneration of the latter erased the historical record, or tried to. At the time of Regiomontanus, the erasure was somewhat schizophrenic. The Muslim influence was too deep and enduring to be wiped out of mind completely. As in a palimpsest, traces of the first record could still be seen. Regiomontanus's *de Triangulus*, the first systematic treatise on plane and spherical trigonometry to be published in Europe, made no secret of having drawn heavily on Arabic sources. That was when Regiomontanus had written it, in 1463. By the time *de Triangulus* was published in 1533, the Arabic sources had in those intervening 70 years been lost or forgotten and almost totally replaced by Greek sources.[12] During the 16th century, Renaissance mathematicians and scientists continued to diminish the Arabic contribution and its originality, attributing much that had been done in Islamdom to Archimedes, Papus and Proclus, while depicting Muslim mathematicians and astronomers as mere correctors. As al-Khwarizmi and ibn al-Haytham faded from memory, Archimedes rose to mathematical divinity.[13] Antonio Codro (d. 1500), a critic of Ptolemaic astronomy who was in Bologna when Copernicus was a professor of Greek there (1496–1500), dismissed Arabic science and mathematics completely, seeing nothing of significance having been done between the Greeks and the Latins, while ignoring or not realizing that his own criticisms of Ptolemy's inconsistencies and flaws had been made long before by sources in Arabic that had been translated into Latin.[14] A century earlier, Petrarch had written plangently that after the Arabs no one was permitted to write on mathematics, they had done it all and their word was final.[15] Arabic sources were still being considered authoritative when Gregorius Reisch (d. 1525) published his encyclopedic *Margarita Philosphica* in 1503.[16]

Testimony of the enduring, though diminishing, European respect for Muslim science is evident in Peurbach's popularization of Thabit ibn Qurra's theories of precession of the equinox. Peurbach's title of his book, *Theoricae Novae Planetarum*, implied that they were new planetary theories – new for the West but in fact more than 500 years old in Muslim astronomy. Treatises by the two 9th century Abbasid astronomers, Farghani and Battani, whom Regiomontanus included in his address to the Paduan scientific community in 1464 on the progress of the mathematical sciences, were published in 1537 in Nuremburg. Regiomontanus's de *Triangulus* used the sine and cosine laws of spherical trigonometry that he recognized as having been invented by Muslim astronomer–mathematicians. A contemporary of Regiomontanus, Piero della Francesca, who died the year Columbus set sail across the Atlantic from Spain, was making modest mathematical advancements based on the works of Khwarizmi, abu Kamil and Leonardo Fibonacci's reworking of the Muslim algebraists in his *Liber Abaci*. Andrea Alpagus (d. 1539) looked to Muslim astronomers in his search for new knowledge.[17] Some Muslim works that had not been translated because they were written after the translation movement eventually did get translated and had their own late influence on Western science, and they were therefore difficult to forget or erase. Such were the works of two 16th century Muslim astronomers, Taqi al-Din and al-Yazdi, who appear to be the sources that introduced decimal fractions to the West.[18]

And then there is the extraordinary case of Leo Africanus, who as a former Muslim was highly regarded in the West for his scientific knowledge. Hasan ibn Muhammad ibn al-Wazzan (d. circa 1550) came from north Africa. How or why he left his religion and his country and became a professor in the University of Bologna is a mystery. It possibly had to do with an Italian merchant community based in North Africa. Leo became a source of late Arabic science. It is speculated that he directly or indirectly communicated Maraghan astronomy to Copernicus, who was at Padua the same time Leo Africanus was at Bologna.[19] Members of the Royal Society in London as late as the 1660s were urging the translation of the Maraghan astronomy of Nasir al-Din al-Tusi. The dye that colored the Arabic legacy of Latin science could not be altogether washed out.

Parity

With the premature deaths of Europe's premier astronomers, Peurbach and Regiomontanus, science was left treading water for a century, or so it seemed if viewed from the mountaintops. At lesser elevations, Europe's scientific potential undramatically continued to grow in the universities that gave sustenance and continuity to scientific study by providing a structure for a scientific community numbering several hundreds and a network providing for the diffusion of their works. From Paris and Oxford to Prague (Regiomontanus regarded Nuremburg the center of European scientific communication because of its centrality of geographic location and European commerce), a scientific society centered in universities was in communication through letters and publications that were made plentiful and available by the printing press, all of which ensured the continuity and enrichment of a tradition that was socially rooted and as much religiously as secularly oriented.

Having accomplished in astronomy and mathematics what Buridan, Oresme, Bradwardine, Sacrobosco and the Oxford Mertonians had accomplished in physics the century before, Peurbach and Regiomontanus initiated what would culminate in the "Copernican Revolution."[20] Published in printed form in 1496, the *Commentary* and *Epitome* were the sources from which young Copernicus learned his astronomy. Since the *Epitome* contained a detailed analysis of Ptolemaic astronomy and incorporated the improvements made to the system by Arabic astronomers, particularly al-Battani and al-Zarqalli, Copernicus learned of their accomplishments through the work of Peurbach and Regiomontanus.[21]

Strong evidence indicates it was through ibn Rushd that Copernicus learned of the relationship between a planet's distance from the earth and its orbital period: the more distant, the longer the period; the closer, the shorter. It was this simple fact of observation, first stated by Aristotle, then Vitruvius (first century B.C.) and Martianus Capella (fifth century AD), with others to follow, that first forced Copernicus to consider rearranging the order of the planets in accordance with that relationship.[22] It was ultimately from the Latin translation of Aristotle's *On the Heavens* (*de Caelo*) that Copernicus knew of the relationship but very likely that ibn Rushd's grand commentary translated to Latin (*Aristotelis opera cum Averrois*

Commentariis, published many times in the 15th and 16th centuries) had some influence.[23]

An answer to the intriguing question was whether, from the influence of Regiomontanus, Copernicus took seriously the idea of a heliocentric universe and if it might be found in one of Regiomontanus's private letters, where he refers to the motion of the earth as having an effect on the observed motion of the stars. A moving earth was anything but a common idea at the time, and though the letter has not survived, it was commented on by a leading astronomer of a later generation who claimed to have seen it and who, as a supporter of the Copernican system, referred to Regiomontanus as the first astronomer to articulate the revolutionary idea of a moving earth. Copernicus could have received the idea from Regiomontanus by way of Novaro, his former student, who later became a teacher of Copernicus.[24] Where Regiomontanus got the idea, or what specific problem led him to it, is as much a mystery as why he did not write more on it. A few leading Muslim astronomers, al-Farabi (d. 950), al-Biruni (d. 1038) and al-Katibi (d. 1272) among others, had considered the idea of a rotating earth as a hypothetical possibility, since, from the point of view of economy and motion, a stationary sun made more sense than a system in which the sphere of the fixed stars was obliged to spin daily at an incredible angular velocity around the earth at the center. In the words of an 18th century Muslim writer, who remarked when first exposed to the Copernican system, "It is the kabab on the shish that turns on the fire, not the fire around the kabab." Those earlier Muslim astronomers who considered the possibility of a moving earth always ended by rejecting it because it was against the principles of Aristotle, Ptolemy and the reality of God's creation.[25]

In the several decades intervening between the work of Regiomontanus and Copernicus, there was much that made the Western and Muslim scientific achievements look like mirror images of each other. Witelo, Buridan, Oresme, Bradwardine, Swineshead, Peurbach and Regiomontanus, even Copernicus and Kepler, can be seen as Latin versions of, al-Kindi, ibn al-Haytham, ibn Sina, ibn Baja, al-Biruni, Umar al-Khayyam, Nasir al-Din al-Tusi, ibn al-Shatir, Ghiyath al-Din Jamshid al-Kashi and Shams al-Din al-Khafri, men inspired by a common source. All were exceptional physicists, astronomers or mathematicians revering the same Greek authorities, working on similar problems, sharing a common point of view on God, nature and Intelligent Design. And all were practicing astrologers who wrote on it and accepted it as the spiritual sister of astronomy, though there was a great divergence in degrees of credence, from the conservative belief that the configuration of the heavens at the time of birth had some influence on character and disposition, to the radical belief that the stars, if properly read, could foretell the future in the events of nature and man. Biruni had cast horoscopes for sultans, as had Peurbach and Regiomontanus for kings. There was nothing unusual in that for Muslim or Western astronomers. Astrology was a rational science; the stars and planets affected the main features of character as they affected nature, the seasons and the tides. Scientists on either side of the Danube or Mediterranean shared the same conviction that rational knowledge complemented scriptural knowledge. Latin scholastics in Paris mirrored their Mu'tazilite forerunners in Baghdad.

For that brief period, from roughly 1460 to 1540, scientific knowledge in the West approximated the level achieved in Islamdom. The difference was that one trajectory of learning was slanting precipitously downward at the time, while the other was soaring upward. Sometime during that period those hypothetical trajectories intersected. Except for the astronomer–mathematicians of Maragha and Samarqand, Islamic science had begun to be continuously less productive centuries before that presumed period of intersection. In this approach to parity, Muslim science's counterparts to Peurbach and Regiomontanus, men whose works represent the summation of scientific knowledge of their time, were three Iranians, two of whom were associated with Maragha, the third with Samarqand. They were al-Qazwini (d. 1283) whose *Aja'ib al-Makhluqat* (On The Natural Wonders Of Creation) is a general compendium summing up the state of natural philosophy and astronomy. Qazwini's student, the Maragha astronomer Qutb al-Din al-Shirazi (d. 1311), who produced general works summarizing the contemporary state of scientific knowledge, the most famous being *Nihayat al-Idrak fi Dirayat al-Aflak* (The Last Word in the Knowledge of the Heavenly Spheres), and al-Jaghmini (d. 1445), who wrote a general commentary on astronomy for Ulug Bey, *Mulakhkhas fi'l-Hay'a*. The works of the two sets of scientists, Muslim and Western, can be taken to establish critical points along the two trajectories of science in their respective societies.

Qazwini's *Wonders of Creation* is a cosmography whose system of the heavens is based on Ptolemy and whose earthly physics are based on Aristotle, as refined by Muslims during the previous several centuries. Written originally in Arabic, the work was later translated to both Persian and Turkish and enjoyed great popularity throughout the Muslim world. Writing from the perspective that astronomy and natural philosophy were manifestations of divine wisdom in the patterning of creation and cosmic order, Qazwini was a popularizer, not a scientist; his aim was to glorify God, not promote enthusiasm for natural studies. Jaghmini, a practicing astronomer in Samarqand, produced something similar to the *Epitome* of Peurbach and Regiomontanus in his *Summary of Astronomy*, a popular hand book of Ptolemaic astronomy that was widely used as a text and often referred to by Muslim astronomers.

Qazwini, Shirazi and Jaghmini were commenting on a comprehensive state of knowledge in a civilization whose scientific endeavor was approaching the end of a creative road, whose foundations and theoretical formulations of nature were fast winding down and would not change over the centuries as the sights of Muslim scientists were lowered and scientific productivity went from creativity to commentary, gloss and recension, settling into a moribund state of comments scribbled on the margins of past works. Peurbach and Regiomontanus, on the other hand, were poised at the threshold of a long period of creativity. Science in Islam was coming to the end of a three-millennia tradition that reached back to Egypt, Mesopotamia, Iran and India; in the West it was approaching a revolution that would transform that tradition and continue transforming it until it would be no longer recognizable as having descended from its ancient Hellenistic and Muslim origins, except for its Arabic numerals, system of decimal notation and the Greek and Arabic loan words that came with the translations in Baghdad and Toledo.

Between 1460 and 1540, there would have been nothing that Muslim and European scientists could not have discussed together and understood, nothing they would have found objectionable in the other's principles of science. Even the *de Revolutionibus* of Copernicus would have given no problem to Muslim astronomers. As a geometrical exercise, it was simply Ptolemy turned inside out and would have been a gripping subject of discussion for those Muslim critics of Ptolemy, from ibn al-Haytham to the Maraghans, and equally for the Ismaili Brethren of Purity, who were as spiritually attracted to the Pythagorean system with its cosmic relations to numbers as were Copernicus and Kepler. A half dozen or so representative scientists, physicians and mathematicians, chosen from the brightest in Islamdom between 900 and 1450, and matched to a like number of their Western counterparts chosen from the middle of the 15th century to the middle of the 16th, would have made for a lively seminar.

Mathematically, team West may have been hard pressed by their Muslim rivals. Before 1500, the sum total of mathematical knowledge in Europe consisted of Robert of Chester's Latin translation of Khwarizmi's *Algebra*, copied out by Regiomontanus in 1456, and Leonardo Fibonacci's *Liber Abaci*.[26] And on the Muslim side, from the 12th century on, the pool from which top Muslim representatives could be selected for this hypothetical seminar was beginning to drain away, while the pool for Western candidates was not yet beginning to collect. If seminar candidates were to be selected from scientists who lived a half century after the death of Copernicus, the Muslim side would have found great difficulty in mustering representatives competent enough to hold their own against the Western side of the table. As put by an eminent historian of Muslim astronomy, the West had "outpaced" their heretofore Eastern benefactors.[27]

Though outpaced, there was still a remnant of creativity in the moribund tradition of Muslim science that refused to disappear with all its traces, despite the reality that most everything that remained of it was in the memory of a miniscule elite. Even during the latter half of the hypothetical period of parity, Islam was producing creative scientists and mathematicians. Science had been a strong and creative tradition for too many centuries for it to give up the ghost without a struggle. In fact, the ghost of science in Muslim civilization was never completely surrendered. Scientific treatises were constantly being written, though they lacked originality in theory, observation, mathematics and instrumentation and were for the most part limited to the practice of time reckoning (*'ilm al-tawqit*) for religious purposes. Mathematicians and astronomers were publishing treatises based on the old traditions right up until the end of the 18th century.

The decline in creative scientific activity was not without brief revivals in mathematics and astronomy, as seen in the work of Qadi Zada al-Rumi (d. 1436), an Anatolian Turk who took over the directorship of Ulug Bey's observatory upon al-Kashi's death, and Ali Kushji, another Turk from Ottoman lands who had come to Samarqand to study and succeeded Qadi Zada al-Rumi as director. These men kept the Samarqand school of astronomy going. Istanbul provided a brief haven after Samarqand's eclipse. Qadi Zada's son married a daughter of Kushji, and their son was father to the Turkish mathematician Miram Chelebi, who in the new capital of the Ottoman Empire continued the mathematical tradition of Maragha's Nasir al-Din al-Tusi and Samarqand's Jamshid al-Kashi. Astronomical craftsmanship in the West and in Islamdom was also closely comparable in the 15th and 16th century. The observational equipment Tycho Brahe was building at the end of the 16th century bears striking similarity to equipment that was used by astronomers of Ulug Bey's observatory in Samarqand.[28]

At the Western frontier of Islam and Christendom, abu al-Hasan al-Qalasadi (d. 1486), a mathematician who was born in Spain and died in Tunisia, to where he immigrated upon his birthplace falling to the conquistadors, kept the high tradition of Muslim mathematics alive with his several works on arithmetical operations, which drew on the mathematical heritage going back to al-Khwarizmi. Qalasadi also advanced the field an important step by introducing symbols to simplify complex mathematical operations, in some instances using short Arabic words or single letters to signify addition, subtraction, multiplication, division, equality, root, an algebraic unknown, and powers of square and cube, and so forth. Though showing some originality in this, he was in fact advancing a tradition begun the century before by mathematicians in Morocco, Algiers and the eastern regions of Islamdom. Qalasadi's works also signify a tradition in decline in that they are commentaries, compendiums, summaries, or summaries of summaries, signs of a tradition losing its vitality but refusing to die.

The optics of al-Kindi, Haytham and Kamal al-Din al-Farisi had faded, leaving no young genius who could have critically engaged Descartes on his *Dioptrics*. Galileo's telescope, which ravaged Aristotelian worship of the unsullied heavens, compounded with Galilean physics and Kepler's laws of astronomy that Newton united in laws governing earthly and heavenly motion, where mathematical formulas factored out God's omnipotent will, would have stopped Muslims cold, as it actually did in the 18th and 19th centuries. By 1600, the al-Kindis, Farabis, ibn Sinas, Birunis, Umar Khayyams, ibn Rushds and al-Tusis, and that flexibility of scientific mind they represent, had been pretty much washed out of the culture. Only a rare strain or two of scientific creativity remained. Mathematics and astronomy had gone the way of optics, physics and mechanics. A rich legacy had come to an end, just as the West was entering its scientific revolution. By the end of the 17th century, the parting of ways in scientific knowledge and the reversal in the balance of military expertise and power had become obvious to both sides. By the early 18th century, Muslim reformers were turning toward the West, consciously and deliberately, even eagerly calling for the adoption of its inventions and translation of its science. Taking from the West would cause severe problems for those who could not let go of that old attitude of superiority. It would also, as we shall see, uproot indigenous institutions to be replaced by alien transplants creating a state of confusion over cultural identity and authenticity that persists most violently to this day.

Notes

1 Henri Hugonnardi Roche, "The Influence of Arabic Astronomy in the Medieval West," pp. 285–286.
2 Cited in John Freely, *Aladdin's Lamp: How Greek Science Came to Europe Through the Islamic World*, p. 141.
3 In fact, the title of his book, *de Perspectiva*, is a direct translation of Haytham's *al-Manazira*, and follows Haytham's order and organization of subject matter, as well as his methodology and sequence of treatment right, down to the amount of space given to each subject. The single instance of deviation was Witelo's theory that light traveled instantaneously, though this contradicted his belief, taken from ibn al-Haytham, that the more transparent the medium the faster light traveled.
4 Albertus was inclined to the simplified system of al-Bitruji, though he was aware of the system's inabilities to account for some observations. Henri Hugonnard-Roche, "The Influence of Arabic Astronomy in the Medieval West," pp. 294–295.
5 *Purgatorio, Canto XVI*, Penguin Classics, London, 1962, translated by Dorothy Sayers.
6 William Dampier, *A History of Science*, Cambridge University Press, Cambridge, 1966, p. 102.
7 P.L. Rose, *The Italian Renaissance of Mathematics: Studies on Humanist Mathematicians From Petrarch to Galileo*, Librairie Droz, Geneva, Switzerland, 1975, pp. 90–117 for an appraisal of their work.
8 Aydin Sayili, *The Observatory in Islam*, Turkish Historical Society, Ankara, 1960, p. 396.
9 Renowned as a professor of philosophy, Bessarion was on the Council of Ferrara-Florence that was delegated to unite the Greek and Latin rites as a means of bringing the Latins to the defense of Byzantium against the Ottomans. Nicholas of Cusa was also a member of the Council. The union failed. Bessarion stayed on in Italy as a diplomat of the Church and was made a cardinal. A few years later Byzantium fell. Two years after that, Bessarion came close to being elected pope (1455). His being Greek cost him the election. John Freely, *Aladdin's Lamp*, pp. 174–175.
10 He had brought an immense collection of 482 manuscripts, which injected a good dose of Greek learning into the Renaissance. Rose, *The Italian Renaissance of Mathematics*, p. 44.
11 Rose interprets Regiomontanus' zeal to translate the Greek original of the Almagest as a way to gain "true astronomy," as if the many important Muslim reforms and parametrical recalculations had not improved anything.
12 Rose, *The Italian Renaissance of Mathematics*, pp. 98–99.
13 Rose, *The Italian Renaissance of Mathematics*, pp. 262–268.
14 Rose, *The Italian Renaissance of Mathematics*, p. 121.
15 Rose, *The Italian Renaissance of Mathematics*, p. 8.
16 A. Hall, *Scientific Revolution, 1500–1800*, Longmans, London, 1971, p. 11 ff.
17 George Saliba, *Islamic Science and the Making of the European Renaissance*, MIT Press, Cambridge, MA, 2007, pp. 230–232.
18 A 15th century Byzantine manuscript written after the fall of Constantinople and appearing in Vienna in 1562 refers to decimal fractions as having been introduced by the Turks who were ruling the lands that had formerly been Byzantium. Rashid Rushdi, *Encyclopedia of Arabic Science*, vol. II, p. 415, note 41. Elements of the decimal system were slowly creeping into Western usage in the 16th century. The mathematical notation of a vertical line setting off numerator from denominator in the decimal fraction system of the Samarqand observatory's Jamshid Ghiyath al-Din al-Kashi was used in the mathematical works of several scientists of that century, most notably those of the physician and mathematician Girolamo Cardano.
19 Saliba, *Islamic Science and the Making of the European Renaissance*, p. 226.

20 Rose, *The Italian Renaissance of Mathematics*, p. 84.
21 Hugonnard-Roche, "Influence of Arabic Astronomy in the Medieval West," p. 298; pp. 298–303 gives a brief but somewhat detailed analysis of Arabic astronomy's contribution to Copernicus' revolutionizing of the Ptolemaic system. Also, Arun Bala, *The Dialogue of Civilizations*, "The Wider Copernican Revolution," for a more general account of Muslim influences on Copernicus. Chapter 13, pp. 145–176.
22 There were of course other arguments for placing the sun in the position of the earth. One was it explained retrograde motion and eliminated epicycles; another was that it did away with the extreme orbital motion of the fixed stars. But these were consequences of making the shift, not the prime reason for making it.
23 *Aristotelis opera cum Averrois Commentariis* contained three texts: (1) the Latin translation of Aristotle's works that William of Moerbeke made from the Greek; (2) the Latin translation of Aristotle that Michael Scot made from ibn Bitriq's Arabic translation from the Greek; (3) and Michael Scot's Latin translation of ibn Rushd's commentary on ibn Bitriq's Arabic translation. See B. Goldstein's article cited above for incisive analysis of the texts in relation to the distance and period relationship and the evidence pointing to ibn Rushd's commentary as the source from which Copernicus learned of it. A spirited discourse over ibn Rushd's principles was going on at the time Copernicus was a student in Italy. "In particular, Achillini (d. 1512), one of the most celebrated philosophers in Italy at the time, published in Bologna an Averroist attack on Ptolemy while Copernicus was a student there." B. Goldstein, "Copernicus and the Origin of His Heliocentric System," p. 225; also, Arun Bala, *The Dialogue of Civilizations* (Chapter 13), "The Wider Copernican Revolution," pp. 145–176, for a more general account of Muslim influences on Copernicus's astronomical thinking. The book gives serious consideration and credence to Chinese influences on Copernicus and the West's Scientific evolution, transmitted by Italian Jesuits in China. Bala also holds that ancient Egyptian sun-worship influenced Copernicus.
24 Edward Rosen, "Regiomontanus," in *Dictionary of Scientific Biography*, vol. II, edited by Charles Coulson, Gillispie, Scribner and Sons, Detroit, 1970–1980, p. 352.
25 E. Wiedemann, "Zu den Anschauungen der Araber uber die Bewegung der Erde," in *Mitteilungen zur Geschichte der Medezin und der Naturwissenschaften*, vol. VIII, 1909, pp. 1–3; A. Sprenger, "The Copernican System Among the Arabs," *Journal of Asiatic Society of Bengal*, 25, 1856, p. 189.
26 Rose, *The Italian Renaissance of Mathematics*, p. 143. But during the next 70 years mathematicians in north Italy would be matching the best of their Muslim counterparts.
27 Professor Saliba wishes to use this term in place of the idea that Muslim science declined. The choice of words comes to the same. European scientists did outpace Muslims after the 15th century; compared to Muslim creative productivity between 800 and 1100, it would be hard to argue that Muslim productivity suffered no decline as measured between 1300 and 1600.
28 For a comparative description in modern Turkish see Sevim Tekeli, "Nasiruddin, Takiyuddin ve Tyho Brahe'nin Rasad Aletlerinin Mukayesi," *Dil ve Tarih – Cografya Fakultesi Dergisi*, Ankara University, Ankara, 1958, XVI, no. 3–4, pp. 301–393.

7 Renaissance and Revolution

Introduction

The Renaissance crowned a four-century ascent in the West's maturation from an intellectually impoverished civilization, clearly and admittedly inferior to its political and religious rivals Byzantium and Islam, to a stature of parity, going on to supremacy. By the time the Renaissance was in its early flowering in Florence, Byzantium was in the throes of falling to the Ottomans, whose empire stood as the leading power of several Muslim powers whose borders collectively extended from east Europe to India and deep into Central Asia, to the western frontier of China. Though Muslim civilization had reached its scientific apogee centuries earlier, politically and militarily it was ascending into the 17th century. Appearances were deceptive. Before the Renaissance was over in the north of Europe, Copernicus, Kepler and Galileo, standing successively on the shoulders of lesser known giants, had prepared the way for the giant of them all, Sir Isaac Newton, during whose prime of creative science the Ottomans were at the Danube, their artillery battering the walls of Vienna. It did not look at the moment to contemporary observers that the West had gained supremacy. But contemporaries are not usually cogent observers of their own conditions.

Springing from the Renaissance, the Scientific Revolution imparted to Western society a heightened appreciation of science that extended beyond the professional academics cloistered in their universities; it brought to the scientific community avid amateurs from a wide variety of professions, from law to commerce, attracting greater levels of social interest and investment in science that ended the century of parity and left Islamic civilization in its wake. Once the methods and mentality of doing science, with its emphasis on precision of measurement and analysis of motion, were translated to military technology, Muslim ascendancy on the battlefield was brought to a dramatic end. This ultimately meant an end to Ottoman triumph: the raison d'être of the *jihadi* state. The Ottomans, the Muslim world's last formidable defense against the global reach of the West, were at the very beginning of the 17th century fought to a stalemate. They still won some wars and took more land, but at the end of the century, after a last gasp of glory before the gates of Vienna in 1683, they were being rolled back in a series of stunning defeats. From then on, the military glory of Islam, like its scientific achievement before, would be reduced to a memory, its wars with the West reduced to defense along shrinking borders on all fronts. The only Muslim border that held firm was the psychological wall blocking out recognition of the West's great transformation. An open mind and border might have allowed scientific and technical transfer. With proper investment and organization, the Ottomans might have restored Islamdom military parity and possibly revived scientific productivity. There is no reason this could not have been possible. The psychological wall was not insurmountable. But for this to happen, the mind that opened with the border would have to be one that embraced more than a handful of reformers at the top of the ruling elite. At the lofty top there were indeed those few who were able to see another horizon beyond that metaphorical wall. However eloquently those somewhat narrow-minded exponents of the "Clash of Civilizations" thesis argue their case, the historical record does not support the idea of a "closed Muslim mind." From a great distance, where the details of historical analysis are lost in the mists of cultural and linguistic differences, it could appear that the defensive mechanisms of a presumably hypothetical collective Muslim mind made for a mighty wall of cultural superiority, bordering on arrogance, crushing all curiosity to look over it, condemning its civilization to defeat, humiliation and despair; a civilization that was informed by a religion, as the thesis contends, that is incompatible with political and intellectual freedom. The thesis overlooks many things, some small but telling and some monumental, especially after the 17th century. It overlooks, for example, that Sultan Mehmet II, conqueror of Constantinople, had a Greek version of Aquinas's *Summa Contra Gentiles*, among other Greek manuscripts, translated to Turkish.[1] It disregards all those Muslims from the early 18th century on who penetrated that proverbial wall and attempted innovative reforms along Western lines.

A few Ottoman leaders had been introduced to Western science through Dutch and French ambassadors presenting gifts of expensive scientific books to the sultan during the last half of the 17th century. These, in translation, had stirred some interest at the very top of government and opened a few minds to the new science; too few for much to come of it. Military defeat was more effective at opening the mind. A decade-long disastrous series of them following the second failed siege of Vienna (1683) alerted some Ottomans that a new reality had come about and that they had better make some adaptive changes – and look to the West to see what they had to change. In the early 18th century, the Ottomans attempted a series of reforms designed to adopt Western military techniques and science. Each defeat lead to a renewed effort to reform. However, the handful of reformers at the head of Ottoman government had little chance against that traditional wall of mind represented, firstly, at court by conservatives whose self-interests demanded preserving the status quo and whose conviction of cultural superiority and contempt for the Christian West resonated through the ranks of the established ulema; secondly, by the military, with its own vested interests in the status quo and sharing the same convictions and contempt; and thirdly, by the street, always ready to fill with riotous, impoverished discontents at the slightest provocation from the lower ulema and military officers.

But even by the early 18th century, when modernizing reforms were first instituted by Ottoman leadership in an organized way, it was getting too late to wrench Muslim destiny from its ill-fated encounter with the modern West. It was not just a problem of reforms and borrowing that might enable Muslims to hold the line against the West, but something deeper in Western culture that would have to be discovered, a way of thinking in regard to the world and one's place in it. The Renaissance and the global voyages, followed by the Reformation and Scientific Revolution, created a secular and individualist turn of mind that was as alien to 18th and 19th century Muslims as it was to 12th and 13th century Latin Christians.

Renaissance

Westerners had been coming to terms with a world wider than their own ever since Gorza and Gerber traveled to Spain in the 10th century. No magic moment suddenly illuminated a new way of thinking. No bell proclaimed the end of the medieval mind and birth of the modern. At some point, the mounting accumulation of experience and number of people who found new ways to accomplish work reached a critical mass. What was once an aberrant variation assaulting venerated tradition became, decade by decade, over the centuries a cultural norm. In Islamdom it was not until the middle of the 19th century that a miniscule gathering of young Ottoman intellectuals was speaking in terms of a change of mind over and above a change in apparel or weapons. Without change of mind and cultural norms, the cadre of reformers with their programs of reform directed from the top were legless bodies groping their way in the dark.

In the West, the men and women of the Renaissance made a virtue of innovation and the experimental, of casting off the old and traditional – like so many gleeful reincarnations of Siger of Brabant, Roger Bacon and their radical bunch. Combined with a heady dose of individualist narcissism, the men of the Renaissance reveled in their rejection of Aristotle and the authority his philosophy still commanded among the academicians. Intoxicated by the sense of their own creative spirit breaking free from the supposed tyrannical shackles of old authority, they were blind to the tremendous intellectual accomplishment their forefathers had made since the 11th century. Faceless humility was no virtue of the Renaissance. Writers and artists saw themselves giving rebirth to the free spirit of classical antiquity. Florentines declared Socrates their honorary mayor. Aristotle, an expression of oppressive authority and scholastic sterility, was replaced with Plato and Neoplatonism as expressions of the creative force discovering and illuminating the inner forms and structures hidden within physical nature. Their break was as much a declaration of intellectual independence cast in art, sculpture, architecture and philosophy as an act of individualist liberation in the name of innovation, experimentation and critical questioning. If humility was an ideal of the previous age, pride was the virtue of Renaissance genius.

Whether Plato over Aristotle, or novelty over tradition, the inversion was not directed against the Church or religion, for the Church and the popes of the

Renaissance were among the greatest patrons of Renaissance genius. Just as popes, cardinals and bishops shared Renaissance values, so also the Church and secular society shared the same world, holding hands in paeans of praise to harmony, beauty, perfection, perspective, aesthetic creativity, measurement, discovery and individualism. Unwittingly, the Church was promoting a powerful and uncontrolled spirit that ultimately undermined its authority, just as its cult of reason had done.

The global voyages that opened unsuspected horizons, of which traditional authorities had been ignorant, fed into the critique of doubt regarding scholastic tradition. The acid of critical analysis seeped into all the cracks and crevices of medieval authority, eating away sinews, ligaments and fundaments. Philological analysis laid bare the fraud of the so-called Donation of Constantine, a forged document of late vintage attesting to Constantine the Great's willing his office and empire to the Bishop of Rome. The fraudulent claim of papal supremacy heaped discredit on a papacy and Church already charged with materialism, sensuality, venality, corruption, nepotism and hunger for power. But who could have foretold the consequences of an unimportant friar in a German town using a cathedral door to publicly air his spiritual problems with authority?

There is more than a touch of irony in the reformist rebellion against the oppression of the humanist, Renaissance-friendly Church. Once Protestantism had been established, its leaders, Luther and Calvin, reacting in the name of divine grace against a theology drenched in free will and reason, were more authoritative and restrictive to human will and the spirit of free inquiry than the Church, in its most oppressive days, had ever been. Theological freedom, sexual laxity and material corruption seemed to hang together. Protestations against Church corruption and worldliness had been brewing long before the Renaissance and Reformation. It had taken many forms, among them the movements of St. Francis, the Cult of Mary, mysticism, pantheism, Albigensians, Waldensians, and the reforms of Wycliffe in England and Hus in Bohemia. Several were bloodily repressed. The Renaissance added an intellectual fuel more shattering to Church unity than the flames of reform and rebellion.[2]

The trauma of schism that ended European religious unity led the Church to tighten up and retrench. Reforms were enacted to counter the Reformation. As the Church in self-defense was casting off its liberal humanism by painting over the Vatican nudes of the Renaissance, censoring books, burning Bruno and showing Galileo its instruments of torture, Protestantism, once Luther and Calvin were buried, became increasingly liberal in its deistic theology of intellectual and individual freedom. Protestant freedom did not open the way to a scientific revolution – the revolution had been in progress for some time – but it did generously foster it once it was in progress. The road to revolution had been bedded, bricked and paved by the Church. Copernicus was a monk; Barberini, Galileo's former friend and ecclesiastical support in his scientific endeavor, was a cardinal, then pope. Mersenne, unofficial secretary of international scientific communications, was a monk. The traceable roots of the revolution ran deep in Church and university scholarship of the 13th, 14th and 15th centuries.

More immediately, those roots are seen in the heirs of that scholarship, in the land of the Renaissance, of Leonardo da Vinci, Girolamo Cardano and Galileo, where Florentine and northern Italian craftsmen and engineers, whose technical expertise, an extension of Renaissance sculpture and painting, was characterized by realism, exactness of measure, experimentalism and innovation. These were the foundations from which would spring the Scientific Revolution and, ultimately, the West's military power, not merely over Islamdom, but over the globe.

The Renaissance began as a northern Italian affair, with Florence at the center. Cities in northern Italy had been spared the depressed conditions of the early medieval period. Unlike their less gregarious French, German or English counterparts, Italian nobles preferred city life to the country estate and joined the wealthy bourgeoisie in trade and small industry, even dropping their aristocratic pretensions and marrying daughters of well-heeled non-nobles. The social cohesiveness of nobles and bourgeoisie in an urban mix of trade, travel and industry proved more conducive to intellectual vitality than the rural isolation of lords devoted to hunting and overseeing vassals on the manorial estate. With the business of exchanging goods, of improving their quality to beat competition, of increasing productivity and of the construction of ships to transport them, came other exchanges of a nature less material than the management of industry and commerce. The production of wealth allowed time for reflection and experimentation. Men of means, urban nobles, bankers, manufacturers, merchants, provided sources of patronage for genius to work out the inspirations of mind and handicraft. In Florence, corporate patronage by the city's leaders took the place of church and royal patronage. Creative in art, sculpture, architecture and literature, but to a lesser degree in philosophical penetration and scientific discovery, the Florentine Renaissance paralleled the one taking place at the time (sculpture aside) in the Ottoman, Safavid and Mogul realms, the last empires of power and cultural prestige to grace Islamdom.

It could be said that the Renaissance in Florence and northern Italy, coming as it did on the heels of the translation and assimilation of the Greco–Arabic legacy, was midwife to the Scientific Revolution. This would be in reference to the Renaissance's emphasis on naturalism, experimentalism, precision of measurement, critical inquiry and humanism. If the rise of medieval Latin science between 1150 and 1500 can be called a renaissance, in the sense that the Greco–Arabic legacy was "reborn," science between the publication of Copernicus' *de Revolutionibus Orbium Coelestium* in Catholic Prague in 1543 and Newton's *Principia Mathematica Philosophae Naturalis* in Protestant England in 1682 was by every measure a revolution, even if it was not called such until the 20th century.

The Italian Renaissance brought together the three strands that went to make that Revolution: mathematics, logic and, the adolescent of the triumvirate, experimentation. Until then, mathematics and logic had existed as separate branches in the study of natural philosophy and were developed independently. Physics, another branch, received little attention after the accomplishments of Buridan, Oresme, Bradwardine, and the Mertonians in the first half of the 14th century. The strands gradually intertwined to form what would be called the "scientific method" in the early 16th century.

The logical component of this weave had been exhaustively worked out by the 14th century scholastics at the universities of Paris and Oxford and was reformulated by Renaissance scholars of philosophy and medicine in the northern Italian universities of Padua and Bologna, where there evolved a method of scientific problem-solving called "Resolution and Composition." Resolution was an inductive process moving from observed facts to their cause; composition was deductive, moving from cause to effect. The observation of nature, a prominent feature of Aristotle's scientific practice that had been lost in the West, or was at most sporadically practiced, was quietly and progressively cutting away the ground upon which rested the logical structures of medieval Aristotelianism. Where this logical component of the evolving method was a product of the universities, the components of experimentation and measurement came from other provenances.

Experiment and measurement came not from the heights of university scholarship but from the practical concerns of 15th and 16th century Italian artists, craftsmen and engineers, men dedicated to precision and artists who believed that the quality of art resided in the precise proportioning of forms. The artist strived for realism and naturalism; the craftsman and engineer for exactitude in the precise working of machines – mechanical clocks, hydraulic lifts and artillery. The emphasis they placed on precision and their practice of experimentation made these Italian craftsmen and engineers the pioneers of an empirical science. That old aristocratic prejudice institutionalized by Aristotle's philosophy and the Aristotelianism that followed, that those who know the general principles of things are nobler than those who know only to make things, that action of mind is closer to God than action of hand, began giving way to the empirical methods of creative practitioners who, versed in mathematical measure, joined the cunning of hand to the creativity of mind.

Leonardo Da Vinci

The spread of knowledge and newly discovered facts facilitated by the printing press and government-funded research on physical problems that were of military interest served to promote the convergence of logic, experiment and mathematics.

Towering over this breed of practical men who dealt with nature and its forces with hand and mind, and who worked and thought outside the university's pressure of scholastic conformity, stands the figure of Leonardo Da Vinci (d. 1519). A self-made genius of humble origin who, unable to afford a university education, spent his youth working, Leonardo embodies the fullest meaning of the Renaissance ideal of 'uomo universale,' the man who, as master of art and knowledge, enjoys the wholeness of life. Craftsman, engineer, architect, sculptor, painter, physicist, anatomist, biologist, mechanician, and visionary, Leonardo united these individual pursuits as parts of a transcendent whole in the Platonic philosophy that was so popular in the Renaissance. As it was for his contemporary Nicolas of Cusa and many Renaissance figures, at the core of Leonardo's spiritual and creative life was a humanist Platonism verging on pantheism: Man was the measure of all things, while his mind and spirit embraced the immeasurable unity of cosmic being.

The times were liberal and Leonardo dared think accordingly. Claiming the moon and planets to be made of the same stuff as the earth, bodies in a vast cosmos that were no more nor less divine and incorruptible than the earth, Leonardo left no doubt about his break with Aristotle and the logical athleticism of the scholastics. In the grand manner of the Renaissance man breaking with tradition, Leonardo unabashedly wrote of having dissected ten corpses with his own hand. As scientist, Leonardo held that the only way to investigate nature was through experimentation and observation of objects in motion. "Wisdom," he included among his many aphorisms, "is the daughter of experiment . . . Experience is never at fault . . . Who would know nature must know motion." The experimental spirit of Friar Roger Bacon was emerging after centuries in the closet.

With Leonardo dawns the scientific method. And so too does the conflict between the university scholar of natural philosophy and the emerging engineer–craftsman as a non-academic scientist. Having no university degree, Leonardo was aware that the educated elite regarded him as a social and intellectual inferior, a craftsman and painter who worked with his hands, an unlettered tinkerer with dirty fingernails. In his own words:

> I know well that because I have not had a literary education there are some who will think in their arrogance that they are entitled to set me down as uncultured – the fools . . . they do not see that my knowledge is gained rather from experience than from the words of others: from experience, which has been the master of all those who have written well.[3]

Leonardo taught himself many disciplines, but for some reason not mathematics. This is odd, for he maintained that mathematics should be applied wherever possible, as it was only in number that knowledge was made certain and exact. His greatest inspiration came from the master mathematician of antiquity, Archimedes, the one upon whom he modeled his own scientific work, but the extant works of Archimedes did not appear in Latin until 15 years after Leonardo's death. Had Archimedes been available 15 years before his death, it would have made little difference, for Leonardo believed that mathematics had to be learned when young, in school and in the university. Poverty had deprived him of that. His lack of mathematics and university credentials pained him all his life. His genius was not an impervious shield against the envy-driven glances and whispers of fellow professionals.

His lack of mathematics did not impede his genius in intuiting analogies hidden in nature's diverse phenomena. Through his knowledge of the hydrodynamic principles of Archimedes and his studies in the propagation of waves in water, he believed that sound and light were also wave-propagated: waves of water, reverberating echoes, transmission of light, these were all analogous manifestations of a single natural principle.

Like his hero of antiquity, Leonardo was a visionary of mechanical contraptions; but unlike Archimedes, whose drawings of pulleys, levers, hydraulic screws and pumps were purely for amusement, Leonardo believed in the practical application of the machines he sketched. His notebooks with their masterfully detailed, precise and realistic drawings of anatomy, of his machines, and of his experiments determining the relationships between loads, momentum, cross section areas of cantilevers and breaking points, all these indicate that the heretofore separately existing elements of mathematics, experimentation and logical deduction drawn from observation were coming together.

By the end of Leonardo's life, a creative community of Italian scientists and mathematicians centered in the universities of northern Italy were reviving mathematical study after a lull of centuries. The center of mathematical study in this little renaissance was Bologna.[4] The more these mathematicians came to absorb the contributions of their Muslim counterparts, the more important discovering the original Archimedes became. This driving passion was not as much for the mathematics – this they had from the translated Arabic sources – but out of the obsession to reach back to the classical original, the Greek source of genius that was at the inspirational heart of the Renaissance.[5]

Filosofo geometra

The Italians who provided printed Latin editions of Archimedes's work were driven by the ideal of precision and measurements in number as a control over mere words. Words were what the scholastic logicians had lost themselves in, as these mathematically minded men saw it. Through their own speculations into nature, they progressively insinuated mathematics into a natural philosophy that became increasingly non-Aristotelian. This in turn gave rise to a variant of the high medieval natural philosopher, the "filosofo geometra," master craftsman, engineer, logician, astronomer and geometer all in one, a Leonardo armed with mathematics.

Adept in many things, the filosofo geometra was natural philosophy's version of the Florentine ideal of the "uomo universali," a jack of all trades and master of them as well. He introduced the gritty labor of experimental work into the ethereal realm of natural philosophy. He was science's version of the Renaissance artist who carefully determined optical perspective; the clinical anatomist who cut into the cadavers of men, women and animals to study and capture in precise drawings the reality of life from the dead; the sculptor who precisely measured, chiseled and shaped the marble block; or the architect–engineer who, in transforming chunks of physical matter into designed constructions, needed to calculate weights, cantilevered loads, arcs, angles, or more grandiosely, the geometry of space in order to capture the magic of heaven's volume under a cathedral dome in Florence or Rome. The accomplishments of these filosofo geometra spanning the 16th century (Tartaglia, Commandino, Cardano, Guidobaldi and Benedetti) bridged the genius of Leonardo to that of Galileo and brought European mathematics up to the highest level achieved by Muslims.

Niccolo Tartaglia (d. 1557) published Latin versions of several works of Archimedes in 1543 and applied mathematics to his studies in mechanics and military science, published as *Nova Scientia*. The growing importance of artillery opened the way to ballistics, a field of applied mathematics in which governments were seriously investing. In his state-funded studies on fortifications and ballistics, Tartaglia made ballistic tables and calculated that an elevation of 45 degrees would provide maximum range. His analysis of trajectories showed them to be evenly curved at all points in a projectile's flight, rather than just precipitously falling to earth upon depletion of impetus. His student, Guidobaldo del Monte (d. 1607), wrote a book on mechanics that was considered to be the most authoritative treatise on statics to emerge since antiquity and was to remain so until the appearance of Galileo's *Two New Sciences* in 1638.[6] The collective work of Commandino and Guidobaldo in statics, mechanics, mathematics, astronomy, falling bodies and the crafting of measuring tools shaped the scientific culture into which Galileo was born. Guidobaldo was a patron and friend of Galileo and figured largely in Galileo's obtaining his first academic appointment, at Pisa in 1592, and then several years later another prestigious post, this one in Padua. They corresponded on scientific subjects right up to Guidobaldo's death just two years before Galileo constructed his first telescope. Galileo wrote of him as the greatest mathematician of his time. In his work on trajectories, Guidobaldo anticipated Galileo's discovery that a propelled body moved in a parabolic trajectory. Guidobaldo had referred to the trajectory as similar to a parabola or hyperbola, the arc of ascent being symmetrical to the arc of descent.[7]

Another great pioneer in applying mathematics to experimental physics was their younger contemporary, Girulamo Cardano (d. 1576). His physics was a mélange of old and new, one part mired in the antiquated Aristotelian theory of motion, the other experimentally rooted in the geometrical school of Buridan, Oresme and the Oxford Mertonians. Trying to resolve the contradictions between them, he studied trajectories, as had his older predecessor, Tartaglia, but where Tartaglia determined the trajectory of a projectile to be a curve of sorts and left it at that, Cardano more precisely described the curve as approximating a parabola, anticipating both Guidobaldo and Galileo. Cardano was one of the few scientists who was still in the last half of the 16th century properly crediting Arabic sources for having given the Latins the mathematics they had. Mathematics, Cardano believed, offered a logical map of life that should be rigorously followed. A divine science whose power influenced life, earthly counterpart to the ancient logic of astrology whose influences came from the stars. Cardano's passionate belief that mathematics ruled the world, together with his addiction to gambling, led to his developing a new mathematics of probability to narrow the possibilities of chance – and avoid his own personal bankruptcy. With a touch of mischievous humor, or misplaced devotion, Cardano, as devout a believer in the stars as in numbers, cast a horoscope for Christ and dared to have it published, becoming perhaps the only man in the West ever to be arrested for casting a horoscope.

While Cardano was studying the trajectory of propelled bodies, Giovanni Battista Benedetti (1530–1590), Venetian mathematician, physicist, philosopher and instrument maker, was speculating on freely falling bodies. His anti-Aristotelian views in mechanics and criticism of Aristotelian cosmology, combined with his application of mathematics to physics and bodies in motion, "heralds Galileo's general overthrow of Aristotelian science."[8] His formal institutional education had ended when he was seven, from which time he was, like Leonardo, self-taught. In his mid-teens, he learned some music and mathematics from his father and geometry from Tartaglia. At 22 he published his first book, *Resolutio*, which was on the solution of geometric problems in Euclid's *Elements*. In his early 30s he was designing fountains, constructing sundials, advising on engineering projects for the ministry of public works and the military, all while lecturing on mathematics and Aristotle and making money on the side casting horoscopes. As a physicist, he advanced a step in the study of falling bodies. Taking a note from Leonardo's analogies on waves, Benedetti thought of a freely falling body in terms of a body sinking in water, giving rise to the name of his concept, Benedetti's "buoyancy theory of fall." He arrived at the theory by way of two disparate sources, one Greek, the other Arabic. The Greek was the Archimedean treatise on bodies in water, translated into vernacular Italian by Tartaglia and published in Venice in 1551. The Arabic source was the medieval theory of impetus going back to the concept of *mayl*. By means of his buoyancy theory, and spurred by a flash of intuition, Benedetti concluded that objects of the same material but different weights fell at the same rate through a given medium. That the rate of fall did not depend on weight was a triumph of intuition overcoming common sense, comparable to accepting a moving earth.

The idea of weight being irrelevant to rate of fall must have been in the air breathed by the northern Italian physicist–mathematicians in the early 1550s: the year before Benedetti published his buoyancy theory in his second book, *Demonstratio* (1554), a mathematician from Brescia, Battista Bellaso, published a book "in which it was asked why a ball of iron and one of wood will fall to the ground at the same time."[9] But with their identity of interest and anti-Aristotelian views and close proximity, being all packed together in a corner of Italy comprising Padua, Pisa, Bologna and Florence, laced together in student, teacher, patron relationships, it is hard to believe that Galileo would not have been aware of Benedetti's work.[10] It was a close community, and its mathematical members were working in the same field of dynamics. Tartaglia's and Cardano's ballistic tabulations and mathematical analyses of trajectories foreshadowed Galileo's studies of motion. Before Galileo aimed his homemade telescope at the stars, he had already mathematized Benedetti's buoyancy theory of falling bodies.

The works of Cardano, Commandino, Tartaglia and Benedetti collectively reflect the growing unity of mathematics, physics and experimentation, while their support by government and wealthy individuals indicates the social importance being given to science and mathematics as a source of military and economic power. Individual investment in science was also becoming a norm. Such was the origin of northern Italy's first specialized public schools, such as the "Piatinne" schools. Dedicated to the study of Greek, mathematics and astronomy, these were public schools founded by the physician and mathematician Tommaso Piatti (d.

1502), whose patronage expressed the importance that science was coming to share with art, literature and medicine during the Renaissance. Cardano, after studying medicine, mathematics and astronomy at the university in Padua, taught mathematics in a "Piatinne" school in Milan. Through the lives of these 16th century Italians of the north, science is seen clearly as coming into its own, having emerged from the chrysalis of the Church and standing on its own secular legs, becoming practically an ordinary profession offering promising careers in government, military, academia and specialized Piatinne-type prep schools

Owing to these post-Leonardo Italians, mathematics in the West had, by the mid-to late-16th century, reached the level achieved by the cumulative best of six centuries of Muslim mathematics, though some algebraic methods that had been in common use by Muslims were known only to a rare few of the best of the Italians. For example, a method of resolution for third degree equations of the form $x^3 = ax + b$ was, independent of Arabic sources, discovered in the early years of the century by Scipione Ferro (d. 1526). Like an alchemist guarding his secret process for making gold, he kept it to himself. Secrecy of discovery was a common practice in the culture of science and mathematics. Before he died, Ferro taught the method to his student, Nicolo Tartaglia, who in his turn preserved it from public knowledge. Decades after learning the solution from Ferro, Tartaglia reluctantly divulged it to his friend Girolamo Cardano, but only after the latter had implored him to vow never to reveal it. Cardano vowed. This was in 1539. But 6 years later Cardano published the solution, giving full credit to Ferro and Tartaglia as the original discoverers. Giving credit did not help. Tartaglia was so furious over what he considered to be a betrayal by a trusted friend that he publicly ended the friendship and insulted Cardano in scathing attacks published for the whole scientific community to read. Not satisfied with having discredited his former friend, Tartaglia challenged him to a duel. Cardano accepted. At this point, another mathematician, Ferrari, famous for his solution of 4th degree equations, defended his friend Cardano's right to publish Tartaglia's method, arguing that knowledge should be public, and offered to duel Tartaglia in place of Cardano.

The duel was held in 1548 in the Church of Santa Maria del Giardino in Milan and was open to the public. The duelists faced off in the presumably packed church and instead of pistols or swords, aimed a total of 62 problems in mathematics, optics, physics and astronomy at each other. The duel of minds ended without a clear winner. The contest's questions and answers, preserved in the archives of Santa Maria del Giardino, stand as a good measure of the level that science in the West had reached by the middle of the 16th century. Five years earlier Copernicus had at last allowed his long-written *de Revolutionibus* to be published. That and the questions and answers recorded in the Santa Maria duel would make a fair gauge in fixing the point in history when the West can be considered to have equaled the highest level achieved collectively by Muslims over their long span of scientific and mathematical mastery.

The mid-16th century can be marked as the time that Western science clearly started to surpass the Muslim achievement, ending the long period of parity that had existed between the two civilizations. Having diverged for centuries, as measured by rates of investment in science, their respective trajectories of cumulative scientific learning commenced an enduring divergence. The West continued investing more and learning more, while Islam continued investing less and losing intimate contact with the impressive heritage their scientists had created.[11] After 1600, the growing divergence between them shifted from linear to exponential.

The drastic consequences this divergence would have on the Muslim world did not become evident to Muslim political leadership until the end of the 17th century. The inability to redress the imbalance would plunge Muslim civilization into a long nightmare of defeat and contraction, accompanied by loss of cultural autonomy and self-confidence, the effects of which presently weigh so heavily, even fearfully, on the whole world, and most humiliatingly on the Muslim half of that world. Catching up to redress the imbalance would prove most difficult. By the time Muslims recognized the reversal of fortune, the reforms they initiated were too feeble to diminish the distance between themselves and their powerful rivals.

As the Renaissance spread northward from Italy, natural philosophers in and out of the Church and universities adopted the new methods of addressing nature. The walls separating the mathematical, logical and experimental approaches to science crumbled as trained mathematicians experimented, scholastic logicians measured and experimented, and empiricists adopted the scholastic's method of resolution and composition, which a century later Descartes would formulate into a method of analysis and synthesis. The global discoveries of the late 15th and 16th centuries ushered Ptolemy's *Geography* out the door, to be replaced by Mercator's new cartographic technique of stereographic projection. Europeans began to perceive that the frontiers of scientific knowledge set by Greeks and Muslims had been reached. Some triumphantly trumpeted their superiority. Leonardo claimed to know more about anatomy and physiology than any Galenic physician and more about physics than any Aristotelian. Paracelsus boasted he knew more about medicine than Galen and ibn Sina combined, and to make the point, this German Swiss physician, otherwise known as Philippus Aureolus Theophastus Bombastus von Hohenheim (d. 1541), publicly, right in front of the University of Basle, burned a copy of ibn Sina's *Qanun al-Tibb*, the standard university medical text used throughout Europe. His burning the traditional authority and renaming himself Paracelsus, that is, above the famous Roman physician Celsus, were acts of self-applauding theater intended as serious symbolism. "Knowledge is experience" was the logo he claimed to live by, and he set out to reinvent medicine accordingly.

Proclaiming that it was experience from the book of nature that was to be followed, rather than blind acceptance of ancient authority, he advised against overtreating the patient and attempted experimentally to find natural substances that induced healing. The body should always be given the chance to heal itself. Rebellious by nature, Bombastus von Hohenheim claimed to have abandoned the authority of Galen and ibn Sina for a new system of medicine based on natural healing and chemistry. Instead of bleeding, purging and sweating to balance the four humors, he treated patients directly with herbal essences, chemicals and drugs that he made in the course of his medical experimentalism. Chemicals such as mercury and derivatives of sulfur had been used before, usually in fatal does. Paracelsus administered small doses and treated his chemicals by oxidation or by rinsing them in water or alcohol. Realizing mercury was lethal, he used it sparingly and in a derivative form. There was as much that was old and harmful in his medical practice as new and beneficial, but by observing the healing powers of the body he opened medicine to new possibilities.

Two years after Paracelsus died, Ptolemy followed Galen out the door.

Copernicus

Contemporary with Tartaglia, Cardano and Paracelsus was the Polish monk who was a self-taught amateur astronomer. While a student of canon law and medicine at the University of Bologna, he seized the opportunity to formally study astronomy. He continued his study in Padua, where he imbibed the heady spirit of the Renaissance and the mathematics of the northern Italians and evidently came into contact with the reform astronomy of the Maraghan School of Nasir al-Din al-Tusi and his followers. Taking the example of the filosofo geometra, he built his own observational equipment, learned Greek and studied Hellenistic and Muslim mathematics, logic and natural philosophy. At some point while studying in Bologna and Padua, the young monk came to the conclusion that God could not have designed the architecture of the universe along the dizzying whirls of Ptolemaic circles and epicycles. The system was too awkward and ridden with problems that were as much astronomical as philosophical and aesthetic. If humans were so cunning in crafting clocks whose precision was not marred by the many complex parts working in unison, whose harmonious movements were geometrically integrated to produce a mechanical marvel, then could not God in all his wisdom do as well?

His mathematical studies led him to Pythagoras and heliocentrism, to a universe where the warming hearth, the source of life, was at the center of the house, filling it with light. The ancient Egyptians, so admired by the Greeks, realized the sun was source of everything. It was only right that the sun, symbol of light, life and truth, should have the honored place at the center. The sun-centered system was spiritually, rationally and aesthetically right. The Greek astronomer Aristarchus had thought so too. Copernicus looked to Pythagoras and Aristarchus as Leonardo had to Archimedes.

Astronomers through the ages had considered heliocentrism or a form of it. It was only logical that they did, given the headaches of Ptolemaic astronomy. With the sun, the planets and the sphere of fixed stars being so far from earth, it seemed simpler to have the earth move and thereby to avoid the staggering velocities that a stationary earth imposed on the universe. But there was an opposing logic whose empirical basis was as overwhelming as the velocities of celestial spheres turning on a stationary earth. That the angle of the fixed stars as observed from earth did not change during the course of a year was proof of a stationary earth. No one suspected that the stars were too far away for any parallax to be detected by the unaided eye. Also, if the movement of the heavenly bodies was really the effect of earthly motion, then the earth would have to be revolving around the sun and rotating on its axis so fast that everything would be swept away. Motion opposite to the direction of earthly motion would be impossible. Something thrown vertically up in the air would land miles away. Birds would be swept away if they tried to fly, trees uprooted. How Copernicus resolved these problems in his own mind is a mystery, though a 13th century Arabic source, *Hikmat al-'Ayn* of al-Katibi (d. 1272), has it that if the earth did move, then the atmosphere would be moving along with it as part and parcel of the earth's motion and birds would not be hindered from flying opposite the earth's direction of motion. Despite his correct reasoning, in the end al-Katibi rejected the possibility of a moving earth because it was not in accordance with Aristotelian principles. No one after him took it up.[12] Copernicus must have come to the same insight as al-Katibi, for how else to explain the possibility of life on a rapidly moving planet?

His dissatisfaction with Ptolemy stemmed from the same sources that fed Muslim dissatisfaction, the system's cumbersome structuring and the equant, though the Maraghan's had finessed the equant out of existence using the double epicycle or Tusi Couple, with which Copernicus was familiar.[13] Lacking Paracelsus's sense of superiority over the ancients, Copernicus at first had no intention of sweeping away the old. His break with Ptolemy was motivated not by a search for a new astronomy but a reversion to a yet older Greek system, the Pythagorean. The Copernican system's planetary models preserved the bulk of the Ptolemaic architecture of deferents, epicycles and eccentrics. Ptolemy, it will be recalled, made the center of the universe an imaginary point he called the eccentric that was a defined distance from the earth. Copernicus preserved the eccentric as the center and merely exchanged the positions of earth and sun so that the center of the universe became the center of the circle around which the earth orbited the sun, namely the eccentric, thereby retaining an essential element of Ptolemy's geocentric system.

The Copernican system was not that much less complex than Ptolemy's in terms of numbers of circles and circles turning on circles. It did simplify Ptolemy in one instance. Since Copernicus had the planes of all planetary orbits passing through the center of the sun, planetary motion in latitude became more uniform and easier to determine. Copernicus was also able to banish the Ptolemaic equant that had challenged the imagination of Muslim astronomers to the limits of geometric invention, especially the 13th and 14th century astronomers of Maragha.

The Maraghan connection to Copernicus has been a subject of considerable research. It centers on the planetary models of Mercury and Venus that Copernicus produced in his *de Revolutionibus*. So close are those complex revisions of Ptolemy to those of ibn al-Shatir, the 14th century Damascene astronomer who was a student of Nasir al-Din al-Tusi in Maragha, that it is highly unlikely that the latter were made independently of the former. The connection is hard to discredit. The diagrams and alphabetical signs Copernicus used to designate arcs and angles precisely match those found in manuscripts of Nasir al-Din and ibn al-Shatir.[14]

Tusi's double epicycle, a sphere rolling in a sphere of double its diameter, was reproduced in drawings of planetary models in a book by Giovanni Battista Amico (d. 1538). The book was printed three times, in Vienna and Paris, and would have been found in the collections of major universities such as Padua and Bologna.

While a medical student at Padua, Copernicus became acquainted with a physician and philosopher named Girolamo Fracastoro (d. 1553), who taught at Padua and had written a book on medical astrology that borrowed from Amico's book. Like Copernicus, Fracastoro was an amateur astronomer searching for an alternative to Ptolemy. His classical hero of choice was Eudoxus, whose theory of planetary concentric spheres Fracastoro used to model his own universe. Copernicus could easily have learned of Maragha reform astronomy from Amico through Fracastoro at Padua.

Copernicus hesitated to publish his revolutionary astronomy. Though the church was still relatively liberal and steeped in the Renaissance, he feared that Biblical literalists would take unkindly to the idea of a moving earth, since that would reduce it to being but one of several planets orbiting the sun, thereby lessening man's cosmic significance in the scheme of divine creation. Aristarchus had been accused of impiety for theorizing a moving earth, and that was back in the theologically less dangerous days of live and let live polytheism, when there was always room for another myth and another god. Oresme had also written of a moving earth without incurring any problem, but that was in the 14th century, when the Church and its authority were more or less securely fixed at the center of Latin civilization. How would the Church react in these bitter days of Lutheran and Calvinist protest? There was no telling. Years before *de Revolutionibus*, Copernicus had published a small treatise called *Commentariolus*, which explained his system and brought him no trouble. On the contrary, as word of Copernicus's new system made its way through the papal corridors and Pope Clement VII came to hear of it, he had his secretary explain it to him and then rewarded his secretary with a valuable Greek manuscript, implying he must not have been too displeased with what he heard. Indeed he was not. Pope Clement told a cardinal about the new theory, and not long after, the cardinal, who happened to know Copernicus, urged him to publish his much larger work on the system. In gratitude, Copernicus dedicated *de Revolutionibus* to Pope Clement and prepared to publish. Yet, cautious to the core, he still hesitated. Finally, when he was 70 years old and close to death, his student George Rheticus implored him to let him arrange for the book's publication while he still lived.

Copernicus by then felt he had little to lose. But to be on the safe side, it was arranged that the book would be printed in Protestant Nurenberg. When Martin Luther got wind of what was going on, he vehemently prohibited any book with such heretical ideas as a moving earth to be published in a Lutheran country. He would have burned the book, maybe along with its author, had he got his hands on them. Fearing for his life, Rheticus, a Catholic, fled the city, leaving the manuscript with Andreas Osiander who was supervising the printing until Luther stopped it. Osiander was a Lutheran theologian with that streak of liberality that, in the next century, would become mainstream Lutheranism (while the Church was concomitantly abandoning its liberalism to take on the conservatism cast off by the Protestants). Had early Lutheranism not inherited a spark of Church liberality in the likes of Andreas Osiander, Copernicus's manuscript might well have been burned, or more probably, its publication further delayed.

Ignoring Lutheran fundamentalism, along with Martin Luther's order, Osiander secretly traveled to liberal Leipzig where, in 1543, he had the manuscript printed. To protect Copernicus from the wrath of Catholics and Protestants, Osiander appended his own anonymous preface to the book, assuring the reader that the new theory presented by the author was merely a hypothetical mathematical exercise that made no claim to represent the reality of the heavens. The hypothesis was neither right nor wrong and neither true nor false, being that the Copernican and Ptolemaic systems were identical in everything but the point of reference from which the motion of heavenly bodies was observed and measured: one point of reference could be substituted for the other without incurring any harm to astronomy and its mathematics.

Copernicus escaped any possible wrath coming from the Church by dying in Frauenburg days after the book's publication.[15] He did not need to die to spare himself grief. Osiander could as well have spared himself the trouble of the disarming preface. The book ruffled no Church feathers. Clement VII gave no notice of it, nor apparently did anyone else in the Church. Owing to the forbidding denseness of its computations, geometric diagrams and spherical trigonometry, probably few but the most dedicated professional astronomers got beyond the introductory chapter. Osiander's preface may have been taken at its face value and the book reckoned nothing more than a mind-wrenching exercise in mathematical astronomy. Only the Lutherans were put out. As for the Church, the book was a non-event, eliciting more yawns than raised eyebrows from any bishop or cardinal who might have opened it out of curiosity or by mistake, only to quickly replace it after a glance at the mass of complex diagrams and mathematics that were crammed in the pages.[16] For the little fuss it raised, *de Revolutionibus* was not a promising beginning to what would later be called a revolution. Luther aside, the only revolution was in its name. Not until over half a century after the book's publication did it become an issue within the Church – but more than the contents of the book, the cause of the eventual uproar may have been the wounded amour propre of the pope, which was more Galileo's doing than it was Copernicus's.

Galileo was a filosofo geometra in the tradition of Leonardo, Tartaglia, Cardano and Benedetti. He studied ballistics, mechanics and hydraulics, built his own equipment and applied mathematics to his experimental studies of objects in motion.[17] He was not a trained astronomer; his interest in astronomy was that of any natural philosopher and mathematician in that astronomy was an important subject in the scientific corpus. His true interest was the physics of motion and the application of mathematics to dynamics.

Falling bodies had been a subject of interest and controversy for two thousand years. Aristotle got the analysis of it off to a bad start with his peculiar theory that a body in motion pushed the air, or the medium in which it was moving, from the front to the rear of the object, whereby the displaced medium acts as the force propelling the body forward. The medium thus both impeded motion by offering resistance to the body in a frontal direction and at the same time acted from the rear to push the body onward. There was much to criticize. The most thorough critical analysis came from John Philoponus in the 6th century, almost a millennium later. According to Philoponus, a freely falling body, or a body impelled to motion, was acted on only by the resistance the medium offered to it. Philoponus remarkably intuited that objects of different weights dropped from a given height strike the ground practically simultaneously. As mentioned in an earlier chapter, the anti-Aristotelian critique of Philoponus had been advanced by Muslim naturalists, particularly ibn Baja (Avempace) and ibn Rushd (Averroes) in 12th century Cordova. Based on the Latin translation of their work, kinetics was advanced by Oresme, Buridan, Bradwardine and the Mertonians in the 14th century. In the 15th and 16th centuries, the study of motion received new importance with the development of gunpowder. Governments in the 16th century were funding ballistic studies and creating university chairs for specialists in the field. This had earned Galileo academic appointments in Pisa and Padua.

As a mathematician working in his specialized field, Galileo received early recognition for his studies of free fall. Standing on the shoulders of giants, from Philoponus to ibn Baja, Buridan, the Mertonians and his 16th century mathematician compatriots, Galileo succeeded in mathematizing the law of falling bodies.[18] His book on the subject of motion, *de Motu*, made his name in the regional science community of north Italy. The international fame he gained, however, was not in this down-to-earth physics, but in his serendipitous encounter with the stars, which brought both him and Copernicus to public attention. Galileo was not an astronomer. Astronomy was literally a world apart from physics. It had been so ever since the philosophy of Aristotle first seized the minds of naturalists and astronomers. Astronomy was of the divine heavens, pure and perfect, with flights of angelic intelligences spinning circular orbs. Physics pertained to the physical and biological, to the living, the rotting and the dying. Its domain of study was the imperfect earth of change, accident, gross matter and imposed linear motion. Galileo's interest in astronomy was nonetheless keen enough, and his reason free enough from the restraints of institutionalized Aristotelianism that he became a Copernican as a young scholar in the 1590s, though not enough of one for him to support the system publicly, in spite of the urgings for him to do so by another mathematician that had converted to Copernicanism, Johann Kepler, who was also not an astronomer.

What drew Galileo to Copernicanism was that he found those who attacked the system to be the ones who knew least about it, while its defenders had studied it in depth and detail. He had experienced enough in the affairs of men and science to know that those who knew absolutely nothing about something were the ones most likely to attack it the loudest. But this was not enough for him to make a public defense of the system. It was a matter of accident that kindled Galileo's passion to defend it. He happened to hear of a new optical instrument invented recently by a Dutch lens-grinder, Johann Lipperhey, that magnified objects so they could be clearly seen at a great distance. It was regarded as a nautical instrument for sighting land and ships at sea. Galileo's inquisitive mind was immediately aroused. Unable to obtain one quickly, and knowing the physics of optics, he built himself a 30-power instrument based on a description of the Dutch device, and in

January of 1610 turned the instrument in pure curiosity to the heavens to see what he could see.

He was not the first to use it as an astronomical device. The English scientist Thomas Harriot (d. 1621) used it to observe the moon in the summer of 1609 and in fact sketched the lunar landscape Galileo was to see six months later. But because Galileo rushed to publish and Harriot was too much of a perfectionist to publish anything he thought to be unfinished, meaning he published very little, Galileo's name has been immortalized as the first to gaze at the heavens through a telescope and see what no living creature had ever seen before. And what he saw was not supposed to be there, the furrowed wounds and imperfections of the scarred lunar landscape with its craters, mountains and valleys that belied the Church-endowed Aristotelian purity of unsullied crystalline spheres God designed to be the flawless architecture of the heavens.

It was not only the physical reality of the moon. The planets appeared as disks, while the stars remained points, all of varying degrees of brightness, proving that the fixed stars were an immense distance away from the earth and the other planets, and were not at all equally distant, fixed in spherical sphere, the firmament, as the Aristotelians believed. The sun had spots, Jupiter had satellites and the Milky Way was composed of thousands and thousands of individual stars. Like a prophet revealing God's eternal scripture writ across the starry heavens, Galileo had it in print in less than two months of seeing it. Another two months and he was the most renowned scientist in Europe, a physicist–mathematician whose fame was made in astronomy. His *Starry Messenger*, dashed off in unadorned Florentine dialect and meant for public and academia alike, was modern history's first scientific best seller, a heavenly voyage as devastating to scholastic Aristotelianism and traditional science, already pretty tattered, as had been the oceanic voyage of Galileo's compatriot Columbus a little over a century earlier.

Where Copernicus had meant to improve on Ptolemy, Galileo's mission was the total destruction of Aristotle. Galileo never mentioned Copernicus or heliocentrism in the book. He did not have to. Aristotle and Ptolemy were the twin foundations of the sciences of earth and heaven. One could not support the system of the universe without the other. They stood and fell together.

Three years after his *Starry Messenger*, Galileo published a book on the sun-spots, again in the popular idiom, and it was here that he, at last, publicly declared himself a Copernican. His declaration and the publicity it aroused over the following two years made Copernicanism an issue the Cardinals considered too important for the Church not to take a stand on. Many of the Cardinals, dyed in the wool Aristotelians, were personally embarrassed by Galileo and his telescope. If the instrument could not be condemned, the theory could.

A Vatican council was held in 1616 to determine whether Copernican astronomy was permissible in view of what the Bible said about the subject. Fearing that a wrong decision by the council would harm both the Church and the free pursuit of scientific truth, and believing it would be a fair and open hearing and that he, the man who had seen the lunar mountains and the four moons of Jupiter, could convince the council to make the right decision by simply presenting the irrefutable proof of a moving earth in a calm and rational manner, Galileo made up his mind to attend. And so, steeled with the confidence of truth winning out, Galileo left his beloved Florence for Rome, in disregard of the advice of friends who, perhaps more familiar with the frailty of truth and reason in a world of institutional politics where preservation of power and vested interest ruled, warned him not to go. But to no avail. Never doubting his powers of persuasion and the irrefutability of science, Galileo traveled to Rome and entered the lion's den.

What was he thinking? That he would be given the awed respect owed to someone who had stepped on the moon? That like Nicolas of Cusa he would overwhelm the council with his brilliance and wit and come off with an honorary cardinalship? More likely, Galileo's self-invitation to defend Copernicanism was seen by the Vatican Council as a brazen challenge from a popular publicist who had stirred the whole thing up and was using science to embarrass the Church, another Abelard too clever by half, sticking his tongue out at authority. What is certain is that Galileo innocently failed to understand the fact that preserving institutional authority in a critical time of dissent and schism left no room for Church authorities to consider impartially the niceties of natural philosophy. Astronomy was not the issue. It was the danger of more schism and the poisonous seed of speculation that might sprout into a more dissent. The preservation of the institution was at stake. Galileo might have taken warning from the fate of Giordano Bruno, whose scientific speculations sent him up in flames in Rome 16 years earlier. Astronomer, mathematician, philosopher, playwright, defrocked Dominican priest, perpetually in and out of trouble and on the run from university to university, city to city, adventure to adventure, religion to religion, Bruno would have had a hard time of it career-wise at any time in any culture. For starters, he believed the earth was, in relation to the cosmos, nothing but a speck of dust moving around the sun in an infinite universe that had multiple suns and earths. Worse, he published these ideas in the form of a dialogue written in popular Italian dialect, exactly the way Galileo would publicize his natural philosophy, broadcasting it to the public at large. This printing press was stirring up a lot of trouble for old institutions. Bruno's dialogue set the Copernican system within an infinite universe with an infinitude of suns and earths. Printed in 1584, it was curiously titled "Ash Wednesday Supper." Ironically, Bruno was to be the ashes, but did the author mean them to be those of Aristotle and Ptolemy?

Bruno anticipated Galileo in another way, by arguing that the Bible was a book of morality divorced from natural philosophy. When the Church protested his ideas, Bruno left the Dominicans, then the priesthood, and finally Catholicism, to become a Protestant, but only to return later to Mother Church, in Paris, where he was teaching and where he thought he was safe from Church authority. Had Bruno remained in Paris, his bold philosophy and cavalier way of changing religions may not have brought him down. But, defying fate, Bruno left the safety of Paris and returned to Italy, where the poisonous atmosphere of schism and Reformation was thick. Venice, he thought, would be safe enough from the Church. There had been good reason for his return to Italy. The chair of mathematics had just become vacant at the University of Padua and he hoped to have it. Unfortunately for him, the chair was awarded to a rival candidate: Galileo. Disappointed, Bruno returned to Venice where he awaited what new adventure in life awaited him.

It was not what he expected. In Venice he was betrayed by a young friend who, supposedly dissatisfied with the mathematical lessons Bruno was giving him, told the Roman Inquisition where to find him. A more probable version of the episode has it that his young friend denounced him in a fit of jealous rage because Bruno was planning to leave Italy and abandon him for a university position in Protestant Germany. Before he could leave for Germany, the inquisitors arrested him and delivered him to Rome, where he was tried and condemned to death. When given the chance to save himself by recanting his philosophical heresy, he replied that since he had not been informed as to what part of his philosophy was considered heretical, he had no idea what it was he should recant. The Vatican inquisitors failed to see the relevance. Bruno could have recanted everything in general or asked what particular points were troubling the inquisitors, but, refusing to be cowered, he did not. Times had changed since Nicolas of Cusa had talked his way out of charges of heresy and been rewarded for it. When the death sentence was pronounced, Bruno, emulating Socrates, told the members of the papal court that perhaps they feared his death more than he did and still refused to recant. When he continued saying things the papal court had no desire to hear, he was gagged, then taken to the Forum and burned at the stake. Sixteen years later it was Galileo in Rome defending science before a council of frightened Vatican theologians who saw danger everywhere in this time of crumbling Church authority.

Galileo should have realized that, given the defensive posture of the Church, the wisest policy would have been to publicly profess orthodox belief or keep quiet, but he was no more able than Socrates or Bruno to keep his mouth shut. To the insecure and frightened Vatican authorities, Galileo was an arrogant, quick-witted nuisance who loved to humiliate those who disagreed with him: another Bruno. In this they were not far from the mark. A master of satirical wit and debate, Galileo would listen patiently until his opponents had finished attacking and disproving his non-establishment science and then would supply them with even more powerful arguments against his ideas than they had been able to muster and finish by disproving them all, his and theirs, in a masterful display of ironic humor, more humiliating than humorous to those whose misfortune it was to debate him. Galileo could not suffer pomposity or fools and was quick to reveal the shallowness of those who did not know what they were talking about. His favorite targets were the Aristotelian logicians of academia who failed to base their knowledge of natural philosophy on experimentation and mathematics. His having publicly made laughing stocks of a number of high personages in the Church and academia had given him powerful enemies.

On the other side of the ledger, he had some powerful friends in Rome, among them Cardinal Barberini, who would become Pope Urban VIII in 1624. Barberini was a friend and fellow scientific enthusiast with whom Galileo could speak freely. And there was Cardinal Roberto Bellarmine, theological advisor to the pope. Five years before the 1616 Vatican Council, Galileo had demonstrated his telescope to the Jesuits in the College of Rome, among them Cardinal Bellarmine. Impressed by the new optical device, Bellarmine had been most encouraging to Galileo. Galileo also must have expected support from those Jesuit and Dominican scientists in Rome who were fully aware of Kepler's discoveries that mathematically related planetary distances from the sun to their orbital periods. By the time Galileo had taken up Copernicanism, the theory, with its Keplerian support, had quietly spread among the European scientific community and received powerful support.

Owing to Galileo's position as Europe's leading mathematician and physicist, the Vatican Council treated him with great respect and listened politely as he argued the case for Copernicus. After a brief deliberation, the Council condemned two books, the first being one by the theologian Paolo Foscarini reconciling the Bible with a moving earth, the second being *de Revolutionibus*. Galileo was dumbstruck. He had presented the truth so clearly to the men of the council: valleys and mountains on the moon, satellites of Jupiter, starry distances explaining lack of measurable parallax, sun spots and a rotating sun. Not only was the decision a slap in his face, but by condemning what was unarguably true, the Church had struck a terrible blow against science and would now be forced to defend a system the scientific community knew to be false. The Church he loved would suffer for its blindness of having rejected the obvious – in effect, the same blind rejection Ghazali had written against five centuries earlier when warning of the dangers of religion's hostility to logic, mathematics and astronomy.

The upshot of Galileo's journey to Rome was that he was instructed not to defend the Copernican theory. Implied in the Vatican Council's vague instruction, however, was that he could teach, write or lecture on Copernicanism as long as he refrained from declaring it the true system of the world. To Galileo, this was meaningless. How could learned men deny the obvious? How could a seeker of natural truth teach, write and lecture on something that was known to be the truth and not present it as truth? How could men who knew the truth of astronomy keep silent? His faith in the decency, honesty and common sense shared by intelligent men must have dimmed a bit that day.

It would have dimmed more had he known that because of his testimony in defense of Copernicanism the Vatican opened a secret file on him. There was no way he could have known that, nor that among the documents deposited in his dossier was one he was never shown until he was forced by the Vatican to return to Rome many years later: a letter bearing his forged signature that bore testimony to his having vowed never to teach, write about or defend the Copernican system in any way, shape, or form. It was this planted letter that almost 20 years later trapped him into publicly recanting his belief in a moving earth.

Returning to Florence a chastened man, he avoided controversy and devoted himself to mechanical problems, of which the Bible had little to say. But Vatican fears and politics made for a tricky sea, and like Bruno, he was too creative and humanistically optimistic to navigate those treacherous waters whose calm belied a tempest in the waiting. In 1624 Cardinal Barberini became pope. Taking heart that a new day had dawned, Galileo returned to Rome to congratulate his old friend and see if the proscription of 1616 against teaching Copernican theory as the true system of the heavens could be lifted. Barberini refused; the political situation was not right, religious war had been raging in Europe since 1618, the conservative faction in the Church should not be antagonized. The pope did nonetheless confide to Galileo that it would have been better all-around had the edict proscribing Copernicus not been issued.

This was encouraging. Galileo must have taken it as an indication, as it most probably was meant to be, that the pope took the proscription of 1616 purely as a political device to foster Church unity and not wholly a rejection of Copernican theory as representing natural truth. More encouraging, the pope allowed him to write a book on Copernicanism, but with the strict proviso not to defend it or in any way put it forth as the true system of the world. The book he wrote would have to balance the Copernican hypothesis with an account of the Ptolemaic system, leaving it open as to which of the systems represented God's construction of the heavens. The book would have to be absolutely neutral and non-committal. These conditions were put into writing and made official by the papal director of censorship.

Galileo worked on the book in Florence for the next several years and entitled it *Dialogue Concerning the Two Chief World Systems*. Modeled on Plato's use of the form, it was an amusingly chatty dialogue among three characters in the spoken idiom of Florence. The idea of using dialogue as narrative may have been suggested to him by Bruno's use of it, though not necessarily. Both Bruno and Galileo, children of the Florentine Renaissance, shared the spiritual Platonism of Leonardo, Michelangelo and many others of their great Italian predecessors. Platonism stood as a flag of opposition to the institutionalized Aristotelianism of the day. Plato and Platonic philosophy were at the heart of Galileo's own approach to nature, where universal forms and ideas became mathematical, and the real was the inner structure that gave form to matter. The name of the select Italian scientific society Galileo belonged to was taken in that spirit: the Lincei, the lynx, whose penetrating sight was a metaphor of the natural philosopher's vision into the inner world of form and spirit hidden within the world of matter, which answered to Galileo's personal philosophy of nature: The physical world was made comprehensible through the world of number and mathematical relationships; God's mind, like nature's grammar, was mathematical, and to the extent that man knew mathematics he partook of the divine.

The dialogue ideally suited Galileo's purpose of comparing two systems, and he used it most effectively. Like Plato, Galileo knew where his dialogue would go and what truth it would produce at the end. Unlike Plato, he was simply to lay out the elements of two opposing world systems without a word that could be construed to favor a particular one of those two. But for Galileo, the evidence of one was so overwhelming, how to be impartial? His urge as a scientist to reveal truth as it was found, undisguised, was too powerful. Once he started writing he was unable to conceal what he knew to be true – equally unable to restrain his playful wit – and so in the joyous light of revealing truth to the world the pope's cautionary instructions evaporated from mind.

The ribald banter of three characters in colloquial dialect going on about the systems of Ptolemy and Copernicus is as humorous as serious science could be made to be. The humor was lost on those churchmen and academicians who might have seen themselves and their arguments mockingly reduced to absurdity. The discourse is carried by a Copernican, a Ptolemaic–Aristotelian and between them a supposedly neutral participant who comments either favorably or critically on the arguments and counter arguments that the two protagonists bat back and forth. (Curiously, the planetary orbits are still described as circles, nothing being said of ellipses or the sun occupying one of the two focal points, all of which was published decades earlier in Kepler's *Astronomia Nova* of 1603.)

The book was finished in 1630, submitted to the papal censors, given the papal imprimatur, published in 1632 and acclaimed by scholars all over Europe as a scientific and literary tour de force. It appeared for a while that Galileo had pulled it off. Barberini himself may have been amused, until someone pointed out to him that the argument of the book was overwhelmingly supportive of Copernicanism and that the character defending the world of Ptolemy and Aristotle, Simplicius, was an ineffectual but well-meaning fool whose arguments, it seemed to the pope after looking closer at the dialogue, sounded suspiciously similar to those he himself had used years ago when as a cardinal he defended Ptolemy and Aristotle in his friendly debates with Galileo. The pope suspected he had been used as the model for the gullible foil; that he had been caricaturized and made to look like a buffoon; and that he had been duped by his friend. Worse, his Jesuit advisors claimed the book would have an effect on Church teaching far more calamitous than Luther and Calvin combined. In rage and humiliation, the pope wanted the book withdrawn. The problem was, it had been approved by the Vatican censors.

It was then that the document was "discovered" bearing Galileo's signature and dated 1616, warning him never to teach, support or discuss the Copernican hypothesis under penalties of the Holy Office. This made it appear that Galileo had obtained the license to publish his book under false pretenses. The license was nullified and Galileo was ordered to Rome to stand trial for "vehement suspicion of heresy." Galileo tried desperately to avoid going to Rome. He kept putting it off, pleading illness and old age, until the Papal Office ran out of patience and informed him that if he did not present himself at once he would be brought to Rome in chains. By now an old man and failing in health, he made the journey and humbly cooperated with the inquisitorial commission. He had not the character of a martyr. He loved life and its sensual pleasures. He would avoid prison and Bruno's fate and say what he must to save himself. After all, it would be only a matter of time until scientific truth won out. Why die for a cause whose victory was guaranteed?

Galileo was dealt with firmly but not harshly. Seeing the old man had lost his health, and with it the cutting wit that had made his intellectual aggressiveness so lethal, the inquisitors treated him cordially; nor did they keep him in prison during the trial. Shown the incriminating document, Galileo denied ever having seen it. The sympathetic commissary general of the inquisition would have settled for a reprimand and release, but there were those in the inquisitorial court who demanded that an object lesson of some severity be made of Galileo. Accordingly, he was found guilty of teaching, supporting and believing the heretical system. His punishment? The same offered to Bruno: recant or be burned for heresy.

He recanted in a public ceremony in St. Peter's Cathedral. The prison term that accompanied his sentence was commuted by the pope to house arrest in his village home in the hills above Florence. Death came a decade later. During that decade he led a comfortable existence with his mistress and daughter, while continuing his scientific work. His publications were now in less controversial areas, mechanics and strength of materials, on which he composed a groundbreaking book based on his research, *The Birth of Two New Sciences*. He continued making heavenly discoveries, right up until his sight failed and his telescope was no longer of any use to him, but he prudently refrained from publicly relating them to Copernican theory.

Galileo's fame came from his telescopic discoveries and popularization of Copernican theory. Beyond this, he made no contribution to astronomy. His field of expertise was mechanics, a subject he made into a science and established as an independent field of study. The emphasis he placed on mathematics, capping the century-long efforts of the Italian filosofo geometra, and his discovery of the relationship between time and the distance traveled by a freely falling body, one of the first fundamental laws of nature to be expressed mathematically, were immense contributions to the transformation of science and methodology that had been developing since Leonardo da Vinci. As Benedetti anticipated Galileo's kinetics of freely falling bodies, Galileo anticipated Newton's law of inertia, that rest or motion was a state that left to itself would continue until acted on by an external force.[19] Riding the crest of mathematical mechanics built up by Guidobaldo, Commandino, Baldi, Tartaglia, Benedetti and additional filosofo geometra of lesser fame and imbibing the teaching of Francesco Barrozzi, mathematics professor at Padua, that mathematics was superior to logic in bridging the divide between divine and natural philosophy, Galileo made himself the new Archimedes. He used mathematics to bridge the divide between the static mechanics of Archimedes and the new dynamic mechanics coming from the ballistic studies of the filosofo geometra. By reducing mechanical dynamics to mathematical principles, he laid the basis of modern physics, "transforming the north Italian Renaissance of Mathematics into the Scientific Revolution."[20] With Galileo, the experimental physics and mathematics of the geometric philosophers that derived from Leonardo and Archimedes merged with the heliocentric astronomy of Copernicus that derived from Pythagoras and Aristarchus. Adding Kepler to this, it only took Newton to complete the equation.

Galileo's mechanics, Tycho Brahe's precision of measurement in observational astronomy and Kepler's theoretical mathematical astronomy merged to form a system of natural philosophy based on the quantitative relationships of mass, weight, distance, velocity and acceleration, sending to the grave once and for all the Aristotelian system of qualitative relationships based on the four qualities and their correlated elements, air, earth, fire and water. Galileo's concept of time and motion was a fundamental break with the past, "a true mutation in ideas" that "altered man's consciousness of a real world outside himself in nature."[21] What the Mertonians had accomplished by their analysis of velocity in the 14th century, Galileo revolutionized by his analysis of acceleration. His experiments on falling bodies, whether performed or imagined, and the dynamics he conceptualized by measuring distance against time, were revolutionary in their transformation of time into "an abstract parameter of a purely physical event."[22]

Tycho Brahe and Johann Kepler

Galileo's trial in Rome dampened scientific enthusiasm in Italy for a century, but it did not halt it altogether. Rather, adopting a low profile, scientists cautiously and quietly continued on with their work in northern Italy. Carried by the momentum gathered through a century of Italian scientific predominance, science could no more be stopped there than it had been in Islamdom when faced by religious opposition. The Church's banning of Copernicus and condemnation of Galileo had some effect in shifting the center of scientific activity from northern Italy to England, France and Germany, but the building momentum of discovery eventually forced the Church, as Galileo was sure it would, to come to terms with its past. In 1822 the Church recognized the validity of a moving earth, and in 1991, Pope John Paul II admitted the Church's injustice to Galileo.

Galileo's mathematics of freely falling bodies produced one of the keys that was to transform natural philosophy. The other was provided by Kepler's mathematics of planetary bodies elliptically orbiting the sun. Both scientists were reaching toward a conception of inertia and the dynamics of moving bodies, where gravity was conceived as a form of magnetic force acting on bodies in motion.[23] Together, Galileo and Kepler held the key to unlock the mathematical physics of inertial and accelerated motion that joined heaven and earth. It would take Newton to turn it.

Kepler's discoveries were made in conjunction with the work of another scientist, the Dane Tycho Brahe, whose many years of excellent planetary observations provided the mass of raw data from which Kepler's mathematical mind would grind out the planetary laws of motion. Without Brahe there would have been no Kepler. Without their collaboration, their names would have been footnotes in the history of science.

A more contradictory and incompatible coupling than Brahe and Kepler would be hard to imagine. Their temperaments made for the oddest couple in the history of science, while their respective scientific strengths and weaknesses fit like a key in a lock. Mathematical Kepler was of humble background, impoverished, and weak sighted. He had an optical defect causing him to see double or triple, odd for a man who would gain fame for his astronomical vision. Brahe was a sharp-sighted aristocrat, proud, peremptory, possessive, and endowed with a princely fortune to build and equip his own observatory and collect a mountain of data but devoid of the mathematical skills to interpret it. Kepler was a devout Copernican, a mystic of astonishing geometric imagination. Brahe was a down-to-earth, unimaginative instrument builder who, unable to break completely with the old astronomy, adopted a mixed system that went back to Heracleides, where the planets orbited the earth, which in turn orbited the sun at the center (see Plate 5).

Brahe was to the manor born, an arrogant overbearing aristocrat with a nose of gold and silver alloy molded to replace the one he had lost in a student duel. Kepler, born into a poor family that was afflicted by all forms of miseries and conflicted by having every imaginable kind of misfit in it – in the middle of his career Kepler dropped everything to learn law and defend his mother, who had been accused of being a witch, and win her acquittal – was a sickly, self-effacing, self-loathing mystic who likened himself to a cowering dog, hungry for affection, obsequiously licking any hand that offered him a bone and snapping at anyone who tried to take it away. To complete the dark picture, he was lovelessly married to a complaining woman whose ignorance was equaled by her lack of respect for her remarkable husband. Unlike his marital relationship, his relationship with Brahe was short, hectic and fruitful[24] (see Plate 6: Kepler and Tycho Brahe at work in the Prague Observatory. Ventura, J. (19th century)).

Kepler, like Galileo, entered his study of astronomy through a side door. Both were mathematicians who got swept up by Copernicanism. For Galileo, it was the telescope that fastened him to Copernicus. For Kepler, it was a Platonic dream of cosmic geometry. Kepler had been from an early age mystically transported by an inspired vision of a marvelous geometrical order of the universe and its heavenly bodies, and from that moment on and the rest of his life, he endeavored to impose his Platonic geometry on Copernican theory.

Kepler had begun his academic career as an impoverished student of Lutheran theology. Upon completing his studies, he was offered an undemanding position in the local town school to teach mathematics. Knowing no more of mathematics than what he had picked up in secondary school, which was enough for the job, and needing the money, he dropped his religious studies to take the position. Though only a low-level mathematics teacher in the provincial countryside, he had enough interest in astronomy to read through and understand Copernicus's *de Revolutionibus*, which few professional astronomers had even done during the half century since its publication. Kepler was immediately convinced Copernicus was right, that it was fitting the sun should reside at the center of the celestial family. Even before Galileo tilted his telescope upwards from the horizon, there were solid physical reasons for a young and open mind to be disenchanted with Aristotelian and Ptolemaic structures. There were the heaven-scarring novas and comets that gave physical reasons for an observer to doubt that the masters of antiquity had it right about the heavens above the moon being pure and changeless, but above all, it was the philosophical beauty of heliocentrism that stirred Kepler's soul and mind. One day in 1595, during his mathematics class, while chalking on the board a figure of an equilateral triangle with an inscribed circle and a circumscribed circle, he was struck by inspiration, as by a bolt of lightning from above: the orbits of the six planets circling the sun were spaced out, one from the other, in a geometric series described by the five symmetrical geometric solids, known also as the perfect or platonic Solids: the tetrahedron composed of four equilateral triangles; the cube with six square sides; the eight-sided octahedron composed of

equilateral triangles; the 12-sided dodecahedron composed of equilateral penta-
gons; and the 20-sided icosahedron composed of equilateral triangles. Kepler was
25 at the time, unmarried, relatively untroubled, his mother unsuspected of being
a witch and his mind lost in the musical harmonies of celestial geometry.

How could it be, he asked himself, that there were six and only six planets in
the starry heavens and, if one included the sphere, the same number of perfect
solids? There was more to this than coincidence. God's mind was mathematical,
and the secret of heavenly creation was geometrical. God structured the heavens
in a complex geometry of spheres and perfect solids, and Kepler, possessing an
infinitesimal fraction of God's mind by virtue of his knowledge of mathematics,
had intuited the architectural secret: the planetary universe was structured by a
series of spheres, each sphere inscribed and circumscribed in and around each of
the perfect solids in a perfect fit of nesting concentric spheres spaced out one from
the other by the geometry of the solid figures. A system of planet-carrying concentric spheres inscribed and circumscribed inside and around the perfect solids,
from the tetrahedron to the icosahedron, was God's cunning way of structuring
the planetary orbits. It was too perfect not to be true. "Geometry," Kepler wrote
in his *Harmony of the World* (*Harmonica Mundi*), "existed before the Creation,
is co-eternal with the mind of God, is God. . . geometry provided God with a
model for the Creation." The geometric architecture of the universe was a reflection of God's mind.

The idea had a long and respectable genealogy going back to Pythagoras and
Plato. Kepler did not believe these geometric forms existed physically. They were
simply forms in the Platonic sense that revealed in geometric patterns the sequential spacing of the planets from the sun that accorded with their orbital periods.
Kepler was not thinking physics; he was formulating the geometry of astronomy,
for the heavens were still the realm of the divine, pure substance devoid of
physicality.

He never abandoned this Platonic epiphany of planetary spacing. Not only
did it mirror the perfect geometric architecture of the divine mind, but some of
the planetary orbits derived from calculations based on the configurations of the
geometric solids came encouragingly close to those established by observation.
The correspondences were way off for the other planets, but any doubts he may
have had were banished by the beauty and elegance of the geometry. It was the
mathematics or else his interpretation of the data that was wrong. Kepler spent
many years trying to prove his epiphany, making countless calculations that
covered hundreds and hundreds of pages, changing one parameter and then
another, tirelessly squeezing observation into the ungiving girdle of theory. Like
any number of scientists driven by a vision of reality that had to be true, he
juggled the figures, fudged data and forced calculations to tally with his inspired
models. Fixing the diameter on one sphere so the model approximated observational values of the planet's orbital distance, he incurred greater discrepancies
on the next. Year after year he obsessively plodded on, calculating and recalcu-
lating, sometimes coming close to what the numbers required, but even his
computational sleights of hand failed to produce results close enough to satisfy

his equally obsessive sense of scientific honesty and the prevailing demands of exactitude.

His *Cosmographicum Mysterium*, published in 1597, records the passionate joy of his inspiration and the agony of his not being able to harness it in mathematics. Like a prophet inspired by a revelatory vision into God's mind, Kepler drove himself all his life to prove the geometric patterns that ruled the planets, and could never understand why his calculations of orbital ratios based on the perfect solids failed to fall in place. Twenty years after his *Mysterium*, long after he had discovered the basic three laws of astronomy and had not long to live, this inspiration of his youth appeared in another book on the perfect solids, *Harmonica Mundi* (*Harmonies of the World*) in which he took literally the Pythagorean idea that the planets were spaced apart in the order of musical octaves. As the planets whirled through the cosmos around the sun, they each set up a pattern of vibration that collectively produced a musical harmony, a divine symphony that was God's mind resonating in sound and number through the universe, but which, as Pythagoras had said, went unheard by man for the coarseness of his soul, corroded over and benumbed as it was by materiality and overindulged appetites. Kepler even composed a musical score of the heavenly symphony based on his computations of the distances of the planetary orbits in their nested spheres around the perfect solids, musical notes and all. Any applause by an audience able to sit through it would come more for its having at last ended than its soaring beauty.

His *Mysterium* and *Harmonica* did not arouse much comment. The Church did not condemn it. Kepler's Copernicanism went unremarked. He was a Lutheran, but many Catholics read him. Part of the reason there was so little fuss over his ideas was that by 1600 the market was being flooded with books on science, good and bad, a great number of them being minimally circulated. In Kepler's Germany alone, the average number of scientific books published yearly amounted to more than a thousand, in which quantities of many would-be controversial books got lost in the shuffle. On occasion, when speculative science ran up against theology, the Church took notice, especially of books by Catholic authors. As a Lutheran, he was out of the Church's reach. It was not until toward the end of Kepler's days, when he published an epitome of his works and ideas in the late 1620s, that he was finally noticed by the Church, condemned and indexed, no doubt in great part because of his analogy between the Holy Trinity, the geometric triangle and his universal harmonies.

His magnum opus, as established by the judgment of scientists and historians, was not what Kepler himself considered it to be. In his judgment his great contribution resided in the *Cosmographicum Mysterium* and *Harmonica Mundi*, geometric and cosmic revelations that were at once spiritual and scientific. Posterity disagreed. For those who came after him, Kepler's fame as a founding father of the Scientific Revolution was his *Astronomia Nova*, the product of his collaboration with Tycho Brahe. This was a work of labor, not love. How could it have been otherwise, having to work as a serf under a great Dane of a tyrant?

At the beginning of their collaboration in the final years of the 16th century, Tycho was the man of his age, one of the greatest names in science, ranking with

Peurbach and Regiomontanus of the previous century. Kepler was a young, self-taught mathematician whose *Mysterium* had only recently made his name known within the advanced circles of astronomical study, namely the Copernican community scattered through Europe. It was enough to have his reputation reach the heights of science's Valhalla, where Tycho Brahe, secluded like a Danish god in his Uraniborg fortress, sleeplessly surveyed the heavens with his legion of astronomers and mathematicians. Kepler was everything the lordly Dane wanted to be: a mathematician–astronomer with a mind brimming over with theories in search of the universal system. To optically afflicted Kepler, Brahe was everything he needed, a sharp-eyed observer who lorded over a magnificent observatory that housed a princely armory of precision-made equipment and a treasure of planetary data collected over the years by his vassal observers. Theirs was a marriage made in hell that opened earth's doors to heaven.

Turning on the twin hinges of Tycho's precision of observation and Kepler's collation of the data, that door grudgingly cracked open and then only because of the obsessive nature of the men pushing it.

Tycho was no less bent than Kepler, only in the opposite direction. The Dane's entry to astronomy, as accidental as Galileo's and Kepler's, had swept him from his intended aristocratic career of soldier–statesman. It happened about the time Kepler was born. As a young boy Tycho observed a solar eclipse. He recorded the event. It was in 1560. He was 14. So astonishing was the eclipse to him that he took to observing the heavenly bodies and recording their positions as a hobby. Precision was important. He would be later appalled to discover how large the discrepancies were between the values of his painstaking observations and those given in the standard tables of planetary positions: as much as 30 minutes (or one-half a degree) of arc and more. The standard tables in use went back to Peurbach, Regiomontanus and the Alfonsine Tables that were ultimately based on the observations of the 12th century Andalusian astronomer al-Bitruji. Observational astronomy was a shambles to Tycho's keen eyes and high standard of precision made possible by his own carefully crafted equipment. Deciding to reform it, and wealthy enough to do it on his own, he built larger equipment to make more precise measurements in tracking the planets. Passion for observing the heavens became passion for precision. The secret of the heavens would be forced out by measurements of starry motion more exact than had ever been made.

Until around 1500, observations within a difference of 20 minutes to 30 minutes of each other were not considered intolerable; half of a degree in variation was thought to be no more than an inconsequential margin of error that had no effect on the working of the system. This was changing during the 16th century. Demands for precision found their way into a broad range of arts and crafts, as well as the technology of the age: the precision of measurement characterized by Leonardo's work and expressed by his anatomical drawings; the requisites of exactness in ballistic studies; the construction of mechanical clocks, with their intricate assembly of springs, gears, wheels and stops; and the demands of maritime commercial interests in need of mechanisms accurately measuring longitudes for global navigation, where an error of half a degree for an oceanic vessel sailing a few thousand miles had too often resulted in men, cargo and ship missing the landmass and sailing off to their doom. By Tycho's time, a mind-set had evolved that was less tolerant of the "close-enough" mentality, back when the sundial had sufficed to meet the daily demands of life.

The equipment Tycho meticulously built, similar to that described by astronomers in 15th century Samarqand, and the care with which he put his equipment to use, produced a remarkable degree of accuracy for the time, within 4 or 5 minutes of arc, a four-fold improvement in observational precision. It was using such a piece of equipment, a sextant measuring 5 ½ braccia (arms) in radius and graduated down to minutes, each minute of arc corresponding to about 1/50 of an inch along the arc, that Tycho measured the parallax of a star he observed that had not been in the heavens before. This was a contradiction to astronomical principle: a new star had been born in the eternal starry heavens where nothing was supposed to change. Young Tycho could scarcely believe his eyes. Had the star not remained day after day he might have passed it off as an optical illusion, but there it remained, and for months he measured its motion, until it finally faded and disappeared. This slight parallax that he measured during the nova's duration proved to him that it was way beyond the orbit of the moon. The celestial virgin had been deflowered.

If the nova did not convince him of it, then the comet he observed in 1577 did, since it too was far beyond the sphere of the moon, that heavenly realm Aristotle had said was of changeless purity where only circular motion existed. It was shortly after this that he abandoned Ptolemy, not for the system of Copernicus, but of Heracleides. Tycho had read Copernicus but could not break completely with the past. Too many convincing proofs stood in the way of his accepting a moving earth. He allowed that the planets circled the sun, but the sun in turn circled the earth, which remained motionless at the center. It was this hybrid system of antiquity that he set out to prove through his rigorous observations.

Brahe's observational skills were legendary. So too were his character and princely style of life in his Gothic observatory of Uraniborg on the island of Hven, given him as a personal fief by the king to keep this living national treasure of an astronomer in Denmark. Tycho staffed Uraniborg with an academy of first class astronomers, including Kepler, whom Brahe had brought from Prague in 1599 to be his theorist-in-chief. Servants and cooks tended to the scientists' daily needs so they could devote themselves like monks to recording the motions of the heavenly bodies. Uraniborg had observation posts under retractable turrets at all four corners of the fortress-like structure. For a quarter of a century before Kepler's arrival, Tycho and his assistants had been making observations of planetary positions that were more accurate than any that had ever been recorded. Unlike the Babylonian, Greek and Muslim astronomers, who would make but a few observations of a planet's position over the course of a year, Tycho and his team made nightly measurements, building up a monumental fund of data that Uraniborg's eccentric astronomer–king guarded with tyrannical jealousy and shared with nobody.

Tycho's Uraniborg project was the most determined, organized, focused and controlled enterprise ever undertaken in the cause of science up to that time.

Like a modern totalitarian with his secret intelligence agencies, Tycho divided up his astronomers in separate units, assigned them special tasks and took possession of their observations as they made them, claiming them his own and his alone, even forbidding his astronomers from exchanging information. One was not to know what the other was doing. Only Tycho would know the overall design of the universe as it emerged from the shadows, thanks to his observers and theorist–mathematicians, whom he considered no more than hired servants, his manorial staff. It was his intent that he would be the one to discover the true system of the world, and that it would turn out to be the hybrid one he espoused.

To Kepler he assigned the planet Mars. This was a fortunate choice. Mars has the highest eccentricity of all the planets, large enough within the limits of Tycho's observational precision for Kepler, eventually, a few years after Brahe's death and a mountain of agonizing computations, to recognize that the eccentricity matched an orbit that was mathematically equivalent to an ellipse.

It rankled Kepler that Brahe limited him to one planet. He wanted to see all the observational data and resolve them into a geometric pattern, the pattern of his Platonic inspiration. Heracleides be damned. In a way, the birth pains of modern science was a contest between two ancient Greeks waged by a mad Dane and an obsessed German.

What most bothered Kepler about Brahe, more than his dopey fixation on Heracleides, was his imperious manner. On one occasion Kepler threatened in one of many bursts of anger to leave Brahe and Uraniborg, but he returned in remorse and, like the fawning dog he described himself to be, begged forgiveness from the old master, who nobly condescended to forgive.

Their relationship struggled through almost two years of contentious collaboration, right up to Brahe's sudden and painful death.[25] His death occasioned a crisis over the ownership of Kepler's computational work on Mars and the table of observations Brahe had given him. Brahe's son and heir laid legal claim to them as belonging to his father's estate, but before they could be locked away, Kepler packed them up with his belongings and like a thief in the night absconded with the lot from Prague.

Kepler's was an epic two-year struggle wrenching a mathematical pattern from Brahe's observations of Mars. Napier's system of logarithmic computation would have helped simplify what was a tremendously laborious task, but logarithms were not to appear until 1614, several years after Kepler's dogged triumph. Two of his three famous laws came out of the Mars observations: that planets orbit the sun in an ellipse; and doing so, they sweep out equal areas in equal times in the plane of their solar orbit. His third law, discovered later, was a surprising ratio, that the cube of the mean distance between a planet and the sun divided by the square of the orbital period was equal for all planets. Known later as Kepler's Laws, these fundaments of the new astronomy were not stated as discoveries. They were not highlighted, as one would have thought, as the successful culmination of a long and hard quest. They were stated most casually as they emerged from the mountains of computations and then dismissed.

Kepler's was a labor of Sizyphus pushing Mars along a circular trajectory that always collapsed into an odd shape that was almost but not quite circular. He kept at it, starting over and over again, computation after computation, page after numbing page, but the orbit resisted falling into a circle. He would try new points on Mars' orbit from the storehouse of Brahe's table of observations and calculate again, repeating the process over and over, folio after folio, but he kept coming back to a disappointing orbit that bulged slightly on the sides, like, as he complained, an egg. Even shaving the data here and there failed to produce the circle.

The shape of the orbit, whatever it was other than grotesque, eluded being a circle by a mere eight minutes of arc. Eight minutes difference was still too high to ignore. At an earlier period, those eight minutes would have been ignored and the divine spheres preserved, but by Kepler's time those minutes had to be accounted for with the same attention to exactitude that an Amsterdam merchant weighed his gold and balanced his debits and credits. Kepler persevered for so long and became so familiar with the numbers and decimal fractions of his endless computations that he at last recognized the discrepancy of eight minutes of arc between the observed trajectory and the obligatory circle to be the eccentricity of an ellipse, as measured by the radius of Mars' orbit. Geometric blasphemy! The Almighty had not used circles and spheres in patterning the universe but an egg-shaped ellipse! Overwhelmed and disillusioned by the imperfection of God's creation, he regarded his discovery as he regarded himself, unworthy, a blemish, a flawed creation. The ellipse was a mockery of his inspired conception of concentrically nested spheres around and within the five perfect solids that the divine should have more worthily followed in designing creation. It was ghastly, aesthetically all wrong. All symmetry was wrecked.

Perplexed and embarrassed, he accused himself of debasing the cosmos, of trading in a perfect sphere for a barnyard egg. How could he tell the world? In Kepler's own words of self-confessed shame, he had cleared the Augean stables and filled them with dung. That he had discovered the basic laws of astronomy counted for little. What counted was the geometric mystery and harmony of the spheres and their octaves of heavenly music.

Kepler never surrendered his belief in the beauty and truth of his geometric harmony of heavenly construction. He regarded it his greatest discovery, writing on it to the end. The laws that he discovered seemed unimportant compared to what he could not prove. He wiped out epicycles, equants, circles, uniform motion around a fixed center and all the complex geometric paraphernalia that Ptolemy, ibn al-Shatir and Copernicus employed to account for planetary motion. Only the eccentric point was in a sense preserved in that the sun was situated at one of the two focal points of the planetary ellipses, not at the center. This too perplexed Kepler. How did God decide on which one of the two focal points of an ellipse to place the sun? Playing dice with the universe? Could God's natural creation be so whimsical?

Kepler's distress is that of a scientist with one foot planted in the old world, the other in the new. It took a generation for circle to surrender to ellipse. Galileo, coming to scientific maturity during Kepler's creative years, appears to have successfully repressed the ellipse. He received from Kepler a copy of his *Astronomia Nova*, but the idea of an ellipse must have been so abhorrent aesthetically and spiritually that Galileo could not accept it. In none of his writings did Galileo ever mention Kepler's ellipse. Planets moved in Copernican circles. A generation after Kepler's death, Halley, Hooke, Wren and Newton in England had no trouble accepting the ellipse. What was repugnant to one generation was quite normal to the next. Darwin would two centuries later experience the whole cycle of discovery, repugnance and acceptance all within his lifetime.

Just as Galileo had come close to formulating Newton's principle of inertia, Kepler had come close to formulating Newton's law of planetary attraction. Kepler was aware that bodies were attracted to teach other, with respect to their proportional magnitudes, and that two bodies, uninfluenced by a third, would come together, each moving a distance inversely proportional to the ratio of their magnitudes, while two bodies of equal magnitude would move equal distances through space until they met. And so he knew that the moon, if released from earth's grip that holds it in steady orbit, would move 53 parts toward the earth, while the earth would move one part toward the moon, in accordance to the inverse ratio of their sizes. Between Galileo and Kepler, all the elements of mathematical physics and astronomy seemed to be there for the making of a universal theory of heavenly and earthly physics. One element was lacking, the introduction of a force that would transform the kinetics of Galileo and Kepler to a system of dynamics.

Newton

The relatively humble origins of Isaac Newton, and the much humbler origins of his predecessor, Kepler, both of them country boys, indicate that the capital invested in schools and universities during the high medieval period was paying off with interest. As payoffs of long-term capital investment, the same could be said of many others in the 14th and 15th centuries who contributed to the rise of modern science, men whose genius, had the opportunity for higher education not been as accessible as it was, would have otherwise been lost to science. Newton was born to a farming family of modest means in Lincolnshire. A university education for Isaac was not in the planning. When it came time for him to assume his duties in running the family farm after his completion of secondary schooling, it became apparent he was as unfit for farm work as he was disinterested in agriculture and animals. He could barely tolerate human beings.

His mother and stepfather (his father had died when he was a child) were able to send him to Cambridge. The Lincolnshire parsonage had a place open at Cambridge for a worthy boy and the family knew the parson. Young Isaac soon proved himself better suited to studying geometry and Latin than tending to cows and planting.

In his first year at Cambridge he devoured Euclid's geometry, a trifling exercise he called it. He then proceeded to Descartes' analytic geometry and optics. When the plague of 1665 struck and the university closed, he left Cambridge and returned home to Lincolnshire where he put his ability to more serious application. The

18-month stretch of leisure that he spent on the farm waiting for the plague to subside was the first of two richly productive periods that marked Newton's career. At the ripe age of 23 and free of chores, he devoted himself to what interested him that summer, the result being his setting down the basic elements of what would eventually be his *Mathematical Principles of Natural Philosophy*, later abbreviated to *Principia*.

Altogether, within those months of plague, he developed his method of differential and integral calculus, or fluxions as he called it, and formulated the Binomial Theorem for solving high order algebraic equations. He studied optics, developed a theory of colors, and, while working on a theory of gravity and intuiting that the force acting between earth and moon was the same force as that acting between earth and an apple falling from a tree, he calculated the force required to keep the moon in its orbit. He was later to reflect on that brief burst of creativity:

> All this was in the two plague years 1665–1666, for in those days I was in the prime of my age for invention, and minded mathematics and philosophy more than at any time since.

In a rare moment of humility, of which there were not many for Newton, he later admitted that if he had been able to see further than most men it was because he was standing on the shoulders of giants. By giants he must have meant Copernicus, Kepler, Galileo, William Gilbert and Descartes, who each in his own way cleared the path for Newton's synthesis of their respective discoveries and reflections on natural philosophy. Kepler had struggled toward a physics of the heavens. And so too Galileo and Descartes. From the vantage point of those towering shoulders, he saw heaven and earth coming together on the horizon, and during those plague-mandated 18 months of leisure, he came close to framing it in mathematics. Endowed with that same inexplicable intuitive power of his great predecessors, Newton intuited that it had to be a set of opposing physical forces interacting to keep the planets in an elliptical orbit. He reasoned that if only one force was acting on a planet, the resultant orbit would be circular. An ellipse indicated something more, a tug of war in the heavens between opposing forces that caused the orbiting planets to speed up and slow down as their distances from the sun varied. One of the forces had to emanate from the massive sun and be inversely proportional to the square of the mean distance between it and the planet. Kepler had likened the force to a heavenly broom, the end of whose handle was fixed to the sun and swept the planets along in their orbit. Thinking in terms of magnetism, he had imagined magnetic hands stretching out from the sun and holding the planet in orbit, something like rays of light. But not light itself, for when cast in shadow during eclipses the planets would have then stopped. No, it was not light, but some physical force, a swirling vortex carrying the planets around in its medium, something like a magnetic force acting at a great distance.

The concept of magnetic force had become popular in 1600 when William Gilbert (d. 1603), the greatest English scientist until Newton, published his *de Magneta*. The book caused a lively stir in Europe's burgeoning scientific community.

Basing his conclusions on years of magnetic experiments, Gilbert reasoned that the earth was a giant magnet and that some form of cosmic magnetism held the planets in their orbits. An early follower of Copernican theory, Gilbert was more advanced than Copernicus in his belief that the stars lay in space at great and differing distances from the earth and were not fixed in a spherical firmament. Gilbert's book appeared while Kepler was in the midst of his epic struggle with the orbit of Mars. In the introduction of his *Astronomia Nova*, published a few years after Gilbert's *de Magneta*, Kepler describes gravity as the tendency of two bodies to unite or make contact, akin to the attraction of magnetic bodies: two stones near each other in space and beyond the reach of a third body, he writes, would come together, the force of attraction being related to the size and density of the bodies. There were also the apparently coincidental phenomena of lunar and solar positions and the earth's tides. For Kepler this was causal, not coincidental: the pull of the moon, and even the distant sun, caused the seas of the earth to rise and fall. Kepler did not know why it was that these forces, magnetic or whatever they were, resulted in tides and elliptical orbits, but he knew there was a causal relationship, a force of attraction proportional to the size and density of the bodies acted upon. It was left to Newton to formulate the unifying mathematical principle, but he knew no more than Kepler why nature acted the way it did. Lacking Kepler's mystical belief in God's musical harmonies that held the universe together in purposeful mathematical relationships, Newton honestly admitted his ignorance beyond the mathematical physics of it all and pretended to no higher metaphysical hypothesis.

Part of the reason Kepler failed to formulate the mathematical principle of gravity was psychological. The idea of a gravitational force acting through empty space between two bodies was too closely reminiscent of the spirits, souls and angelic intelligences of medieval animism. To the secularized natural philosophers of the 17th century, positing a force acting at a distance through space seemed to be a return to the teleological system of scholasticism, the hand of the unmoved mover. The concept was too filled with medieval mystery to be taken seriously as science. Even Newton had a problem with it, though the fact that gravitational attraction could be expressed in mathematics blunted somewhat the charges of medievalism and magic fingers acting at a distance. But even with that, the empty void would have to be filled with some material substance through which the forces were transmitted, some light airy ether in order to provide the requisite tactile medium to carry their effects, as sound travels through air or waves through water. How could force be transmitted otherwise? Aristotle's nature abhorring a vacuum lived on in the ether.

Renee Descartes (d. 1650) advanced Kepler's theory of an ether-filled spatial vortex. His shoulders were among those that Newton stood upon to view the new universe. Descartes anticipated Newton's researches in optics, mathematics and in formulating a mechanical philosophy. His development of analytic geometry, that is, the application of algebra to geometry, and his graphic system of plane and cubic coordinates used to map out algebraic equations in two and three dimensions, known after him as Cartesian, were each a profound contribution that helped

Newton in his mathematization of force, matter, motion and distance. Descartes' mechanics were in fact a close pre-statement of Newton's law of inertia, while his philosophical popularization of Bruno's infinite universe made the idea of cosmic infinity more acceptable in learned circles. Nonetheless, a cold and lonely infinitude of dark empty space presented a universe even more unsettling than the Copernican one that had the sun usurp the favored place of the earth. Going from the comfortable, cozily enclosed universe of Ptolemy and Copernicus, where either earth or sun was at the center, securely wrapped in the starry firmament at some finite distance, to the unfathomable specter of a limitless universe where earth was but a speck lost in a soulless void, was like a fall from grace.

Fearing the fate of Galileo, Descartes withheld publication of his cosmology and tempered his Copernicanism by positing a sleight-of-hand system in which the earth moved but did not move. It moved around the sun; but it did not move with respect to its spatial envelope. What moved was the vortex. Hence both Copernicus and the Church-enforced Ptolemaic system were right. Spheres and epicycles gave way to vortices. To account for the motions of the planets and their satellites, 14 of them were required, all whirling around in a cosmic storm, the resulting balance between their inward centripetal and outward centrifugal forces giving them their elliptical orbits. Descartes had broken enough with the past not to be loath to speak in terms of Kepler's ellipses, though the crudity of Descartes' cosmic physics is a clear measure of the difficulty natural philosophers still had with the idea of forces acting through the vast planetary distances of empty space without embarrassing themselves with analogies of spiritual forces and angelic intelligences guiding the planets. Newton simply stated the facts through the mathematics of gravitational attraction and let it go at that. He assumed no responsibility to frame hypotheses beyond what the equation said. He was not a metaphysician.

Practically the whole of Newton's scientific contribution can be traced back to those 18 months of the plague years when he was free of academic course work and his mind was free to search where it would. Nor was he in a hurry to share the fruits of that productive respite in the country. He tucked his calculations on gravitation in a drawer, where they were to remain until forgotten or lost. There was a reason Newton was in no rush to publish. A brush with institutional resistance to new ideas had early in his scientific career caused an abiding wound that tempered his collegial relations with fellow professionals. It had to do with an early treatise of his on light. During that 18-month stretch of discovery on the family farm, young Isaac had done much research in optics based on the work of ibn al-Haytham and Descartes.

Rainbow theory had been a subject of intense interest in Greek and Muslim science. Grosseteste, Bacon and Witelo had resuscitated it. In 1611, the archbishop of Spoleto, Antonio de Dominis, published a theory of the rainbow that was little different than the explanation of primary and secondary reflections found in ibn al-Haytham's *Book of Optics* and Kamal al-Din al-Farisi's early 14th century recension of it. Kepler had also studied light. Descartes carried the science forward in his *Dioptrics* by incorporating into the theory of light what he termed its refrangible nature. Newton went further. Where ibn al-Haytham had beamed a ray of light through a small hole into a camera obscura in which were suspended water-filled glass spheres to simulate moisture in the air, Newton passed a beam of sunlight through a glass prism suspended in a dark room. He was not the first to produce the resulting spectrum of colors on a screen, but he was the first to explain them. Following his intuition, he then recombined the spectrum by passing it through a second prism, an experiment that would come to rank among the top ten most critical in the history of science. Newton's other optical experiments were designed to determine if light propagated itself through the subtle ether in waves or particles. He concluded that light sometimes acted as a wave, sometimes as a particle.

Some years later, after he had graduated and was teaching mathematics at Cambridge and was a member of the Royal Society, he submitted his optical researches for publication, fully confident they would be well received. When they were met by something between critical reserve and cold opposition, young Newton was stunned, then disheartened. Like Galileo and his run in with the Church, he had expected the world of science to rejoice in its embrace of fresh discovery backed by experimental proof. Galileo and Newton failed to understand that the large majority of scientists regarded their cherished theories as a capital investment, personal property to be protected. Having invested their lives in gaining the knowledge that came with those theories, upon which they hinged their prestige and by extension their honor, they could not bear to see them go down. Letting go of what they deemed precious intellectual property was especially galling when the "new" that brought down the "venerated" was the work of a young newcomer. Newton would later become just as possessive, protective, small-minded and combatively jealous over discoveries that he regarded as personal property.

There was also an esthetic repugnance regarding Newton's optical discoveries that diminished their appeal and cooled acceptance: light, a symbol of truth, purity and divinity, was now cast as nothing but a mongrel mix of mundane colors. The idea was as repellant to optics as was Kepler's egg to astronomy. Robert Hooke, head of the Royal Society at the time, took the lead in discrediting Newton's research. Newton never forgot or forgave Hooke. Withdrawing within himself, he maintained an outer visage of Olympian imperturbability and from then on refrained from publishing for many long years. His bitter entry into the jealous world of science thus resulted in a prolonged period of self-isolation to preserve the tranquility of mind he needed for his research.

During this period, his prime of creativity, Newton occupied himself in a number of divergent pursuits, among them theology, the Bible, and alchemy, to which he gave long and serious attention. He built an oven in his garden in Lincolnshire for alchemical experiments and filled many notebooks in his small, neat and crowded script with descriptions of his experimentation and theological musings. Around 1680, his interest returned to mathematics and natural philosophy, by which time he was a distinguished professor at Cambridge, holder of the prestigious Lucasian chair of mathematics and a highly respected member of the Royal Society. It was during this period of renewed interest in physical astronomy that his book on the theory of gravity, the *Principia*, for many historians and scientists the world's greatest scientific work, came almost accidentally to be written and published. The story begins in 1685 with a visit the young astronomer Edmond Halley paid to Newton in Cambridge.

A lively discussion on gravity and planetary motion had been taking place in London among members of the Royal Society, in particular among Halley, Hooke and Christopher Wren, whose minds were busy with the discoveries and theories of Kepler, Gilbert and Galileo. Halley had come up from London with the sole purpose of asking Newton what he thought the curve of one planetary body orbiting another would be if the force of gravity between them was inversely proportional to the square of the distance. An ellipse, Newton responded without a moment's hesitation. Astonished at Newton's quick reply, Halley asked how he knew it would be an ellipse. Newton told him he had worked it out some 20 years ago during the plague. Halley excitedly asked to see the calculations. Newton searched, was unable to find them and promised Halley he would work them out again and send them to him in London.

One wonders why the members of the Royal Society would not have known that the planetary orbit was an ellipse. Almost 80 years had elapsed since Kepler published his *Astronomia Nova*. What was missing was not the curve of the orbit, or that the gravitational attraction of bodies varied inversely as the square of the distance. What was missing was the equation that related the gravitational force to distance and mass. This, thanks to Halley's visit, was what Newton now provided, initiating a second burst of creativity paralleling the first of 1665–1666.

The *Principia Mathematica Philosophiae Naturalis*, which included Newton's computational method of calculus, was as difficult to digest as Copernicus's *De Revolutionibus Orbium Coelestium*. It took 40 years of discussion among scientists before its implications could be fully appreciated. One of these implications, quickly realized, was the reunion of the principles of heaven and earth after their separation by Plato and Aristotle. The *Principia* established a single universal physics and would remain the new system's primary authority for over two centuries. Newton's masterpiece would also provide the classical paradigm for new physical sciences that emerged in the 19th century – electromagnetism, optics, heat, mechanics – and would endure as a scientific marvel defining the mathematical structure of the universe and foundation of classical modern science until the rise of relativity and quantum mechanics in the early 20th century. Generations of scientists after Newton understood physics to be a matter of working out the details of natural phenomena in the light of Newtonian mechanics, while the Newtonian system of universal gravitation that maintained balance and harmony in the organization of the planets would be the guiding paradigm of the Enlightenment.

Newton's drawing of the separate strands of discovery together into a single theory completed the basic structure of what would be called the new science, as opposed to the old of Greek, Arabic and Latin. To the old scholars, the new science would have appeared alien, disembodied, a soulless shell of mathematics, divorced as it was from any moral or spiritual system. God and soul had no place in it. Where God's mind had integrated and preserved the universe in Greco–Muslim and Latin natural philosophy, the preserving force was now a cold and comfortless deism, a lifeless mechanics of force and matter.

The displacement left even Newton cold. He feigned to reject the total abandonment of the divine from the system, claiming that God's preserving hand was needed to keep the clockwork universe tuned and running in accordance to the laws God had created for the intricate machine. But why a preserving hand was necessary to keep in tune a creation by a perfect creator was impossible to explain. Newton, who feigned no hypotheses on the nature of gravity and why things were as they were, namely, how forces were transmitted across vast distances to keep the planets in orbit, had nothing to say here either. At the same time, he could not accept the metaphysical implication of the mechanistic universe created by his mathematical principles and the natural philosophy that flowed from them. Newton's physical marriage of heaven and earth in a mechanical universe left God out in the cold, at the margin of the universe, a shadow whose only job was to oil the machinery, rewind the spring and keep the wheels turning.[26]

In sum, Kepler transformed Copernican astronomy from a mathematical hypothesis to a mathematical certainty. His laws of astronomy, synthesized by Newton with Galileo's law of freely falling bodies, produced a universal theory relating force, mass, distance and acceleration. The synthesis so transformed science that it was no longer in the old Greco–Muslim–Latin league. Science was no longer the old ball game in which all three divisions in the league knew the ground rules and could play the game, where four balls was a walk and individual players could be traded and their skills and ideas of game strategy fit easily in with those of any of the other teams and team members. With Kepler and Galileo, it was another game in another league in another universe. With Newton, the curve of scientific divergence would shoot off like an arrow into deep space, severing all contact, commonality and correspondence that had bound the three successive traditions together, where, given a common language, one tradition could be comprehended by and contained within the other. Western science had grown beyond the contours, patterns, principles and design of its progenitors and was setting its own rules, its own goals along its own path. When in the mid-17th century Western science began trickling into the Muslim world, it was unrecognizable.

When Sir Isaac Newton died in 1724, a handful of Ottoman reformers had just returned to Istanbul from an extended study tour in France in an attempt to discover the secret of Europe's new-found strength on the field of battle. The Ottoman delegation turned out to be the Muslim world's first engagement with Europe's new science. It was for a few Ottomans in the delegation an easy, self-evident equation: science equaled power. But easy as it may have been for those few, for the collective body of Muslim political and religious culture, taking possession of the new science was to be a long and, even up to today, unfulfilled task.

There would be many attempts at taking possession. It would at first glance appear to be a simple matter of Muslims reclaiming their old heritage of scientific supremacy and getting on with the job, especially with Europe on the attack. But the new science was so transformed from the natural philosophy of the

Greco–Muslim heritage that had been brought to Europe by the Latin translators half a millennium earlier that its patrimony was no longer recognizable. Digestion of it would go on for centuries in the Muslim world, from the time of Newton's death to the present day, reform by reform, one attempt aborted after the next. Even with the West's rising power bearing down on the countries of Islam, making their cultural and political survival incumbent upon Muslims learning as quickly as possible the science and technology of the West, even this did not, has not, forced the issue of assimilation so that Muslim society could join the international brotherhood of science as a full and respected member.

More daunting than the intellectual challenge of mastering modern science and its application, it was to be the formidable combination of socio-religious resistance, the same that had, in its time, turned a cold face to the indigenous scientific movement of the earlier centuries of Islamic civilization, and, more urgently, the unwillingness of political leaders to invest adequately in the education and institutionalization of science, that would make life so difficult for reformers and science so difficult to acculturate. In the earlier period, the sciences had come from an extinct civilization, the Classical and Hellenistic Greeks, and from defeated civilizations, Persian and Indian. In the modern period, the sciences were coming from a rival civilization that, for a millennium, had been Islamdom's cultural inferior. That rival had suddenly become strong, expansive and very much a threat. Learning from the West would present Muslims with a Gordion knot of psycho-religious conflict. The refusal of Muslim rulers to invest in the future of their people would guarantee the knot would remain securely tied.

To this we now turn.

Notes

1 Gutas, *Greek Thought, Arabic Culture*, Routledge, New York, 1998, p. 174.
2 Significantly, while the followers of Luther and Calvin were tearing up ecumenical unity in the West, a Shi'i reformation of sorts, centered in Azebaijan and expanding through Iran and southern Iraq, was doing the same at the heart of the Muslim world. As in the West, the religiously explosive period was one of artistic and architectural renaissance, and it was the last of those movements that gave a certain out-of-phase parallelism to the two civilizations' patterns of development between the 8th and 16th centuries. In this instance they were temporarily in phase, but unlike the Protestant Reformation in Europe, the Shi'i reformation in Iran and the eastern fringes of the Ottoman Empire held no unintended consequences of intellectual expansion. In Islamdom it resulted in a geo-religious division innocent of the secular intellectual trends that came with the critical examination of religion in the West. This also profoundly deepened the parting of the ways between the two civilizations. The balance of intellectual and material power was from then on falling away ever more tellingly from East to West.
3 Eugene Rice and Anthony Grafton, *The Foundations of Early Modern Europe 1460–1559*, Norton, New York, 1994, p. 21.
4 Fredrich II, "Stupor Mundo" of the mid-13th century, had given to the University of Bologna Latin translations of the Arabic and Greek works on mathematics that had been made in Palermo under Fredrich's patronage. Rose, *The Italian Renaissance of Mathematics*, p. 145.
5 Rose, *The Italian Renaissance of Mathematics*, p. 103. The Latin translation of Archimedes in Greek completed the assimilation of the Greco–Arab corpus of mathematics,

an accomplishment comparable to that of Peurbach and Regiomontanus the previous century in astronomy.

6 Rose, *The Italian Renaissance of Mathematics*, p. 222. The innovative craftsmanship of the filosofo geometra is shown in Commandino's "reduction compass," a drafting instrument with two pivoting arms connected by an adjustable micrometer that was used for transferring and comparing dimensions and geometrical proportions. He also devised an elaborate instrument that divided circular arcs into degrees, minutes and seconds, and on into thirds, and even fourths, an order of precision that approximated the mural quadrant of the Samarqand observatory during the time of Jamshid Ghiyath al-Din al-Kashi, a century earlier. In the same order of technical craftsmanship, Guidobaldo invented an instrument for drawing ellipses and other conic sections.

7 Rose, *The Italian Renaissance of Mathematics*, p. 228.

8 Rose, *The Italian Renaissance of Mathematics*, p. 156.

9 Stillman Drake, *Dictionary of Scientific Biography*, vol. I, edited by C. Coulson, Gillispie, Scribner and Sons, Detroit, 1970–1980, p. 605. In his third book on the subject, *Speculationum* (1585), Beneditti wrote that the speed of a freely falling body would be impeded by the medium it was falling through, the impedance depending on the object's surface area.

10 Benedetti lived 60 years, not overly long, but a full and productive life. Years before his death he had cast a horoscope for himself that predicted he would die in 1592. On his deathbed in 1590 he recalculated, found a small error in the original computation and died satisfied he had got it right.

11 The great works of Muslim science and medicine were in fact more alive in the West than in Islamdom. Consider the pulmonary system of blood circulation described by ibn al-Nafis. While in the East ibn al-Nafis and his theory were lost even in memory, in the West the work of the 13th century physician was avidly pursued and communicated to the scholarly community. The Spanish anatomist Michael Servetus almost certainly learned of ibn al-Nafis's description from a northern Italian physician named Colombo when they were both at the University of Padua. Colombo learned it from another Italian, Alpago, a physician who lived over 30 years in Damascus in the early decades of the 16th century, during which time he collected a great number of Arabic medical texts, many of which he translated. Servetus's description of the pulmonary system of circulation reads in his 1555 edition of *Fabrica* as if it were a free Latin translation of ibn al-Nafis's Arabic text. Servetus's Latinized description of ibn al-Nafis's pulmonary system was quickly transmitted to European medical schools and given serious study. Joseph Schacht, "Ibn al-Nafis, Servetus and Colombo," in *al-Andalus*, Madrid, vol. xxii, 1937, pp. 317–336.

Another case, in the 1590s, Thomas Harriot (who preceded Galileo by some months in observing the moon through a telescope), having studied Haytham's optics from a Latin edition made in 1572, solved Haytham's special problem of rays reflected from a curved mirror. Then he went on to investigate other optical phenomena the Arab scientist had not included or adequately explored in his book – and this at a time when ibn al-Haytham had faded to a dim memory on the East. Centuries after Nasir al-Din al-Tusi's reform astronomy at Maragha had somewhat faded from Muslim memory, members of the Royal Society, in the last half of the 17th century, were discussing plans to translate al-Tusi's astronomical works.

12 Alois Sprenger, "The Copernican System of Astronomy Among the Arabs," *Journal of the Asiatic Society of Bengal*, 25, 1856, p. 189.

13 More than the equant, the contradiction that drove Copernicus from Ptolemy's geocentric universe to the sun-centered system of Pythagoras and Aristarchus was the relationship between a planet's orbital period and its distance from the center around which it orbited. The more distant a planet from the center, the longer the period. This had been known since antiquity through observation. But the relationship failed to hold for

Mercury and Venus, the "inner planets" in the Ptolemaic system. In a geocentric universe, Mercury and Venus are the closest to a stationary earth, followed by the sun, Mars, Jupiter and Saturn, in that order. But during a part of an earthly year, the inner planets were observed to be on the other side of the sun. Assuming a sun-centered universe, the solar years of Mercury and Venus were in order with the "outer planets," and the period to distance ratio held. It also resolved the problem of retrograde motion. He had been led to this by his readings of writers of antiquity, particularly Vitruvius in the first century B.C. and Martianus Capella in the fifth century A.D. in addition to Aristotle and the commentary of ibn Rushd: "The doctrines of Averroes were very much at the center of philosophical discussion when Copernicus was a student in Italy (1496–1512). In particular, Achillini (d. 1512), one of the most celebrated philosophers in Italy at the time, published in Bologna an Averroist attack on Ptolemy while Copernicus was a student there." Bernard Goldstein, "The Origins of Copernicus's Heliocentric System," *Journal of the History of Astronomy*, 33, 2002, pp. 219–235.

14 For this see: George Saliba, *Rethinking the Roots of Modern Science: Arabic Manuscripts in European Libraries*, Georgetown University Press, Washington, D.C., 1991 and his *Islamic Science and the Making of the European Renaissance*, MIT Press, Cambridge, MA, pp. 193–232, where on p. 200 he provides a detailed presentation of the evidence for the matching diagrams. Also Arun Bala, "The Wider Copernican Revolution," in *The Dialogue of Civilizations*, Palgrave Macmillan, 2006, p. 121, pp. 145–176.

15 A copy just off the press in Leipzig is said to have reached him the day he died: a heartwarming ending to a lonely life devoted to science. However, the journey from Leipzig to Frauenburg in Poland was, even by express coach, a few days longer than the time that elapsed between the date of publication and the author's death.

16 A study by Harvard historian of astronomy Owen Gingerich shows how very little it was read: Owen Gingerich, *The Book That Nobody Read*, Walker & Co, New York, 2004.

17 See F. Minazzi, Galileo, *Filosofo Geometra*,Rusconi Libri, Milan, 1994.

18 For a brief analysis of the treatment of motion from Aristotle to Galileo see Edward Grant, "Aristotle, Philoponus, Avempace and Galileo's Pisan Dynamics," *Centaurus*, II, 1965, pp. 79–95. This is based on an earlier, groundbreaking study on the connection between ibn Baja and Galileo: Ernest Moody, "Galileo and Avempace, the Dynamics of the Leaning Tower Experiment," *Journal of the History of Ideas*, 12, 1951, pp. 163–193; 375–422.

19 Working from the theory of impetus and impressed motion that went back to John Philoponus, ibn Sina, abu'l Barakat al-Baghdadi, ibn Baja, Buridan, Oresme and the Mertonian School and was developed by Tagliatti, Cardano and Benedetti, Galileo reasoned that a moving body's impetus was consumed by the resistance of the media in which it moved and that, in the absence of resistance, its velocity would remain forever constant. As Galileo put it, a ball rolling on a perfectly frictionless and infinite horizontal plane would never come to a stop. His account of rolling balls down inclined planes was a thought experiment: I.B. Cohen, *The Birth of a New Physics*, Norton, New York, 1985, pp. 95, 117–121.

20 Rose, *The Italian Renaissance of Mathematics*, pp. 280–294 for Galileo's contribution to the Scientific Revolution.

21 Cohen, *Birth of a New Physics*, p. 95.

22 William Gillespie, *The Edge of Objectivity*, Princeton University Press, Princeton, NJ, 1966, p. 52.

23 Their analogy of gravity and magnetism came from another groundbreaking work: William Gilbert's *de Magneta*, where the earth was likened to a huge magnet.

24 For a fascinating account of their collaboration see Arthur Koestler, *The Sleepwalkers*, Hutchinson, London, 1959, and many other editions.

25 The Dane's death was as bizarre as had been his life. For a long time, his death was ascribed to a burst bladder caused by having held his water too long during a banquet, but recent chemical analysis of his remains (he is interred in a tomb in Our Lady Cathedral in Prague where he died) indicate the cause was mercury poisoning, probably caused by his prosthetic nose. When a law student in Germany (secretly studying astronomy), he lost the bridge of his nose in a dual in the dark of night, and he had various prosthetic noses made (reputed to be an alloy of gold and silver but recently discovered to be made of brass) in order to cover his disfiguration.

26 The system's secularist divorce from the creator was, on the other hand, in the eye of the beholder. Eighteenth century scientists did not allow their religious beliefs to run into the pages of their scientific treatises. It was a matter of keeping things in their proper place. As heaven and earth came together, science and religion drew apart, and they were treated apart, professionally and publicly; but in the privacy of one's mind, natural philosophy could continue to be the mirror of God's mind. The new system and the philosophy that framed it could be given a religious content. There would have been nothing in the new science inherently abhorrent to the religious mind, nothing that could not have been resolved by allegorical interpretation to agree with scripture. A decade and more before Newton was born, Galileo had argued this with passionate eloquence in a remarkable letter he wrote to the Grand Duchess Christina when, seeing the dark shadows of the Vatican's inquisitors closing in around him, he thought to reach out for help: "Letter to the Grand Duchess Christina" in Stillman Drake, *Ideas and Discoveries of Galileo*, Anchor Books, New York, 1957.

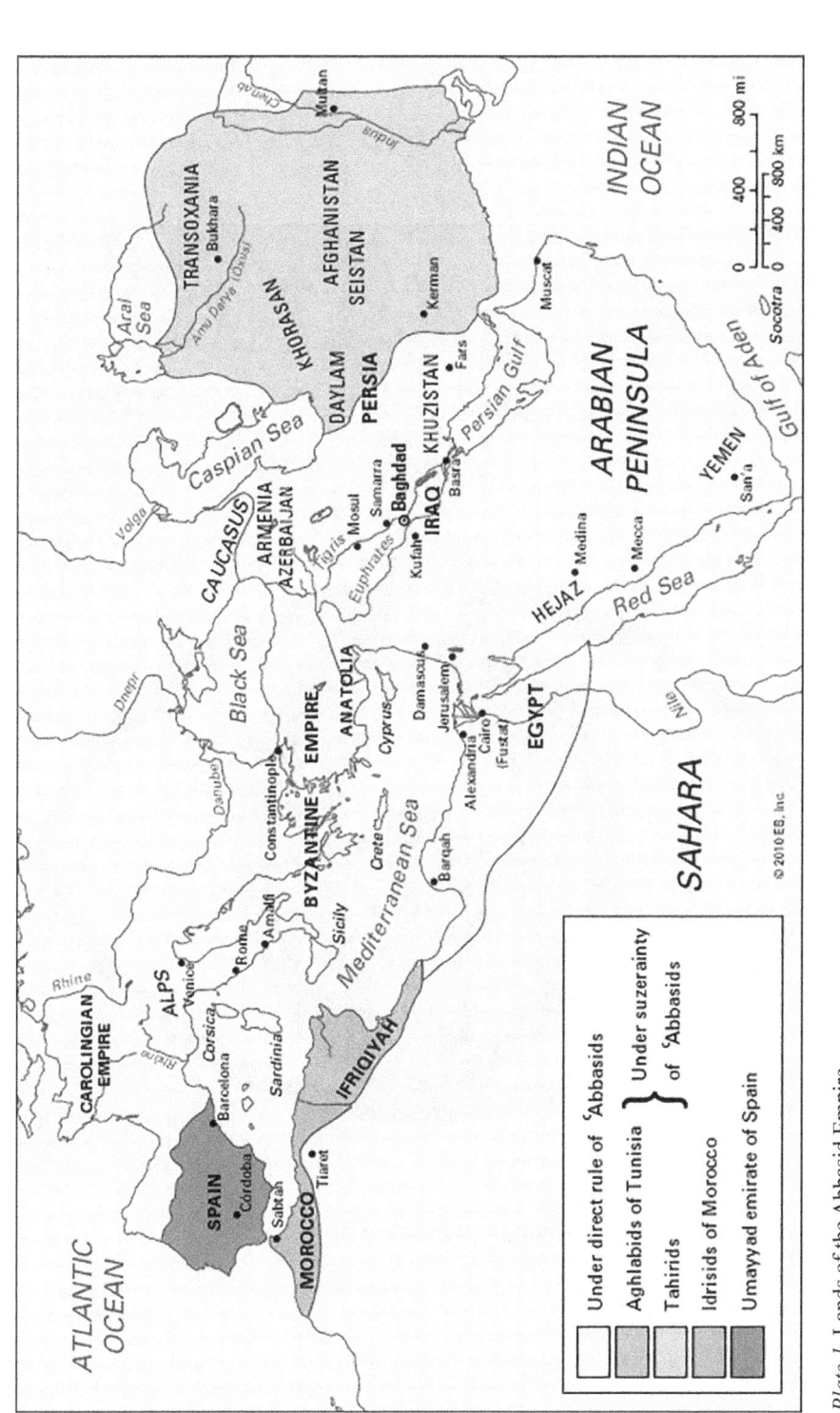

Plate 1 Lands of the Abbasid Empire

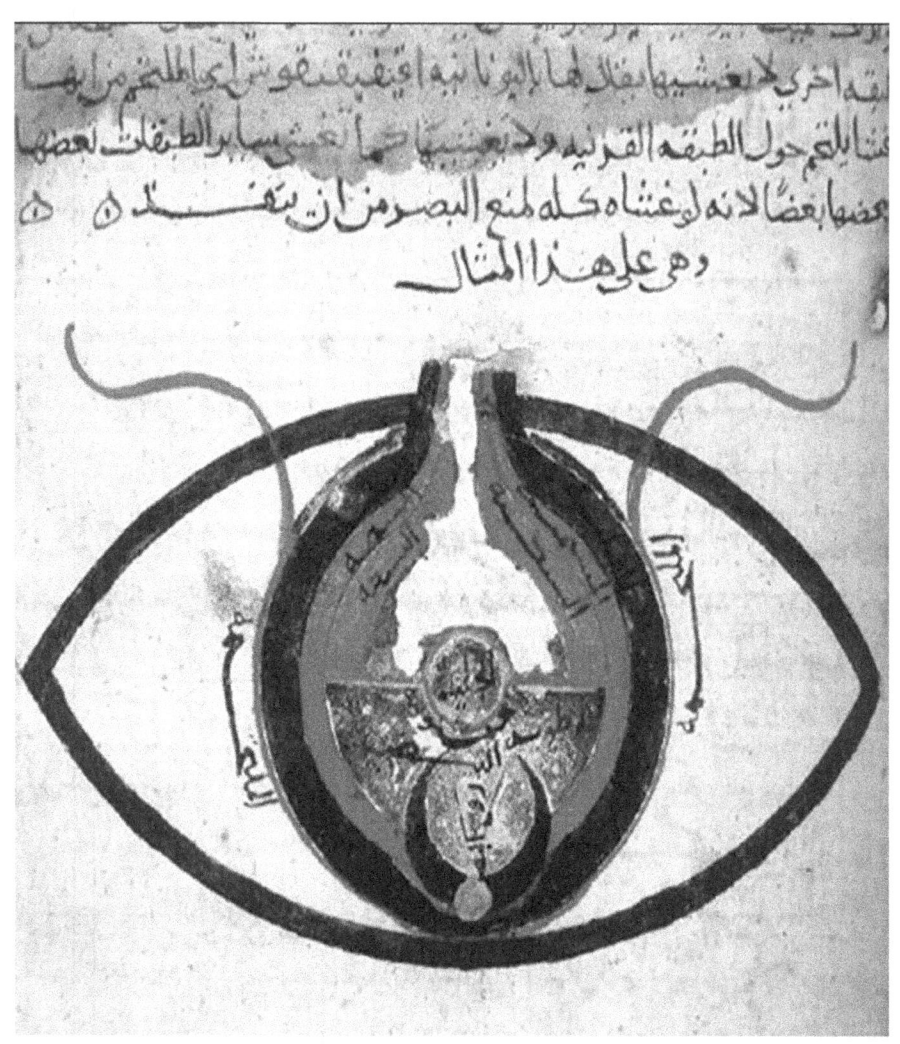

Plate 2 Anatomy of the eye according to Hunayn ibn Ishaq

Plate 3 Takyuddin and other astronomers at the Galata observatory founded in 1557 by Sultan Suleyman, from the Sehinsahname of Murad III, c. 1581 (vellum), Turkish School, (16th century)

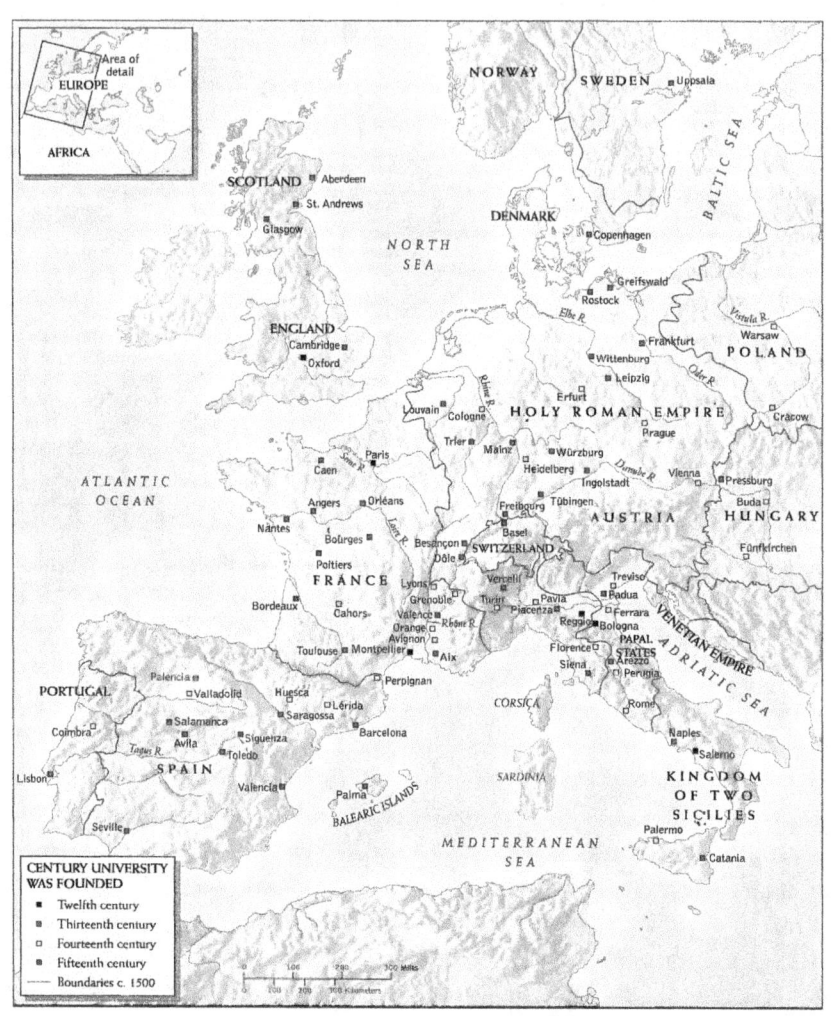

Plate 4 Map of the Spread of Latin Universities in Europe, 1200–1600

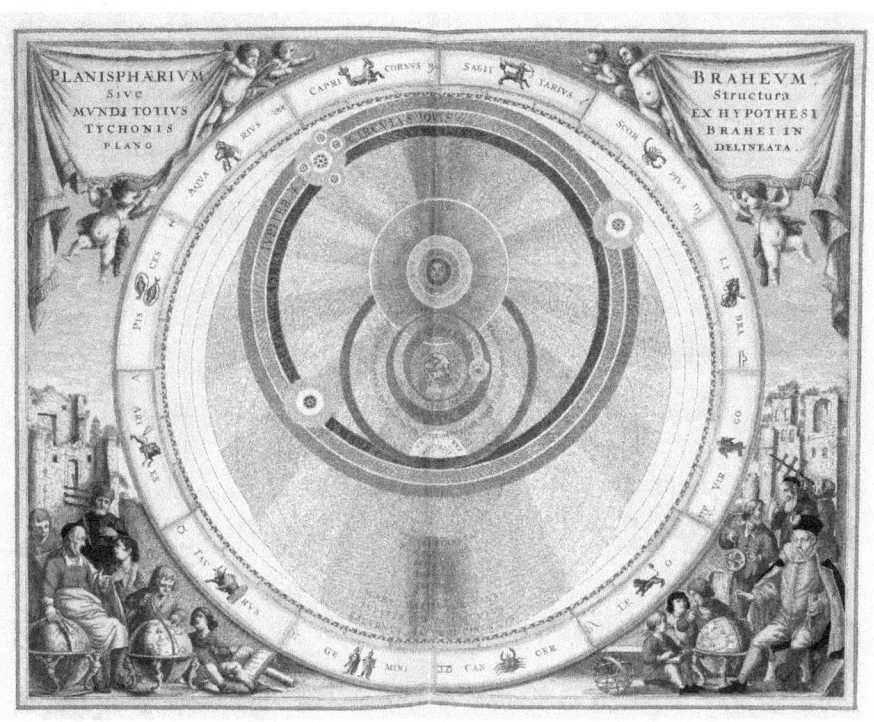

Plate 5 Map showing Tycho Brahe's System of Planetary Orbits, from "The Celestial Atlas, or The Harmony of the Universe" (*Atlas coelestis seu harmonia macrocosmica*) pub. by Joannes Janssonius, Amsterdam, 1660–1661 (hand colored engraving)

Plate 6 Kepler and Tycho Brahe at work in the Prague Observatory. Ventura, J. (19th century)

Plate 7 Map of Christian Constellations, from "The Celestial Atlas, or The Harmony of the Universe" (*Atlas coelestis seu harmonia macrocosmica*) pub. by Joannes Janssonius, Amsterdam, 1660–1661 (hand colored engraving), Cellarius, Andreas (c. 1596–1665)

Plate 8 Map of Muslim Empires: Ottoman, Safavid and Moghul, 1500–1600

Plate 9 Napoleon Bonaparte (1769–1821) visiting the plague-stricken of Jaffa, 11th March 1799, 1804 (oil on canvas), Gros, Baron Antoine Jean (1771–1835)

Part III

From Muslim empires in rise and fall to Western ascendance

8   Military ascendancy

1258–1600

Between 1100 and 1600, the West was advancing stage by stage, on one front or another, in its millennial contest with Islamdom. Western improvements in technology and food production led to increasing population and urbanization, with the subsequent emergence of an expanding market economy whittling away at serfdom and the self-sufficient, agriculturally based manorial feudalism it supported. Universities proliferated with increasing urbanization and the new wealth that was accumulating through international trade and urban industry. What was taught in the universities by the late 16th century had gone beyond the knowledge gained from the translation of Arabic and Greek texts. Rising centralized monarchies were sending ships around Africa to India and westward across the Atlantic to begin the world's first global empires, while society was being internally transformed by a Renaissance, then a Reformation accompanied by a Scientific Revolution. Meanwhile, Islamdom was also undergoing some fundamental changes, some of them not altogether positive.

In the 11th century, the Fatimids in Egypt unleashed the destructive force of tribal Bedouin invasions across the coastal plain of north Africa in vengeance for political dissonance in Tunis. Cities and towns were destroyed, farms laid to waste, orchards and groves cut down and trees uprooted. What had been green became desert. Writing centuries after the event, the historian ibn Khaldun, who was born in Tunis, could claim that north Africa had never recovered from the devastation of the Bedouin storm. Two centuries later, it was the Mongols to the east leveling cities. On the positive side, the transformative influence of Islam turned city-destroying conquerors into proud rulers of post-Mongol Muslim empires, Ottoman in the West, Safavid in Iran, Moghul in India. Of these, the Ottoman was the most powerful and most relevant in terms of interaction and intellectual exchange with Europe. The Ottoman Empire lasted six centuries: several times the normal run of Muslim dynasties.

The Ottomans gave Sunni Muslims their last taste of conquest and glory. The conquest of Constantinople was a momentous event in the march of Ottoman arms under the banner of Islam. It crowned the glory of earlier Turkish triumphs, the Ghaznavid Sultan Mahmud's conquests in India at the end of the 10th century and the Seljuq Sultan Alp Arslan's Anatolian victory at Menzikert in 1071. The Ottomans would go on to penetrate into the heart of central Europe, but their failures against the walls of Vienna, in 1529 and 1683, would preserve the capture of Constantinople as the jewel in the crown of Turkish conquest.

Constantinople gave the Ottomans a status and prestige above all other Muslim powers, and it also provided them with a mighty base along the southeastern flank of Europe from which it could launch invasions deeper into Christian territory. Campaign after campaign, the Ottoman Empire expanded: Greece, Croatia, Bosnia, Albania, Slovenia, Montenegro, Herzogovina, Moldavia, Wallachia, Bessarabia, Hungary, Podolia, Crimea, the Black Sea and Caspian Coast; all fell before the devastatingly effective Janissary infantry and heavy field artillery.[1] Ottoman conquests balanced the Muslim loss of Andalusia.

In 1517, taking time off from conquests in Europe to campaign against the Shi'i Safavids in Iran, Sultan Selim I (1512–1520) added Syria, Iraq, Palestine, Arabia, Egypt, Libya, Tunis and Algeria to the empire.[2] What had begun as a precautionary measure to protect the army's southern flank while fighting the Safavids ended in the Ottoman annexation of the Arab world. Considering that 50 years before the fall of Constantinople the Ottomans had been decisively defeated by Timur, who took the Ottoman sultan prisoner, parceled out Ottoman territories and made it look to all the world as if the young empire had come to an end in the manner by which so many other Muslim imperial dynasties that had risen and fallen within the span of a century or less, the Ottoman recovery and resumption of conquest in every direction was an extraordinary feat that reversed the dynastic pattern seen in the previous seven centuries of Muslim history.

With the fall of Hungary to Sulayman (1520–1566: known as The Magnificent in the West, The Law Codifier in Ottoman historiography), only Vienna stood in the way of an Ottoman advance from the Danube to the Rhine. Reinforced by support from western Europe, the defenders of Vienna were able to hold out against the Ottoman army. The Danube would remain the border between Europe and the empire. The conquests were enough for Sunnis across the Muslim world to regard the sultan as the worthy caliphal successor to the best of the Umayyads and Abbasids. Protector of Mecca, Medina and Jerusalem, ruler of Constantinople and half of Europe, of North Africa and the Arab speaking regions of Islamdom, the Ottoman sultan was hailed as the legitimate successor to the political leadership of the Prophet, symbolized by the purported cloak of Muhammad that was worn by the sultan on religious occasions. (The cloak supposedly came into Ottoman possession with the defeat of the Mamluks of Egypt who in turn had come into possession of it when a relative of the Abbasid Caliph executed by the Mongols escaped from Baghdad with it and found refuge in Mamluk Cairo.)

For six centuries Ottoman armies and navies battled against the Christian powers of Europe, the first three of which were remarkably successful. Ottomans were always at war, if not against the Byzantines whom they destroyed, then the Venetian or Portuguese navies or Hapsburg and Russian armies, and sometimes against a coalition of two or three of them at once. When not warring the West, it was the Mamluks and Safavids on the eastern border. Raiding was how the Ottomans began, conquest was how they grew to an empire, and holy *jihad* their raison d'être. The warring ghazi was the Ottoman ideal throughout the empire's existence: It endures as the principal image and the title that modern secular Republican Turkey has bestowed on its practically deified founder, Mustafa Kemal.

What was the secret of Ottoman power and longevity that defied the century-long generational pattern of conquest, efflorescence, and decadence, with possibly a last flourish of brilliance before extinction that made such a strong impression on ibn Khaldun that he based a socio-economic philosophy of history on it? The source of Ottoman strength and durability, even in adversity, is of course found in the empire's early institutions. Administrative and military landholding institutions were inherited from the Seljuqs of Anatolia, whose divided emirates the Ottomans defeated and absorbed. Ottoman rulers patterned themselves and their military organizations on the model of the Mongols, fellow tribal horsemen who had gone from pastoral nomadism to conquest and world rule. Several institutions the Ottomans created themselves. Previous Turkish dynasties had relied on Iranian statesmen to set up and run the centralized administration and preserve the realm from the recalcitrance of the tribal army chafing at the constraints imposed by the government's central control, which was forcing the tribes to accept a settled life in an organized state based on agriculture, trade and taxes. Control of the military tribesmen had been an endemic source of instability for earlier Muslim dynasties. Another source of trouble had been the rivalry among the leading chieftains who resented the ruler's power and were jealous of each other's position at court. To avoid this, the Ottoman sultan and his trusted advisors established the *devshirme*, a collection of Christian youths from the Balkans. Every few years, several thousand Christian boys were taken as slaves to serve the sultan. It was strictly illegal to enslave Christians of the subject population. They were "People of the Book" with legal status and were to be protected and left free in their religion. Employing a clever twist of casuistry, a *fatwa* was issued claiming it to be legal: since the Balkan people had not become Christian until long after the time of the Quranic revelation that laid down the law, it was permissible to enlist their boys into the sultan's service as slaves.

Since the Ottomans wanted strong, simple, unspoiled country boys, they were collected in villages, then taken to the capital where they were converted to Islam and taught Turkish. The brightest of them were entered in what was called the "inside service" to became palace interns. Educated in history, literature, religion, composition, calligraphy, politics, government and courtly formality, the boys served in the palace as the sultan's personal slaves and rose through the courtly ranks in serving their padishah, their father-king, until they were ready to be assigned responsible positions of government. Many of them ascended up the administrative hierarchy to become provincial governors, ministerial viziers and grand viziers, in which capacity they often led Ottoman armies to battle. As elite palace servants whose material needs were adequately satisfied by the sultan, it was reasoned that these converted slaves would serve the sultan more faithfully than members of the Turkish aristocracy, whose loyalties were conflicted by vested interests, devotion to family, personal ambition, petty jealousies, court intrigues, not to mention their sense of privileged entitlement. The *devshirme* would avoid that. Its members would have one interest in life, one purpose: to serve and obey their royal father.

Those of the *devshirme* boys who were not selected for palace service were sent to the Anatolian provinces, where under military supervision they were Islamized, Turkified and rigorously trained in the martial arts in preparation to entering the ranks of the *Yeni Sheri*, or Janissaries, meaning New Troops. This was the sultan's elite slave infantry corps, his crack troops who were sent in to finish off the enemy and whose name for generations sent chills of fear through Europe. Martin Luther thought them to be a divinely sent punishment for Europe's wickedness. A century after Luther's inverted praise of them as God's scourge, institutional corruption had so rotted the system that the Janissaries turned out to be more the scourge of the empire than of Europe. By the 18th century, it became clear to leaders on both sides of the Danube that the Janissaries had lost their fearsome edge. Some Ottomans perceived it as going deeper than just Janissaries: that a serious reversal in the power relationship between the Ottomans and the West had somehow occurred. The reversal could hardly have been more dramatic in its decisiveness.

In their days of military glory, the Ottomans showed themselves to be quick learners, at least in things to do with war. Wrought iron cannons had been in use since the end of the 14th century, but they were technically crude and not too effective against heavy stone walls. In the 1420s and 30s, cast-iron muzzle-loading siege cannons were developed in Europe. The Ottomans were soon to master the technology and improve on it. By the 1440s they were producing huge artillery pieces of cast-bronze. One of the largest ever made played a part in the siege of Constantinople in 1453, at which time the Ottoman artillery corps was the world's finest. It will be recalled that Greek Fire, a flaming mixture of sulfur and pitch blown by bellows from cannon-like tubes mounted on the prows of ships, had saved Constantinople from the navy of Islam's first empire in the early 8th century. Now its more deadly descendant gunpowder, a product of Chinese technology, destined the city to become the capital of Islamdom's last empire. Western technology that could be applied directly to the battlefield was perfectly acceptable to the Ottomans. Non-military technology and the science that went with it would be, on the surface at least, another matter.

To the east, the Safavids in Iran and the Moguls in India also stood strong. An English traveler who visited the Ottoman, Safavid and Timurid Hindi realms in the very late 16th century thanked God the Muslims were occupied fighting among themselves, for he feared the armies of Europe would not stand a chance were the Muslims ever to unite. It was a fair assessment of how the balance of power appeared. Not to be outdone by the Ottomans, the Safavids, under the greatest of their rulers, Shah Abbas (1586–1628), builder of Isfahan, also developed an artillery corps. An Englishman who had come to Iran as part of a diplomatic embassy to establish a commercial agreement between the Shah's government and the British, and who had experience in the technology of cannon casting, was employed by Abbas to supervise the manufacture of artillery and train the new corps. Another Ottoman–Safavid parallel was the young Christian boys who were taken as slaves from Georgia in the Caucasus and converted, trained and formed into a corps serving the ruler. This was the "Shah Seven," Turkish for "He Who Loves the Shah,"[3] a Safavid version of the Ottoman *devshirme*. The Moguls in India instituted military, landholding, tax and administrative systems that were similar to those of the Ottomans and Safavids (see Plate 8).

At the height of the Ottoman, Safavid and Mogul empires, royal patronage produced a splendid Islamic renaissance in painting, architecture, poetry, and historiography, personified by Sultan Sulayman (d. 1566) for the Ottomans, Shah Abbas (d.1628) for the Safavids and Akbar Khan (d. 1605) for the Hindi Moguls. Each ruler was an icon of power, cultural elegance, artistic refinement and religious rectitude. Later generations would look back with nostalgia at this period of grandeur, when Muslims were masters of their world and the world around them, setting their own styles of life, dress and cuisine, confident in their culture, education and institutions, feeling no need for foreign languages and unburdened by the shame of backwardness. Had there at the time existed a sense of modernity that was bound to go global, they would have been the ones to project the model and set the standard.

The imperial cities Istanbul, Isfahan and Delhi (and to this could be added Samarqand of the Timurids and Uzbeks) were bejeweled with monumental mosques, palaces, mausoleums and public buildings from whose glazed tiles set in floral and geometric patterns shined the brilliance of Muslim genius. The grand bazaars, displaying the goods of trade from China in the east to North Africa in the West and from Russia and Scandinavia in the north to central Africa in the south, reflected in abundance the more mundane levels of well-being: perfumes, spices and fruits of every kind, leathers, fabrics, furs, dyes, ivories, porcelains, objects in gold, silver, lapis and jade. The public baths, gardens, fountains, mosques and mausoleums reminded subjects of their ruler's munificence. Small industry flourished in a market of indigenous manufacturers and skilled craftsmen.

Tastes of the royally supported artistic renaissance were sometimes too much for the ulema to take, but the rulers' power and prestige overruled the opposition, as they had during the Abbasid caliphate's days of strength. Depicting the human form in painting or sculpture is proscribed but out of fear that the power of imagery might detract from God's glory in the believer's mind, but this did not stop the potentates of the age from having their likenesses produced in color on canvas. As if to show his independent power and support of high civilization, Sultan Mehmet II, conqueror of Constantinople, broke with custom and invited Bellini from Florence to do his portrait. Religion and tradition could not deter Sultan Sulayman from having his portrait done. Akbar's portrait in India depicted him contemplating a rose, from which the Mogul dynasty received its name, the Rose Dynasty, signifying its artistic sensitivity, spiritual depth, transcendent composure and aesthetic inner peace. Akbar's was an Islam of distinct Indian and Buddhist influences. Miniature painting flourished in all three empires, the masters of course being Persian, the Florentines of Islamdom. Monuments, mosques and mausoleums were architectural triumphs commemorating the glory of rulers who built for their immortality. The Mogul ruler Shah Jihan (d.1658) chose pure white marble for his beloved wife's resting place in Agra, the Taj Mahal, an enduring reminder of

Mogul splendor. Jihan's own mausoleum was to be an exact replica in black marble facing his wife's from across the long reflecting pond, but his son had him deposed and cast in a dungeon before it could be built.

Akbar deserves special mention for his daring and imaginative attempt to synthesize Islam and Hinduism, the religion of most of his subjects. Religious synthesis, he hoped, would provide a stability to his realm that went beyond coercion by military power. Akbar enforced religious tolerance, observed Hindu religious feasts and customs, appointed Hindus to high office, used them in the military, and encouraged the building of both Hindu and Muslim schools. In deference to Hindu custom he gave up eating meat and expected Muslims in his court to follow. His example of tolerance was intended to bring the followers of the two religions closer together, for reasons of state, but Akbar's innovations, anathema to the Sunni ulema, failed to go far beyond his royal court and ended with his being overthrown in a palace conspiracy instigated by his conservative son. Aurangzeb's sharp reaction against his father's tolerant policies easily undid the innovations, as far as they had gone. Reforms that challenge the cultural and religious mainstream to take root and break through the hard ground of tradition would have required generations of enforcement by strong leadership, as suggested by the history of the modern Republic of Turkey, whose secularist surface even now must allow for the emergent response of deep religious and cultural sentiments that for a century were suppressed by the guardians of the republic. What survived of Akbar's attempt at a Hindi–Muslim synthesis was the new language he and his scholars created from Arabic, Persian, Turkish and Hindi. Intended to unify his realm in a common tongue, Urdu grew to become an important language of poetry, religion, literature, history and learning. It survives as the national language of present day Pakistan.

When Shah Jihan had been thrown into his dungeon, he was able to procure a broken piece of mirror with which, when held at a certain angle, he could catch the barest glimpse of a corner of his wife's Taj Mahal mausoleum through his barred window. Akbar's broken mirror and the abortive end of his innovations make an apt metaphor of the failure that was about to befall the whole of Islamdom. What Shah Jihan saw in his fragment of mirror was the image of magnificence of an autonomous Muslim state. The time was rapidly approaching when there would be no more Taj Mahals or autonomy. Autonomy of the Muslim state, even as the glow of glory still lit the near horizon, was soon to slip away. Already in the reign of Shah Jihan's son European merchant colonies were being established on the Indian coast and expanding, initiating the economic processes that would soon undermine the Mogul empire's independence. Muslim power was shining deceptively bright, even as its foundations were being quietly but relentlessly cut away, European cargo ship by cargo ship, colony by merchant colony firmly digging in along the urban shores and harbors that traced the traditional maritime trade routes that fed Muslim prosperity. Buoying the innocent-looking economic activity of Westerners in Muslim lands in the 17th century was a less visible foundation of power, namely European science and technology, the effects of which were to be devastatingly manifested to Muslims before the century was out.

Nothing more traumatizes a civilization's collective self-esteem than wars lost. The trauma of defeat was all the more severe for Muslims who for the best of a millennium had the West on the defensive, whose founding heroes swept over two empires in little more than a decade, and who had come to take their superiority of arms, and therefore superiority of religion and civilization, as axiomatic, an incontrovertible truth manifested by God acting in history. The roles were about to be dramatically reversed. The inferiors became victors. While Muslims had been celebrating Ottoman additions to the Islamic Abode of Peace, and Safavids and Moguls were building proud empires of their own that glorified the vigor and brilliance of Islamic civilization, fateful events were quietly drawing the noose around the proud Muslim world.

In 1492, shortly after Grenada fell ending the last vestige of Muslim rule in Spain, Columbus sailed under the Spanish flag for India across the Atlantic in search of a maritime route for those exotic spices that affluent Europeans, in rising numbers, were demanding to enliven their daily fare: coriander, cardamom, ginger, nutmeg, cinnamon, clove, pepper, mace, saffron and camphor. The Portuguese had monopolized the route to India around Africa; so Spain sailed in the opposite direction. While Vasco da Gama opened the way to a Portuguese empire in Africa, India, the Arabian Sea and the Persian Gulf, Columbus pioneered a Spanish empire in Mexico and the Caribbean based on slave labor. What had begun modestly as a Portuguese explorative venture to the Canary and Verde islands and southwards along the western coast of Africa in quest of gold and slaves grew into a vast international commercial enterprise that gained astounding riches by procuring the spices that so pleased the European palate, and culminated in the emergence of Western maritime empires that stretched around the world, embracing the west coast of Africa, the Persian Gulf, India, Indonesia, Malaysia, the Philippines, the islands of the Caribbean and the land masses of the western hemisphere.

Before Bartholomieu Diaz and Vasco da Gama rounded Africa, there were two main routes by which Eastern spices reached Europe. One was by sea from either Malacca or Malabar to Suez on the Red Sea, from where the cargo was then transported overland to Alexandria and shipped to Venice or other European ports. The other route was by sea and land from the Spice Islands to the Persian Gulf, then overland to the east coast of the Mediterranean, usually to Beirut, from where the cargo was reloaded onto ships bound for European ports, usually Venice. Venice, whose vessels were the main carriers, grew rich, but then declined when the new oceanic routes were discovered. So while the Ottomans had been gloriously taking Constantinople and eastern Europe to the Danube, and the Arab region to the Tigris, and besieging Vienna to the pounding of drums, cannon fire and triumphant cries of God's greatness, western Europe's maritime powers were quietly taking over the world's waterways and ensconcing themselves along the eastern shores of Islamdom.

Vasco da Gama's rounding of Africa established a direct maritime route between India and Europe, thereby cutting Egypt out of the lucrative customs duties on

Eastern goods that had, until then, been unloaded from ships at Suez and cara-vanned to Alexandria, Rosetta or Damietta for transshipment. That became a thing of the past. The Portuguese were, at the same time, building fortresses along the coasts of Africa, the Persian Gulf and India, monopolizing trade from Malaysia, Indonesia and India, and driving out Muslim navies and merchant ships, as would the Dutch and British drive out the Portuguese a century later. No longer would Muslims carry the trade of the Indian Ocean, Arabian Sea, Persian Gulf and Red Sea, which had been private lakes of theirs for eight centuries. European cargo ships and merchant colonies were everywhere in the 17th and 18th centuries: along the coasts of Africa, in Alexandria, Cairo, Damietta, Rosetta, Suez, Sidon, Beirut, Damascus, Tripoli, Izmir, Istanbul, Aleppo, Basra, Baghdad, Hormuz, Shiraz, Isfa-han, Bombay, Goa, Calcutta and east to China – isolationist Japan allowed in only the Dutch, who were given special trading privileges.

Other factors that had nothing to do with Europe were meanwhile undermining the Ottoman economy. Agricultural land was contracting because of poor farming methods and mismanagement of irrigation systems. As mentioned earlier, in the 11th century the Fatimids in Egypt had sent large Arab tribes across North Africa on a mission of vindictive destruction that laid waste to agricultural and urban life. Trees were uprooted, orchards and farmlands destroyed, cities plundered, turning the once fertile soil of North Africa into desert. Mesopotamian agriculture had been in precipitous decline since the Abbasid caliphate's implosion in the late 9th and 10th centuries.

Diminishing revenues from agriculture, coupled with losses in trade revenues because of European monopolies of maritime commerce, led the Ottomans and other Muslim governments to debase the coinage to make up for the shortfall. Debasement was economically disastrous, especially for the Ottomans, whose precious metals were being drained away because of the high inflationary pres-sures caused by the great influx of Spanish silver from the Americas. In what has been called the Great Price Revolution, commodities were sucked out of Ottoman lands in return for cheap silver. A vicious cycle ensued: the more silver came in, the higher inflation became and the more the Ottomans debased their currency. The Ottoman economy was by now too dependent on trade with the West for its cur-rency not to be affected by Europe's economic swings. By the middle of the 17th century, the Ottoman government had fallen into a state of perennial insolvency.[4]

Not having enough currency to cover expenses, the rulers resorted to various stratagems that ran counter to good government and central control of the empire. High offices and provincial governorates were sold, tax collection was farmed out to wealthy individuals for lump sum payments, all of which ran the risk of putting government positions in irresponsible hands. Anyone buying the right to collect taxes in a province would make sure to get his investment back with as much profit as possible, even if it meant raising taxes and oppressing the peasants more than they already were. The expense of irrigation repairs was considered money out of pocket. Investment and development with an eye to the future became victims of the quick fix. Diminishing revenues often resulted in long delays in salaries paid to government officials or reductions in salaries, ripening the conditions for bribery, administrative irresponsibility, oppression of office and every other form of corruption. Similar conditions corroded the military institutions. This was true for Iran and India as well.

By the early 18th century, the Safavids and Moguls had all but disappeared. The Ottomans, having suffered disastrous decades of defeat and territorial loss, seemed to be following right behind. It took a bitter swallow of pride for Ottoman leaders to look to the West with new eyes and institute reforms based on that which they perceived to be the sources of Western power. It was either that or go down before a despised enemy grown suddenly superior. However, as will be seen, even two centuries of importing Western science, technology and institutions failed to halt Europe's territorial expansion and economic domination and may have in fact hastened the Ottoman march to join their Safavid and Mogul cousins into the annals of lost empires.

Notes

1 Colin Imber, *The Ottoman Empire*, Palgrave, Basingstoke, 2002, pp. 11–86 for a detailed chronology of military success and reverses up to the middle of the 17th century.
2 Geographically between the two great Sunni empires, Ottoman and Hindi Moghul, arose a third Muslim power that was Shi'i, the Safavids, centered in Iran. Though Turkish in origin, the Safavid ruler and his tribesmen soon became Iranianized upon conquering the country, as had Arabs, Turks and Mongols before them. Before achieving power over Iran, the Safavid ruler had led a confederation of seven Turkish tribes united in a mystical order of warriors for the faith. They believed their order to have been founded by an early 15th century pious mystic, Safa al-Din, hence the dynastic name Safawi, as pronounced in Arabic, Safavi in Turkish and Persian. The Sufi order had originally been of Sunni persuasion, but to win recruits in the heavily Shi'i region of Turkish Azarbayjan to where the tribes had migrated, a descendent of Safa al-Din switched to Shi'ism, his tribesmen following. To its enemies and outsiders, the Safavid order of Sufis was known as the Kizilbash, Turkish for redhead, because of the towering red conical headgear that members wore, crowned with 12 tassels symbolizing the 12 Shi'i imams.

  The seven Turkish tribes of the Safavids were quickly absorbed and Iranianized. Iran in whole or in part, had been ruled by a dozen or more Sunni and Shi'i dynasts since the 10th century, but none left a trace as deep and indelible as the one made by the Safavid dynasty. In 1501, a year as decisive in Iranian and Muslim history as it has been divisive, the Safavid leader Ismail, a descendant of Safa al-Din and chieftain of the seven tribes, declared himself Shah and his dawlah Shi'i. Henceforth, Iran's religion would be the cult of betrayal, martyrdom, expiation, sorrow, suffering, persecution and, ultimately, triumph when the Hidden Imam would reveal himself and announce the end of time and the wicked would be punished for their persecution and denial of the family of Ali and his descendants through Hasan and Husayn. Unlike earlier Shi'i dynasties, of which there were many in Muslim history, the Safavids ruthlessly forced the Iranian populace to convert, persecuting, executing or driving out those who refused. Many Shi'i dynasties had ruled in lands where the large majority of the population was Sunni, such as the Fatimids in Egypt and Palestine, but the rulers left the populace to their religion and in general made no changes that affected the populace. Until the advent of the Safavids, a state policy of forced conversion and expulsion had never occurred in Islam. By ironic coincidence, when it did occur, the Christian rulers in newly unified Spain were at the time doing the same thing, just as brutally, to Muslims and Jews – a less savory instance of historical parallelism in the familial pattern of Muslim and Western civilizations.

3 Sevmek, to love; Seven, he who loves or lover.

4 Dan Diner, *Lost in the Sacred: Why the Muslim World Stood Still,* Princeton University Press, Princeton, 2009, pp. 96–101; Omer Lutfi Barkan, "The Price Revolution in the 16th century: A Turning Point in the Economic History of East and West." *International Journal of Middle East Studies,* 6, 1975, pp. 3–28.

9   Prologue to decline

The past as future

A word should be said about personal identity and the formative power of historical consciousness in Islamic civilization in order to appreciate the difficulties that
Muslims experienced in their interaction with Western knowledge during the
period of defeat and territorial contraction. For without some understanding of the
internalization of culture and history on the personal level, the psychological elements underlying the refusal by many Muslims in positions high and low to accept
the reforming innovations would be missed, making the stern resistance to change
look irrational, to the point of willful self-destruction, which it was not.

If by civilization is meant an ordered way of life informed by religious, political,
and legal institutions, and elevated by a literary and artistic esthetic of mind and
craft of hand, then there was little that could be called Muslim civilization during
the first century of Islam's history, beginning with the first revelation of the Quran
around 611/612. Islam was born into a primitive Arab society that could perhaps
boast of its tribal warrior heroes and the oral poetry that praised them and its energetic middlemen of the camel caravan industry, but it could point to little that it
had of high civilization. During the century following Muhammad's triumph over
pagan Mecca and the spread of Islam through Arabia, Muslims achieved astounding success under their tribal chieftains, who could be barely held in control by
their caliph at the best of times. They had no time to produce anything more than
a rough-draft copy of local Byzantine and Sassanian regulations needed to organize people into an ordered society capable of keeping the peace, collecting taxes
and maintaining an economy.

Arab energy was consumed in maintaining order, holding the constantly expanding empire together in the face of tribal factionalism and rebellion and in interpreting the Quran and organizing Prophetic traditions to formulate a system
socio-religious law. The Islamic venture was pretty much an all-Arab enterprise
until the end of the century of conquests, upon which the Arabs were consumed
by their own military success with the fall of the Umayyad caliphate in Damascus.
During that first century, and more so in succeeding ones, new peoples, non-Arabs
living in the conquered lands, came into Islam through conversion, bringing with
them inherited traditions that reached back to antiquity. Conversion was in fact
incidental to cultural contribution. Many were the unconverted Jews and Christians who contributed to the Muslim appropriation of the rational disciplines of

high civilization. Mass conversion of the indigenous people of Syria, Iraq, Iran and Egypt did not come until centuries after the conquests, by which time the assimilation of the traditions of high civilization, and the Arabization of the people, were well under way. Iran would accept Islamization but preserve its national and linguistic identity, and pass it on to those Turks and Mongols who conquered Iran. Accepting Islam, Iran accepted Arabic as a language of religion and knowledge and became a primary contributor to Islamic civilization from the 8th century on, but it never surrendered its language or pre-Islamic identity. Parallel to the Arabization of self-identity and language that occurred in these old lands of high civilization, from Mesopotamia to Morocco, indigenous traditions that insinuated themselves into the community of believers came to be perceived as genuinely Islamic.

By means of the patronage of the Baghdad caliphs and their wealthy Iranian viziers and the intellectual work of Christians, Jews, Pagans – and those of them who had converted to Islam – the seminal strains of civilization's millennial traditions of science, philosophy, medicine and mathematics were infused into the upper strata of the Muslim social body. The cross-fertilization of their cultural traditions enriched the soil from which grew the new civilization. The new was, in a sense, the recasting of the old in the religious norms of Islam and the Arabic language, so that children of the converts reared in the evolving civilization and religion shared, as much as any Arab Muslim, in the ultimate triumph of Muhammad or sorrows and defeat of Ali and Husayn, as a historical experience of their own personal religious life. As deeply as did any Arab descendant of the tribal conquerors, they experienced the uplifting triumph of good defeating evil when hearing or reading of the heroic battles that their Prophet had won in Medina against his stonehearted Meccan adversaries. Their sharing in the sorrows, challenges and triumphs of the beloved Prophet, and their faith in Islam and the glorious Quran, transcended genealogical origins and national identity. The memory of Muhammad, an illiterate Arab shedding God's light in a dark, heathen and primitive place, of the Ka'ba and Mecca, of the Hijra and Medina, the Battle of Badr: these were to every Muslim, of all ethnicities, holy portals through which God gave life and eternal memory to His last prophet, bearer of His last revelation.

The glories of primitive Islam in its Arab setting were the glories of all Muslims. Even the high-browed Iranians accepted this. In later times, when the Ottoman *devshirme* brought fresh recruits into Islam from the newly conquered Balkans, these boys, once Turkified, Islamized, educated, and infused with the glories of Muslim history, were subject to the same religio-historical emotions as their Arab coreligionists to the south. Their dedication to *Din ve Devlet, Religion and State,* was the bedrock of their purposeful existence that fused their souls to Muhammad, Mecca, Medina, and the Quran.

The completeness with which the *devshirme* recruits identified themselves to Islam is truly remarkable. Even *devshirme* recruits who came into Islam when already grown and mentally formed young men were able, with evident ease, to abandon their ancestral home, their culture and Christianity, to devote themselves body and soul to Islam, sultan and empire. In becoming Muslim, converts assimilated not just a religion but a civilization and the historical heritage it embodied, going back to the Prophet receiving his first revelation in the cave at Hira. The *devshirme* boys, though born eight centuries to a millennium after the event, gloried in Muhammad's triumph as fully as they did in the triumph of Ottoman arms that carried the prophetic message into the heart of Europe, against their former coreligionists. Like any Arab Muslim, their consciousness of self was infused by the Islamic drama of historical fulfillment. Their sense of a former ethnicity and religion was dissolved and expunged by the informing power and spiritual passion of Islam. No less than the pounding blood of Arab Muslims, their pulse quickened with righteous fervor when reading or hearing of Muhammad's persecution in Mecca, of the boycott of his clan by the greedy and wicked Sufyanids, of his flight to Medina and the victory of the Prophet with his small but brave band of companions at Badr, 300 defeating a Meccan force three times larger. The triumphs of Arab tribes bound as one under the banner of Islam were their triumphs too. The history of the fall of Sassanid Iran to Arab warriors became a glorious event, even to Iranian converts, since the conquest brought the light of Islam to extinguish the night of pagan ignorance, the *Jahiliyya*. The alchemy of Muslim faith turned national defeat into a spiritual triumph.

The glory of those early times, when enemy armies were annihilated in a single battle, empires knocked off in a few years, a world conquered in a century, became the Muslim's cultural and historical heritage. In accepting Islam, one accepted history, for just as the Quran was spiritual proof of Islam, history was empirical proof of the truth of Muhammad's prophecy. If this were not so, then how to explain the triumph of Muhammad and the triumphant armies of Islam that followed his own successes? Obviously, Muslim triumph was the work of God's abiding favor of the believers in order that the final revelation would be communicated to the world.

The Muslim historical experience was opposite to the experience of early Christianity where history was adverse to the infant religion, whose only victory was the spiritual triumph of the self over evil through suffering and surrender of wealth, through martyrdom, not triumphs or glory in the field of physical battle. The primitive Christian belief that the world is evil and corrupt, that personal victory is achieved through self-denial and martyrdom, may intersect with some aspects of Sufism and Shi'i belief, but not orthodox Sunni belief.[1] For the Shi'i, the Arab conquests were not proof of divine favor in the way they were for the Sunni. They were conquests carried out by illegitimate rulers, usurpers, tyrants and enemies of God. Opting for Shi'i Islam may have put some Iranians at ease in accepting the religion of a people they regarded as primitive destroyers of the Sassanid dynasty, ending Iran's proud tradition of civilization, since it was evil usurpers and not true Muslims who conquered Iran.

Unlike earlier Shi'i dynasties that had come to power in regions that were heavily Sunni, the Safavid was the first to extirpate Sunni Islam from the dynastic realm, root and branch – or at least to try. Previous Shi'i dynasts dared no such thing, leaving the Sunnis to their four schools of legal interpretation. Even the revolutionary Fatimids of the Sevener (Ismaili) branch, once they had ensconced themselves in Egypt and Syria, settled down and built their own capital of Cairo apart from the vast Sunni majority. Keeping their distance, they proceeded to govern their Sunni subjects without coercive propaganda, forced conversion or any form of religious intervention. The Shi'i Fatimid dynasty, with its government and religious institution, ruled Egypt but was not of Egypt. When the Fatimid line ended after some two centuries and Egypt returned to Sunni rule, it was as if the Fatimids had never been, except for the new capital, and al-Azhar and several other grand mosques and their architectural modeling. The same could be said of the Buwayhids in Iran and Mesopotamia. The Shi'i Safavids were something different. In so far as the vast majority of Muslims in Iran were Sunni, Shah Ismail's revolutionary institutionalization of torture, exile and execution amounted to a perpetual political and religious split in the Muslim community. Hence the bitterness of the centuries of wars between Ottomans and Safavids, and on Iran's other border, between Safavids and Uzbeks.

Shah Ismail was a young man when he and his warrior tribesmen took over Iran. His youth may have rendered him less averse to applying the violence required to impose his revolution, but the annals of history are in no short supply of revolutionary rulers more advanced in age than Ismail who did not refrain from extreme violence in their efforts to remake the world as they wanted it. Behind his religious revolution was a coldly calculated political ideology: to establish an Iranian identity that conformed with the Shi'i tribes upon which the ruler's military power was based and that defined the new dynasty in its confrontation with the neighboring Sunni powers. But the Safavid success presented Shi'i theologians with an embarrassing dilemma, one not too dissimilar from the one Christian theologians faced after Constantine's conversion, when the blood of martyrs was no longer needed to seed the Church. Self-interpreted as a persecuted community of martyrs for centuries, the Shi'i sect, or rather its leaders, now had to address the uncomfortable and unfamiliar condition of having become a dominant religion of state.

Sunnis, on the other hand, like the Romans of old, for whom God and military triumph marched hand in hand, gloried in the conquests of the Arabs, and later of the Turks. Not until the 18th and 19th centuries, as land after Muslim land was falling under European control, would Sunni thinkers search their souls to fathom an explanation for this inexplicable reversal of fortune, as if the train of history had jumped its divine track. Defeat and occupation were not in God's plan. Even the Shi'i sect with its suffering cult of oppression was ill prepared to accept defeat by the West. Islam had indeed incurred its share of civil wars and losses – Spain, and the ignominious end of the Abbasid caliphate by the Mongols – but defeat and loss were assuaged by the Seljuq success in Anatolia, the defeat and expulsion of the Crusaders by Salah al-Din and the Mamluks, and above all, by the conversion to Sunni Islam and the subsequent emergence of triumphant empires, the Ottomans and Hindi Moguls. The Sunni historical collective consciousness left little space for the memory of defeat.

Except for those very few Muslims who seriously read the critical works of the great historians, such as al-Tabari (d. 923), who faithfully and honestly preserved reports of vicious strife and calumny within the small Muslim community at the time of the Prophet and just after,[2] the believers eagerly accepted the purified version of history: The early Muslim community lived in divinely inspired harmony and triumph, owing to the radiant goodness of Muhammad. With the passage of time, the harmony and triumph shined ever brighter; the defects dissolved in the idealized image of a perfect society in oneness with itself and God. It became an act of faith that early Muslims had been the most excellent Muslims of all, their goodness a consequence of their being close to the Prophet, who was idealized as the "Perfect Man" who radiated goodness like a nourishing sun. The excellence endowed in Muslims by virtue of their being sincere believers assured Islam of conquest and triumph against the evil and ignorance that opposed God's true religion, just as the Prophet had triumphed over those hard-necked Meccan hypocrites and evildoers who opposed him. Muhammad's struggle and ultimate triumph became a metaphor for the historical destiny of Islam.

The defeats in the modern era were thus theologically explained away in the same way the Old Testament prophets explained to their people the coming of the Assyrians and Babylonians; or Martin Luther explained the relentless Ottoman advance into Europe; or American TV evangelicals explained God's hand in bringing down the Twin Towers. When Muslims had been true Muslims, God granted them power. Defeat came because they had fallen away and been corrupted by adopting alien practices. They had forgotten the ways of the first Muslims, and until they purified themselves and returned to God and the ways of the early community, they would suffer defeat and humiliation.

In the latter part of the 19th century, a more sophisticated theology of cultural and political decline would be articulated: Christian nations were strong because they had abandoned Christianity to become more Muslim in their beliefs and practices than Muslims in theirs. In the process of abandoning Christianity for secular humanism, Christians took on beliefs, values, morals and work ethics that were rooted in early Islam and that characterized the first generations of Muslims, the *salafi*. European Christians were Christian in name but Muslim in spirit. While Christians were leaving Christianity and becoming strong, Muslims were leaving Islam and becoming weak. Only by going back to true Islam could Muslims become strong again. By reforming and taking on modern ways, Muslims would in fact be returning to the Islam they had abandoned. The argument would become popular in modern apologetics.

The theme is fraught with irony. Islam was born in a merchant society, the Prophet was a Meccan merchant, the Quran is laced with terms and phrases from its commercial environment. In Medina, where Muhammad sought refuge from the hard-hearted Meccan leaders of commerce who opposed him for greedy commercial reasons and evil pagan practices, he became a head of state bearing all the responsibilities of the growing community of believers: chief administrator, head of treasury and tax collecting, military commander-in-chief, lawgiver, statesman and diplomat. He fought at the head of his army, was victorious against great odds in one battle, was defeated, wounded and thought to be dead in another, but in the end emerged triumphant. Prophet, military commander and wise lawgiver, Muhammad combined the roles of Moses, David and Solomon all in one. The first

Arab dynasty was led by the merchant-elite of Mecca, erstwhile opponents of the Prophet, whose caliphs conquered from France to China. Trade, conquest, jurisprudence, civil order, responsible government and war for survival were the embryonic sinews of Islamic society. The honest, sober, God-fearing and hard-working merchant ready to do *jihad* for his faith, an abstraction of the religion's social genesis, was the epitome of the good Muslim, an image drawn from the historical memory of the Prophet himself. Wealth was considered a blessing from God; holy war was a duty. Islam needed no Augustine to argue the goodness of material wealth or the just war and its rightful desserts. Conquest, trade and religious rule were bred in the bone.

The Arab conquests had placed Muslims in the central position of the Afro-Eurasian landmass, between the West and China, in which position they dominated world trade. The irony is that after a millennial struggle between Christendom and Islam, dominance was at last won by a civilization whose scripture advocated surrender of material wealth, withdrawal from the world, forgiveness, meekness, and the otherworldly pacifism of turning the cheek for another slap. The reversal of fortune was mind-boggling. The Prophet of Western civilization had died nailed to a cross, a humiliating, ignominious end for someone who would be claimed to be God; Muhammad, an illiterate orphan, had died a success in everything he did, from commerce and law to running a state and expanding it. The contrast between the two could not have been greater.

How was it, then, that a civilization whose adopted prophet threw merchants and money changers out of the temple, advised followers to give up all their possessions and follow him, because a rich man had as much chance passing through heaven's gate as a camel through the eye of a needle, and preached that the believer should trust in God to provide for them just as He provided for the lilies in the valley which toiled not, who taught that trade and toil were wasted energy since the world was going to end soon – how was it that a civilization following such an otherworldly, dead-end, loser religion could win the world by military conquest, industriousness and global trade? Did not Saint Paul preach that the material world should be abandoned and Saint Jerome that seldom if ever can a merchant please God?

The contradiction between the scriptural basis of the West's religion and the practice of its followers was as momentous as it was puzzling. To Muslims the answer is simple: Westerners had abandoned their religion. By shedding themselves of Christianity they became dominant in the world, and the virtues that made them dominant were the very virtues of Islam that Muslims had forsaken. While Muslims shed their essential virtues of industry, trade, military discipline, social justice and quest for knowledge, and as a consequence lost their dominance in the world, Westerners shed their religion and in its place acquired the virtues of true Islam. What should have been Islam's historical destiny was usurped by the West, Christians becoming the true Muslims and Muslims surrendering to the passivism, rejection and mystical otherworldliness that was the scriptural essence of Christianity.

The same reversal occurred in science, according to the argument. The Quran and the Prophetic tradition are filled with verses and sayings of the Prophet urging believers to seek knowledge, to look to the heavens and nature as signs of God's wisdom and power, while in Christianity nature, like the material world, is dismissed to emphasize the miraculous: walking on water, raising the dead, curing the blind and diseased and multiplying fish and loaves. In true Islam the only miracle is the Quran. Islam is a religion of science, not of miracle. The Quran is saturated with allusions to natural phenomena and commands for people to observe and think. But as Westerners freed themselves by throwing off their world-denying religion with its rejection of material pleasures and its miracles that discouraged the pursuit of natural knowledge, they walked away from their intellectual passivity and learned Muslim science, just at the time that Muslims were abandoning and forgetting all that they had accomplished in science, flinging themselves into the morass of miracles, mystical saint worship and superstition, sinking into the swamp of mental torpor that doomed them in face of the energies of the New West, Christian in name but Muslim in virtue. The challenge was for Muslim society to go back to those pristine days of Mecca and Medina and recapture the essence of true Islam so that Muslims could get back on track and become what God prepared and intended them to do.

The response to meet the challenge presented by this upstart civilization of the West was prescribed by earlier crises Muslim society faced. The prescription followed the same pattern as that which ibn Taymiyya and ibn Qayyim al-Jawziyya preached when Islamic belief, as they perceived, faced the crisis of magic, superstition and the occult sciences in the 13th and 14th centuries. It was the same formula as that which motivated the reformist puritans that went by the names Murabitun, Muwahhidun and Wahhabi. Deep in Muslim consciousness was that idealized vision of those few years in the past, in early Islam, when everything was right. Regeneration was only possible by returning in spirit to those early days, to relearn and infuse into society the recaptured spirit of those traditions of the pious ancestors. Society, government and institutions had to be remodeled on those traditions. This was the remedy prescribed by religious reformers who grappled with the confusing and contradictory threat of a covetous and devouring West. The tribal Wahhabis of Arabia preached and fought for an enforced rebirth of the pristine past of Mecca and Medina in the 18th century. More peaceably, modernist reforming shaykhs of al-Azhar in the next century also based their arguments on a return to the past, as did some secularist reformers, while giving the argument a distinct modernist twist. This response of harking back to an idealized picture of a bygone earthly communal paradise as a divine prescription to rehabilitate present dislocations had become, as it were, genetically encoded in the culture, a relentless therapy for whatever ailment afflicted society at whatever time.

This same response arose as a prescription for reform as soon as it was perceived that the Ottoman destiny had taken a wrong turn. For secular-minded reformist Ottomans, the idealized image of pristine Medina was overlaid by another idealized image of the past, of the days back when Ottoman institutions were functioning as they had been designed to function. In 1635, Sultan Murad IV asked his official scribe Kochi Bey to draw up a report analyzing the causes for the military's lackluster performance against those armies of the West, the Hapsburg army in particular, which Ottoman armies had for centuries been accustomed to defeating. This is the earliest of over 2 ½ centuries of reports that attempt to understand and analyze the changed conditions brought into relief by Western success and how to meet the challenge. The first attempts focus on internal factors: the lassitude of religious morality, institutional decay and administrative corruption. It is not until a century after Kochi Bey's memorandum that embassies would be sent to Europe to investigate the changes that had taken place there, in the heartland of the enemy. A century after that, in face of much resistance, Ottoman reformers would be incorporating Western civil and military institutions into the empire. Imported legal codes and industries, even modes of Western dress, furniture and lifestyle, would sweep away a millennium of Muslim tradition. A century of reform and the empire itself was swept away, divided up and parceled out by the very people the Ottoman reformers had been for two centuries struggling to emulate.

It had been only a matter of time. Western power, now that it had the upper hand, was not about to tolerate any Eastern state strong enough to compromise its interests. Reform was encouraged by Western powers; their bankers, entrepreneurs and industrialists grew rich on it. But Western-encouraged reform was supposed to go only far enough to keep order in the Muslim states, to keep them afloat and safe for commerce, investment and Western-financed development; just enough to keep "moderate" leaders securely in power to serve as partners and protectors of Western interests, not to become rivals or threats to those interests, not to become strong enough for independence.

The Ottomans at least outlasted by two centuries their sister rivals, the Safavids, Uzbeks and Moguls, but the price Ottomans paid for those centuries was severe. Stage by stage, reform by reform, the empire's reformers, with an eye to the West, imitated, borrowed, innovated, adapted and secularized in a desperate effort to survive. The halcyon past of pristine Islam held no allure for these 19th century government reformers, who had traveled to Europe, knew its languages and to some extent had become Westernized. In their zeal to reform, they practically secularized and reformed Islam and Ottoman institutions out of existence, readying the empire for the European carving knives. Religious reformers and conservatives would claim that it was the aggressively imposed secularization by Westernized reformers that killed religion and the soul of the empire. The West merely carved up the corpse on the banquet table of Versailles.

The fate of the empire was a lesson that 20th century religious reformers could point to in their battle with secular nationalist reformers who looked to the West. The inspiration of reform and the way forward should not be the West, they continue to claim, but Islam of a thousand and three hundred years ago. The interaction between the West and the peoples of Middle Eastern countries, from the time of the Treaty of Versailles to the present, has not lessened the bitterness of the lesson nor diminished its receptivity among Muslims. To be sure, the bitterness has grown deeper, for in the eyes of Middle Easterners, whether Sunni, Shi'i or Christian, Western power has imposed more than one Versailles upon them.[3]

The appeal of the past is a salve to the wounds of defeat and political failure. Many civilizations in time of troubles look to a perceived brighter past for the inspiration it takes to continue. Muslim civilization has a more persuasive history than most to fall back on; so persuasive, in fact, that the past has been seductively over-idealized enough to become a trap. It can beguile believers who dwell too long on it. They can become blinded by its brilliance. So eagerly do they grasp to bring it back, its spirit slips through their fingers.

Notes

1 The Christian's belief that the end of the world will be preceded by the return of Christ, parallel to orthodox Judaism's belief that prior to the re-establishment of Israel a Messiah will be sent by God, finds its Shi'i analog in the Hidden Imam, the twelfth and last in the male line of Ali and his son Husayn. Having disappeared in the 9th century, the Imam will return to lead the persecuted followers of true Islam to victory over the evil usurpers responsible for the persecution and deaths of Muhammad's descendants through Ali and Fatima. The conquering caliphs, abu Bakr, Umar, Uthman and the dynasty of Umayyids (661–750), who according to Shi'i political theology usurped power that rightly belonged to the family of the Prophet through Fatima, Ali, Hasan and Husayn, were enemies of true Islam, more pernicious and dangerous than infidels. Their punishment would come when God gathered the souls on Judgement Day.

2 For instance, the Satanic verses where the Quran mentions the "daughters of God" to win over the polytheist opposition in Mecca; Muhammad's sending out raiding parties in Medina during the holy month; A'isha's lost necklace and her being accused of adultery and the sorrows this caused Muhammad; the enmity between A'isha and Ali; and the murder of Uthman and civil war between Ali and Mu'awiyya.

3 For a thorough account of the abiding damage and dislocation the Treaty of Versailles caused in the Middle East see David Fromkin's, *A Peace to End All Peace: The Fall of the Ottoman Empire and the Creation of the Modern Middle East*, Henry Holt and Company, New York, 2009.

10 Military misfortune and the beginning of scientific and technical transfer 1600–1722

An early taste of defeat at the hands of a Western power that hinted of worse to come was the Portuguese naval victories in the Arabian Sea at the beginning of the 16th century. Then in 1571, the Mediterranean fleet was demolished at the Battle of Lepanto, an event often taken as the turning point in Ottoman military might. However, led by the efficient grand vizier Sokolli, a Christian convert who had come into Ottoman service as a slave in the *devshirme* system, the government vigorously responded to the defeat by rebuilding the lost fleet and maintaining Ottoman naval power in the eastern Mediterranean. The Ottomans continued their offensive against the West, but the victories failed to come as they had previously. In a long series of battles with the Hapsburgs lasting from 1593 to 1606, the sultan's army could do no better than hold their own. The long war, having drained the finances and manpower of both sides, ended in a stalemate. For the Ottomans, so accustomed to victory, this was equivalent to defeat.

The transformation that the West had been undergoing for centuries was beginning to tell on the battlefield. In the treaty ending the war with the Hapsburgs, signed at Sitva Torok, the Ottomans agreed to relinquish the diplomatic nicety that for over a century had given precedence to the sultan in his correspondence with the kings and emperors of Europe. From 1606 on, the sultan was no more than the royal equal of Europe's many monarchs, a humiliating comedown for the Ottoman government in its diplomatic relations with the West.

While the Ottomans were struggling to hold their own on the western front, the Safavid empire had reached its ephemeral zenith under Shah Abbas the Great, who was at the time buying rifles and artillery from Europe and hiring European military experts to strengthen his military. This was another ominous sign for Muslim fortunes. In military technology and international trade, the West began casting its shadow on Muslim regimes. In 1598, in the midst of the long Ottoman–Hapsburg war mentioned above, the British government sent two English brothers, Antony and Robert Sherley, to Shah Abbas in Isfahan to explore the possibility of trade and a European–Iranian alliance aimed at engaging the Ottomans on both their western and eastern front. An alliance never materialized. But before returning to England, Antony Sherley, a man of military experience, stayed on to instruct the Shah's military officers in the art of casting canons. His brother Robert remained in Iran, married a girl given to him by Abbas and in 1608 was sent to England on a mission that resulted in the establishment of Anglo–Iranian diplomatic and commercial relations. England was at the time establishing itself in the Persian Gulf, where its merchants were setting up trading colonies along the coast and elsewhere in the Muslim east. These would be the economic beachheads that with the coming of the English Industrial Revolution would devastate the local industries of the host countries across the Middle East and Asia.

Advanced technology in the weapons industry and international commerce were already in the early 17th century preparing the way for manifest Western dominance in the next. Blinded by their sense of superior power, Muslim rulers could not fathom the consequences of their actions. Instead of having his military officers and craftsmen learn the technology of metallurgy and how to cast cannons and master the chemical and industrial processing of gunpowder, building foundries and producing his own weapons, as the Ottomans had indeed done two centuries earlier, Shah Abbas, a Muslim leader called the Great and reigning at the apogee of his dynasty's political power and cultural efflorescence, chose to have it done for him by foreigners of a country whose merchants were busily engaged in founding the East India Company in preparation to laying claim to Eastern trade. Without realizing it, Muslim rulers began their surrender to Western technology at the height of their power. Abbas the Great set the precedent, and the precedent became a pattern that became all the harder to break as the West got stronger.

The Ottomans put up a hard, century-long battle until they, too, succumbed. The Ottomans continued to have little to cheer about after the defeat at Lapanto and, later, the stalemate that ended at Sitva Torok with its step down in international prestige. A war with Poland in 1620 also ended in stalemate. Had the Thirty Years War (1618–1648) not diverted Europe's attention, it could have been far worse for the Ottomans. Weakness on the battlefield was reflected internally by the growing discord in Ottoman institutional life. The *devshirme* slave recruitment system was rapidly breaking down in the early 17th century. The Janissaries, having become more interested in forcing donations from the cash-strapped sultan than in expanding or defending the empire, became unruly to the point of insubordination. Whenever their demands went unmet, they turned rebellious and overturned their giant soup kettles in sign of their discontent and began rioting for what they considered their due, which was often no more than their back wages. On occasion, a sultan and his grand vizier would be overturned after the soup kettles. A precedent was set with the assassination of Sultan Osman II in 1622.

The royal court was another debilitating element in the Ottoman structure of government. The court was riven by personal jealousies and rival cliques intriguing to win the sultan's favor. A *devshirme* faction of converted Balkan Christians was opposed by a faction of old Turkish families of the landed elite, who resented that high government positions went to the converted outsiders, as the old families saw the *devshirme* men. The Inside Boys of the *devshirme* were intended to avoid such rivalries, but over the two and a half centuries of the institution's existence, the *devshirme* officers had devolved into just one more divisive faction. The women of the royal harem were another.

The mothers, wives and favored concubines of the sultans plotted against each other, forming parties within the court in favor of one or another of the many princes to succeed the reigning ruler. So influential did the women become for a time in the affairs of state that Ottoman historians refer to it as the period of Harem Rule. Rivalry, intrigue and jealousy had always existed at court and in the palaces, but it was not until the 17th century, when the sultan's prestige and power as lord of the world was wavering, that their effects came so disruptively to the surface. The impecunious state of the empire poisoned the air of the ruling circles. Troops and government officials went unpaid for extended periods. Graft and corruption became compensatory payments for survival. A malaise spread through and corrupted the ranks of the once-disciplined military and civil bureaucracy.

Soaring inflation caused by the influx of silver from the New World wrought economic havoc. When the government debased the coinage in response, raw materials and goods manufactured in the empire were sold outside for a stronger coinage, a classic example of Gresham's Law. The empire was being drained of the precious little capital it had left from centuries of warfare. The conquests having ceased, no new riches were being brought in, nor was wealth being produced. Fortunes were made and lost, but this was merely a matter of material possessions changing hands, not a production of new wealth through expanded industry, trade and agriculture. With economic decline sapping the strength and integrity of central government, a rebellion broke out in 1623 in eastern Anatolia under the leadership of local governors. A young sultan had just come to the throne, Murat IV. Reaching maturity a decade later, he addressed himself to the intestinal disintegration of the old institutions and initiated a program of reform, the first in a long series of progressing reforms that would last right up until the end of the empire.

Murad's reforms were outlined in a treatise that he had asked one of his advisors to draw up on the causes of decline. Kochi Bey, an Albanian *devshirme* recruit who had risen in Ottoman service from slave convert to the sultan's chief advisor, knew little of the changes that had transformed Western Europe. Consequently, his report focused on changes in Ottoman institutional and personal integrity rather than external changes beyond Ottoman control. He assured Murad that Ottoman institutions were far superior to anything in the West. The problem, he maintained, was that Ottoman institutions were no longer functioning correctly. The moral fiber of men, not the institutions, had changed. The remedy was to appoint honest and competent officials; end the factionalism and favoritism that disrupted courtly life; put down the bandits roaming the countryside; eliminate unruly elements in the military and tighten up discipline; the sultan should more closely oversee military and government affairs, as had earlier sultans; and taxes, maintenance of agriculture and the miserable conditions of the peasantry should be reformed to bring conditions back to the high standards of earlier times. The old institutions were to be refurbished, not new ones instituted.

It was a program of reform that met with general approval in that nothing really had to be changed. And nothing was. Twenty years later the same diagnosis and program of reform were restated, this time by the famous writer Hajji Khalifa, known also as Katib Chelebi, who composed a long memorandum in response to

Sultan Mehmet IV's query as to why government was unable to meet expenses. Muslims, he advised, needed to study the rational sciences. In light of the changed circumstances prevailing in the empire, old prejudices that had opposed those sciences had to be swept away so that the benefits of geometry, astronomy and geography could be accepted.[1] This looked to be a step in the right direction, but it was not easily taken. In general, Ottomans who pondered the reversal of power and growing insolvency had difficulty comprehending the nature of the changed circumstances. Attention was focused on internal conditions. Awareness of conditions outside the empire was vague. It would not be an exaggeration to say that Ottoman knowledge of Europe was totally lacking.[2] The same state of ignorance existed throughout the whole of Islamdom. No Muslim country maintained a permanent embassy or consulate in Europe or knew anything of significance about Europe until well into the 18th century. What knowledge of Europe was brought into the empire by the *devshirme* men, whose origins were rustic and who were recruited young, which did not amount to much. Occasionally the Ottomans sent a diplomatic mission to Europe, but it was always for the immediate purpose of arranging a truce or peace treaty, not to become familiar with Western society or to learn anything. Diplomatic matters between the Ottomans and Europe were settled in Istanbul, where European countries with interests in the empire had permanent representatives. A parallel situation existed in Iran with the Safavids. When Shah Abbas desired a commercial treaty with England, he sent an Englishman to London, not an Iranian embassy to evaluate and explore.

The Ottomans sent a diplomatic mission to Vienna in 1665–1666, the very time young Newton was down home on the farm in Lincolnshire developing calculus, pondering universal gravity and passing light beams through prisms. The Ottoman mission included 150 officers. It included a remarkable Ottoman traveler and writer who, uncharacteristically, showed keen intellectual interest in the West. Evliyya Chelebi may have been the first Ottoman to have had such an interest, or at least to have expressed it in writing. In his travel book, *Safar Name*, Evliyya enthusiastically described Vienna's libraries, hospitals and surgical practices, several of which he witnessed. His is the first recorded Muslim insight recognizing that Western medicine had advanced way beyond the brutishly and lethally primitive level that Muslims had witnessed at the time of the Crusades, when Latin surgeons cured headaches by cutting a hole in the skull, or infected legs by smashing them off with an axe.

At St. Stephen's Hospital in Vienna, Evliyya saw a surgeon cutting and opening a soldier's skull with vices and clamps and exposing the brain to remove a bullet that had been lodged in it. A powerful concoction of wine and opium was used as an anesthetic to keep the patient under, and it did. Evliyya describes the surgeon as wearing a kind of white handkerchief over his mouth and nose, possibly the first historical reference to a surgical mask. He describes the operation in caring detail and was much impressed by Viennese medical expertise. Once the bullet had been removed and the damaged part of the brain cleansed, the brain's protective covering was replaced, then the clamps released and the surgically severed section of scalp put back in place and the skin sutured tightly at the incision by using hungry giant leaf cutter ants whose jaws held the edges of the incision firmly together, their heads being severed from their bodies the moment their jaws clamped down. According to Evliyya, the patient survived.[3]

The account was the first crack of light in the formidable wall of self-complacent superiority that Muslims had built around themselves. But the literate elite who read Evliyya's *Book of Travels* took it as an amusing travelogue filled with imagined and exaggerated stories. Who could believe the mighty empire, even if not as mighty as in the days of Sulayman's reign, was in any way inferior to the states of Europe? There was no compelling reason to doubt Ottoman power. The empire was still strong enough to dispel any doubts of Ottoman superiority. Some battles had been lost, but others won. Crete was seized in 1669 from the crumbling empire of the Venetians and added to the sultan's far-flung lands, along with some Polish territories not long after.

So emboldened were the Ottomans by their conquest of Crete that old aspirations of major conquests came back to life, sending a surge of giddy vigor through the leadership who decided it was time for another go at Vienna. The Ottomans had reason to believe that where they had failed in 1529 they would succeed in 1683, for this time they would be allied with France, traditional enemy of the Hapsburgs and friend of the sultan ever since the reign of Sulayman the Magnificent. The idea of a joint Franco–Ottoman attack on the Austrian Empire seems to have been hatched in the halls of Versailles by the Sun King's foreign minister, who had the French ambassador in Istanbul assure the sultan's government that the time was right, the Hapsburgs were in no condition to resist a joint Franco–Ottoman attack. Vienna would fall like an overripe apple. Louis XIV, the ambassador claimed, was fully behind the plan. And so he may have been.

The sultan and his ministers were assured that once the Ottoman army invaded, the French would attack from the west, drawing a large part of the Austrian army away from the defense of the capital. The lure of a glorious triumph was seductive. The more the sultan and his advisors considered it, the more tempting Vienna became, to the point that the Ottomans did not insist that the French put their commitment in a formal agreement bearing the Sun King's signature. The Ottoman leadership's will to believe that the French would join in an attack on the Hapsburgs was as powerful as was the temptation to take Vienna – too powerful to be restrained by any rational analysis of the prospects of success and the consequences of failure. The foregone conclusion of a military triumph left no room for doubt that the French might not abide by the unsigned agreement; critical analysis of the French position in Europe that could influence or curtail French actions was lost in the visceral response to economic, institutional and military problems: more conquest. Like that of Constantinople, the conquest of Vienna would invigorate the empire, raise it to new levels of glory and prestige, reignite the ghazi spirit of old, and resolve the fiscal problems.

In retrospect, the second siege of Vienna looks to be a delusional act of desperation by a misled leadership of a declining empire in the throes of fiscal chaos, military insubordination and gross administrative corruption. Nonetheless, the

Ottomans came close to pulling it off, and probably would have had Louis XIV sent his army against the Austrians. Whatever fantasies the French king might have entertained before the Ottomans headed for Vienna, upon cooler reflection he realized he could not act. His assisting Muslim Turks to seize a capital city of Catholic Christendom after the Protestant Reformation, and the terrible wars of religion that accompanied it, would have been too blatant a betrayal of the French king's position in the eyes of the Church as Christendom's foremost defender. In addition, from a purely political perspective, France would have been ill served by upsetting the balance of power in assisting to replace on its eastern border a weak Hapsburg empire with a victorious Ottoman one. The Ottomans would have known this had they known something about Europe and European diplomacy.

By not having permanent embassies in Europe, the Ottoman government assured itself of remaining in ignorance, not only when it came to connecting the dots of international European diplomacy and the shifting dynamics of alignments and configurations in Europe's balance of power politics, but also of Europe's advances in science and military technology. Consequently, French intentions and the state of Hapsburg strength and weakness were badly miscalculated, resulting in a failed siege, fiscal bankruptcy, and a disastrous European counterattack that cost the Ottomans heavily. Nonetheless, that the siege came so close to succeeding in spite of French inactivity is testimony to the ability, motivation and energy that still endured in the sclerotic sinews of the empire's civil and military leadership, amidst its many crippling ailments.

The second siege of Vienna stands as the empire's last grand gesture of greatness. But the European reaction it provoked was catastrophic for the Ottomans. From then on it was downhill. As soon as the Ottoman army had retreated from Vienna to winter quarters in Belgrade, the pope declared a Holy Crusades against "the Turk," whose defeat before the gates of Vienna was correctly interpreted as failing strength. The time had come, the pope declared, for the Turks to be driven out of Europe by the combined forces of Christendom. Several European leaders took up the pope's call for this latter-day crusade. In a series of wars between 1692 and 1699, Ottoman armies were battling the Hapsburgs in Hungary, Bosnia and Serbia; the Poles in the Ukraine; Venetian fleets in the waters of Dalmatia, Albania and Greece; and the Russians in Crimea, Moldavia and Wallachia. The battering defeats were at last brought to an end with the treaty of Karlowitz, signed in the last year of the century.

More poignantly than ever, responsible Ottomans at the head of government realized that the glory days were over; that the reason for the devastating loss was not just because of the Ottomans becoming weaker but of Europe becoming stronger. The defeat at Vienna, and defeats that soon followed, deeply wounded Ottoman pride and dignity as they had never been wounded before. The last several years of the 17th century came as a terrible awakening to reality. Statesmen who contemplated the events of 1683–1699 and the attendant internal discord, who knew well how bad conditions were in the Ottoman state, realized that their destiny was no longer of conquest but of simply holding on to what they still had of the empire. And to do that, they also realized, would require making some changes.

The treaty of Sitva Torok at the beginning of the century had put an end to the accepted idea of Ottoman supremacy; the treaty of Karlowitz at the end of the century marked the beginning of territorial contraction and the idea that the empire was now locked in a struggle for survival. If Evliyya Chelebi's description of Viennese medicine was a chink in the wall of Muslim complacent self-satisfaction and arrogance, Karlowitz was a cannon blast blown through it. Its effect on the quarreling factions of the court in Istanbul was to cast the ruling establishment into a divided and confused funk of defeatism and escapism, with some of the elite grimly resolved to institute reforms. These few realized that not only had the old institutions become ineffective and destructive, but Europe had somehow transformed itself to a higher level of military ability.

Since military battle was the one field of contest in which Ottomans and the West openly competed for all to see and make judgments, the obvious conclusion was that the secret of European power resided somewhere in its military organization and weaponry. That there were deeper transformations underlying Europe's military ability had not yet penetrated Muslim consciousness. Accordingly, there was an attempt to incorporate certain European methods into the military by order of the sultan as advised by a miniscule faction of reform-minded ministers and high military officers courageous enough to call for change and adopt from the West what they thought would strengthen the empire in defending it against the infidels. The new reformist outlook regarding the West was an inversion of that spirit of an earlier time of strength and confidence, when sultans had their portraits done by Renaissance artists without a hint of embarrassment and when artillery technology was taken from the West, not for defense but further conquest. In the changed circumstances, taking from the West would, in the minds of many Ottomans, be a source of wounded pride and religious betrayal, in addition to its being a threat to the interests vested in the military's institutional status quo.

The reformist ministers tended to be men of the *devshirme*. Such was grand vizier Husayn Pasha Koprulu, descendant of a 17th century *devshirme* grand vizier whose origin was Balkan Christian and who, as a high Ottoman officer, fathered a family that would through the generations provide the sultans with a dynasty of outstanding grand viziers who were able to preserve the empire, more or less intact, through most of the 17th century. With the sultan's blessing, Husayn Pasha surrounded himself with like-minded officers and undertook reform of the army and navy. He commissioned another *devshirme* officer, Husayn Pasha Mezzomorto, to the grand admiralship of the navy, authorizing him to execute whatever changes he thought the navy required. Protected by the grand vizier against the formidable conservative factions opposed to change, Mezzomorto, being originally Italian and having some familiarity with European naval innovations of the 17th century, overhauled the navy along European lines. He had a new fleet of galleons built with sails in place of oars, a European innovation going back to the early 17th century. He also created a more rapid and efficient chain of command, instituted a new supply system and pay structure that were placed under the autonomous authority of the navy and organized a new recruitment and training program for both sailors and officers, whose rank and advancement were to be in accordance to ability rather than family connections. This latter reform was one of the chief reasons for the conservative resistance to Mezzomorto.

In recognition of the economic reality undermining the empire, the grand vizier Koprulu ordered new industries to be built so that Ottoman products could compete with European manufactures, which were running local crafts and industries out of business. This was not an overly ambitious aspiration. At the time, around 1700, with proper direction, investment, training and explorative missions to Europe, there was no reason the Ottomans could not have competed. Had that direction been taken and heavy investment in it been made, and Koprulu's and Mezzomorto's reforms taken hold and been advanced by reformist successors, Muslim history would have been different. But the ruling establishment was too divided between reformers and opposing conservatives for the ship of state to keep an even keel. As would happen to reformist governments over and over again during the whole of the 18th century, the conservative faction at court, composed of a loose alliance of old noble families, leading ulema and high military officers, all fiercely hostile to innovation and feeling threatened by Europeanized reforms, prevailed upon the sultan to dismiss the reforming grand vizier, whose reforms, it was claimed, were perverting the empire with unholy importations from the lands of the unbelievers. The sultan feared enough for his life and throne to bend with the wind. With Koprulu went Mezzomorto and his cadre of reformers and with them both went the naval and industrial reform programs. Central leadership was too weak to make a committed stand for reform. Things would have to get far worse before a firm commitment was made, but by then it would be too late.

All governments, absolutist no less than democratic, have their parties and rival factions operating within the state structure in one form or another, and the closer to the center of power, the more cutthroat they seem to be. Such mutually hostile parties can serve to prevent government from extreme or ill-considered action, but if balanced, they can induce paralysis and prevent or undermine any action at all. Whenever the sultan appointed reformist ministers, usually at a time of crisis that came upon the heels of military defeat, the conservative party would wait for its chance to swing into action to undo the reforms. Paralysis at the center assured that Ottoman fortunes would continue to spiral downward. In turn, further decline assured that reformers would increasingly consider Western knowledge and expertise to be not only of value but necessary for survival. It was no coincidence that appreciation of Western knowledge was making its way into the empire at the time Ottoman elites were becoming aware of the limits of their power in facing the West. As in the Abbasid period and Caliph Mansur's stomach ailment in Baghdad, medicine led the way in this latest transfer of knowledge, this time from the transformed West back to Islamdom.

During the last few decades of the 17th century, sultans had begun recruiting European physicians to care for them and their families. This was even while ibn Sina's medical compendium, which had been issued in 60 Latin editions between 1500 and 1674, was still being used as a standard text in European medical schools. Ibn Sina's *Qanun* was equally available to the Ottomans, but the elites preferred European medicine, and so while European medical students read ibn Sina, wealthy Ottoman families followed the sultan in having their ailments treated by European physicians, whose training was still in part based on the Latinized *Qanun*. Evliyya Chelebi's admiration of the surgical practices he witnessed in Vienna in the 1660s was likely a reflection of the elite's attraction to Frankish physicians. In the early 18th century, a Swedish doctor was appointed as personal physician to the sultan's mother. This could have been a consequence of the close Ottoman relationship with the Swedish king, Charles XII, in their common struggle against Peter the Great over Poland. The Swedish doctor possibly may have accompanied Charles when the Russians defeated the Swedish army at Poltova in 1709, sending Charles fleeing from Poland to seek refuge with the Ottomans. In spite of politics, religion and cultural prejudices, European physicians found a warm welcome in the Ottoman court and upper strata of Istanbul society, members of whom opted to give up Razi and ibn Sina long before European universities did. Not even the ulema's opposition to European physicians dissuaded Muslims from going to them, if they could afford to.

How is it that Ottomans could prefer European physicians but not act to learn European medicine? If the question could be answered it would explain much about the tardy and clumsy way Ottomans engaged the critical process of reform. As for the question of medicine, only the wealthiest Ottomans went to European physicians. The wealthy looked upon them as skilled servants from abroad, tolerated experts of little more importance than those European craftsmen who made the clever mechanical clocks Ottomans so adored. And then there were the issues of religious sensitivity and public opinion. The idea of founding a European-style hospital or medical institute to train young Ottomans in European medicine was out of the question. A hospital would have been a public affront to mainstream Muslim sensitivities, an open admission of a religious failing. The prejudice, however, was against Europeans, not Christians as such. When it came to medicine, religion was no issue.

In fact, the native physicians of Istanbul came largely, if not solely, from the city's Jewish and Christian minorities. By long tradition, Jews and Christians were favored as physicians in Muslim society. In the 8th, 9th and 10th centuries, Abbasid caliphs had Jewish and Christian physicians. Ibn Maymun (Maimonides) had left Andalusia in the 12th century and in Ayyubid Egypt became personal physician to the Ayyubid ruler, Salah al-Din's son. Jewish doctors often attended to Ottoman sultans. The conqueror of Constantinople, Mehmet II, had an Italian Jewish physician, who eventually converted to Islam in order to avoid any trouble with the ulema.[4] Following the expulsion of Jews from Christian Spain, Jewish physicians of Spanish, Portuguese and Italian descent became common in Ottoman society. Sulayman the Magnificent's chief physician was one of them. Local Jewish physicians of Spanish origin enjoyed a favored status with the sultan and Ottoman elite. This was because many of them had received their medical training in Europe or had remained in contact with Europe and learned the medical advances being made there.

Sometime around the middle of the 17th century, the Sephardic Jews lost their favored position. A possible reason that has been given to explain this loss is that the immigrant families from whom these doctors came were in time acculturated into Ottoman life in Istanbul, causing them to lose their ties with both Europe and European medicine. Their place was taken by Istanbul's Greek Christians. Seeing how the Ottoman elite were coming to prefer European over their own physicians, members of the Greek community traveled to Italy to study medicine knowing that a lucrative career would be theirs when they returned. A Greek who had been trained in the medical school at Padua became the personal physician of the *devshirme* grand vizier (1656–1661) Mehmet Pasha, the first of the Koprulu dynasty of grand viziers. European medical knowledge and ability were his passport to high Ottoman office, even though he remained Christian. Religion no more stood in his way than being Jewish had hindered Maimomides's medical success in Egypt centuries earlier. Some Muslims did in fact learn Western medicine; not in Europe but from either European or European-trained Greek physicians in Istanbul. These would have had patients a social rung or two beneath the wealthy elite.

By the early 18th century modern European medicine was making its entry into Ottoman society, that is, the practice of it, not the clinical science itself. That was not to come until the 19th century. There was one exception to this: Umar Shifa'i, an Ottoman physician who learned medicine from European physicians in Istanbul. In 1704 Umar published an eight-volume encyclopedia on chemical cures based on the work of Theophrastus von Hohenheim.[5] Theophrastus, it will be recalled, renamed himself Paracelsus for his having advanced beyond, as he claimed, the medical frontiers of Greeks and Muslims. Umar Shifa'i's popularity as a physician and his claim that his book was a translation of Paracelsus's book indicates that Western medical knowledge was to some extent penetrating Ottoman society. This was concomitant to Husayn Koprulu's and Mezzomorto's military and industrial reforms. The wall between western Christendom and Islamdom looked for a moment to be coming down. However, it remained.

While the forces of conservative reaction were putting an end to the work of Koprulu and Mezzomorto, the sultan was succumbing to conservative pressure on the medical front. A royal decree was issued forbidding the medicine of Umar Shifa'i and requiring Muslim doctors to take an examination that tested the religious orthodoxy of their medical practices. The royal rescript also prohibited foreign physicians from practicing anywhere in the empire. This appears to have been in reaction to the practice of Western medicine penetrating down into the non-elitist levels of Muslim society. Responding to the complaints of the local Muslim doctors who were losing out to the competition, the ulema spoke up to protect them, declaring Shifa'i's medicine to be non-Islamic.[6]

While the New Medicine surreptitiously continued to be practiced and European physicians were introducing modern medicine in the course of following their well-remunerated careers treating the ills of the elites in the up-scale quarters of the city where they were out of sight of the pious ulema, another beam of European knowledge was shining through the chinks of that mental wall.

The first notice so far recorded the introduction of the new astronomy to Muslim society came with the travels of an Italian aristocrat and began in a roundabout way with a broken heart. Pietro della Valle (1568–1652) of Rome, crushed to the point of considering suicide when rejected by the woman he loved, took the advice of his friend and physician, Schipiano, and made a pilgrimage to Jerusalem to heal his dying spirit, which was quick to recover in the holy city. This was in 1614. A year later, Pietro traveled to Damascus, Aleppo, Istanbul, Baghdad, Isfahan, all the while studying Arabic, Turkish and Persian, and collecting a large library of manuscripts and keeping a detailed diary in the form of long and frequent letters to Schipiano, his dear friend who had at the time cultivated trade relationships with the West. By 1616, Pietro had shaved his head, grown a beard and was dressing in the Eastern style in order to mingle more easily with the people, which led to his marriage to a young Armenian–Syrian Christian girl he described as a high spirited, intelligent, fearless and of exceeding beauty.

In Iran, Pietro met a cosmopolitan society of intellectuals from all over the Muslim world, a number of whom were devoted to mathematics and astronomy. One of them, Zayn al-Din, was vaguely aware of scientific developments in the West and wanted to learn Latin so he could read the new science in the original. In 1620/1621, while in Goa, Pietro did what he could to satisfy his friend's passion for learning by writing from memory, in Italian and Persian, an account of Tycho Brahe's hybrid version of the Copernican system, with diagrams. Why Brahe's and not Kepler's is a mystery, especially since Galileo, a fellow Italian, had been popularizing Copernican astronomy for five years before Pietro left for Jerusalem.

What influence Pietro's epistle had in the hands of Zayn al-Din remains in the shadows of history and serves as no more than an indication that an early form of the new science was finding its way eastwards through a broad variety of channels, from individual initiative in the early part of the 17th century to courtly sponsorship in the later decades of the century.[7]

In some small measure, Copernican astronomy was being introduced into the educated circle of the Ottoman court. In 1638, the court received a printed copy of the French astronomer Noel Durret's (d.1650) astronomical tables, which were based on Brahe's and Kepler's Rudolfine Tables. The book, in Latin and most probably a gift of the French ambassador to either the sultan or his grand vizier, may be the first recorded Muslim encounter with modern European astronomy. More than its content, the book's print excited some interest among a few Ottoman officials who wanted to explore the possibilities of printing being used. But formidable conservative opposition put an end to any thought of importing such a satanic practice, for not only was printing religiously repugnant, it would put the copyists, who came from the ulema, out of work. Beneath the rock of religious protest against innovation can be found the worm of economic self-interest. It took more than 20 years before the book was translated. Sometime between 1660 and 1664, Durret's astronomy was translated to Arabic, still the language of science throughout the Muslim world. The translator was a court secretary, Tezkireci Kose Ibrahim Efendi.

How was it that this court secretary happened to know Latin? The answer explains much about the small group of Ottoman officials who during the 17th and 18th centuries became the reformers demanding Ottoman assimilation of Western science and technology. Ibrahim Efendi came from Szigetar in Hungary.

He apparently entered Ottoman service either through the *devshirme* or, as seems more likely, as an educated man who crossed over and made his way independently to Istanbul to enlist his talents in the service of the sultan's government, perhaps to escape the European wars of religion, or because of some personal problem. His passport to Ottoman service was conversion; his rise in the service was his intellectual ability. He would not have been the first or last European to seek Istanbul and Ottoman service as a refuge from European intolerance, or as a chance to make a new life after having fallen into trouble back in his native home. Learning Arabic, and of course Turkish, perhaps also Persian, would have been part of his education in becoming an Ottoman. Ibrahim also learned some astronomy along the way. His rhymed translation of Durret's book indicates the thoroughness of his cultural transformation. Turning science in Latin into rhyme in Arabic, he entitled his translation: *Sajanjal al-Aflak fi Ghayat al-Idrak* (*Mirror of the Heavens in the Purpose of Comprehension*)[8]

By 1660 there must have been a small collection of astronomy books on the shelves of the palace library given by Western ambassadors to the sultan. The reason Ibrahim chose to translate Durret's book and not another appears to have been because of the mystical element it contained. Entitled *Novas motuum caelestium ephemerides Richeliane* and published in 1637, Durret's work is mainly on astrology from the angle of mysticism, hermeticism and geometry and includes a discussion on the concept of infinity and medical astrology. The astronomy of Copernicus, Brahe and Kepler are mentioned without much detail. The book's astronomical tables would also have attracted Ibrahim. As a work explicating modern astronomy, Durret's book was of marginal importance, though as a royal gift it must have looked to be a magnificent piece of work[9] (see Plate 7).

Those interested in Ibrahim's Arabic translation of Durret's book amounted to a handful of court scholars. One of its readers, the palace chief astronomer, scoffed in derision when coming to the description of the Copernican system. The earth moving around the sun! He credited Europeans for having no end of "fanciful innovations." But upon advancing deeper into the text and reassessing it as a whole after examining the Rudolfine tables of observation, he declared it the equal of Ulug Bey's tables of Samarqand, the highest praise his knowledge of astronomy could reach and requested that a copy of it be made for his personal use. What elicited the astronomer's admiration was not the new theories of planetary organization but rather what was familiar to him, the tables of observations that Ibrahim had translated and arranged in columns in the form of the traditional Muslim *zij*.

Copernicanism failed to interest the court astronomer. It did interest some others. Several of Ibrahim's associates who read his Arabic version persuaded him to make a Turkish translation of the introduction, which contained an account of the improvements made in observational astronomy from Hipparchus to Kepler, with Ptolemy, Battani, Zarqali, Alfonso the Wise, Peurbach, Regiomontanus and Copernicus in between. This Ibrahim completed. In his Turkish translation and commentary, he praised Copernicus as a superior astronomer, who realizing the difficulties that Ptolemy's system posed to observational astronomy, resolved them by having the earth move around the sun. Tycho Brahe was another great astronomer who "made observations with numerous excellently made instruments and began to correct the *zij* of Copernicus." Ibrahim names Kepler (Kepelyarus) as another.

Ibrahim apparently accepted Copernicus. His translation had a limited circulation and no discernible effect in arousing courtly interest in European science. He gave a copy of his Arabic translation to the chief military judge in Belgrade, where Ibrahim had gone as a member of the military campaign in the Ottoman war against the Austrians, the war that ended with the 1666 diplomatic mission to Vienna that Evliyya Chelebi accompanied, leading to his description of Vienna's hospitals and medical practices. Twenty years after having been given Ibrahim's translation of Durret's book, the chief judge had another edition of it made, but again, only because of the value of the tables.

While they dismissed the new theory as a Frankish fantasy, Ottoman astronomers could still appreciate the value of Durret's tables and critically evaluate them as being at least as good as, if not better than the best of all Muslim observations. It was only to this limited but practical extent that Ottoman astronomers were receptive to Western advances. There was no compelling reason at the time for Muslim astronomers to regard Copernican theory as anything more than a simple hypothesis – one of those "fanciful innovations" of which Europeans were thought to be so fond. As Ottomans understood it at the time, they were preserving the wheat of practical observation while casting the chaff of Frankish whimsy to the wind.

European texts of modern science sporadically arrived at the Ottoman court as ambassadorial gifts to the sultan. A few years after Kose Ibrahim Efendi finished his translation of Durret's astronomy, the Dutch ambassador in Istanbul gave the sultan a copy of the monumental 12-volume atlas of Willem Janzoon Blaeu (d.1638), cartographer for the Dutch East India company. Blaeu, who had learned astronomy under Tycho Brahe, was determined to become the world's leading mapmaker after the death of Gerhard Mercator in 1594. His *Atlas Novus*, published in 1603 in six volumes, contained the best and most up-to-date maps of the world, and it was the first atlas to represent the world in an eastern and western hemisphere. Willem's two sons expanded the work to its monumental 12 volumes, which, renamed *Atlas Major* and published in 1662 in Amsterdam, was the most expensive work produced in Europe in the whole of the 17th century. It was indeed a royal gift.

More than a world atlas, the immense and splendidly done work was an encyclopedia of Western science up to 1660. The Dutch king's gift to the sultan remained seven years on the royal shelf before a translation was at last undertaken. Durret's work had remained 20 years before being translated, giving an idea of the little importance Ottomans ascribed to Western knowledge – though Jansoon's 12 daunting volumes in Latin may have had something to do with the long delay. When a translation was finally begun, it was for some reason not Ibrahim Efendi who was entrusted with the task but a Syrian from Damascus, abu Bakr Efendi al-Dimashqi, a mathematics teacher in Istanbul who knew Turkish and somehow had come to know Latin, possibly through Dominican or Franciscan missionaries in search of prospective converts from the Syrian Christian community. Over a 10-year stretch, 1675–1685, abu Bakr Efendi was able to translate nine of the 12 volumes.

The massive, sturdily bound tomes, with their clear orderly print and finely executed and detailed maps of the world laid out in precise Mercator projection and brilliantly illuminated, must have impressed the translator, for he interjects favorable comments in the narrative of his translation and confesses to admiring Blaeu's work, an admiration that clearly included the artistry and technical industry that went into producing it. The Arabic title he gave the translation would suggest he found his work a pleasant experience: "Victory of Islam and the Joy of Editing the Atlas Major." Some of his comments have a defensive tone. Many Muslims, he writes, are involved in mathematics, astronomy and the physical sciences, which are very much alive in Islam. Of the two main branches of science, astronomy and geography, Bakr Efendi claimed that it was only in the latter that Europeans had gone beyond Muslims, and it was this, their knowledge of geography, that had enabled the West to cause so much trouble for Muslims – an obvious allusion to the maritime powers having seized the western hemisphere and much of the eastern one, encircling the world of Islam and taking over the trade that had until recently been carried by Muslims.

Damashqi rejected the systems of Copernicus and Brahe that he found described in Blaeu's *Atlas*. They were contrary to scripture. He accordingly reduced Blaeu's three pages of text comparing the Ptolemaic, Copernican and Tychonian systems into a single short paragraph. Holy scripture, abu Bakr al-Dimashqi wrote, proved the Ptolemaic system to be the correct one.[10]

In spite of the books of science making it over the wall, resistance held strong. As an awareness of European science dawned in the minds of some of the Ottoman elite, and reformers attempted to employ Western methods to turn military and economic decline around, the ulema took notice. In 1717, the shaykh al-Islam, head of the Ottoman religious institution, issued a *fatwa* forbidding the deceased grand vizier's large library to be made into a tax-free religious endowment because it contained books on astronomy, philosophy and history. Books on these subjects were also to be banned from all public libraries. The vizier's library was therefore donated to the palace library.[11]

It took another military disaster before the ulema and the conservative faction in the sultan's court gave way to another reformist government.

Excited by the territorial gains made in the war concluded by the Treaty of Karlowitz in 1699, Austria and Russia rushed once again to the attack in one more of the many wars that Austria and Russia would wage against the Ottomans. Fredrick the Great would later characterize the incessantly warring Russians and Ottomans as the one-eyed fighting the blind. In this war, the blind were able to defeat the one-eyed, and almost take Peter the Great prisoner. The blind, however, were not so fortunate on the Austrian front. Allied with Venice, the Austrian army advanced eastwards into Serbia and Bosnia to take Belgrade. The long war ended in 1718 with another Ottoman humiliation. The Treaty of Passarowitz gave the Hapsburg empire all of Serbia and its capital Belgrade, territory that had been in

Ottoman possession for 330 years. Bosnia, on the other hand, with its large Muslim population, was returned to the Ottomans. Adding salt to the Ottoman wound of defeat, the Austrians demanded there be free trade between them and that the merchants of either country be allowed free entry with extraterritorial rights. As it was unlikely that many Ottomans would be setting up business in the Hapsburg realm, what this meant was that merchants of the Austrian empire would be free to trade in Ottoman lands and have consular protection, that is, be protected by the laws of Austria, a form of diplomatic immunity for foreign merchants.

It was a one-way agreement. Except perhaps for some Jews, Greeks and Armenians, merchants of the Ottoman empire were of no mind and in no position to establish trade anywhere in Christendom. Another galling humiliation for the Ottomans was the right Austria demanded, and received, to station consular representatives in the empire, meaning spies and agents to work toward alienating the empire's Balkan Christian subjects from the sultan's government in order to have them serve Hapsburg interests. This gave the Austrians the same capitulatory privileges that France had received from Sultan Sulayman in 1535, back when the empire was powerful and sultans granted capitulations as favors given by the strong to the weak, rather than as demands made by the victorious on the defeated.

Though not registering the same level of catastrophe that Karlowitz did, the treaty of Passarowitz was enough to shock even the conservatives out of their complacency. For the moment they retreated in sullen silence to their corner, leaving the way open for another round of reformist leadership. It took contracting borders, loss of maritime trade routes, and European consular protection of the religious minorities before the reformist impulse could gain another go. The reforms set in motion by Passarowitz went much further than the previous ones. Sultan Ahmad III and his reforming grand vizier, who was also his son-in-law, Damad Ibrahim Pasha, bit the bullet of change and went straight to the heart of the matter, to the West itself, in order to learn what had to be learned for the empire to survive. Whatever it was that had made Europe so powerful had to be discovered and adopted.

The diplomatic mission sent to Paris in 1721 was an open admission that the Ottomans, at least some faction of them, had come to accept the bitter but necessary pill of a European tutorial. Because of the friendly relationship that existed between the Ottomans and France, going back to King Francis I and Sultan Sulayman in the early 16th century, Paris was chosen as the destination of the Ottoman embassy of exploration that would unlock the secrets of European power.

The initiation of reforms by Ahmad III and Damad Ibrahim Pasha gave every appearance that Ottoman leadership was determined to change and innovate. An imperial printing press was set up following the diplomatic mission to France, an Arabic font was made for it and books of science were translated and printed, while among the ruling elite an interest in French fashions and amusements was aroused. What is more remarkable, descriptions of the courtly life of Versailles seized the imagination and developed into a surreal imitation of Versailles along the banks of the Bosphorus, which, if nothing else, provided a moment of escape from the distress induced by the growing sense that the empire had lost something of its soul, spirit and muscle. The ruling class's obsessive preoccupation with luxury, amusement and ostentatious display made its greatest impression through the mindless frenzy of cultivating tulips, for which the reformist movement gained its name: *Lale Devri*. The Tulip Period.

Notes

1 Taner Edis, *An Illusion of Harmony: Science and Religion in Islam*, Prometheus Books, Amherst, NY, 2007, p. 59.
2 On this see: Bernard Lewis, *What Went Wrong? Western Impact and Middle Eastern Response*, Oxford University Press, New York, 2002, p. 7.
3 He also describes an operation in which a false tooth carved from ivory was implanted in the gum cavity of a freshly extracted tooth. The implant held firm as the gum healed around it. Evliyya was equally impressed by the clinical garments and hats the doctors wore while operating, and he concluded that medicine as practiced in Vienna had advanced even beyond the level the Greeks had achieved. John Livingston, "Evliyya Chelebi's Account of Surgical Operations in Vienna," in *Abhath*, American University of Beirut, 1970.
4 Bernard Lewis, *The Muslim Discovery of Europe*, Norton, New York, 2001, p. 227.
5 Lewis, *Muslim Discovery of Europe*, pp. 230–231; A. Adiver, Osmanli *Turklerinde Ilim*, Remzi Kitabevi, Ankara, 1982, pp. 141–143
6 A contemporaneous Western counterpart to the conservative reaction against innovative medicine that was taking place in Istanbul is Lady Mary Wortley Montagu's frustrating efforts to introduce England to the Turkish practice of inoculation against smallpox. Conservatives and many physicians opposed it. In London, an assembly of physicians, surgeons and members of the Royal Society was invited in London to hear Lady Montagu out. In the middle of a long and cantankerous debate she exclaimed in exasperation, "I had rather be a rich effendi with all his ignorance than Sir Isaac Newton with all his knowledge," and then and there grabbed the instrument and inoculated her daughter in front of the astonished assembly, among whom sat none other than Sir Isaac himself, by then the Royal Society's most famous member. M.W. Montagu, *Turkish Embassy Letters*, University of Georgia Press, Athens, Georgia, 1993.
7 Della Valle's letters were published posthumously by his sons in three volumes, the first in 1652, the second in 1658 and the third in 1663. An English translation in two volumes by G. Havers was published in 2010 as *Travels of Pietro della Valle*. A succinct account of Pietro's travels by Caroline Stone is found in Aramco World, January/February, 2014, pp. 20–27.
8 Ekmeleddin Ihsanoglu, "Introduction of Western Science to the Ottoman World: A Case Study of Modern Astronomy (1660–1860)," in *Transfer of Science and Technology to the Muslim World*, edited by E. Ihsanoglu, A.H. de Groot, Istanbul, 1992, p. 69.
  For Noel Durret, Ibrahim Tezkireci and an analysis of the text and its translation, see Avner ben Zaken, "The Heavens of the Sky and the Heavens of the Heart: The Ottoman Cultural Context for the Introduction of Post-Copernican Astronomy," *British Society for the History of Science*, 37, March 2004, pp. 1–29.
9 Ben Zaken, "The Heavens of the Sky and the Heavens of the Heart," pp. 2–6.
10 Ihsanoglu, "Introduction of Western Science to the Ottoman World," p. 80.
11 Adnan Adivar, *Turk ilim*, p. 139; Ahmet Evin, "The Tulip Age and Definitions of Westernization," in *Social and Economic History of Turkey* (papers presented to the *First International Congress on the Social and Economic History of Turkey*) Ankara, July, 1980, p. 136.

11 The Tulip Period

Damad Ibrahim Pasha was an able and energetic civil servant of humble origins. Neither *devshirme* convert nor favored youth of an influential family, he had come to Istanbul from the provinces as a young man in search of a career. His rise exemplifies the possibilities open to a man of talent. Keen of mind, Ibrahim gained a lower-level position in the central administration and ascended steadily through the ranks, eventually reaching a position high enough to come to the notice of the sultan. This brought him the rank of pasha, then marriage to the sultan's 13 year old daughter. With the royal marriage came the honorific title "damad," son-in-law. The office of grand vizier followed shortly after, which he was holding when the war against Austria occurred. The decisive defeat made him realize that the Ottoman way of doing things would either have to change, or the empire would perish. He also understood the problem to be more than a simple disparity in military strength. There was something unseen behind the West's strength.

After he had negotiated the Treaty of Passarowitz, Ibrahim prevailed upon the sultan to initiate some fundamental changes and was given authority to do what he thought had to be done. His first step was to appoint a committee of 25 scholars and scribes to make translations of books Europeans considered to be of the greatest significance in their cultural heritage. Books, not military weapons, would reveal the secret of European power. With this began the third phase in the translation cycle that would bring scientific, philosophical, technical and medical knowledge full circle into Islamic civilization. The first phase had been Greek and Syriac to Arabic, then Arabic to Latin, and now it was to be Latin and French to Arabic and Turkish. Previous translations in this third phase had been of a passive nature: books given as gifts to the sultan that laid around for decades before being translated. Damad Ibrahim's Committee of 25 was an active, purposeful effort to seek out critical books for translation. Another sign of the direction and dimension of Ibrahim's spirit of reform was his endowment of a madrasa in which he ordered mathematics to be taught, a subject rarely if ever offered in a religious school.[1] It was a small step, but it was a beginning.

In order to determine what structural changes should be undertaken that went beyond existing Ottoman institutions, and how to go about making them, Ibrahim tried to learn as much as he could about Europe from various European ambassadors at court. Consulting so many different European sources that were for political, religious and economic reasons contradictory or hostile made for more confusion than clarity about Europe. This was Ibrahim's introductory lesson that Europe was not an integral whole representing a single outlook formed by religion and a particular set of institutions but was as divided and contentious as the world of Islam or the Ottoman court.

Concluding that little could be learned of Europe's inner sources of strength without experiencing Europe directly, he obtained the sultan's blessing and took the unprecedented step of sending small embassies to a few European countries, and a major one to France. Owing to their having a mutual enemy in the Austrian Hapsburgs, the Ottoman Empire and France had enjoyed friendly relations for two centuries. Also, the Ottomans believed France to be the most powerful country in the West and its cultural heart. Embassies had been sent to Europe before, but only to negotiate treaties, never to learn anything. Just the acceptance of the possibility that infidel countries had something worth a Muslim's effort to learn was an intellectual revolution in itself.

Damad Ibrahim's intuition that behind Europe's superior military weaponry, organization and combat ability must exist institutions and knowledge was a part of his committee of translation and his having mathematics taught in a madrasa. Reforming indigenous institutions, he concluded, had to go hand in hand with learning the sources of Europe's strength. One had to complement the other.

Ibrahim put much importance on the mission to France. Its success depended on finding the right man to lead it. This meant someone open-minded, observant, quick to penetrate and comprehend what would be totally unfamiliar, a man whose mind was not closed with pretensions of superiority but who was at the same time not easily intimidated. He would also have to be able to get along with foreigners, infidels. His knowing French would have been a great asset, but this was asking too much.[2] Lacking knowledge of French, Damad Ibrahim's envoy would require a translator, another veil that could obscure the reality that the ambassador was seeking.

Ibrahim found his man in a former *devshirme* officer. His name was Yirmisekiz: 28, the number of his Janissary regiment. His formal name was Mehmet Chelebi, but he became popularly known by the number. He was a military man who had transferred to civil administration, ascended the ranks, and in the same manner that his patron Damad Ibrahim had come to the attention of the sultan, Yirmisekiz came to the attention of Ibrahim. Such were the career possibilities open to any bright and able young man, whether a European convert or born into the religion. His experience in the military and government, but above all his strength of character and courage to initiate reforms, decided Ibrahim to appoint Yirmisekiz as chief of the embassy to France with the mission "to visit fortresses, factories and works of French civilization and education, and report on those which might be applicable," a mission as large as it was vague. He set sail from Istanbul to Marseilles in November, 1720, and reached Paris five months later.

Yirmisekiz and his entourage were as strange to the French as the French were to them. According to Yirmisekiz's account of the trip, when they arrived in France, people flocked to see the visitors in their fairy-tale clothes of long flowing robes with wide sleeves and voluminous head wear, as if aliens from another planet had arrived. As the embassy traveled the barge canals and rivers on its way north to Paris, crowds of country people lined up along the banks. So great was the crush of excited spectators that many were pushed over the banks and into the water. French troops that had been assigned to guard the embassy were obliged to shoot their rifles in the air to disperse the frenzied crowd, unnerving the embassy. Outside the Palace of Versailles, people stood shivering in the rain for hours just to catch a glimpse of the strange looking embassy, which must have appeared to the country folk more wondrous than a traveling circus.

Yirmisekiz's report portrays a lively account of the affair.[3] It describes the excitement of the crowds, the beauty of France, of Versailles, the splendor and magnificence of its court and of Paris. His praise is such that the French ambassador in Istanbul who was later to read the report commented that the author's descriptions were highly unusual, as praise of anything foreign went against the high regard the Ottomans had of themselves and could have offended the sultan and grand vizier who might well have interpreted it as a slight on Istanbul and the sultan's court:

> The Turks are so filled with their grandeur and the pride of their princes is so delicate that it is extraordinary and even surprising that Mehmet Efendi had dared speak of the beauty of France and the magnificence of our court in the terms he used in a report to be read by his master.[4]

Yirmisekiz apparently had enough confidence in the integrity of his superiors that he could write honestly and openly, proving Damad Ibrahim had made a sound choice.

Reminiscent of Evliyya Chelebi's praise of what he had witnessed in Vienna 60 years earlier, Yirmisekiz described Paris's hospitals, operas, theaters, industries, bridges, canals, locks, roads, gardens, zoos, the status of women, and particularly the Royal Observatory in Paris and its scientific instruments. Where the traveler and writer Evliyya was most fascinated by the medical practice in Vienna's St. Stephen's Hospital, the ambassador in Paris was much taken by the royal observatory. He reports that its construction was begun by the Sun King in 1667 in connection with his creation of the Academy of Science and then completed in 1672.[5]

The observatory is described as a tall stone edifice of three floors, each containing many rooms and housing various objects of recent invention for scientific research. The scientific apparatus on display in the observatory's various chambers excited the ambassador's admiration. There were geometrical devices, immense spheres set on iron pedestals, some kind of a complex mechanical device that was used to determine phases of the moon, and other instruments to observe the stars, many of which "are unknown to us." Several of the machines exhibited to him and his subordinates were sufficiently complex to elicit expressions of awe. One of these was an instrument used to determine solar and lunar eclipses, described by

Yirmisekiz as consisting of a system of circular plates with the degrees engraved along the perimeter of the plates:

> When the circles commence to turn, a needle-like pointer with a round button the size of a small coin at the end extends toward the sun and moon, and when it covers part or all of one or the other, one knows the month and duration of an eclipse. Another circle indicates the year, month and day of an eclipse. It is a wondrous thing. The French are much proud of it and tell me there is no other like it and only King Louis XIV's generosity made its construction possible, as this prince knew the expense of science and bestowed gifts and honors on inventors and put their machines in the observatory for students to use.

In another chamber of the observatory was an array of burning mirrors with concave lenses "the size of one of our great dining tables," whose power the French scientists demonstrated by showing how it could melt lead and consume a block of wood in flames the moment it was placed before the mirrors. His description of an "astonishing" machine which consisted of little more than a sophisticated system of pulleys to lift great weights indicates the extent that Muslims had lost contact with their Archimedean tradition of mechanical expertise. The machine would have hardly impressed those Muslim scientists and builders of cunning mechanical contrivances in bygone centuries, but for the ambassador: "Truly, I saw so many marvelous things that not even a treasure would suffice to acquire them."

Of all the marvels Yirmisekiz witnessed in his two visits to the observatory, it was of course the telescope that attracted his greatest interest:

> Among other things there was an apparatus for observing stars and planets, the glass lens of which is as large as a barber's mirror. It is fitted in a casing of laminated white metal and resembles a well pump. Its length is 50 axe lengths and is fitted in a hollowed out gutter-shaped conduit to which it is firmly fixed. Another glass [lens], this one smaller, is fixed at the end. In all, it has two glasses. The telescope is moved up, down, to the sides, and forward or backward by an involved system of pulleys and weights that enables a single man to position the large instrument in any direction he may wish.[6]

Yirmisekiz may have been the first Muslim recorded to have peered through a telescope and seen a heavenly body close up. His reaction echoes Galileo's surprise over a century earlier. He saw Venus rising over the horizon as having the shape of a three-day moon tinted in red lilacs, while the moon itself appeared so big the viewer could have drowned in it. Strangest of all, it looked to have the soft and spongy texture of bread owing to its caverns, valleys and the play of light and shadow thrown across the rough surface of its elevations and depressions. "But Glory be to God," he exclaims in mock response to his hosts' having some fun at his astonishment, "I did not see the trees and water that some of the Frenchmen claimed were there!"[7]

It can only be surmised what thoughts raced through his mind that never made it into his report as he peered through the huge polished barrel of the telescope aimed at the night's sky, but there can be no doubt Yirmisekiz was aware that what he was witnessing went beyond even Ulug Beg's table of observations.

In regard to modern astronomy, Yirmisekiz's report raises many questions, more by virtue of what he does not report rather than what he did. For example, he writes of what European scientists had determined with great precision concerning the orbit of Jupiter, of the precise and detailed explanation he received on this planet's orbit, of the planetary model he saw with Jupiter's four satellites orbiting it, each equidistant from the other, and of how the model marvelously indicated where the four satellites would be, right to the degree and minute of any day, for "as Jupiter makes its orbit, its satellites make theirs, praise be to the most high for whom nothing is impossible!" He has no doubt Jupiter orbits the sun. Yet nothing is said about the earth doing the same, a fact that would have been worth mentioning as one of the fundamental European astronomical discoveries. In view of his discourse on Jupiter, he could not but have known of the theory. During his visits to the observatory, the French would have made a point of informing him of it if they least suspected he was of another opinion – French scientists took great delight in impressing, surprising, dumbfounding and shocking, literally, Muslims with their knowledge. Did Yirmisekiz not mention a moving earth because he was unsure whose unfriendly eyes might see his report and bring accusations of heresy raining down upon him? The report was more of a non-confidential memorandum meant for anyone to read, not just the sultan and grand vizier. He sent a copy of it to the French ambassador, who, after commenting on it, had it translated to French. Possibly the report was read by more Frenchmen than Ottomans. Not until a quarter century after the embassy returned to Istanbul was it published.

Another revolutionary invention that attracted the ambassador's interest was the printing press. This was not as unfamiliar as the telescope. Products of the printing press had been seen by Ottomans in the literary gifts presented by European ambassadors to the sultan over the years. Presses had been in fact used by the religious communities of Armenians and Greeks in the empire since the 17th century. Printing had been shunned by Ottomans not so much for religious reasons as economic and esthetic. It would put copyists out of work; manuscripts with flourishes of calligraphy were a thing of beauty.

Yirmisekiz realized the power of both telescope and printing press, and of the two it was the press he would bring back to Istanbul. This would prove to be the most immediate and tangible result of his mission to France. Yirmisekiz would probably have brought a telescope back as well, if the funds had been provided and one had been for sale. The far lesser expense of the press must have been a factor if there was any choice in the matter. Another factor was his son, Sa'id, who accompanied the expedition. Sa'id was at the time a young man in his twenties. He is reported to have ventured out into the foreign country beyond the protocols of diplomacy, engaging French society, studying French, making acquaintances, and, more than his father, who gained the fame, setting the direction of Ottoman reform. It was he who, with the support of his father and Damad Ibrahim, was the driving force in breeching the tradition of the written word and had books printed instead of their being laboriously copied out by hand, copy by copy. From the young man's experience in Parisian society, Sa'id believed that real reform would begin with the spread of literacy through the printed word.

For a literate culture that considered certain words to be sacred and the act of writing them a sign of devotion, it was no easy matter to legitimize in Muslim society a Western contraption that, noisy and unclean, mechanically put words to paper. Sultan Bayezid II had issued a ban on printing in 1485.[8] Adding to the difficulty, many of the ulema made a living as copyists, a profession that the press would quickly destroy. To help get over the hurdle, Ahmad III requested the shaykh al-Islam at the top of the Ottoman religious institution to issue a *fatwa* stating that printing did not contravene the holy law in act or spirit. Sixteen members of the high ulema in Istanbul were persuaded to profess themselves in favor of the press. Those of the ulema opposed to it were generally calligraphers and copyists who claimed that printing the script of revelation was an affront to God and scripture, that writing by hand was a tradition going all the way back to Mecca and Medina and was the way God intended.

The last thing the sultan wanted was to have trouble with those who decided what was and was not religiously acceptable. The ulema spoke for the Muslims of the empire and without their cooperation, or at least grudging passivity, change would be impossible. And so a compromise was worked out. The number of presses allowed to operate in the empire would be limited and the printing of anything touching on religion or religious subjects would be prohibited. Books on subjects such as geography, astronomy, mathematics, physics, medicine, language and history could be printed; copyists of the ulema would be employed as proofreaders. This gave innovative reform enough of a semblance of religious legitimacy to disarm at least temporarily the conservative opposition. Once the compromise was agreed upon, a *fatwa* was issued legalizing it. An Arabic font was then prepared.

The translations and books printed were meant to expand the horizons of an educated elite. But there was no plan, or implementing structure, to introduce young students to new knowledge in preparing a new generation to carry reforms to another level. Just printing books was not going to do it. Unsure of how to implement deep reform, and how far they could go before the sword of tradition came down on their necks, the reformers proceeded cautiously. They knew that the ulema would certainly have resisted introducing printed material and the subjects related to it in the madrasas on the grounds of endangering the faith of young Muslims. Instituting new schools was out of the question. The sultan and grand vizier were not about to take on the religious establishment.

The ulema's endorsement was needed for reform to progress. A glimmer of hope that the more educated members of the ulema would join the reformers came with Esad Efendi (d. 1722), a gifted proofreader, and one of the very few of the ulema who knew a European language. Esad Efendi, later known as Analema Alim for his scientific work, translated Aristotle's physics from Greek to Arabic in preparation for its being printed on the new press.[9] A translation of Gilbert's magnetism or Galileo's physics would have better served the purpose, but Aristotle, whose works had been given several Arabic translations a thousand years earlier, was at least a start. Esad Efendi's treatises on the telescope and microscope were considerably more fitting for the task at hand. Approximating the role played by Latin Church scholars during the rise of science and philosophy in the West, Mulla Esad Efendi embodied grand vizier Damal Ibrahim's dream come true, a member of the ulema engaged in scientific work.

To make the likes of Esad Efendi become the norm among the ulema would have meant a revolutionary reform of education in the religious institution, but that was not about to happen: secular studies had no place in the madrasas. Loathe to provoke religious discord, the reformers were between the proverbial rock and a hard place. The idea of instituting a state system of education that would parallel and eventually marginalize religious education was at the time far beyond reformist vision, which at its best was blurred. The overseers of reform were groping: none of them saw far enough ahead to lay out a plan with a comprehensive agenda of translation. This was unfortunate. A methodical program of translation of the basic texts of modern science, from *de Revolutionibus* to *Principia*, would have set the ship of state on a sound course of reform. The gap between Islam and the West in the early 18th century was still manageable. Mastering the *Principia* would have brought the Ottomans up to the scientific frontier. Newton was still alive at the time of the 1720 mission to France. The scientific frontier had not progressed beyond the point that Newton had brought it, but it was not going to remain at that point for long.

With a printing press having been brought from France, a competent man had now to be found to run it and supervise the translations. The responsibility called for a man knowledgeable in Western languages and of the West itself. In addition to his being favorable to reform, he had to be technically able and an upright Muslim acceptable to the chief *mufti* who was given the last word on the appointment. The man was found in Ibrahim Muteferrika (d. 1745), a Hungarian convert to Islam.

Like the translator Ibrahim Kose Efendi of Szigetvar half a century earlier, this Ibrahim was one more of those anonymous Europeans who emerged from the shadows of a Christian background to be given a new name, a new religion, a new life, and gain fame as a Muslim in Ottoman service as a dedicated reformer of the empire in its defense against the West, the former abode. So completely had these men become Ottomans that even their European Christian names went down the memory hole. The integration of these young men into Ottoman life in name, religion, language, culture, personal identity and devotion to imperial Ottoman service is truly astonishing. How to explain this voluntary surrender to oblivion of everything they were before going Ottoman?

The open-armed acceptance into the middle and upper reaches of Ottoman society must have been both attractive and rewarding even during this not so affluent period in Ottoman affairs. The attractions of becoming Ottoman were many, and many were the Westerners who left their past lives behind to join Islam and begin again. The price of entry was conversion, learning Turkish, perhaps also Arabic and Persian. A young man from the West with talent and languages, preferably with some military experience, and no qualms about converting, had an open career in Ottoman service. One might speculate that the cuisine, rich in fish, fruits, vegetables, spices and sweets, the delights of polygamy, concubinage, not to mention the sunny weather, the natural and architectural beauty of the seven hills of Istanbul surrounded by Bosphoros, Marmara, Black Sea, Mediterranean, the Golden Horn, and, in terms of career, the possibility of becoming a pasha, must have had their attractions for men of robust sexual energy and ambition, and a bent for adventure. The mystery is not so much why a man would stay to be absorbed and indelibly remade once he had crossed over, but why would he cross in the first place, being not so sure what awaited on the other side? The Ibrahim from Szigetvar remains a mystery. In the case of Ibrahim Muteferrika a little more is known.

A Hungarian source claims he was a poor 18-year-old student at the Calvinist College in Kolosvar in Transylvania when in 1692 he was captured by an Ottoman raiding party that had crossed the Danube. Being of humble origins, he went unransomed, was taken to Istanbul and sold into slavery.[10] A Turkish source dismisses the story of his capture and enslavement and focuses on the religious strife that was raging in Hungary and Transylvania in the late 17th century. According to the Turkish version, Ibrahim was not a Calvinist but a Unitarian from the town of Erdel, in Transylvania, where Unitarianism was widespread. Transylvania was at the time being taken over by the Hapsburgs, who were forcing the Unitarians to become Catholic. It was this that decided Ibrahim to cross over. The Unitarian repudiation of the trinity and uncompromising belief in the absolute oneness of God, with Christ being a human prophet and in no way divine, made Ibrahim's conversion to Islam an easy journey. (St. John of Damascus, a millennium earlier, it will be recalled, had considered Islam so like Christianity that he thought it to be an anti-Trinitarian heresy, while al-Ghazali thought of Sufism as a form of Muslim Christianity.)

Ibrahim wrote a brief autobiographical treatise around 1710, *Risale-i Islamiyye*, in which he reveals his anti-Trinitarianism. As a youth he secretly read books against the Catholic doctrine and discovered in the Bible references to the coming of Muhammad, revelations, he claimed, that were excised or falsified by believers in the Trinity. According to his treatise, he fled Transylvania because of Hapsburg religious oppression in order to seek refuge with the Ottomans, after which, discovering the close doctrinal kinship between Islam and Unitarianism, he freely converted.[11] His choice of the Muslim form of Abraham for his new name would have provided another bridge taking him from his old to his new religion in the monotheistic family.

Knowing Hungarian, and Latin from his religious studies, he was made an officer in the *Muteferrika* corps, or special service branch, by which name he came to be identified in his new life, in the same way Yirmisekiz did. He quickly learned Turkish and Arabic, Shari'a law, and Islamic literature. His knowledge of languages gave him a boost in Ottoman service; his intelligence and personality gave him another. As a middling but keen officer in the bureau of imperial correspondence he came to the attention of Damad Ibrahim, who introduced him to Yirmisekiz and his son Sa'id. Ibrahim Muteferrika, they agreed, had the right qualifications, and he was given full and absolute control over the new press, in which capacity he became one of the most important Ottoman writers in the last two centuries of the empire's existence.

With the exception of mechanical clocks, the press was the first invention adopted from Europe that was unrelated to the army or navy, and the first significant step toward Ottoman modernization. Emblematic of the deepening understanding of the West and its sources of power, the new press was to lead the way toward Ottoman intellectual reformation, with Western books on geography, science and technology now sharing honors with Western military achievements as subjects of interest. Printing, Ibrahim Muteferrika claimed in a treatise praising the press, not only avoided the mistakes copyists were prone to make but speedily produced many copies of a book and served to spread literacy and education. He saw the press as the spearhead of intellectual revolution, where new ideas would work their subtle change in building a consensus toward reform and renewal.

In physical plant and intellectual output, the press was a modest affair. Established in his house, it operated for 18 years, 1727–1745, and printed a total of only 16 different works in 20 volumes, six in the exact sciences (none of them groundbreaking works of the Scientific Revolution), the remainder in history and geography. The other presses that were planned to follow never materialized. Ibrahim's miniscule budget and few operating assistants reflected the financial problems afflicting the Ottoman state, with its treasury perpetually empty, leaving government officials and the military going often months and more without pay. It also reflected the unwillingness to invest adequately in intellectual reform. Meaningful intellectual reform would have required financial reductions in other areas. To accomplish the goals envisioned by Yirmisekiz's embassy to France, and by Ibrahim's printing press that resulted from the embassy, many more embassies and printing presses would have been necessary, and scores of Ibrahim Muteferrikas. The investment in energy, money and men was not enough to wrench the empire from its ill-omened tracks and set it on a new course.

Ibrahim's press remained a government project alien to the Muslim populace. It created no broader appeal beyond government to excite private initiative for literary or political purposes. No intellectual radicals had yet come on the scene to broadcast their ideas. A market for printed books had little chance of emerging. There were reasons for this. Ottomans, and all Muslims, valued the written word. Writing was a sacred act that penetrated deep into Muslim culture. Calligraphy was a respected profession associated with religion. The beauty of penmanship was a prized art, of which there were several defined forms. The writing hand was moved by a God-given human soul. What moved a machine?

It was not that the printing press was all that unfamiliar to the Ottoman elite. Ibrahim's was not the first press in Istanbul; it was only the first to use Arabic characters. Jews fleeing Spain and seeking refuge in Muslim lands had brought one with them to Istanbul as early as 1493; the Greek community was later to possess one, as was the Armenian Church. While for Jews and Christians the machines were permissible, for Muslims the press remained for centuries blithely ignored, a foreign invention of questionable legitimacy, even sacrilegious, submitting as it did the script of revelation, Arabic, God's sacred language, to the tortures of a soulless machine invented in the devil's workshop by infidels.[12] God would not want his name stamped out by a clanking machine.

Printing would not easily overcome the resistance to change. In any case, the reading public was not large enough to sustain a private press. Ibrahim's hope of spreading literacy did not happen.

The Tulip Period's press did not much change things, but it was a precedent. Just getting the press sanctioned, even conditionally, was a major accomplishment when considering European reports of Ottoman attitudes regarding adoption of European invention. It goes without saying that any judgments made by Western observers on the practices and prejudices of Easterners should be taken with a large grain of salt; but the commentary left by Europeans on Ottoman disregard of useful mechanical inventions should at least be considered as part of the historical record and critically reviewed, if for no other reason than to see how these observations by outsiders depict not only their subject but themselves. In 1560, at the very height of Ottoman power, the ambassador of the Holy Roman Empire in Istanbul, Ghiselin de Busbecq, a not unsympathetic observer of Ottoman institutions, had this to say:

> No nation has shown less reluctance to adopt the useful inventions of others; for example, they have appropriated to their own use large and small cannons and many other of our discoveries. They have, however, never been able to bring themselves to print books and set up public clocks. They hold that their scriptures, that is, their sacred books, would no longer be scriptures if they were printed; and if they established public clocks, they think that the authority of their muezzins [callers to prayer] and their ancient rites would suffer diminution.[13]

The implication that public clocks would cause the muezzins to suffer, analogous to the printing press putting copyists of the ulema out of business, which reflects both the vested interests and influence of the ulema and the government's disinclination to innovate counter to them, is a piece in the social pattern of Ottoman ruling style, that is, to leave well enough alone and not rock the boat for the sake of keeping tranquility, similar in purpose to the Millet System of keeping harmony among the religious minorities by leaving them to their communally autonomous selves. There is also an argument to be made about putting *muezzins* and copyists out of work and the esthetic quality of the written word. But purely on the level of technological and scientific innovation, some Ottoman acts do not belie ambassador Busbecq's observation. It has been mentioned in another context that in 1574, while Tycho Brahe was building his Uranoborg observatory on the island of Hven, Sultan Murad III ordered the demolition of Istanbul's only observatory, which, only just completed, had been modeled on Ulug Bey's in Samarqand. A naval ship approached the towering structure along the coast and, with its artillery, obliterated the masonry to the cries of triumph by the crowd gathered to watch, as if a great naval victory had been achieved. In fact, to those who believed it a sin to pry into God's secrets by observing the heavens, which was the basis of the *fatwa* obliging the sultan to issue the order of destruction, it *was* a victory. The court poet Ala al-Din Mansur celebrated the event in verse, claiming the observatory had become useless and was needed no longer as it had achieved its purpose of correcting Ulug Beg's tables. In all fairness, this was not a general attitude in the Muslim world, nor was it a general Ottoman one. The Mogul ruler in Delhi had a large observatory in the Samarqand tradition built around the time the Istanbul observatory was being destroyed. The observatory's mural quadrant was sturdily constructed and so finely calibrated it was capable of measuring the position of heavenly bodies with an accuracy of up to six minutes of arc, about as good as Tycho Brahe could do.[14]

In regard to the Ottomans, they indeed recognized the inventive skill of Europeans in making complex mechanical devices and marveled at the beautiful timepieces European merchants in Istanbul imported from Europe. The clocks were in great demand by those who could afford to buy one. But as for Ottomans learning the craft to make and sell them for profit, or of financing local skilled craftsmen to master the art and set up shop, the European watchmakers who had set up shop in the Galata quarter of Istanbul had taken care to monopolize the industry. In this they acted no differently than their national counterparts in global commerce who fought wars to control trade and drive Ottoman and all other ships from the seas. The drive to monopolize and control locally and globally gave no chance for Ottomans to cut into the trade, the effect of which, in Western eyes, was to make them look indolent. Hence the uninformed judgments such as Voltaire's when in writing to Fredrick the Great in 1771 he describes Catherine the Great as an enlightened despot waging war against benighted tyrannical Turks:

> It is now sixty years since they (Ottomans) have been importing watches from Geneva, and they are still not able to make one, or even regulate it.[15]

Ottoman entrepreneurs attempted in the early 18th century to enter the textile industry, but this too was undermined by a combined force of European merchants, textile manufacturers, and the capitulatory privileges granted to Europeans by the Ottomans in former centuries.[16] To Westerners, looking through the glass darkly with their cultural prejudices filling in what they could not see clearly, industrial entrepreneurship was beneath Ottoman interest and dignity. Here is the French traveler Etienne Savary writing on Egypt in the 1780s:

> Here the Turk smokes pipes of jasmine and amber all day, in a fragrant place, thinking little without ambition; activity, the soul of all our talents, is unknown to him. He peacefully enjoys what nature offers, what each day presents to him without worrying about the next day.[17]

A similar criticism is found in a report by the British Consul in Istanbul written a decade or so after Voltaire's and Savary's observations, and this at the very time Ottoman leadership was supporting a large program of reform:

> Traveling, that great source of expansion and improvement to the mind, is entirely checked by the arrogant spirit of their religion, and by the jealousy with which intercourse with foreigners . . . is viewed in a person invested with an official position . . . Thus the man of general science . . . is unknown; anyone but a mere artificer who should concern himself with the founding of cannons, the building of ships or the like, would be esteemed little better than a mad man . . . They like to trade with those who bring to them useful and valuable articles, without laboring of manufacturing.[18]

European observers were outsiders looking in, through many veils, and could be accused of the very arrogance they ascribed to Ottomans. The aspirations of men like Damad Ibrahim, Yirmisekiz, Ibrahim Muteferrika and their scores of assistants and the many reformers who followed them through the whole of the 18th century, seem to have totally escaped European notice. Foreign visitors saw what fit their preconceptions. Even the notable accomplishments of Ibrahim Muteferrika failed to make an impression.

Of his many scientific pursuits, Ibrahim Muteferrika considered himself above all a geographer. The productions of his press reflect this. His first project was to print modern maps of the Black Sea and Sea of Marmara for the navy. This was a continuation of the naval reforms of Grand Admiral Mezzomorto and Grand Vizier Husayn Koprulu at the beginning of the century. Ibrahim's concentration on geography would appear to have been influenced by the empire's ignorance of it. Ottoman knowledge of the world was manifestly abysmal when compared to what Ottomans had known of it in the glory days of the 16th century. Grand admiral Piri Reis had a map of the New World that was apparently made by Christopher Columbus. Piri Reis himself made a surprisingly exact map of the western hemisphere. That knowledge had been lost during the next century. The explorations of Columbus, Vasco da Gama and Francis Drake were at most only vaguely, if at all, known to even the most educated Ottomans of the early 18th century.

Paralleling the anecdotes about Ottoman disinterest in science and in making their own mechanical devices are stories about the birth of geographical knowledge. Most infamous of these mostly apocryphal stories is the purported Ottoman accusation of its French ally to have allowed the Russians to transport their fleet overland across France from the Atlantic to the Mediterranean after the Russian navy appeared unexpectedly in the Aegean and sank the Ottoman fleet in the early 1770s. A variant of the story has the Ottomans protesting to the Venetians for having allowed the Russians to sail from the Baltic to the Adriatic. The point is the same, that the strait between the sea and the ocean, Gibraltar, bridge to the Islamic conquest of Spain and named after the Berber Muslim conqueror Tariq, was unknown to the rulers of Islam's most powerful state. Another anecdote that made the rounds among the European diplomats in Istanbul was the Ottoman admiral who, having been ordered to conquer Malta, set out with the imperial fleet and after months of sailing the Mediterranean in futile search of the island returned to Istanbul to report "Malta yok!" There is no Malta. A documented account reports that a grand vizier at the end of the 18th century refused to believe military supplies could be delivered by sea from India to the Red Sea coast.[19] As late as 1800, with the French army in occupation of Egypt and threatening Syria, and the British and Ottomans in alliance trying to force the French out, the British Consul in Istanbul could write that no one in the Ottoman navy "has the least idea of navigation and the use of the magnet."[20]

Anecdotal and exaggerated as the stories and reports must be, if any of them conveyed even the least shadow of reality, Ibrahim Muteferrika had good cause to emphasize geography and the arts of navigation. He knew that if the Ottomans were to survive the challenge of Europe they had to know the world, and it was to this end he devoted his literary career and publishing efforts: to expand Ottoman knowledge of the world through geography and history and of the universe through physical science and astronomy. His emphasis on geography may in fact lend some support to those stories ridiculing Ottoman ignorance of it.

His press was authorized to print Western works as well as Muslim classics. The idea was to inform Muslims about the West while reviving their interest in the intellectual contributions made by Muslim civilization in the past. Accordingly, Ibrahim decided to print the 17th century geography *Jihannuma* (*Book of the World*), by Katib Chelebi, known as Hajji Khalifa in the Arabic tradition. The chapter Ibrahim wrote and added to his printed edition of Katib Chelebi's world geography was the first piece of purely Ottoman literature written to introduce Western science to Ottoman readers.[21]

Ibrahim edited and printed a second work of Katib Chelebi, *Tuhfat al-Kibar fi Asfar al-Bihar* (*The Great Gift in Sea Voyages*), to which he added a section of his own that presented the first physical descriptions of the western hemisphere to appear in Ottoman literature since Piri Reis's book.[22] He then wrote his own book on the western hemisphere, *Tarih al-Hind al-Garbi al-Musamma bi Hadis-i Nev* (*History of the West Indies called The New World*). An important source of Ibrahim's knowledge of western science came from a Latin book of relatively recent vintage, *Institutiones Philosophicae*, by the Frenchman Edmond Pourchot (d. 1734), published in Paris in 1695. The book, which could have been a gift of the French ambassador or a book Yirmisekiz or his son Sa'id had brought back from Paris, contained accounts of Gilbert's magnetism, Galileo's physics and Descartes' theory of vortices. An even more recent and specialized source than Pourchot used by Ibrahim appears to have been William Whiston's *The Longitude and Latitude Found by The Inclinatory or Dipping Needle*, published in London in 1720. Ibrahim wrote and printed two treatises on magnetism, the more important of the two being *Fuyuzat Miknatisiye* (*Enlightenment of Magnetism*), published in 1732.[23] The major historical works he printed included a history of Timur, one of Egypt, a universal history written by Katib Chelebi, a history Ibrahim wrote himself on the Ottoman conquest of Bosnia and several Ottoman chronicles.

In addition to his work in geography and history, he translated from Latin to Turkish the cosmographical work of the Dutch geographer and astronomer Andreas Cellarius (Keller): *Harmonica Macrocosmica*, also known as *Atlas Coelestis*, published in Amsterdam in 1661. Like Blaeu's monumental *Atlas Major*, a copy of Keller's *Cosmography* may have reached the court as a gift from the Dutch ambassador.[24] In his translation, Ibrahim provides critical remarks and categorizes the rival systems as Aristotelian–Ptolemaic, Pythagorean–Platonic–Copernican, and Tychonian, rather than the traditional categorization of Ptolemaic, Copernican and Tychonian.[25] The translation was a major addition to the slowly growing scientific literature in Turkish and Arabic that, little by little, was exposing the Copernican system and modern physics to a miniscule society of Ottoman readers. Exposure, Ibrahim hoped, would lead to the spread of knowledge, then to acceptance, and acceptance to learning and revival.

To everything he printed, whether editions of Muslim works or his translations of Western works, he added comments and chapters of his own that included material he had garnered from other sources in order to expand knowledge of the subject and enhance the reader's interest. Encouraged by his having incurred no rebuke from the ulema for his pro-Copernican bias in the chapter that he had added to his edition of Katib Chelebi's *Jihannuma*, guarded as the bias was, he expressed his Copernicanism more openly in his translation of Keller's *Cosmography*. Copernican theory is interpreted as the historical development of scientific thought from Pythagoras to Aristarchus and Ptolemy, culminating in Copernicus and Tycho Brahe. To paraphrase Ibrahim: the Christian church and ignorant people at first opposed Copernicus; this was because his system was beyond their common sense; with the development of more advanced astronomical equipment and more precise observations, the system became more credible; even a Catholic Cardinal, Nicolas Cusa, accepted it, and so too the celebrated scientist Descartes, who explained the system's transmission of forces through his philosophy of vortices; as for Tycho Brahe's system, it was a compromise between the Ptolemaic and Copernican.

Ibrahim recognized that heliocentrism might not sit well with some of the ulema. Hoping to avoid complications, he presented a version of the defensive preface that Rheticus had added to Copernicus's *de Revolutionibus* when preparing it for publication, that the Copernican system was an exercise in geometrical astronomy and had nothing to do with the way things really were. Whatever system one followed, Ibrahim wrote, it was only astronomy and did not increase or diminish one's religion. As for the systems themselves, one was the mathematical equivalent of the other, no one of them having any bearing on the purity of faith of a believer, whatever astronomy he chose to follow. And, then, quite as had Galileo in his dialogue on the two world systems that got him in trouble with the Vatican, Ibrahim argued the advantages of one system over the other, claiming that belief or disbelief in any astronomical system was not a religious matter: the new system should be considered on its merits. Science does not contradict anything in religion. The creator who made the universe is the same who revealed himself to the prophets. Religion stood above and apart from astronomy. And though Muslims considered the Ptolemaic system the true one, the others should be examined if for no other reason than for the benefit to be gained in the pursuit of knowledge. Ibrahim concluded that a critical study of a sun-centered system would illuminate astronomy through the process of detecting and testing the evidence that proved or disproved the system.

While trying, though not too successfully, to avoid any overly expressed commitment that the Copernican system was the proven one, Ibrahim reviewed the ulema's arguments in opposition to it, then did his best to clear away any contradictions between God's word in scripture and a moving earth by stating that what was meant by a moving earth in astronomy was movement in space, while what was meant in scripture by a stationary earth was permanence in existence. He then explained the seeming contradictions between sense experience and an earth in motion: the lack of observed stellar parallax was because of the great distance of the earth from the stars, and the reason that trees were not uprooted was because the earth's gravity kept the air around it and everything on it in place.[26]

Ibrahim was spared the trials of Galileo. The Vatican's coercive apparatus had no Muslim counterpart and the ulema had no equivalent of a Jesuit order or inquisitorial court to threaten, torture and roast suspected deviants to the established orthodoxy. The ulema had no authority to do any of those things, only the power to declare heresy and recommend that the political authority burn books deemed repugnant to religion and exile suspected and declared heretics. Ibrahim had the support of the sultan and grand vizier, but this was no guarantee of protection against the ulema's wrath, which was usually expressed through the quietly applied coercive power of persuasion to have public discourse marginalized on subjects of which the ulema disapproved by banning and exile: no torture, no threats of it, no burning of flesh, only paper. There could be no Muslim counterpart to Bruno's burning or Galileo's forced disavowal of Copernican theory. As it turned out, none of the ulema appeared to be the least disturbed by Ibrahim's discourse on a sun-centered system.

Ibrahim kept the press running practically single-handedly until 1747, turning out on the average a volume a year for the 20 years of its existence. It was shut down for lack of support. A single press run by practically one man producing 20 volumes over as many years and then being shut down was an inauspicious beginning of Ottoman regeneration. Remarkably, Ibrahim kept the press operating long after the bloody events that ended the lives of the men who gave birth to the Tulip Period. Ibrahim Muteferrika and the printing press were the period's most enduring symbols.

Demise of the Tulip Period

Pioneers of change like Husayn Koprulu, Mezzomorto Pasha, Damad Ibrahim, Yirmisekiz and Ibrahim Muteferrika were a rare breed. They and the few scores of men assisting them were trying to bring change to a society cast in the sullen lack of curiosity that comes with massive illiteracy, ignorance, poverty, insecurity and traditions shrouded in the aura of religious authority. Outmoded, sclerotic and corroded in corruption, the institutions that had once given the government and the military the means to become a world power had become fossils of a past era of physical growth and fiscal soundness. Those institutions were, however, still internally sturdy enough to resist changes that would jeopardize the positions of those whose livelihood and prestige were hinged to them. As far as these old institutions were concerned, change, if it was to be, would be restricted to cleaning up and reordering what was in place, not replacing what existed with innovations.

The effort to modernize was a labor of Sisyphus. The Ottoman boulder had a high mountain to go and the conservatives would be waiting to kick it back down at the first chance – and a chance was always coming: a military defeat, a lost war, a humiliating treaty, another province lost, as though European expansion and the conservatives were in alliance against the reformers. In spite of failure, defeat and despair, the Ottoman Sisyphus would rise from despair and push harder. The cycle would continue for two centuries, one set of reforms more extreme than the one before, until world war and the unbridled zeal of reform left the empire a heap of ashes.

The Tulip Period has been interpreted as a stage on the way to Ottoman imperial collapse: a time of breakdown in traditional values, an irresponsible hedonist escape by Ottoman leadership from the despair of defeat and economic collapse.[27] The long years of defeat on the battlefield that began with the failure to take Vienna in 1683, combined with the mountainous financial burden incurred by the siege and the losing wars and the subsequent loss of faith in the erstwhile invincible institutions that were now faltering in the face of growing Western power, cast the leadership into an angst of doubt that the empire could survive another war with Europe. War had to be avoided. The sultan appointed viziers who followed that policy. In this defeatist atmosphere, as the narrative goes, the ruling elite avoided the harsh reality that Ottomans had lost the ghazi warrior spirit embodied by the first generations of Muslims and by the first ten Ottoman sultans. Rather than facing up to the fact that they were no longer able to shoulder the duty of expanding the Abode of Islam into the Abode of War – the raison d'être of empire – they surrendered to cynicism and moral turpitude, recklessly plunging into a long banquet of voluptuary self-indulgence where Western modes of luxury and vice became common: French-inspired pleasure palaces along the Bosphorus, Western dress, partying into the night, loud music and the proliferation of coffeehouses and brothels.

Mehmet Chelebi's descriptions of Fontainebleau and Versailles, with the fountains, sprawling gardens, picnics, games, nocturnal musical balls and all the amusements of refined extravagance of aristocratic indulgence gave the Ottoman ruling elite a fanciful glimpse of French high society at play. He had brought sketches of the French chateaus and palaces he had visited. These had then been embellished by titillating descriptions of the marvels of courtly life at Versailles. From this fantastic facsimile of a paradise on earth emerged the Tulip Period's frivolous side of serious reform. With one part of the elite surrendering to the embrace of decadence for one last banquet before the deluge, and another consisting of a few strong-willed reformers working to save the empire, the period presents a portrait of schizophrenic malaise at the highest level.

Sultan Ahmad III and Damad Ibrahim had a foot in each camp. They favored change but enjoyed the frivolity of the banquet too much. Leading the way to voluptuary escapism, the sultan and his grand vizier had European painters come to Istanbul to do their portraits and decorate the walls and rooms of the pleasure palaces that they built along the choice headlands at the tip of the Golden Horn, which offered an unobstructed view of the natural beauty where geographical Europe met Asia. Other pleasure palaces sprouted up along the banks of the Bosphorus and Golden Horn, imitations of Fontainebleau and Versailles, with gardens, fountains and expansive lawns for the partying elite to forget themselves, their responsibilities and the troubles of their declining empire. Extravagant banquets with wine, music, poetry readings, dancers, singers and the wives of high officials draped in fashionable French gowns were nightly events in the sprawling gardens where the cream of the dominant order frolicked among the fountains and the tulips.[28]

Particularly the tulips. Specially cultivated tulips were grown in abundance, gardens and gardens of them. A sudden craze for tulips had seized the court and spread to the whole of those privileged strata of Istanbul society that had the means to purchase and cultivate the bulbs. Cultivation of tulips became an art form. No expense was denied on them, nor on the palaces and gardens they adorned. Exotic strains were prized and eagerly sought. So serious a business did tulip cultivation become that appointment to high office was on occasion awarded to a horticulturist who produced a strain of exceptional beauty. Experimental secrets were guarded as jealously as wives and daughters. By 1726, 839 different kinds of tulips were being cultivated, each with a fanciful name, such as "Beloved," "Roman," "Crown of the Kaiser." Some varieties sold for a thousand gold ducats. Damad Ibrahim eventually had to set price controls.[29] European ambassadors and foreigners residing in Istanbul also caught the fever, some of whom sent prize bulbs to Europe as examples of Ottoman horticultural prowess.

Once the window to the West had been opened, though ever so slightly, all sorts of novelties flew in. Yirmisekiz and his son Sa'id had returned from Paris with European furniture and dress that served as models for the wealthy, who began having furniture in the French design made or imported. In the homes of some high officials, French-style sofas and chairs began replacing the low lying divans and cushions of Ottoman custom. Traditional attire was cast aside for men's trousers and women's gowns. Defying Muslim religious tradition, the well-to-do had portraits of themselves painted by European artists who, like the clockmakers, textile merchants and tailors, flocked to Istanbul to cash in, as they would a century later to Cairo when Egypt was reforming. Not only the rich were affected by the Tulip Period's levity of spirit and laxity of morals. Popular innovations of foreign provenance sprouted up in the more modest quarters of the city, such as coffee houses and taverns. Brothels are said to have flourished. Mirroring the new taste in fashions of the Istanbul elite, Parisian society began taking an interest in things Turkish, but this had its origins in the large Ottoman delegation that had attracted so much attention in Paris at the beginning of the 1720s.

The same excess that went into tulips and pleasure palaces was to be seen in the evening extravaganzas in the gardens of those waterside villas. In imitation of the nightly celebrations at Fontainebleau and Versailles, fireworks lit the evening sky, while music and song wafted across the waters of the Bosphorus and Golden Horn. Gardens were illuminated by colored lanterns, and for an added touch of light fantasy, candles were glued to the backs of turtles, which ambled slowly over the grounds, casting their beams across the delicate splendors of the many tulips. One such party in the spring of 1726 was brought to an abrupt conclusion when the ulema, upset by the nightly exhibitions of wasteful, if not sinful, revelry during such depressed times, prematurely announced the sighting of the new moon that began the religious fast.[30]

On the less frivolous side of the Tulip Period's extravagance, Damad Ibrahim had bridges, public gardens and aqueducts built; schools and mosques were repaired. But he never sent a second embassy to Paris to advance on the first. Nor were scholars sent to Europe to bring back a telescope and other technical equipment or scientific texts. No missions were sent to study industrial and military technology or to explore Europe's universities. Once the press had been accepted and Ibrahim Muteferrika put in charge, Damad Ibrahim failed to provide funds and staff for it to make a difference. A second press never materialized. Government failed to invest the resources that could have transformed printing and translation from an ephemeral innovation to an institution, though there was money enough for tulips, palaces and banquets. In the end, the frivolous defeated the serious. An opportunity had been lost, just as the one that had begun with the energetic start of Koprulu and Mezzomorto at the very beginning of the century.

How different might the world have been had Ottoman leadership more diligently applied itself to reform at the time, two centuries before the empire crumbled. The industrial revolution was still half a century in the future; European technology was growing in complexity but nothing more complicated than the refracting telescope. European science was still small-scale and inexpensive, individual men researching by themselves or in private associations, performing table-top experiments at home or a university laboratory. Science and technology were at the time easily within the grasp of the Ottomans, and once assimilated and accredited as Ottoman, and thereafter Muslim, acceptance would have diffused throughout the world of Islam, including the Arab provinces. Ottoman Istanbul, Islam's window on the West, could have served Islamdom in the manner that Peter the Great's St. Petersburg was at the time serving the Russian empire. The chief *mufti*'s acceptance of the printing of scientific and technical books would have been the thin edge of the wedge opening the door to special technical and medical schools, whose graduates, men striving for change, would have yearly added more reformers to the few of the Tulip Period. A century after the end of the Tulip Period, the Ottomans would in fact institute reforms in more or less this fashion, but by the 1830s Europe had transformed itself many times over, making the gap all the more difficult to bridge. Had the grim seriousness that would characterize reformers of the next century been shared by those of the early 18th century, the condition of the Muslim world would most certainly have been very different from what it is today. As it was, things had to get much worse before reform was undertaken more seriously. In the meantime, the West kept progressing, and the gap in scientific knowledge and technical expertise kept widening, jumping ahead even more rapidly with each generation from the late 18th century to the present.

In retrospect, the Tulip Period might be seen as the last best chance for the Ottomans to have wrenched destiny from its tracks and to have avoided the horror and humiliation that has been most of the Muslim world's condition for over a century. Against this, one might argue that the Ottomans of the time could have had no idea of the importance of science. Many educated Europeans themselves at the time had no idea of its social importance, nor of the fact that they had passed through a significant scientific revolution. Since technology and science had not yet come together as a defined couple in symbiotic relationship to radically transform the world, how could the Ottoman leadership have understood the power of science? In answer to this, Ibrahim Muteferrika's choice of translations and publications clearly shows an Ottoman awareness of science's importance. In a book he wrote to the sultan who came to power following Ahmad III's murder, which ended the Tulip Period, he urged many reforms, among the most urgent of them being science and science education.

The Tulip Period's end came quickly. In 1730, the Ottoman army was defeated on the eastern border by a Turkish tribal confederation, an even more humiliating defeat than the failure at Vienna. The Turkish chieftain Nadir Shah had succeeded to power in Iran after the Safavids suffered an extended period of dynastic weakness and, following the millennial pattern of rise and fall of Muslim dynasties, succumbed to a tribal assault of primitive Afghan warriors in 1722. It was an inglorious end to a dynasty that had been in full glory less than a century earlier. When the Afghan tribesmen proved incapable of building on Safavid precedents to create a strong central government, Nadir took advantage of the political instability and, following another centuries-old pattern, led his tribes from the eastern steppes into Iran, which he ruled as Nadir Shah. Following yet another old pattern, the new ruler and his Turkish tribesmen were soon Persianized, as had been the Turkish tribes of the Safavids, and before them the Il-Khan Mongols, the Seljuq Turks, and earlier yet, the Arabs who had settled in Iran during Umayyad and Abbasid times.

The Ottomans had been observing events in the east. Damad Ibrahim saw an opportunity to alleviate the empire's chronic lack of revenues by plundering Iran in its moment of weakness. He discarded the peace policy that the Ottomans had been following for 30 years and prepared an army of invasion. But preparations were not fast enough. By the time the army crossed into Iran, Nadir Shah had stabilized the country and was prepared. Instead of an easy victory, the Ottoman army suffered a crushing defeat – just what the opponents of Damad Ibrahim's reforms had been waiting for. The military defeat catalyzed an amalgam of many disparate elements of discontent whose widely divergent interests were for the moment submerged in a call for blood. The opposition struck as one against a government blamed for life's many travails – high taxation, inflation, increasing unemployment, poverty and corruption – all simply summed up in the rousing cry of *din ve devlet*, religion and state, meaning the sultan's ministers had betrayed both.[31]

Adding to the explosive force of the social time bomb ticking away in the streets of Istanbul were the large numbers of peasants who, crushed by taxes, had given up trying to scratch an existence out of the soil and had come streaming out of their villages into the capital, where their grievances became one with those of the unemployed urban masses. Poisoning the atmosphere beyond the usual norm of envy and sense of social injustice were the pleasure palaces with their evening parties of music and feasting, their ostentatious show of wealth and foreign ways, the fuss over tulips that looked as sinful as they were incomprehensible to the hungry and homeless. The Janissaries were seething for having been set upon by Damad Ibrahim's military reforms that implied they were in dire need of discipline and training, as if they were in some way below European standards. Equally unhappy were the lower ulema who resented the un-Islamic luxury and behavior of the rich and the upper crust of the ulema who were in league with the reformers and who saw a satanic shadow in the printing press that put out books in God's script. A faction in the sultan's court, a welter of conspiracy and intrigue at the best of times, was only too happy to put the blade of defeat to the necks of the leaders. As long as the empire remained at peace and suffered no new catastrophes, Damad Ibrahim had been able by craft and cunning to play one party off against the other and quietly carry out his reforms. Now he and his reformers were branded as traitors and the perpetrators of defeat. Bonded by their common fear that innovative change would mean loss of status and position, an array of cliques came together: *Devshirme* and Janissary officers; chief *mufti* and high ulema; a Turkish nobility jealous of the *Devshirme* "outsiders" enjoying high office; and a harem riven with rivalry among the sultan's mother, wives and favorite concubines who each acted through the princes and viziers to have their favorite ascend the throne.

As soon as news of the defeat in Iran reached Istanbul, the streets erupted in violence. Rioting led to random destruction and looting, then to conflagration and an orgy of murder. The poor sections went up first, followed by the well-to-do. The burning, looting and bloodletting mounted daily. The police and troops, having at first stood by, joined the raging mob upon seeing the paralysis of government. When the directionless rage of populist vengeance reached a certain pitch, a popular front led by lower and mid-level ulema and Janissary officers coalesced in the name of Islam, justice and political legitimacy. The cry went up for Damad Ibrahim's head. The nominal leader of the uprising was a Janissary officer from Albania, Patrona Khalil, whose troops had taken early to the streets to join the waves of unemployed and impoverished marching alongside religious students and their madrasa teachers, whose robes and turbans gave the uprising a religious hue.

Patrona Khalil and his troops provided some direction and organization to the mob, but the orgy of pillaging, looting and torching raged unabated for days. Shops were stormed, followed by homes of the well-to-do, then the villas and pleasure palaces along the Bosphorus. Prized tulip beds were uprooted. Fleeing from their burning homes, the owners were slaughtered in the streets, their women and daughters raped, butchered and thrown in the Bosphorus by mobs.

Sultan Ahmad III, a somewhat passive but accepting figure in regard to the changes that his grand vizier was trying to make, was at first paralyzed by fear and indecision. When at last realizing that the violence was beyond control and that to save himself his son-in-law had to be sacrificed, he ordered Damad Ibrahim's decapitation. The head was delivered to Patrona Khalil and fixed on a pike and paraded through the streets in triumph, the torso thrown to the mob and torn to pieces. One head failed to quench the blood lust of the rioters. Ahmad III's came off next.

Patrona Khalil was now the military master of the empire. A new sultan was chosen from among the many princes. Raised to the throne, Mahmud I compliantly appointed Patrona Khalil as grand vizier and chief of the military and then immediately began plotting his execution. Allowing a few months for the new grand vizier to feel secure in his authority, the sultan duped him into coming to a meeting in the palace without his armed guard. His corpse departed in twain.

With Patrona Khalil gone, the upheaval was quelled and a semblance of domestic peace gradually returned to the city, part of the price of which was discrediting the reformers, including Yirmisekiz and his son. Ibrahim Muteferrika survived owing partly to the respect he enjoyed as a scholar who wrote on the virtues of Islam. Also, he was not too closely associated with the court reformers. His press was far from the palace and had printed nothing that could be considered objectionable by the ulema during the few years it had been in operation. This helped him overcome the lower ulema's opposition to keeping the press going. It was in fact during the decade following the violent end of the Tulip Period that the press reached its greatest productivity and Ibrahim accomplished his greatest and single most enduring literary legacy in support of reform: publication of a book outlining the path to renewal that the empire must follow, *Usul al- Hikam fi Nizam al- Umam* (*Rational Principles in the Political Organization of Nations*).[32]

Depressed by the bloody events of 1730 and what they portended for the future of the empire, Ibrahim composed and published the tract in 1731. Dedicated to Sultan Mahmud I (1730–1754), in whose hands the destiny of the weakening empire had fallen, Ibrahim urged the new sultan to continue with the reforms of the previous government. In the tradition of Kochi Bey's treatise composed for Murad IV a century earlier, Ibrahim analyzed Ottoman decline in terms of the empire's institutional corruption. Both Kochi Bey and Ibrahim urged reform of indigenous institutions, but Ibrahim went a step further by holding Europe up as a model for Ottoman reform. European state organization and institutions should be studied, and if deemed advantageous to the Ottoman state, adopted. It had taken a century of reformist reflection to break through the wall.

Ibrahim's treatise was explicit regarding what had to be done to arrest decline and transform weakness to strength. His program was a mix of new and old: adopt suitable European methods and institutions that drew on the practical reforms of Husayn Koprulu, Mezzomorto and the Tulip Period government and the traditional reform of refurbishing existing institutions. The latter provided a moral guide and framework for reform going back to the treatise of Kochi Bey, which in turn drew from the historian ibn Khaldun's political philosophy, and before that from Seljuq vizier Nizam al-Mulk's *Book of Politics*, the Siyasetname Rectification of the old was to proceed alongside innovation. Considering the murderous events of the year before, Sultan Mahmud must have read Ibrahim's list of what must be done with a chill of misgiving:

- Laws that were no longer compatible to the times had to be reformed.
- Government had to be reorganized and a system instituted to prevent incompetent people from being appointed to office.
- The government must enforce social justice.
- New ideas and methods should not be rejected.
- Ignorance of modern technology weakened the empire and had to be changed.
- Military discipline had to be restored.
- Ignorance of the outside world had to be turned around.
- Bribery, waste of state wealth through corruption, mismanagement and incompetence had to be reformed out of existence.
- Government must institute new education.

For the first time in Islamic literature since al-Farabi's classic Platonic-inspired discourse (The Virtuous City) in the 10th century on the forms of government, democracy was discussed in a serious way as one of the three basic modes of political organization. Monarchy and oligarchy were the other two. Some Western states, Ibrahim writes, have elements of secular democracy wherein sovereignty resides in the people. In spirit and institutional organization, these are more rationalist than religious. Of the three systems, he writes most favorably on democratic popular sovereignty, with its assemblies, legislature, advisory councils and elected representatives. Democracy was compatible with religion as long as the ulema were the popular representatives and the Shari'a the eternal law and constitution of state.[33]

Ibrahim sees strength in the democratic form of political organization. Governments of such states oversee the well-being of their subjects and open the way to new organizational forms. Some European states have revolutionized themselves by their support of science. The place they have given science education has enabled them to innovate and reorganize their military and battle capability. The most important science in the realization of political power, from which comes military power, is geography and everything related to geography, such as determining locations and putting an advantageous location to good use. America was discovered not simply because of its geographical position relative to the European states bordering the Atlantic, but because of the investment the governments of those states put into science. The opportunities of having a good geographical location are lost without also having the science to exploit them. Western power comes not from civil and military organization alone, but from knowledge of geography and science.[34] This knowledge was the basis of the global explorations that gave the West new possessions, new wealth and the command of the seas, allowing it to take over world commerce through its control of the south seas and its archipelagos and the coastal cities in India and China, all of which effectively put the Muslim world under siege. The West advances while Muslim countries, knowing nothing of the causes behind the reversal in power, stagnate. How to reverse the reversal?

According to Ibrahim, a vital part of the answer would come with the discovery of the sources of Western power. One of the basic sources is expressed in the title he gave his book: rational laws and principles of government. Rationalization of law, government, economy, education and an integrated world outlook form the foundation of Western organizational ability. It is from this same rationalization of life that comes the West's inventive skill and military power. Ibrahim is careful to add that this does not mean political principles and laws derived by reason are superior to those derived from religious sources. Superiority comes from the respect and obedience given to the laws derived from those principles, whether religious or rational. What good are religious laws, divinely perfect though they may be, if they are not followed? His arguing for the rationalization of law (because the religious ones were not being followed) is in effect an appeal to abandon religious law where it did not coincide with reason.

As for the military power of the West, this comes from a confluence of many sources, among them good government, respect for law, social justice, education, mechanical inventiveness and creative intelligence. Military ability is an end, not a beginning of a strong state. Peter the Great's reforms in Russia are offered as an example of rational reform turning a weak country into a strong one. Peter learned about the West and made his new laws and built his new institutions accordingly. The Ottomans can do it too. To accomplish it, Ibrahim calls for a New Order: a Nizam -i Jadid, the very term that would be used at the end of the century to define a new program of reform. Ibrahim anticipates those reforms in both name and substance. Government should be rationalized, the budget balanced, education promoted, the sources of Western power learned and adaptations of them made accordingly. It would not be as easy for Muslims as it was for Peter to use the Christian West as a model, but not doing so would imperil the empire's existence. The West is growing stronger and preparing to attack any moment:

> Let the Muslims cease to be unaware and ignorant of the state of affairs and awaken from the slumber of heedlessness. From now on let them be informed of the condition of their enemies. Let them act with foresight and become intimately acquainted with new European methods, organization, strategy, tactics, and warfare. This will be possible only by ending the state of slumber and indifference, dropping sheer fanaticism with regard to learning of European conditions, stopping tolerance of laziness, indifference and carelessness toward corruption and difficulties in the state and commonwealth.[35]

The program Ibrahim mapped out was to be the road reformers would take for the duration of the empire's existence and beyond. When the empire had expired and been interred, Mustafa Kemal would take Ibrahim's road map to its limit by secularizing the state founded on the Western republican model.

A political scientist has called Ibrahim Muteferrika the John Locke of the Ottomans.[36] Ibrahim and Locke were near contemporaries who gave penetrating thought to civil society and political rule following grave episodes of regicide, and were, each in his own historical context, spokesmen for liberal government. The difference was that Locke was legitimizing in principles a present political condition brought about by an historical process, while Ibrahim was proposing the initiation of the historical process to reach that condition. In fact, the process had already begun with the Tulip Period and been aborted, and for that reason his program was all the more daring – and its chances of being resumed not bright.

The threat to reform came equally from above and below, the proverbial rock and hard place. While the collapse of the reforms of Husayn Pasha Koprulu and Mezzomorto had shown that palace intrigue could be a lethal enemy of change, the storm that swept away the reforms of Damad Ibrahim had shown the street from below to be even more lethal, when Janissary officers gathered into their hands the leadership of furious mobs, madrasa students and their religious faculty and marched them to the palace. The ulema attached to the mosques in the popular quarters of Istanbul were the representatives of the people who had no illusions concerning the government's concern or capacity to alleviate their growing hardship. More ominous for the cause of any reformist party with innovationist inclinations was the willingness of the traditionalist elite to join forces with the impoverished masses and their religious leaders from the lower ranks of the ulema, reinforced by the muskets of the Janissary infantry and the leadership of its officers.

After the Tulip Period, Westernized reform was vilified by its elitist and populist opponents as being anti-religious, socially subversive and serving the interests of the West, whose hired lackeys were traitors posing as reformers. Reformers came to realize that before any sultan could again tread the road of reform, the social problem would have to be ameliorated, the ulema leadership coopted, the Janissaries reined in and courtly factions brought under control. A sultan would have to take the sword in hand.

Force-draft reform by political order from the top was to be the only source capable of opening Muslim society to the new ways and knowledge of the civilization's millennial rival. Given the conditions of a shrinking economy and trade dominated by Europe, if changes were to be made, they would have to be imposed from above, by state authority, for there was no chance of a rising stratum of society, buoyed by trade, banking and industry, affecting change from below. There was no system of private secular education from which it could come. Nor could it be expected to come from the traditional institution of education at the center of society, the madrasa: the ulema had shown collectively through the centuries that they were not about to catch fire for science and philosophy the way medieval churchmen did in the Latin West. Al-Ghazali, who warned against the pitfalls of logic and science through uncritical belief in reason, was as close to an Aquinas or an Albertus Magnus that Muslim theologians would come. Yet even if the opposition to reform were to be smashed and the most drastic reforms put in place, the Western nations were so powerful, their appetites so whetted and expansionist instincts so focused on inheriting what they regarded rightfully theirs, that no reforms might have been enough to save the empire. The European powers were sure to do what they could to undermine the very reforms some of them encouraged, lest the Ottomans become strong enough to defend their empire. In fact, the European factor can be seen already at work during the height of the Tulip Period.

A French Huguenot named Rochefort, who had fled France for Istanbul for the same reasons of religious intolerance that Ibrahim Muteferrika fled Transylvania, desired to benefit both his fellow Huguenots and the country that had given him refuge. Rochefort proposed to Damad Ibrahim that skilled Huguenots seeking religious security be settled in the Ottoman provinces of Moldavia and Wallachia where they would set up textile and other factories so that the empire could be economically self-sufficient and not have to import textiles and other manufactured goods from Europe. Rochefort's plan included building a military school in which Huguenot officers would train Ottomans in the science of modern warfare, which included an engineering school to educate and train young Ottomans in the sciences and mechanical crafts of manufacturing. The source of modern industrial and military technique would be transported right to the heart of the empire. Damad Ibrahim had responded positively and discussed Rochefort's plan with Ahmad III. The idea of having skilled technicians and engineers building up industry, and French officers serving in and training the Ottoman army, appealed to the sultan. But the plan was nipped in the bud.

As soon as the French ambassador in Istanbul, Bonnac, learned of what was being planned, he scotched the whole affair by warning the conservative factions in the sultan's court of Rochefort's ambitions and portrayed his grand project as a European conspiracy. Rochefort, Bonnac told them, was pressing the sultan to open up the empire to European Christians, infidels. Ambassador Bonnac's intrigue was successful. The conservatives at court put pressure on the sultan to forbid Christian Huguenots entering the country. The sultan ordered Damad Ibrahim to drop the affair and that ended it. Complicating all the more an already complicated grid of opposing forces, came now the spoiling current of European ambassadors who were keenly insinuating themselves in the thick of Ottoman courtly intrigues and rivalries. As Ottoman reformers searched westward for models, European meddling muddied the political waters. The mercantilist French government, any European government, had no interest in seeing the Ottomans develop industries that competed with their own, particularly ones that would diminish their profits in the lucrative textile trade with the Ottomans.[37] European economic self-interest would continue to be a perpetual factor of increasing importance in the complex algebra of Ottoman reform for the duration of the empire's existence.

The fate of the Tulip Period brought into relief the disparate forces aligned against reform. From within and without, reformers were trapped in a web of contradictory forces: the palace factions from above, the street from below, with the ulema and Janissaries reacting from both above and below. Between society's and government's own internal contradictions, and the fist of European-controlled global trade closing around the central lands of Islam, with Western economic self-interest now playing into the courtly oppositions that weakened Ottoman government, reform would continue to be a treacherous enterprise at the best of times. Until the next government to undertake a serious program of reorganization and innovation came to power in 1789, scientific and technical change desultorily limped on, devoid of overall direction and noticeable effect.

Although Ibrahim's press came to a halt in the mid-1740s, the spark of intellectual interest it kindled in the classics of Islam's brilliant past, and in analyzing the problem of cultural decline, continued to glow. Following up on Ibrahim's annotated printed edition of Katib Chelebi's geographical works, and his own work analyzing decline in his *Rational Principles of Political Organization*, Shaykh al-Islam Pirizade (d. 1748) translated into Turkish ibn Khaldun's neglected historical masterpiece, the *Muqaddima*, or *Prologue to History*, in which economic justice is the glue of social solidarity: the fundament of a state's strength.

Notes

1 Ahmet Evin, "The Tulip Age and Definitions of Westernization," in *Social and Economic History of Turkey, 1071–1920*, edited by H. Inakcik and O. Okyar, Meteksan, Ankara, 1980, pp. 134–136. For a critically acute student paper that answers Evin's arguments on the innovations introduced by the Tulip Period, and brings into question that there was in fact a "Tulop Period" that could be marked off as something uniquely different from the decades preceding 1720, see Rachel R. Fry, "Gardens in the Air: A Reconstruction of the Ottoman Tulip Period," *Student Publication 103*, Fall 2013, Gettysburg College.

2 The task of diplomatic translation and communication with Europe was left to the empire's Christians and Jews. It was they who knew Europe's languages and handled diplomatic correspondence, in particular the Greeks of the Fanar district of Istanbul, the Phanariates as they were called.

3 Yirmisekiz's report has been translated to French by Julien-Claude Galland as *Le Paradis des Infidels*, Francois Masperio, Paris, 1981. For a general account of Yirmisekiz's mission and the Turkish influence it had on the French see Fatma Muga Gocek, *East Encounters West: France and the Ottoman Empire in the 18th century*, Oxford University Press, Oxford, 1987, pp. 72–75.

4 *Paradis des Infideles*, Paris, 1981, p. 236.

5 Yirmisekiz mistakenly writes that the king had the observatory built for his famous astronomer, Cassini. This was Jean-Dominque Cassini (1625–1712). Originally an Italian, he became a naturalized Frenchman when Louis XIV appointed him director of the new observatory after it had been completed. It was as director of the observatory that Cassini discovered four of Jupiter's satellites, the cause of his fame. When he died, his son Jacques (1677–1756) inherited his position and it was this Cassini who was director at the time of Yirmisekiz's visit.

6 The observatory in fact possessed two telescopes, both made in Rome, one of which was 34 feet long, the other 100 feet. *Paradis des Infideles*, p. 150.

7 Trying hard as he could, he did not see Saturn or more than two of the five satellites he was told were there, the first discovered by Huyghens in 1671, the last four by the elder Cassini between 1665 and 1684. *Paradis des Infideles*, p. 151. The French astronomers leading the tour, Jacques Cassini no doubt among them, were obviously enjoying the show they were putting on for their foreign guests. Similar humor in a similar show would be expressed by French scientists in Egypt at the end of the century. The French

298   *Muslim empires to Western ascendance*

were as proud of their scientific accomplishment as they were patronizing to Easterners.

8 Dan Diner, *Lost in the Sacred*, p. 83.
9 Gutas, *Greek Knowledge, Arabic Culture*, p. 175. Arabic was still considered the language of science.
10 A Hungarian nobleman residing in Istanbul, Rakoczi Ferencz, to whom Ibrahim had been assigned as a translator early in his Ottoman career, claimed to have received the story from Ibrahim himself through conversations they had over the years. Niyazi Berkes, *The Development of Secularization in Turkey*, McGill University Press, Montreal, 1994, p. 71.
11 N. Berkes, "Ibrahim Muteferrika," in *Encyclopedia of Islam*, vol. 3, pp. 996–998. And by the same author, "Ilk Turk Matbaasinin Kurulusu ve Ibrahim Muteferrika, Hajjetepe Universitesi Edebiyat Facultesi Dergisi, vol. 10, part 1, 1993. Also in T. T. Kurucusu, *Belleten*, vol. 26, 1962, pp. 715–737.
12 For the early history of the printing press in the Ottoman Empire, Fatma Muga Gocek, *France and the Ottoman Empire in the 18th century*, Oxford University Press, Oxford, 1987, pp. 108–115.
13 Lewis, *Muslim Discovery of Europe*, pp. 232–233, citing Busbecq's Letters, pp. 213–214.
14 Muhammad Abdus Salam, *Scientific Renaissance in Islam, Singapore*, 1994, p. 18.
15 Lewis, *Muslim Discovery of Europe*, p. 234, note 16.
16 Gocek, *France and the Ottoman Empire in the 18th century*, pp. 106–107.
17 Quoted in Nina Burleight, *Mirage: Napoleon's Scientists and the Unveiling of Egypt*, Harper Collins, New York, 2007, p. 16.
18 Muhammad Abdus Salam, *Renaissance of Science in Islamic Countries*, World Scientific, Singapore, 1994, p. 21.
19 Writing on this, Bernard Lewis notes that like the Ottomans of the time there is no lack of geographically challenged leaders in the West, but Western nations have a standing and professionally trained diplomatic corps to check the ignorance of political leadership. *Muslim Discovery of the West*, p. 154.
20 Salam, *Renaissance of Science in Islamic Countries*, edited by H.R. Dalafi and M.H. Hassan, World Scientific, Singapore, 1994. p. 21.
21 Ekmeleddin Ihsanoglu, "Introduction of Western Science to the Ottoman World: A Case Study of Modern Astronomy (1660–1860)," in *Transfer of Science and Technology to the Muslim World*, edited by E. Ihsanoglu, A.H. de Groot, Istanbul, 1992, pp. 80–85.
22 Ibrahim's edition was later translated to English and published in London (1831) as *History of the Maritime Wars of the Turks*, Publishers for Oriental Translation Fund, London, 1831.
23 A. Adivar, *Osmanli Turk Ilim*, Remzi Kitabevi, Ankara, 1982, p. 148; N. Berkes, *Development of Secularization in Turkey*, McGill University Press, Montreal, 1994, p. 46.
24 Ibrahim's translation of it appeared under the title (in modern Turkish transliteration of Ottoman Turkish) *Mecmuat -i Hey'et-i Kadime ve Cedide (Compendium of Old and New Astronomy)*. Berkes, *Development of Secularization in Turkey*, p. 46; A. Adivar, *Osmanli Turk Ilim*, Remzi Kitabevi, Ankara, 1982, p. 151.
25 Ihsanoglu, "Introduction of Western Science to the Ottoman World," pp. 84–86.
26 Ihsanoglu, "Introduction of Western Science to the Ottoman World," pp. 81–83.
27 Berkes, *Development of Secularization in Turkey*, p. 27.
28 M. F. Gocek, "East Encounters West: France and the Ottoman Empire in the 18th century," in *East Encounters West*, Oxford University Press, Oxford, 1987, pp. 72–75.
29 A. Evren, "The Tulip Period," p. 140.
30 A. Evren, "The Tulip Period," p. 140.
31 Ahmet Evren interprets the beginning of the revolt as occurring before the defeat: "The Tulip Period," p. 141.

32 Translated to French by Baron Rewicski as *Traite de la Tactique*, Traitnern, Vienne, France, 1769

33 Berkes, *Development of Secularization in Turkey*, p. 42.

34 Berkes, "Matbaasinin Kurucusu," pp. 717–718.

35 Berkes, *Development of Secularization in Turkey*, p. 44.

36 Berkes, "Matbaasinin Kurucusu," p. 719.

37 von Hammer, *Geschichte der Osmanische Reiche*, Hartleben, Pest, Germany, 1827– 1833; vol. XV, p. 351; F.M. Gocek, *France and the Ottoman Empire in the 18th century*, pp. 106–107.

12 Toward a new order

A perverse logic dogged the cause of Ottoman reform. When the empire was at peace, the urgency to reform abated with fading memory of the last disaster. The reformist temper rose only with defeat in another war. But reform could only be applied during times of peace. Then when war came and the empire suffered another defeat, the conservatives used this to claim that the reforms were not only useless but induced weakness in that they undermined the empire's venerable institutions. Doing away with old Ottoman institutions for alien new ones guaranteed defeat, they argued. Between the anvil of conservative resistance and the hammer of Austro–Russian expansion, reformist leaders were obliged to navigate a tricky course: keep the borders at peace while facing the wrath of conservatives, whose careers and interests were tied to those old and cherished institutions that the reformers wanted to diminish or replace. The conservative argument was that the venerable institutions were alive and strong; they only needed refurbishing. The argument held some validity. There was yet some vigor left in the old institutions, as it seemed.

In wars with Russia and Austria between 1736 and 1739, the Ottomans exchanged losses and gains, and at the end could feel they had withstood two powerful Western foes while holding the eastern border against Iran. After a shaky beginning, the Ottomans defeated Austria at Banja Luka and, with the Treaty of Belgrade ending the war in 1739, regained the territory that had been lost at Passarowitz in 1718. Azov in the Crimea was regained from the Russians, and Ottoman rule was reestablished in Bosnia and Serbia. Russia and Austria were thereafter, for a time, diverted from their ambitions in the Balkans and the Crimea by wars in Europe. The War of Austrian Succession (1730–1748) and the Seven Years War (1756–1763) gave the Ottomans the longest period of peace in their warlike history. The gains of 1739 and the long respite that followed seemed to vindicate conservative philosophy. Reform was put on ice. No major initiatives were undertaken for another half century.

While the West continued to be transformed through technology, industry and expanding capitalist production, the Ottoman state was being wracked by inflation, unemployment, plague, food shortages, marauding gangs ravaging the countryside and powerful local lords seizing control of large areas in the provinces as their private tax-exempt property. The need for major reform was apparent to everyone in a position elevated enough to see over the wall of self-delusion and denial. Ottoman leadership, conservative and otherwise, generally agreed on the need to reform. The question was how and what to reform. The contrary positions expressed themselves in reports written by Ottoman ambassadors who had led missions to Europe. Yirmisekiz's report was so positive about France that the French ambassador wondered if it would not cause Yirmisekiz some trouble. A generation later, in 1748, an Ottoman ambassador sent to Vienna describes the instruments of science and technology he saw in the royal observatory and other institutions as no more than toys and playthings appropriate for frivolous men with infantile minds. It did not matter what the Austrian scientists showed the ambassador. His mind had been made up long before he left Istanbul for Vienna. Western infidels had nothing to teach the Ottomans. This double-mindedness in leadership on how to reform and what models to accept either killed or crippled attempts to change.

Following the Tulip Period there were some reforms, but they were haphazard, almost incidental, and absent of any overarching program or unity of vision. In 1732, a French nobleman and military officer crossed the Danube in search of Ottoman employment. Claude-Alexandre Comte de Bonneval (d. 1747) was as remarkable in his way as a noble military officer in Ottoman service as Ibrahim Muteferrika was as an Ottoman scholar-reformer. Bonneval had fought for Louis XIV with distinction in the War of Spanish Succession (1703–1713), then had a nasty disagreement with the Sun King, left Versailles in a rage and found service with Eugene of Savoy, with whom he also had a falling out. He then served the Venetian Republic. What trouble he had there is unclear, but he was forced to leave. He then traveled to Ottoman territory where in Sarajevo he presented himself to the governor. A European military officer willing to cross over and convert always had a career with the Ottomans, whatever the mischief he was leaving behind him. Impressed by the Frenchman's aristocratic bearing and military credentials, the pasha of Bosnia sent him on to Istanbul. Sultan Mahmud I was also impressed by Bonneval. Seeing he had the possibility of a fruitful career, Bonneval learned Turkish, converted, taking the name Ahmed, upon which the sultan gave him the high rank of pasha. In effect, Ahmed Pasha Bonneval was a voluntary, middle-aged, side-door *Devshirme* recruit starting at the top.

Commissioned to revitalize the old artillery corps, the former count became known as *Humbaraji* (Bombadier) Ahmad Pasha. The artillery corps had been an object of reform ever since the days of grand vizier Husayn Koprulu. Following Koprulu, Damad Ibrahim had introduced European training methods and some Western weapons to the corps. Under Bonneval's supervision, a cannon foundry and a factory for producing gunpowder and modern muskets were built. To assist him, three young French officers, and several Irish and Scottish ones, were placed under his command. They too had fled their countries of origin for one reason or another to accept Islam and start a new life in Ottoman service, and it was these few desultory officers entering the Ottoman military that helped to keep the germ of reform alive. Ibrahim Muteferrika printed a new, European-oriented military drill and instruction manual for the artillery officers.

The mix of Europeans was not altogether a happy experience. The Irish and Scottish officers were disgruntled at having a French aristocrat suddenly thrust upon them as their pasha. But being proud military officers trained in the strict methods of the professional armies built and cherished by the British and continental absolutist monarchs of the period, they managed to put aside their jealousies to make a concentrated effort at reforming the Ottoman military in the spit and polish image of the disciplined professionalism of their own officer corps training. Besides, if they wanted to hold on to their well-paying positions they had to make the effort.

The Europeans taught courses on military science, designed smart, close-fitting European-style uniforms for the Ottoman soldiers and officers, introduced them to modern European weapons, drilled their troops rigorously and, mother of innovations, had them snap to attention and salute their superior officers.

Bonneval devoted himself to Ottoman service as fervently as any Balkan Christian boy recruited into the *Devshirme*. About a decade after Rochefort had submitted his proposal to Damad Ibrahim to modernize the empire using Huguenots, Bonneval submitted a memorandum to Mahmud I outlining an ambitious plan to modernize the whole military establishment from top to bottom, with himself, Ahmad Humbaraji Pasha, as commander-in-chief. Based on French and Austrian models, Bonneval's plans called for a new method of recruitment, strictly enforced discipline, more efficient lines of communication, a training school to produce responsible officers educated in tactics, military science and engineering, a medical corps to ensure the health of the troops, the prompt disbursement of pay, and a system of pensions to build morale and enhance recruitment. His list of reforms and innovations – in some matters reflecting the reforms Ibrahim Muteferrika listed in his book to the new sultan – offers an idea of the condition of the Ottoman military that less than half a century earlier had been on the verge of taking Vienna.

It was one thing to draw up a rational plan of military reform and quite something else to implement it. Bonneval's plan remained as such, since the Ottomans who supported Bonneval were vigorously opposed by the conservative factions. The most vigorous challenge to Bonneval's plans came from the Janissaries. To overcome Janissary recalcitrance, Bonneval proposed having them divided up into small units and dispersed throughout the empire. When the Janissary officers learned of what Bonneval was up to, they successfully intrigued against him through their contacts in the imperial court. Ever mindful of what had happened to Ahmad III, Mahmud I had no desire to stand up to Janissary discontent. A change of grand viziers was the sultan's way of showing a change in policy. The new grand vizier, an Italian convert to Islam who disliked the French aristocrat, cut funding for Bonneval's artillery corps and factories, bringing the reforms to an abrupt end.

Two years later, 1734, in a moment of conservative distraction, yet a new grand vizier authorized Bonneval to build a military engineering school for training artillery corps officers. Reform was on again. The military engineering school Bonneval designed was a far more ambitious project than the officer training school of Damad Ibrahim's reforms. Called the *Hendesehane*, (the word would come to designate a school of engineering in Turkish and Arabic), it was built in Uskudar, just outside of Istanbul at the time. The curriculum included geometry, trigonometry and the study of ballistics relating target distances to muzzle velocities and firing elevations, which would have introduced student officers to the basics of Galileo's and Newton's physics.

The problem was where to find willing students for the new school. Janissary officers were the most likely candidates, but willingness was not a quality that accurately described their military dedication to service. Humiliated by the implied inferiority of their having to be students sitting at the feet of Westerners, even though the Europeans had converted to Islam, the Janissary officers, many of whom were themselves European in origin, proved to be impossible students. If Western sources are to be believed, they resisted mathematics as if it were poison, but this could have been a problem of their age as much as their recalcitrance. An alternate source for students was the corps called Bustanji, or palace guards, who, it was hoped, would be more amenable to learning. These showed promise in the beginning. One student, the son of a religious judge, was said to have improved the design of a quadrant for artillery gunners during the course of his study in the new school. To supply the students with textual material, Bonneval and his European officers composed a treatise on trigonometry and translated it to Turkish. Austrian Count Montecuccoli's (d. 1680) famous book on military science, *Memorie della Guerra* (Venice, 1703), was translated from its original Italian into Turkish at this time.[1]

As with earlier attempts to reform, the school was cut short before it had a chance to root itself into the military culture as an authentic Ottoman institution of science and technology. Its demise had to do with Janissary discontent. In 1736, in the midst of the war between the Ottomans and the Austro–Hungarian Empire, the Janissaries angrily demonstrated against having Bonneval's trained corps of bombadiers participate in the war, lest its performance embarrass them. With neither grand vizier nor sultan brave enough to face down the Janissary officers, the bombadiers were withdrawn from the field and Bonneval transferred in virtual exile to Kastamonu. With Bonneval gone, the bombadier corps and the military engineering school affiliated with it were left to wither in limbo. In 1750, barely a ghost of itself, the *Hendesehane* was shut down.

Bonneval died in 1747, the same year as Ibrahim Muteferrika. With them went two symbols of unfulfilled reform. The flicker of promise given by the latter's printing press was snuffed out as willfully as the former's short-lived artillery corps and military school. The military, the only institution Ottoman reformers could imagine to have any use for science and technology, stood in firm opposition to change.

Following the war with Austria and the extended period of peace that ensued, the exigencies of reform subsided "and the Ruling Class settled into its stupor, again assuming that the supremacy of Ottoman ways was keeping the enemy at bay."[2] Corruption, nepotism, bribery, sale of offices and favoritism flooded back, as the ruling class that benefited from these practices continued to resist change in the name of true religion with its ancient traditions and institutions. Yet the urge to reform never disappeared. Sultans and grand viziers kept the embers alive, at the margins so not to attract attention and antagonize the opposition. Sultan Mahmud I brought in Polish experts to build and operate a paper factory to meet a rise in public demand for books, a rise that appears to have been the result of an intellectual awakening nurtured during the Tulip Period and sharpened by Ibrahim's press. The sultan did not send educated young men to Poland to study the technology and learn the industry so they could build and operate the factory themselves and then build more factories as they were needed and pass the knowledge on, improving on it with experience. Instead, the Poles were brought to build the factory, and having built it, they left. Technological transfer was minimal. The reformist party was too intimidated by the conservative opposition to send student missions to the West. Young men going abroad to study on their own was out of the question.

In 1748, very shortly after the artillery engineering school and the printing press had been shut down, Mahmud I was stirred by the same urge to discover the secrets of European strength that had moved Ahmad III and Damad Ibrahim to send Yirmisekiz to Paris. This time the embassy was to Vienna, where the Ottoman delegation saw versions of what Yirmisekiz had seen in Paris a quarter century earlier. The dissimilarity of the reports emanating from these two missions, to which allusion has already been made, reflects starkly the contradictory impulses clashing within even the reform-minded party of the ruling elite. Where Yirmisekiz had openly admired what he saw in France and wrote positively about it, Mustafa Hatti's report conveys the traditional air of contemptuous superiority:

> At the emperor's command we were invited to the Observatory, to see some of the strange devices and wonderful objects kept there. We accepted the invitation a few days later, and went to a seven- or eight-storey building. On the top floor, with a pierced ceiling, we saw the astronomical instruments and the large and small telescopes for the sun, moon, and stars.
>
> One of the contrivances shown to us was as follows. There were two adjoining rooms. In one there was a wheel, and on that wheel were two large, spherical, crystal balls. To these were attached a hollow cylinder, narrower than a reed, from which a long chain ran into the other room. When the wheel was turned, a fiery wind ran along the chain into the other room, where it surged up from the ground and, if any man touched it, that wind struck his finger and jarred his whole body. What is still more wonderful, is that if the man who touched it held another by the hand, and he another, and so formed a ring of twenty or thirty persons, each of them would feel the same shock in finger and body as the first one. We tried this ourselves. Since they did not give any intelligible reply to our questions, and since the whole thing is merely a plaything, we did not think it worthwhile to seek further information about it.
>
> Another contrivance which they showed us consisted of two copper cups, each placed on a chair, about three ells apart. When a fire was lit in one of them, it produced such an effect on the other, despite the distance, that it exploded as if seven or eight muskets had been discharged.

The third contrivance consisted of small glass bottles which we saw them strike against stone and wood without breaking them. Then they put fragments of flint in the bottles, whereupon these finger-thick bottles, which had withstood the impact of stone, dissolved like flour. When we asked the meaning of this, they said that when glass was cooled in cold water straight from the fire, it became like this. We ascribe this preposterous answer to their Frankish trickery.[3]

The embassy of 1748 returned from Vienna with little upon which reform could be based.[4]

While the idea of reform remained alive and occasionally resulted in actions of minimal success, Western scientific knowledge continued finding Ottoman ports of entry, usually, as had been the case since the 1660s, through European ambassadors giving scholarly scientific works as gifts to the sultan, the Dutch being the most generous in this regard. Aware of the keen interest in geography kindled by Ibrahim Muteferrika, a Dutch ambassador gave a copy of the *Geographia Generalis* by the Dutch scholar Bernhard Varenius (d. 1676) to Sultan Mahmud I. Published in Amsterdam in 1650, Varenius' *Geographia* presented a synopsis of geography and astronomy that was current in Europe at the time the book was written and so did not go beyond the science contained in Durret's, Blaeu's and Keller's works that had previously been presented to the sultans and translated. A reformist grand vizier of the Koprulu family had Varenius's book translated into Turkish in 1750, adding one more 17th century classic of European geographical and astronomical knowledge to the Ottoman bookshelf. The translator, Osman ibn Abdulmennan, a palace scribe, apparently of Bosnian origin, had previously translated from Latin a botanical work of unknown name, suggesting there may have been a number of European scientific books making it to Istanbul that were translated but whose names have not been recorded. Varenius's book presented a brief synopsis of Copernican theory but followed Ptolemaic models in the astronomical sections of his geography. In a revealing marginal comment, the Ottoman translator expressed his preference for the Copernican system, claiming that it was more reasonable to assume that the earth went around the sun than the other way around, since when roasting kabab on a skewer it is the skewer, not the fire, that turns.[5] The analogy, with its absence of equivocation, surprise or fear of religious contradiction, might suggest the Copernican idea was not unfamiliar to the translator. Had knowledge of it disseminated through certain circles of the court during the century that had passed since Ibrahim Kose Tezkiraci first praised the heliocentric system in his translation of Durret's tables? A likely source of dissemination would have been Bonneval's military engineering school, with its European teachers whose instruction in Newtonian physics would have included something about modern astronomy. Knowledge of these could have spread to some extent beyond the classrooms and training fields. Almost certainly there must have been unrecorded paths that modern astronomy entered the Ottoman Empire, for how to explain the writings of the scholar–mystic Ibrahim Hakki?

Several years after Osman's translation of Varenius, a provincial religious scholar wrote a book in which he presented the Copernican system as God's system. Ibrahim Hakki's *Marifetname* (*Book of Knowledge*), published in 1757, was the first piece of autonomous Muslim literature to defend Copernicus. Previously, modern science had come into the empire between the covers of European books, presented to the court or translated for the military engineering school. Ibrahim Hakki found it somewhere else. He was not an astronomer or a member of the imperial court, but a mystical shaykh from provincial Erzurum. His casual acceptance of a moving earth is more surprising than how this provincial shaykh came to know of Copernicus. Being of mystical orientation with the mystic's use of allegory and metaphor in interpreting the Quran, he may have been free of the scriptural literalism of Sunni orthodoxy; but how a Sufi in Erzurum learned anything of European science is as much a mystery as is large portions of his book's mix of old and new in its mystically interpreted integration of scriptural creation, astronomy and natural history.

Ibrahim Hakki follows the medieval pattern of dividing creation into three orders: plant, animal and mineral. Quranic verses substantiate the order and purpose of the cosmos as God's gift to humankind's earthly existence. Three branches of science make possible a rational understanding of creation and its purpose: geometry and astronomy relate to the heavenly level; anatomy and psychology to the human, where man anatomically mirrors the universe in microcosm. The third branch is mystical and relates to human consciousness of the cosmic whole, the divine truth that resides in the heart and mirrors the divine in the human mind. The author's descriptions of God's creation and the movement of the heavens alternate between passages of fantastic flights of Sufi imagery and down-to-earth hardboiled science: the moon moves across the sky in a carriage of topaz with 300 handles, an angel at each; but heavenly movement is around the sun.[6]

Saturated with metaphor upon metaphor of mystical obscurity, the *Marifetname* makes for a mind-bending cocktail of science, Sufism and Ismaili falsafa. The astronomy comes in the last of the *Marifetname*'s four volumes. Comparing Ptolemy and Copernicus, Hakki repeats his Ottoman predecessors: it is not the system that is essential to belief but acceptance of divine creation. As God's mind is rational, his creation is rational, and of all possible systems of the heavens, the Copernican is the most rational, with the largest body, the sun, at the center and the planets going around it, for what is more rational in regard to bodies in motion than the smaller orbiting the larger? He prefaces his section on Ptolemaic astronomy with Ghazali's caveat that there is no contradiction between the mathematical sciences and religion and that those who in the name of Islam deny what is obviously true pervert both religion and science.

The heart of Hakki's section on astronomy is the nine "matters" (*madde*) that he lists as underlying the change from the old to the new astronomy, which has "won honor and respect." But apparently not from all. The narrative in the first *madde* makes it obvious that there had been heated controversy, literally, on the subject. Those who have accepted the Copernican system have had their homes and courtyards burned with wax and fire by those who blindly follow the old

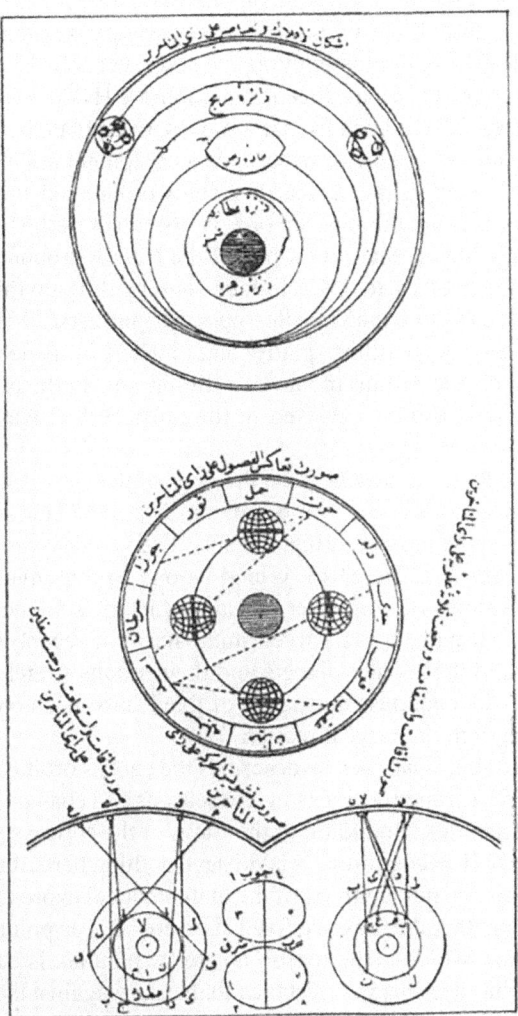

Figure 12.1 Astronomical diagrams of the Ptolemaic and Copernican planetary systems in Ismail Hakki's *Marifetname*. Diagram from page 147 of Ibrahhim Hakki's *Marifetname*: Ottoman manuscript given to Columbia University by the Turkish government and digitalized by Google (https://babel.hathitrust.org/cgi).

The diagram is the first illustration in a Muslim source of the Copernican system. The top illustration of the diagram shows the planetary positions according to the latest astronomers (al-muta'kharin), with sun at the center, followed by Mercury, Venus, Earth, Mars, Jupiter and Saturn. Notice the oblate shape given the earth with the curved line indicating axial rotation. Newton accounted for the oblate shape because of the earth's rotation, which would cause the planet to bulge out slightly along the equator, with respect to the great circles around the north and south poles. Below the diagram is an illustration of the earth's positions as it orbits the sun through the seasons as designated by the zodiac.

astronomy. Hakki refers to those who support and strive to verify the heliocentric system as "pure-hearted, hard-working men at whom stones are thrown to dirty their names and who are treated with malice and revulsion for their believing that the earth moves." And there have been, to paraphrase Hakki's narrative, those who since ancient times believed that the earth moves. Even Plato believed it. Scientists have recently made observations with precise equipment and have experimented to determine the laws of nature, and as their study progressed, the heliocentric view emerged and has become the favored one, well-known as the "New Astronomy." Belief and trust in the planetary motions of the New Astronomy is not related to religion and is not contrary to any religious command. If anything, believing in a false astronomy is no more than heedlessness. The universe, whatever its shape or form, whatever the composition, quality and motions of its heavenly bodies, has nothing to do with the requirements of religion and faith. It is enough just to believe that the universe was created in the most perfect way by the almighty master craftsman.

The second 'matter,' or *madde*, is on the laws of the New Astronomy: Scholars of the New Astronomy say that the bright sun is a fixed star at the center of the universe. Mercury, the closest, orbits the sun. Then comes Venus, completing an orbit every six months. The earth, joined as one to the enveloping concentric spheres of the elements of fire, water and air, orbits the sun once a year in a great circle. Saturn's four and Jupiter's five "small stars" of recent discovery are then mentioned. Beyond these outer planets and their moons stretches the vastness of open space with "its countless thousands of fixed stars, each one a huge sun-like body transmitting light in every direction."

In the third and fourth *maddes* he describes the earth's orbital and axial motions and the underlying astronomical reasons for seasons and changes in daylight hours. The fifth *madde* he titles "Founding of the Laws of the New Astronomy and Their Winning Esteem and Recognition." Hakki says nothing here of the planetary laws of Kepler and Newton, nor anything of the mathematical expressions of their laws, but he does explain that the massive fixed stars are seen as points because of their enormous distance, which is the reason no stellar parallax is detected during the earth's annual orbit; it is this vast distance that argues against the firmament rotating daily around a stationary earth:

> It would require the speed of a bullet for the starry sphere to turn a revolution in 24 hours, or 300,000 times greater and more violent than the speed of the earth turning. . . . Thus it is only correct that any reasonable and intelligent person should accept the earth's motions.

The speeding bullet is Hakki's rational equivalent to Osman Abdulmennon's shish kabab analogy.

Section six of Hakki's final volume is on retrograde motion as viewed by the New Astronomy. The lower planets go faster in their orbits around the sun than do earth and the outer planets. Heliocentricity explains the appearance of retrograde motion and simplifies the movement of planets by doing away with epicycles and planets stopping, going backward and then forward again. The seventh and eighth sections are on questions believers of the old astronomy have regarding the new system. If the earth moves, why can it not be sensed, why is not everything blown away? Ibrahim Hakki adduces the standard analogy of likening the moving earth to a ship whose cargo moves as one with the ship without its motion being sensed. Earth and its cargo move as a ship through space or as fish move with the motion of the sea or birds with the wind. Nothing is blown away or uprooted because of the enveloping ethereal sphere that conceals earth's motion from the senses. This ethereal matter moves with the planet but does not impede its motion of rotation or the motion of falling bodies. The motion of rotation imparts itself to the enveloping atmosphere, so a stone will fall vertically, just as on a moving ship a stone dropped from the top of the mast will fall vertically parallel to the mast. To someone observing the ship from shore, the stone will fall in a curve because of the combined motion of ship and stone:

> People who refute this in the name of religion have misunderstood religion. There are a great number of judgments in religious books which according to our view are not esteemed to be religious commands, as in the Torah where the moon is called a big candle when in truth the moon receives its light from the sun. The Torah and other religious books describe Earth as flat, but they mean 'spread out' and they do not in any way contradict the planet's sphericity.

Ibrahim Hakki's melange of modern science and divine creation with its fantastic embellishments appears to be a reconciliation of science and religion designed to appeal to a broad range of Muslims of varying levels of education. In traditional fashion, he divides people into two general categories, the *khass* and the *'amm*, the elite and the common, each of which has a particular level of understanding. As a popular mystical leader and educated man, Ibrahim Hakki took both constituents as his own and addressed them both in his book, delivering to both the same message, namely a restatement in terms of modern astronomy of al-Ghazali's guiding principle that logic, mathematics and the mathematical sciences are religiously needed and anyone opposing them in the name of religion irreparably damages religion by cloaking it in ignorance.

Hakki's Copernicanism caused him no trouble with religious authorities. His mysticism did. He was interrogated by a council of ulema formed by the anti-mystic *mufti* of Erzurum, and when pressed to defend himself, Ibrahim produced his *Marifetname* and successfully challenged the interrogators to find anything heretical in it.[7] Whatever effect his brush with the *mufti*'s council may have had on him, he avoided the subject of modern astronomy altogether in his next two books. In the one he wrote just before his death, he repudiates his earlier philosophical approach to astronomy as a path that leads to the weakening of faith. His long silence on astronomy after the publication of his *Marifetname*, followed by his repudiation, may indicate a Galilean capitulation to the pressure of religious orthodoxy.

Confusing and contradictory though it may have been in its imaginative and mystical weaving of modern and medieval knowledge, the *Marifetname* was undeniably popular, as it was printed between 1825 and 1914 a total of nine times, four in Cairo (1825, 1839, 1841 and 1863) and five in Istanbul (1867 and four more times over the following half century, once every decade right up to the last gasp of the Ottoman empire with World War I.)[8]

Ibrahim Hakki was the first writer to describe and advocate Copernican astronomy to readers beyond the privileged circle of the courtly elite. His description of the conflicts between those who believed Copernicus and those who rejected the new astronomy, with houses being burned down, indicates Copernicanism was gaining ground among Ottomans. The popularity of his *Marifetname* also indicates a broad dissemination of the theory among the empire's reading populace. For more than half a century his book would be the only one composed by a Muslim to support Copernicanism, giving it a deservedly important place in the early assimilation of Western science in Ottoman society, even with its mystical orientation.

The last half of the century saw a quickening of scientific influx. The year the *Marifetname* appeared, 1757, a new sultan ascended the throne, Mustafa III. Having informed himself of the telescopic discoveries of new planets and satellites through a number of translations and Ottoman sources that would have been available to him by then, Mustafa asked the French ambassador for a book on the new astronomy. When his request was relayed to Paris, the *Academie Royale de Science* sent him several books on the discoveries of Kepler, Descartes and Newton. The astronomical tables of Jacques Cassini were also delivered via the French embassy. An extension of the work of J. D. Cassini, Jacques' father, the tables were an important piece of mid-18th century astronomical literature and included tables of logarithms for simplifying computations. First published in 1754 as *Astronomical Tables of Sun, Moon, Planets, Fixed Stars and Satellites of Jupiter and Saturn*, Cassini's book arrived in Istanbul several years later and was translated to Turkish in 1767, an indication of the accelerating pace of transfer. Another indication of this is Mustafa III's reformist vizier, Mehmet Raghib Pasha (d. 1763), who promoted the Turkish translation of the chapters of Newton's *Principia* that Voltaire had recently translated to French.[9] The French embassy had also sent Alexis Claude Clairaut's addended publication of Cassini's tables, in which Clairaut (d. 1765) had worked out the three-body problem of gravitational attraction and orbital motion, a problem which had exercised European scientists for a century.[10] The Turkish translation omitted this most critical part of the work; not that it mattered: no Ottoman was commissioned to study any of these works once they had been translated. Transfer was one thing; assimilating the content was another.

Not to be outdone by the French, the Dutch ambassador presented the sultan with a copy of Hermann Boerhaave's medical work.[11] Translated in 1771, the gift brought into Turkish the anatomical and physiological knowledge of the circulatory system discovered by Harvey in 1628.

Mustafa III cautiously resumed the innovations of the Tulip Period on a reduced scale. Embassies were sent to Europe on short missions of industrial and military information gathering. By the 1770s, ambassadors were sending reports back to Istanbul on Europe's trade and industry. On another Muslim frontier of transmission, an Iranian ambassador sent to England around this time saw the industrial revolution in full swing and wrote on the machines he saw in textile mills, in foundries, in sheet metal and cannon factories, with their furnaces, steam engines, hydraulic pumps and spinning jennies that turned out manufactured goods "faster than a hundred hands," bringing great wealth and power to the British empire. The Iranian ambassador's report signaled that the mental wall between Islam and the West was becoming more porous. In 1788, the French Consul in Istanbul reported that the *Encyclopedia Universelle*, a rich source of all the branches of science presented in a readable and popular style, had been translated to Turkish.

Yet for all this, it should be noted that after over half a century of reform, of embassies to the West, of reports on Europe's trade and industry and translations of scientific material, the Ottomans still had not begun to produce students of science and technology. No missions were sent to report on the great universities blanketing the European continent. "The sultans had the power and means to hire technologists from abroad, they did not have the power to produce their own technologists from the ulema-dominated educational system."[12] The lack of initiative, or to put it more generously, the glacial rate of growth of scientific and technical appreciation, cost the Ottomans dearly.

A crushing blow against reform occurred in the midst of Mustafa III's revival of innovation and translation. In 1769, the Ottomans were drawn into another war with Russia that ended the long period of peace. Catherine the Great had designed a grandiose scheme for burying the Ottoman Empire once and for all, a partition plan that was to live subliminally in the mind of European diplomacy and come to diplomatic life periodically in the 19th century, to be at last realized by Britain and France as victors at the end of World War I. According to Catherine's plan, Austria would receive Bosnia, Serbia, Herzogovina and Montenegro; France would inherit Egypt and Syria; Venice would take Crete, Cyprus, part of the Greek mainland (Morea) and Dalmatia; while to Russia would fall the protectorship of a resuscitated Byzantine Empire comprising everything else: Thrace, Macedonia, Bulgaria, Romania, northern Greece, a large piece of Anatolia including the Black Sea coast, and of course, Constantinople, ruled by a neo-Byzantine emperor, for which role Catherine had groomed her grandson, appropriately named Constantine. As for the Turks, they could go back to being nomads and peasants in what was left of central and east Anatolia or trek back to Central Asia where they came from.

To draw the Ottomans into the grand war that Catherine intended to end in the empire's quadruple partition, the Tsarina declared a protectorate over the Crimean Peninsula, knowing the Crimean Khan would appeal to the Ottomans for support, which he did. Confronted by this Russian provocation, the conservative hawks at court demanded that Ottoman honor be preserved, and so Mustafa III's peace policy evaporated in the ensuing fever of thundering drums and trumpeting cries for war. As unrepentant and unrealistic as they were opportunistic, the conservatives were forever barking and snapping at the heads of the reformist–innovator government with cries of *jihad* and ghazi, and the empty slogans of a vanished glory in their pushing the financially strapped empire to war – as though they were in collusion with the empire's covert and overt enemies in the West who were impatiently waiting for the kill. French diplomats, keen to have their Ottoman friends check Catherine's expansionist ambitions, joined the conservatives in urging the sultan to go to war, just as they had in 1683 against Austria. The disaster that had followed the failed siege of Vienna was about to be repeated.

Fearing the worst at the beginning of the war, Mustafa III and his reforming ministers, with grand vizier Raghib Pasha in the lead, took the unprecedented step of inviting experienced military officers from Europe to reform the Ottoman army and navy. Many of these officers came from France, whose policy it was to help the Ottomans defend themselves against Russian seizure of territory, a good portion of which was coveted by the French themselves. Previously, European officers who had come on their own to offer their services had been obliged to accept Islam, but now with officers being actively enlisted from Europe through the Ottoman embassies, conversion was waived: a revolution of sorts in Ottoman relations with Europe. The most famous of these officers to come to rescue the Ottomans from the clutches of Catherine was Francois Baron de Tott (d. 1793), a Hungarian nobleman turned French whose colorful military career rivals that of Comte de Bonneval's.

A military officer, de Tott had been involved in a Hungarian uprising and sought refuge in Paris where he enlisted in the French army. Having risen to the rank of general, he was sent to Istanbul in 1755 to learn Turkish and act as military advisor to the French ambassador. His marriage to the ambassador's daughter elevated his career to the highest stratum of French diplomacy and intrigue. Appointed inspector general of the French trading companies in the Levant, he was given the secret mission of gathering information on the commercial and military situation in Egypt and the Red Sea, where the French were expanding trade. He was to report on the possibilities of French navigation and commerce in the Red Sea and of digging a canal and connecting it to the Mediterranean. De Tott's spying did not go unsuspected. The conservative faction suspected every Westerner hired in by the reformers as being spies and kept a sharp eye on them. No longer having to convert made them all the more suspect. The reformers on the other hand were so keen on having European technical help, especially now at this critical moment of war against Russia and Catherine's ambitious plans of carving up the empire, they were inclined to turn a blind eye to any evidence of their guests' duplicity.

The war with Russia was dire. The Ottoman fleet had been totally destroyed at the Battle of Chesme, taking down with it Raghib Pasha's naval reforms; Egypt, Syria and the Hejaz, including the holy cities of Mecca and Medina, had been taken over by the Mamluks in Egypt, whose leader Ali Bey al-Kabir was being encouraged and aided by the Russian fleet in the eastern Mediterranean. Faced by a defeat that threatened to be as disastrous as any the Ottomans had so far suffered, the government reopened Bonneval's defunct military engineering school and put Baron de Tott in charge of it. The immediate objective was to train bombardiers for a corps of rapid fire artillery in defense of the Straits and Istanbul. The Ottoman leadership was so panicked that it allowed de Tott to dress in his French uniform.

At the Baron's insistence, the French officers who arrived in the middle of the crisis to serve under him in training and instructing the Ottoman military were also allowed to dress in their national uniforms. So now it had gone so far that unconverted Europeans in alien uniforms receiving high salaries from an impoverished Ottoman treasury were lecturing Ottoman officers on military science and inflicting European-style drills on sullen Janissaries being marched over the parade grounds in step, while the Russian army and navy advanced on Istanbul and the Ottoman fleet rested at the bottom of Chesme bay. Not since Timur's defeat of Bayazid on the plains of Ankara 375 years ago had Ottoman fortunes look so bleak. In their panic, the reformers had surrendered to the demands of Baron de Tott. The breach of allowing non-converted Europeans to serve in the military and wear their own uniforms was legalized by the *fatwa* of an obliging Shaykh al-Islam. To the reformers it was a bitter swallow of pride and dignity. To the conservatives it was an act of treason not to be forgotten.

The first task challenging the reformers and their French officers was the formation of a modern military corps. But what to create it from? After lengthy debate it was decided to avoid the Janissaries. They were beyond reform. And they would have been sure to rebel against the new military that Baron de Tott was designing for the sultan. Any one of de Tott's innovations would have set them off, whether the military engineering school he supervised, the French-style uniforms he designed, or the modern French weapons he ordered for the new troops. The new weaponry required a *fatwa* of its own. But what to base it on? The Quran and Prophetic Hadith were searched. The best the Shaykh al-Islam and his assistants could come up with was a principle formulated as "al-muqabala bi'l mithl" – confronting with the like – meaning it was permissible to oppose the enemy with his own weapons (in which context the French became the enemy, but that part of the logic was ignored). Reformers also argued from historical precedent. The Arab use of Byzantine Greek Fire in the 8th century legalized the Ottoman use of European gunpowder and artillery in the 15th century, and both of these instances served to legalize the continuation of military borrowing. Greek Fire had in fact prevented the Arabs from taking Constantinople from the Byzantines; now Russian Fire was threatening to take that same city back from the Ottomans to reconstitute the Byzantine state. Whatever had to be done to prevent it would be done, and religion would simply have to justify the means. It was either that or be trampled by Russia, and then humiliated all the more for it being a woman, Catherine, who would preside over partition and extinction of their empire, as she seemed destined to do.

Russian armies conquered the Crimea, invaded Moldavia and Wallachia, took Bucharest and were poised to march on Istanbul. A Russian army invaded the Caucasus region to the east; to the west Russian agents were instigating a revolt among their Orthodox brethren in northern Greece. The Russian fleet mastered the Ottoman Mediterranean coast. The Ottomans were staring extinction in the face. What saved them from it was Catherine's overwhelming success. The specter of her successful offensive that threatened to leave Russia in control of Istanbul, the Balkans and the straits between the Black Sea and the Mediterranean sent Austria and Prussia into a feverish scramble to find a diplomatic end to the conflict and restore a workable balance of power. One was found: partition, but not the one that Catherine had planned for the Ottomans. Another country was chosen as the Christmas turkey. In compensation for Catherine's withdrawal from Ottoman territories, she and the two emperors would partition Poland instead. The Ottoman turn would come later. Catherine was amenable: the Pugachev peasant rebellion was at the moment raging in Russia and had to be addressed. As it was, the tripartite partition of Poland gave the Ottomans another century and a half of slow death.

The Treaty of Kuchuk Kaynarji, drawn up in 1774, stands with Passarowitz and Karlowitz as signposts along the long road of Ottoman contraction and collapse. The Crimea and other territories north of the Black Sea were lost, and in return for Russian withdrawal from Moldavia, Wallachia and the Caucasus, Catherine demanded the right to build an Orthodox Church in Istanbul and to have legal protection of Orthodox Christians in the empire. This was to be a loss more debilitating than territorial loss, as it gave Russia the legal right to intervene in the internal affairs of the empire. It cut a slice of Ottoman sovereignty away. The millet system that had for centuries provided confessional autonomy and kept the peace among the empire's minorities was in dissolution. Coming under the protection of European powers, Christians of the empire now began to be viewed by Muslims either as subjects of divided loyalties or agents of Western powers, with dire consequences for religious confraternity. Between the Capitulations and the assumption by European powers of protective rights over the empire's religious minorities, the Ottoman political body was being eaten away from within.

Sultan Mustafa III abdicated in disgrace, vindicating the conservative charge that reformers who based their infernal innovations on foreign methods, institutions and personnel were in bed with the West. The price of imitating foreigners was defeat and loss of more territory. Accusations of cultural, religious and institutional betrayal drew the curtain down on reform once again. Branded as traitors and enemies of religion, the reformers were lucky to escape with their lives. Following the fate of Bonneval's engineering school and bombardier corps, and Ibrahim's printing press, de Tott's military corps and training school with its French officer–instructors were left to wither and disappear.

The reformists went down. But not without a bold statement on the existential reality that the Ottomans were facing. Historian and statesman Ahmet Resim (d. 1783) accused the hawkish conservatives of taking the empire into war and losing it because of their blind resistance to reforms. The unhappy reality, he explained, was that the old Ottoman spirit of the ghazi warrior was a romance of the past. The only way to stand up to Russia in these times was by reforming and maintaining a policy of peace. Religious fervor, so dangerous and destructive in the present conditions of Ottoman military weakness, had to be put aside. For survival, the Ottomans had to follow a policy of peace and reform. It was a message that brought the historian little popularity outside of the small group of reformers with whom he was associated, and one that the conservative protectors of the status quo were unwilling to accept.

And so just as the rejectionists of innovative reform had awaited their chance to exploit adversity, so now did the reformers await for the next reformist prince to become sultan, in this grim pattern of war, defeat, territorial loss, mutual recrimination and renewed reform that depressingly repeated itself in a contracting spiral of desperate struggle against the fateful specter of unstoppable European power and expansion.

The driving force of the next reformist government was grand vizier Khalil (Halil) Hamid, who, under Sultan abd al-Hamid I, carried on in the spirit of Husayn Koprulu, Damad Ibrahim and Raghib Pasha. In spite of the new sultan's having spent a stultifying half century cloistered in the seraglio surrounded by a legion of women who indulged his every whim and having been deprived of any administrative or military experience when he came to the throne, abd al-Hamid proved to be an energetic and innovative ruler. In 1782, by which time the flames of accusation and counter-accusation over the Treaty of Kuchuk Kaynarji had died down, he appointed Khalil Hamid his grand vizier. Before resuming his predecessor's reforms, Khalil revived Bonneval's old scheme of reducing the disruptive potential of the Janissaries by having them divided up into separate units stationed far from each other. He also intended to do away with the decrepit military landholding system and build a new military institution along Western lines. The grand vizier was planning serious change.

One step toward putting his plans into action was the re-opening of the military engineering school (*Muhendishane-i Humayun*), with Baron de Tott in charge. The Baron wanted to create and train a French-style rapid fire artillery corps, but the school (located in the arsenal and so also referred to as the naval engineering school) had only around 15 students, most older than 30. Hoping to disarm their conservative critics, the reformers attached members of the ulema to the school. Some of them obligingly took mathematics courses since a rudimentary course in it was taught in Istanbul's leading religious schools, but this hopeful sign of enlisting religious students was wiped out when a group of conservative ulema visited the school, declaimed mathematics as belonging to *falsafa* and warned the students that their French teachers were leading them to heresy.[13]

Under the supervision of Baron de Tott, several French officers, graduates of France's most technically demanding military schools, taught physical geography, mathematics, hydraulics, mechanics, fortifications, ballistics, artillery design and navigation. One Ottoman, Gelenbevi Ismail Efendi (d. 1791), who had written on computation of fractions and logarithms, taught mathematics at the school, indicating that some progress in assimilating science had been made since the days of Bonneval's school.[14]

The school encountered major but not insurmountable problems. One was communication. Lectures were given in French. A French-speaking Istanbul Armenian, standing next to the French lecturer, translated the lectures into Turkish sentence by sentence. Another obstacle the school encountered was the overbearing aristocratic manner of Baron de Tott. Even in his own memoirs he is seen arrogantly belittling the student officers he was given to teach, calling them gray-bearded ship captains and 60-year-old school boys. The problem was that de Tott was expecting young, bright students eager to learn, whereas a good number of the students who were admitted to the school were older officers who had connections and thought that going back to school would be good for their career and might even be fun. One anecdote de Tott thought particularly amusing was the reply he got from an older Janissary officer when asked how many degrees were in a triangle: "It depends on the size of the triangle." To a good teacher this would not have been a stupid comment to be held up to ridicule, but an opening to produce a geometric proof. In his commission to construct and supervise a foundry for casting cannon and to teach Ottomans the craft, the Baron found more anecdotal material to diminish his students' technical competence and self-respect. A century earlier, Ottoman artillery had been pounding the walls of Vienna; now a Hungarian-born baron spying for the French was teaching Ottoman officers how to make a canon and ridiculing them for the effort. By the late 1770s, rather than Europeans in Ottoman service having to adopt the religion, dress, manners and culture of their hosts, Ottomans who came into contact with Europeans began adopting European ways. A few Europeans went Ottoman.

The naval engineering school de Tott had been in charge of since its opening in 1776 employed several exceptional European officers, among them a Scotsman who converted to Islam and acculturated to become an Ottoman, Campbell Mustafa Aga. The French and other European officers teaching in the naval school supervised the construction of new shipyards to produce warships that matched the maneuverability and firepower of the Russian ships that had obliterated the Ottoman fleet at Cheshme. To help alleviate the textbook problem – there were no adequate texts in Turkish for students in the new schools – two French engineering naval officers, Lafitte and Truguet, learned Turkish and wrote books in the language on fortifications, naval warfare and military science in general. Their books, as well as other French textbooks translated into Turkish, were printed on the French Embassy's press in Istanbul and published in the early 1780s, providing Ottoman naval officers with at least some up-to-date reading on their subjects of study. This helped, but there never seemed to be a sufficient number of texts in Turkish, nor an adequate number of instructors who spoke Turkish. Most lectures in the naval school were in French and communicated to the students through Turkish translations made, as mentioned, by French-speaking Armenians.

The naval school students, instead of living in their own quarters in the city as they had previously, resided inside the arsenal where they were trained by French instructors and kept under firm discipline. Attempting to solve one problem produced others. Few students were willing to live in an arsenal and subject themselves to regimented training and strict study of science and technology under infidel Europeans. What the reformers with their French officers and technical experts were trying to do was so alien to Ottoman and Muslim culture, so distant from the experience and understanding of the youth and young men, that no rush of spirit could be generated, no youthful dedication to heed the call of duty to God, country and emperor and take up learning the new science, as it would be with the Japanese youth a few generations later. Connections, nepotism and seniority precluded a classroom of young minds eager to learn. Finding good students for the new schools was like pulling teeth. Baron de Tott's manner did not help.

Given the challenges confronting reformers, it is not surprising that insufficient numbers of officers were competent enough to properly sail the ships that the new arsenal was building in accordance to European standards. Too many older officers continued to be appointed through the generational traditions of bribery, favoritism and nepotism. For many reasons, cultural and personal, technical training and professional discipline fell short of replacing what the historian Ahmed Resmi had analyzed as the root of decline: the lost Ottoman ethos. In other words, self- confidence. On the other hand, innovative reform and religion were not immiscible, whatever the conservatives said. One of the Ottoman teachers in the naval school, the abovementioned Gelenbevi Ismail Efendi, who had studied trigonometry and written on the subject, was a religious judge. If one or a few could do it, why not more? The answer to that is found in the lack of adequate funding, and hence infrastructure, afforded the new schools.

The naval school was the centerpiece of grand vizier Khalil Hamid's reforms. According to Gianbatista Toderini, a Venetian priest who was in Istanbul from 1781 to 1786, the school looked as though it was on the way to becoming a success, though he admitted it had a long way to go. Toderini's optimism was based more on what he saw in the school's scientific and technical library than in the classroom. The work he wrote on Turkish literature offers informative observations on the contemporary Ottoman scene, one of which focused on his visit to the naval school where he was favorably impressed by its European nautical equipment, atlases, maps and books.[15] He reports seeing a telescopic quadrant, a compass of Galileo, a French-made theodolite, terrestrial globes, an armillary sphere, astrolabes, an English-made octant, trigonometric tables and Turkish translations of French works, among them LaLande's astronomical tables, in addition to ballistic tables, books on geometry, and translations of scientific books written in European languages other than French. The teachers at the naval school, according to Toderini, knew English, French and Italian and had at their disposal in the library the best books of European marine science.

The library possessed other scientific texts that Toderini failed to mention, such as translations of Western medical books and Count Raimondo Montecuccoli's leading work on military science, *Memoria della Guerra*. A few years after Toderini's visit, the French consul wrote in 1788 of having seen, among many other notable works in the naval school library, a Turkish translation of the French universal encyclopedia, leading him to believe that the expense the government must have put into translating such a large scientific and technical work augured the dawning of a new era in Ottoman intellectual progress. The consul was sure that the Ottomans had embarked once and for all on a cultural voyage that would bring them to art, science, commerce and progress: "Effects proportionate to such exertions have been already visible in the good order and discipline of their armies and have been extended to their dress."[16] Toderini did not quite share the French Consul's upbeat assessment. Impressed as he was by the naval school's library and equipment, Toderini's critical view of the qualifications of the school's students and their devotion to study coincides more with Baron de Tott's dour sarcasm, for he did not see in the school a bright Ottoman future. Its students, numbering around

50, he writes, are "sons of sea captains and Turkish gentlemen," of whom only a few took their studies seriously.[17]

Looking at the Ottoman condition and the world through a window that opened to a view that was diametrically opposite to that through which the reformist historian and statesmen Ahmat Resim peered, the historian Ahmet Vasif (d. 1806) offers a conservative insider view of the reforms and Ottoman dependence on Europe:

> It is evident that the imperial state would not demean itself and condescend to learn unfamiliar martial science from those who belong to the same kind as its enemies. . . . Success or failure depends upon God. The beliefs of Christian nations are contrary to this, as they follow the doctrine of a school of philosophers according to whom the Creator has no role in particular matters such as war.

Vasif goes on to chastise Western nations for believing it is not God who decides victory but the side that has superior material means to make war. This presents a problem to Vasif: "How is it permissible to attribute victory merely to the perfection of the means of warfare and defeat to shortcomings in those means?" Logic and logistics had no place in the equation of war: man proposes, God disposes. Montecuccoli's rational discourse on war had nothing to offer but unbelief.

What sounded like perfect reasoning to Westerners and Ottoman reformers, that superior weaponry, numbers, material support, logistics and organization would give victory over an enemy inferior in these categories, was to the more traditionally conservative mind a betrayal of religion and all that the empire stood for. Ottoman institutions in their uncorrupted form stood for God and religion. Taking from the West and accepting that God's will was secondary to superior arms were heretical acts. To the conservative mind, at home with a theology that subjugated cause and effect to a tenuous probability determined by divine will, the view was perfectly reasonable. And in a strictly political sense, leaving the divine to the side, Vasif's view on the reforms had both feet firmly planted in a reality that could be accepted as rational by Westerners as well: that it was fatal for the Ottomans to accept help from the French who were not at all friends but were acting for their own advantage, consequently, "Trusting Christian states is utterly impermissible."[18] Events would prove Vasif to be right.

It would certainly have been better for the Ottomans to strengthen themselves rather than relying on foreigners. The problem was how to do it without them? In the absence of an Ottoman institutional and intellectual reformation, how would reformers begin to transform a state without first resorting to outside help, in hope that change, induced by a jump start from an external source, would take on a life of its own? Though the conservatives were convincing in arguing that Westernized reform was a strategy to divide, weaken and conquer the empire, they could offer nothing more than the oft-tried and failed attempt at reforming the old institutions. On the westward-looking side of Ottoman reform, the thought of sending batches of students to the West to study was at the time too radical for even the innovators to contemplate. The idea was still so unthinkable, so far out in space that even the idealism, energy and rebelliousness of youth failed to capture it: no young Ottoman went on his own to a European country to learn the language and study science or military engineering.

The Franco–Ottoman relationship was a diplomatic orphan. It was popular with neither the conservatives, who were waiting their chance to set things right with God, nor with many of the French progressive intellectuals of the Enlightenment, such as Voltaire, Diderot and Condorcet, who on the one hand saw Catherine as an enlightened and progressive despot, and on the other saw the sultan and all Ottomans as backward religious fanatics filled with ignorance and superstition. With French diplomatic and popular attitudes being decidedly anti-Ottoman, there was a great deal of antipathy to the relationship from that side. The sentiment was multiplied on the Ottoman side. De Tott's being a suspected spy was harmful enough. What followed was worse, and it once again proved the conservative argument that it was better to cut off an arm than accept the helping hand of a European power.

Choiseul-Gouffier was appointed French ambassador in 1784. Several years earlier he had written a book on his travels in Greece. Greece was part of the Ottoman Empire. Caught up in the romanticism of the time, he added a new preface to the book in which he deplored the servitude of the noble Athenians and Spartans to "the stupid Muslim." If that was not embarrassing enough for someone who would soon be France's ambassador to those stupid Muslims, he wrote a second book, this on the contemporary state of the Ottoman empire. In this book, he encouraged the French to compensate for their loss of Canada and India to the British by taking Greece, the Aegean, Egypt and the Red Sea from the Ottomans. Such were the friendly sentiments of the ambassador of the country upon which the sultan, grand vizier and reform party had placed their trust and hope for support. In spite of these views, the French government, bankrupt and just a few years away from revolution, regarded him as an excellent ambassadorial choice for Istanbul. His views in fact counted for little. It was his influential connections in the court of Louis XVI that gained him the appointment, and so off went Choiseul-Gouffier to Istanbul. The conservative faction in the Ottoman court could not have asked for a better ambassador to aid their cause, as though Versailles and the conservatives of the Seraglio were in collusion to undermine the reformers.

Before leaving for Istanbul, Choiseul-Gouffier had the good sense to send agents to Germany and Austria to buy up copies of his books. But to no avail. The conservative faction in Istanbul knew the ambassador's views long before he arrived. It was the kiss of death for Grand Vizier Khalil Hamid. Could the conservatives have known of the new ambassador's views and not the grand vizier? Not likely. Khalil Hamid was simply too keen for French support to risk affronting the French government by refusing the appointment. Even a contemptuous spy contemplating the empire's partition was acceptable. The drowning man grabs the tail of the serpent.

With the ambassador came a new crop of French experts: military technicians, astronomers, geographers and cartographers, as well as poets, painters, and a number of non-descript travelers who had come to look around and report back. Also accompanying the new ambassador was a French noble, Duc Charles de Montmorency, who through agreement between the French and Ottoman governments had been commissioned to reorganize the Ottoman army and navy and establish a command headquarters on either Crete or Rhodes. It may have made military sense, but to the conservatives at court it smacked suspiciously of a French plot to subvert the empire by placing the command center of the army and navy under the authority of a French officer on a strategic island far from the capital. Little by little, from Comte de Bonneval to Baron de Tott and Duc de Montmorency, Ottoman reformers, seemingly having given up on themselves, were in desperation putting their destiny in French hands, as untrustworthy as those hands were suspected to be.

Combined with ambassador Gouffier's dream of carving off the Greek corner of the empire for French commerce to make up for France's recent overseas losses to Britain, Montmorency's mission set the conservatives to sharpening their swords. Reform was becoming a nightmare of contradictions. As if Khalil Hamid was not in a bad enough situation with the apparently subversive missions of de Tott, Gouffier and de Montmorency, something else was exciting conservative suspicions. Officer Truguet was at the time on a French ship making a commercial treaty with the recalcitrant Mamluks, who continued to rule Egypt more or less autonomously after the revolt of Ali Bey had been put down and his army forced from Syria and the Hejaz. What was a French officer doing negotiating a treaty with the Mamluk usurpers of Ottoman power in Egypt? And what about the mysterious Baron de Tott, inspector general of the French trading colonies in the Levant? Did the Ottoman opposition know that the French general of Hungarian origins who was taking over the Ottoman military by his reforms and new organizations was at the same time carrying out a secret mission for the French government that aimed to take over Ottoman commerce in the east Mediterranean and Red Sea? The grand admiral of the Ottoman fleet, Ghazi Hasan Pasha, knew. Leader of the conservative faction and a jealous rival of Khalil Hamid Pasha, Ghazi Hasan believed he, not Khalil Hamid, should have been appointed grand vizier – adding one more dash of personal ambition and jealousy to the witches' brew of internal Ottoman politics.

Personal animosity made it all the easier for the grand admiral to believe the grand vizier was treacherously weakening the empire by letting the likes of de Tott, Choiseul-Gouffier, Truguet and de Montmorency, all of whom had agendas that served Christian France more than the Islamic empire, take over in the name of innovative reform. The grand admiral trusted Europeans to help the Ottomans as little as did Ahmet Vasif the historian. Europeans who came to help the sultan were either out to help themselves or factions in the French government. That was to be expected. The treachery was not in the Europeans, but those who brought them. Yet, Ghazi Hasan Pasha knew the situation was not quite that simple. A competent admiral, he was aware of Ottoman deficiencies in face of the West and supported reform and what the French were doing to rebuild and reorganize the navy. But he knew not to trust them. He knew they were using the reforms as an entry to attain their own strategic and commercial goals. Ghazi Hasan saw clearly the dilemma of reform.

What it came down to was that Ottoman reformers were depending on foreigners whose inimical designs on the empire had even been published. Secret Western plans to seize the empire and its seas could be read as the next chapter in the West's economic domination, already in large part achieved through the flood of European goods inundating the local market and sweeping away local industries. Direct occupation would be the next stage, if the powers could but agree on who got what. The French and Russians had already drawn up plans to resolve that. It was no secret to the Ottomans that it was only intra-European rivalries that kept the powers at bay. As soon as the balance was upset, one power or another, France or Russia, would make for the prize: Russia by land invasion from the north and east or France by cunning diplomacy and duplicity, and failing that, invasion by sea, probably taking Syria and Egypt. Before the end of the century Bonaparte was to use them both.

The Ottoman relationship with France was fraught with too many contradictions to last. In 1785, while the French were reforming the Ottoman army and navy and carrying out their secret missions, Catherine the Great's government annexed the Crimea. To Ottomans of all political views, it appeared her grand project of partition was once again on the table. This offered an opening for the opponents of Western-oriented reformers with their weak-kneed policy of peace. Determined to bring down Khalil Hamid and end the French connection, Ghazi Hasan Pasha and other opponents of the government were quick to unite the socially diverse components of their conservative coalition and go into action. Their first demand was that the sultan give them the grand vizier's head. Abd al-Hamid at first hesitated. Too much had been invested in the reforms to throw it all away. But when the streets filled with Janissaries, religious students and the impoverished, a scene all too familiar for the sultan not to know what he had to do to save himself, he gave the order. A *fatwa* was issued declaring Khalil Hamid a traitor and he was summarily dispatched. On his publicly displayed headless torso was draped the customary placard stating the victim's crime: "Enemy of God and State." A number of other reformers shared his fate. Compared to the crisis that had ended the Tulip Period, it was a relatively peaceful affair, but the effect was the same in that it put a crimp in the reforms, proving once again that reformers stood alone and fell alone.

The conservative government that now came to power was headed by a Georgian convert who had entered Ottoman service as a young man. His abiding hatred of the Russians tempering his mistrust of the French, he had the latter stay to finish the work that the previous government had contracted them to do, though a deep suspicion chilled the relationship. Relying on a French devil was worth the risk if it helped defend against the greater Russian devil, and so the reforms limped on under a grudging conservative leadership, at least for as long as peace lasted, which was not for long. The Georgian grand vizier's hatred of the Russians was too powerful for him to wait for the reforms to have effect. In 1787, he and his conservative party pushed the reluctant sultan into declaring war against Russia over the Crimea. As always, the time could not have been more adversely chosen.

On the one hand, the newly built fleet was away in Egypt where Ghazi Hasan Pasha was bringing the Mamluks under control and fiscally reorganizing the province. On the other hand, a war against Russia would be an open invitation for Austria to occupy Bosnia and Serbia, especially given the Ottoman army's poor performance of late. There was no rational hope in the world of their even coming to a draw in a two-front war. But reason was no contender against anti-Russian passion, conservative valor, and Vasif's popular belief that God, not military power, decided victory. Any rational analysis would have easily predicted the outcome.

Waiting until the Ottoman army was engaged against the Russians in the east, Austria, as expected, invaded from the west. Something else could have been predicted by rational analysis. The Hapsburg's had recently signed a treaty of friendship with France, the two countries having dropped their centuries-long enmity and become allies in order to counter the growing power of Prussia, which under Fredrick the Great had upset the balance of power to become a threat to both. The treaty was an astounding reversal in the world of European diplomacy. One of its many clauses required neutrality if one or the other went to war with a third country. Austria invoked the clause and forced France to withdraw its military and technical experts from Ottoman territory, as their presence was interpreted by Austria as a breach of the treaty since the Ottomans and Austria were at war. On the brink of bankruptcy, with the country sinking into social discord, the anxious government of Louis XVI complied, too insecure to risk irritating its new ally. Orders were issued for the French engineers, technicians, teachers and military advisors to pack up and leave Ottoman territory immediately. Ottoman reform once again came to an abrupt halt. The new arsenal and artillery and naval engineering schools and their institutional affiliates became hollow shells. While the conflict's progress was predictable, its end was a classic demonstration of war's unpredictability.

By the spring of 1789, the Ottoman armies had been pulverized and swept like dust from the fields of battle. The Austrian army had taken northern Moldavia, Bosnia, Serbia, Montenegro and Kosovo, had marched from Belgrade to Nis and was awaiting orders to move on down the undefended road to Istanbul. The Russians had meanwhile seized southern Moldavia, Wallachia, the Crimea, parts of the Caucasus and were advancing along the northern coast of the Black Sea toward the Sea of Marmara and Istanbul. Another Russian army poised in Moldavia formed the southern thrust of the Russian pincer aiming to close on the Ottoman capital. Nothing stood in the way of the enemy armies and the imperial capital. Catherine's dream of partition and pushing the Turks back to the steppes of central Asia appeared to be at hand. And then, at the moment of Austro–Russian triumph, just as the five-century old empire of the sultans seemed at its end, news of the French Revolution thundered through the imperial halls of Europe, bringing the continent's trembling monarchs together to save their thrones and crush the virus of revolution before it spread across the borders.

Austria, closest to the virus, was the first to sue for peace, in return for which all territories it conquered were returned to the Ottomans. The Russians fought on, won several more smashing victories, and could no doubt have gone it alone to Istanbul. But again, European international diplomacy intervened. Britain, Prussia and Austria, united in the Triple Alliance to fight the French Revolution, pressured Russia to back off. If the Ottoman Empire was to be devoured, it would be an international banquet jointly agreed upon; not some unilateral plan of an empress presiding over the distribution of Ottoman provinces the way she presided over the partition of Poland, slice by slice. Thanks to the French Revolution and the Triple Alliance, the Ottomans emerged from the jaws of imminent destruction, though the war had been destructive enough in its waste of troops and treasure.

The long, expensive and unnecessary war into which the conservative government had dragged the unprepared empire ended in 1792 with the Treaty of Jassy. In return for withdrawing from Moldavia and Wallachia, Russia was allowed to annex the Crimea and maintain suzerainty over the Georgian Caucasus, while Ottoman government was restored in the cities and provinces of the Balkans up to the Danube. The century-long series of wars ending with the treaties of Karlowitz, Passarowitz and Kuchuk Kaynarji had come like knockout blows, each one more devastating than the one before. Jassy, though no victory, had come as a gift from heaven. The empire, staggering but still standing, its self-esteem and confidence sorely bruised if not close to crushed, awaited the blow that might be the one to end it all.

The conservative warriors had been cowed. The imperial stage was set for yet another round of reform in preparation for that next blow that must one year or another come, from Russia, from Austria, or from both at once. Perhaps it was only the conservatives who were not surprised when it was the longtime ally and supporter of Ottoman reform who delivered it.

Notes

1 A. Adivar, *Osmanli Turk Ilim*, Remzi Kitabevi, Ankara, 1982, pp. 161–163. One of the count's famous dictums on success in modern warfare reforms was "Money, money and money" – the insufficiency of which would unremittantly hamper the Ottoman government's attempt to modernize.

2 Stanford Shaw, *History of the Ottoman Empire*, Cambridge University Press, Cambridge, 1976, vol. I, p. 242.

3 Bernard Lewis, *The Muslim Discovery of Europe*, Norton, New York, 2001, pp. 231–232.

4 Aware of Peter the Great's accomplishment in building a modern navy, the sultan sent a delegation to St. Petersburg several years later. The result was of as little consequence as the embassy to Vienna. "Relation de l'ambassade," *Journal Asiatique*, 8, 1826, pp. 118–125.

5 Ihsanoglu, "Introduction of Western Science to the Ottoman World," pp. 86–87.

6 *Marifetname*, Ibrahim Hakk, Istanbul, 1756, pp. 144–151; Ihsanoglu, "Introduction of Western Science to the Ottoman World," p. 91.

7 The story is related by a descendant of Ibrahim's and may be somewhat romanticized. MesihIbrahim Hakkioglu, *Erzurum Ibrahim Haikioglu*, Istanbul, 1973, p. 91.

8 An edition transliterated from the Arabic script of Ottoman Turkish to the Latin script of modern Turkish was made by Turgot Ulersoy and published in Istanbul in 1978.

9 Giambattista Toderini, *Idee generale de la Turquie et des Turcs*, vol. I, Paris, 1787, p. 118.

10 A child prodigy, Clairaut became a member of the *Academie* while still a teenager. At 13, he had presented his first paper to the *Academie*. He was at the forefront in disseminating Newton's theories of gravity and light throughout continental Europe.

11 Herman Boerhaave (d. 1738), theologian turned mathematician and physician was at the time Europe's most famous physician.

12 Lewis, *Discovery of Modern Europe*, pp. 224–225.

13 This according to a letter of ambassador Choiseul-Gouffier, cited by Frederic Hitzel, "Les Ecoles de Mathematique Turque et L'aide Francaise," in *Histoire economique et Sociale de L'Empire ottomane et de la Turqu (1326–1960)*, edited by D. Panzac, Peeters Press, Paris, 1995, pp. 812–825; and also chapter 2 in *Integration et Transformation des Savoirs*, F. Hitzel, Louvain, Paris, 1995, pp. 813–825.

14 The texts used for mathematics and mechanics were the most recent works published: Etienne Bezout's *La Theorie generale des equations algebriques*, D. Pierres, Paris, 1779; Jean-Francois Callet's *Les traits elementaires d'arithmetique et de mecanique*, Paris, 1774–1775; and Charles Bossut's *Les cours de mathematiques a l'usage des ecoles militaries*, Paris, 1782. For details on the school's curriculum, teachers and students see Frederic Hitzel, "Les Ecoles de Mathematique Turques et l'Aide Francaise (1775–1798)," *Histoire economique et sociale de l'Empire Ottoman et de la Turquie (1326–1960): Actes du sixieme Congres international tenu a Aix en-Provence*, vol. 8, 1995, pp. 813–825; also Hitzel's La France et la modernization de l'Empire Ottoman a' la fin du XVIII eme Siecle," for French scientific and technological aid to the Ottomans between 1784 and Bonaparte's invasion of Egypt.

15 *Litteratura Turchesa*, 2 vols., Venice, 1787. For a French translation: *De La Litterature Des Turcs*, transl. A. Cournand, Chez Poincot, Paris, 1789.

16 Cited in Niyazi Berkes, *The Development of Secularization in Turkey*, McGill University Press, Montreal, 1994, p. 60, note 11. The optimistic appraisal of French-directed reforms by a French consul may not have been free of national self-congratulation.

17 *Litteratura Turchesa*, vol. I, Venice, 1787, pp. 177 ff, and in French translation: *de la Literature des turcs*, vol. I, Paris, 1789, pp. 161–167. Also Lewis, *Muslim Discovery of Europe*, p. 236. It should again be stressed, the views expressed above are those of outsiders looking in.

18 Cited by Berkes, *Development of Secularization in Turkey*, McGill University Press, Montreal, 1963, p. 66.

13 The new order

Selim III ascended the throne the year of the French Revolution. The revolution and subsequent beheading of the French king meant little to the Ottomans. They had seen many royal heads roll. But thanks to the fear that the terror and its execution of Louis XVI put into the hearts of Europe's monarchies lest their own subjects do the same to them, the revolution saved the Ottomans from massive land losses. In the longer run, the Revolution would prove to be a mixed blessing to Ottoman fortunes, as the Revolutionary Republic unleashed imperialist aggressions against the empire that had been only planned and contemplated during the days of the monarchy, those grandiose schemes sketched out in reports by the king's ambitious ministers and diplomats who saw in the Mediterranean shores of the Ottoman empire compensations for territorial losses to the British in India, the Caribbean and North America. During the monarchy, diplomacy was based on an inter-European "Balance of Power." From the French perspective, this balance was integrally related to preventing Austria and Russia from swallowing the Ottomans before France had a chance to claim its fair share. The idea in practice served to restrain the expansionist impulses of Europe's great powers and allowed France the pretension of posing as the ally and protector of the sultan and his possessions, rather than being one of the predators waiting for the kill. When French revolutionary armies broke the balance, the Ottoman Empire went from an eastern sideline of diplomacy to a central target in the maelstrom of European powers fighting for supremacy, where it remained until its demise with World War I.

The storms generated by those contending forces exerted ever more profound pressures on the Ottomans to reform, inducing a series of programs, each one more radical than its predecessor, that wrenched the Ottomans from their institutions, their legal system, and their millet system of preserving harmony among the empire's many religious and national minorities, costing them along the way their provinces and their sovereignty.

Undoubtedly distressed by the sorry fortunes of previous reformers, but realizing that the choice was reform or certain extinction, Sultan Selim III, soon after signing the Treaty of Jassy ending the war, began improvising a new set of reforms based on those that had preceded him, but more far-reaching in depth, scope and rigor. Selim went back to hiring French military officers to teach and train a new military corps, and French experts to advise the government. By 1795 there were over 600 Europeans, mostly French, employed in the arsenal, military camps and training schools. No longer secluded in barracks and military posts, they lived in the city and freely mingled with the people, enjoying the public amusements, while they themselves, in their military uniforms and European dress, presented an amusing show to the Istanbulis. Only the conservatives of Istanbul were not amused. To them, it appeared that the new sultan was delivering the empire to revolutionary French officers disguised as friendly reformers, and they were not far from wrong.

Selim was the first sultan to give his reforms a name: *Nizam -i Jedid* (*Cedid* in modern Turkish), New Order, a term originated by Ibrahim Muteferrika and resonating with implications that raised the hackles of conservatives, traditionalists and reactionaries of every hue, for just the word "new" conjured in the conservative mind a specter as menacing as the Russian army.

The plan to create a new military force was the origin of the New Order. In fact, the military was the New Order, but as other innovations followed, the name was applied to the whole period of Selim's reforms.[1] Selim's New Order came into place between 1789 and 1806, reform by reform, extending from the military to diplomacy and government. Of these, it was diplomacy that met with the most success. Ottoman diplomacy was practically an oxymoron until 1793, when Selim established the first permanent legations in the most important European capitals, Paris, Vienna, London and Berlin. Before then, Greeks living in the Fanar section of Istanbul had been assigned to handle diplomatic correspondence and act as interpreters for the Ottomans in their relations with European governments. Diplomatic missions had previously been temporary affairs, a delegation going to a European city on a particular mission and then returning to Istanbul. There had been no permanent embassies. Selim's purpose in establishing them was two-fold: to learn directly from Europe and to integrate the empire into international diplomacy as a means of using European rivalries and "balance of power" to protect the empire from the territorial ambitions of any single power or alliance of powers. Unwittingly, his reforms helped navigate the empire right into the heart of the storm.

Selim's European missions were meant to be the spearhead of reform. Ambassadors and their assistants were ordered to learn the languages and "useful sciences" of their host countries and submit reports about their institutions, economies, governments and military, whatever was deemed useful and important. Ottomans being ordered to learn non-Muslim languages was something of a revolution in itself.

A report from Selim's ambassador in Vienna reveals the relatively penetrating observations Ottomans were soon able to make regarding European society. After being there only a year (1793) during which he studied Austria's military, government and socio-economic organization, abu Bakr Efendi reported that trade, industry and agriculture were supported by banking institutions, a regulated system of taxation and the people's freedom of life, freedom here including political as well as economic rights. The government used tax revenues to build roads, maintain a postal system and support the mining industry; government acted to protect and promote the interests of the people. Another feature the ambassador found remarkable was the absence of scriptural law and religious obligations regulating daily life. He marveled that religion seemed to have little to do with social and political relationships and the strength that came of them. Incredibly, lack of religious orientation in life did not make the people weak or corrupt. Rather, the people had a strong sense of honesty, duty and responsibility. This and their industriousness, efficiency, good organization and above all, education and mental ability, made their government, society and military strong. The government served the people and the people supported the government.

Such observations, absent in the reports of Yirmisekiz and Mustafa Hatti earlier in the century, show not just how much and in what direction Europe had developed in the course of the 18th century, but more to the point, how adept some Ottomans had become in looking beyond the glitter of royal palaces, military parades and museums to detect the social, economic and political principles from which European strength derived. Abu Bakr implied in his report that the sultan would have to change the ways of Ottoman government if the empire was going to compete with Europe. Here the ambassador makes an argument that had been made before and that would become a leitmotif of 19th and 20th century Muslim reformist apologetics: since the Ottomans of olden days had all these modern European virtues and laws, imitating Europe would not be imitation as much as it would be restoring the principles by which earlier Ottomans had lived; adopting European institutions and principles would bring the Ottomans back to their own. Taken up by Arab reformers in the next century, the argument would be extended to claim that the early institutions that had existed in Medina under the Prophet, or in Damascus under the Umayyads, or in Baghdad under the Abbasids, were in essence the same as those of contemporary Europe. Reclaiming one's own was not borrowing or imitating.

As the experts and the Ottoman ambassadorial reports arrived from Europe, Selim's reformers and their French advisors mapped out a set of goals and procedures that would define the New Order. Their vision covered a broad range of areas: economic, financial, administrative, political, educational, scientific and military. Rather than leaving his viziers to carry out the reforms as had earlier sultans, Selim took the lead and involved himself in daily decisions and details in order to show his commitment and encourage his fellow reformers. His involvement precluded any chance of his saving himself in exchange for the head of a grand vizier in the event of a conservative-driven populist uprising. It was a guarantee that if any heads were to roll, his would be the first.

Unlike most of the sultans in the 18th century, Selim was physically active, mentally alert and knew something of the world, even though, like them, he had grown up in confinement within the walls of the imperial harem that he never left until becoming sultan. As a young prince, he had written a small book on artillery and had collected some Western scientific books and instruments. He had a Venetian doctor who kept him informed on the West and its threat to the empire. By the time of his accession, Selim had some idea of what needed reforming. Furthermore, he knew that to be successful he would need a cadre of courageous and capable men around him, men as determined as himself to enforce reforms in face of the accusations of betrayal and heresy bound to come from the opposition.

With this in mind, he quietly and patiently surrounded himself with about 20 such men whom he could trust completely: companions, former slaves, lower-level officers, men of intelligence, daring and loyalty who were outside the contending circles of power, who had no courtly or family ties other than to their padishah – father sultan – who had adopted them and whom they would follow to death.

Selim cautiously collected these men over a three-year period so as not to raise suspicions, then appointed them one by one to key positions, all the while feigning the outer visage of a pliable young prince of the House of Osman whose wits had been dulled by too many stultifying years immured in the harem, where every vice and luxury known to man were his for the taking. He dissimulated well. When his men were in place and he dropped his mask, the conservatives at court, members of the royal family, the ulema, the military and the bureaucracy were taken by surprise. The effete dullard had become a lion. Or, as the Fatimid Caliph Hakim put it after playing the same game eight centuries earlier in Cairo, the lizard that he had been called in contempt had become a dragon.

The men he had chosen came from modest, even humble origins. The leading one was a former slave of Selim's, Mustafa Rashid. Mustafa was wholly devoted to Selim and his reforms, but uncomfortable with the sultan's decision to rely so heavily on the French, so-called allies who had proven themselves not only a weak reed but duplicitous. Mustafa was certain France coveted the empire's Mediterranean provinces and would betray Selim at the first chance. He would have preferred Britain as a partner in reform, a country that had shown no interest in taking over Ottoman lands, but Selim was set on France. France had been involved in Ottoman reform for over half a century and new reforms could build on what the French had already done. The sultan would follow precedent.

Two other leading members of Selim's inner cadre of reformers were also former slaves: a Circassian, Kuchuk Husayn Pasha; and a Greek, Yusuf Aga.[2] Mahmud Efendi was a fourth leader in Selim's reformist cadre. An intelligent and energetic official of humble background, he had served as scribe to the Ottoman ambassador in England before being selected to join Selim's reformist avant garde. While in England he had taught himself English and French, in addition to geography, history and politics. It was his knowledge of Western languages that escalated his career to a high position in Selim's new Bureau of Translation.[3]

What was unique in Selim's New Order was its institutions. New ones were funded, but without the old ones being abolished. The old were kept, but apart from the new, a situation which contributed to a confusing dualism. Selim thought it safest to reform without touching or disturbing anything. Abolishing the old would have caused violent turmoil. Expectations were that the new would grow while the old withered and died. But the old did not die. The cultural soil of the Middle East does not easily let old things die a quiet death. The old remains in the soil often to reappear unexpectedly, threatening what had replaced it. The consequent coexistence of old and new that Selim's New Order unintentionally institutionalized became a prelude to the dichotomy that would run through Ottoman society as secularist reform became more extreme with the growing impact of the West during the 19th century. In Selim's time, the most obvious feature of this social fracturing was seen in the military. Even though the Janissary institution was deemed ineffectual and obstructive, it was too much feared and too influential for Selim to terminate it, and consequently it lived on, distant from the New Order Troops, but nevertheless a threat to innovation.

High religious officials on the other hand, had no problem recognizing the dire conditions that called for innovative reform.[4] Within the ranks of the ulema resided reformers whose ideas were most radical for the time. One of the leading proponents for an army of new troops was the Shaykh al-Islam.[5] Too often has religion and its ulema enforcers been misidentified as the opponents of modernizing change. In the succeeding reign of Sultan Mahmud II, a high ranking member of the ulema wrote a memorandum advocating a reduction in the high lifestyle of the upper classes, to which he himself belonged. The resources gained thereby should be used to initiate a government program in industry and trade. The upper classes, he wrote, must be driven to engage in these endeavors that they looked down upon as beneath them. They must be stripped of their aristocratic disdain of commerce, manufacturing and profit-making. The government must set itself to forcing the wealthy classes to be productive and forcing itself as well. Toward this purpose, the government must restrict the building of wasteful palaces and luxury homes, reduce salaries, encourage interest in industry and commerce, and restrict imports to nurture home-grown industries, with a three-year tax exemption for new enterprises. Industries should be built outside Istanbul in the countryside where wages would be less than in the capital.[6] The memorandum reveals that Ottomans in high positions, within and outside the religious institution, realized drastic changes for the better must be and could be made, if only the leadership was willing to transform the flab of upper class waste to the lean of modern economic enterprise. Obviously, the economics of reform were fundamental to success or failure. The mulla who authored the memorandum clearly knew it was not simply a matter of God's will. Determined and severe action had to be taken to save the Islamic Empire from the West. God helps only those who help themselves.

The shaykh's message of reducing the lifestyle of the elite to fund reform fell on deaf ears. When one reads of the contortions that Selim and his top reformers went through in searching for revenues to fund the New Order corps – initially a modest affair of a few thousand troops and a score of officers, the French weapons and officers required to arm and train the troops being the big expenses – a strong impression is made that the reformers found no way to finance reform the way it should have been. All available revenues were earmarked. Nothing new to tax was found. Everything that could be taxed was already being taxed. Raising taxes beyond the present level would have been counterproductive. The tax burden was already at the crushing point. The empire was in wretched economic condition. Its crafts and industry were smothered by cheap European imports; its agriculture was contracting for neglect of maintenance and for natural reasons; and a wealthy class of old families and large landowners refused to pay taxes or condescend to invest in trade and manufacturing. After weeks and months of searching for sources of revenue, of analyzing the economy, debating, compromising and reviewing again and again the empire's incoming revenues and expenses, Selim and his top men took to seizing lands in different provinces, and taking funds from the Holy Cities foundations. A new tax on grapes from Greece was levied, and on walnuts sold in public markets. Taxes on wine, raki, spirits, coffee, textiles and sheep slaughtered in Istanbul were raised.[7] But the class of powerful family landowners who had amassed great wealth amidst the impoverished masses whose plight was worsening were left untouched. The debilitating economic condition that resulted from the grossly maldistributed wealth in Ottoman society made ibn Khaldun's socio-economic philosophy of history a popular read among the reformers.

The New Order ran on the proverbial shoestring. Recruits for its military corps were drawn from homeless boys in the capital and the peasantry of Anatolia, origins as humble as those of the leading reformers. Lifted from their bleak conditions to become the sultan's favored troops, they may not have resented their French-style uniforms. New armaments factories that were built and maintained under French supervision supplied them with modern rifles and artillery. Existing Ottoman armaments factories were a shambles, as useless and dangerous as the Janissaries. The steel used for casting cannon in the old foundries was inferior. Occasionally a cannon would blow up with lethal results. The gunpowder was rank with impurities and too often failed to fire. Bullets were made without strict adherence to standards. A soldier in battle might find his ammunition did not fit his weapon, the result of corners being cut to cheat the government by contractors who thought the government was cheating them. Rather than tearing down the old factories and building anew, the old were left to function, old and new existing in a nightmare of parallel universes of technique and experience, like the Janissaries and the New Order troops.

As for the new, the French republican government dispatched 70 master foundry experts to build a modern cannon foundry and gunpowder factory. French munitions experts were brought in to design and supervise the new munitions factories. They were also to train Ottomans who were supposed eventually to take the place of the highly paid French experts and teachers, but the Frenchmen did not fancy losing their lucrative positions overly soon. British military experts were also hired. Under British direction a second bullet factory was built, to the chagrin of the French foundry and munition experts. French petulance was further exacerbated when the Ottoman government went to the British for the machines and equipment to manufacture the new armaments. Britain's Industrial Revolution had been in progress a generation. British machines were known to be the best. So while at war with each other, France and Britain were uneasily joined in building the new Ottoman army and navy. On occasion, the Anglo–French conflict was transported to the new armament factories in Istanbul. The contradictory situation was dramatically resolved when France invaded Egypt, a treacherous act of war against the country France was supposedly allied to, not to mention the hypocrisy of attacking an ally that was paying the French generously for their technical assistance in a program of modernization. Again the conservatives were proven right. And so too was Ingiliz Mustafa on the reformist side. He had always figured the French a poor choice compared to the British. And so it was that once again French engineers, experts and officers were sent packing, this time by order of the sultan, who luckily survived the conservative wrath brought against him by his erstwhile ally's treachery. Such were the hazards of Ottoman reform.

Bonaparte's Egyptian campaign highlighted the pitfall of contracting a European power for the work of reform. To depend on the West was to build on sand, and dependence was becoming addictive. The reformers saw no other way. The old institutions seemed to them beyond redemption. New ones were necessary, but how to build them without advice, learning and training from Europe? Europeans were necessary to start the process that would eventually become progress. As the French experts were leaving Istanbul in the summer of 1798, British and Swedish replacements were arriving right behind them, calculating the profits to be made by helping the Turks stand up to those who would seize their lands – lands, of course, they had seized from others centuries earlier, in the casino of history.

The French departed but they left something behind. Selim's French-directed resuscitation of de Tott's military school, the *Muhendishane*, produced an educational forum that, at least on paper, took students through a four-year sequence of science and technology in a much more systematic and rigorous way than had Bonneval's and de Tott's earlier versions of the school. The school's curriculum was impressive. In addition to lessons in French and English and military science courses, student officers of the New Order military corps studied a rigorous array of basic and advanced mathematics and physics courses that included geometry, trigonometry, algebra, conics, differential and integral calculus, statics and dynamics, physics, ballistics, Newtonian mechanics and astronomy.[8] The fourth year was devoted entirely to science and technology. Gifted students who desired to go on to advanced studies after graduation could study optics, advanced mechanics, and complex variables. The age of entering students was recorded and surprisingly low, between 10 and 12, quite a change from Baron de Tott's 40 and 50-year-olds. The reason given for admitting them at this early age is that the reformers wanted students whose minds had not yet been so hardened by years of Quranic memorization and religious catechisms that critical thinking would have been impossible. Selim donated his own collections of Western scientific books to the new military school's library, which was said to have contained some 400 European books, most of them, like the school's teaching staff, being French. While the New Order program of reforms was overall pretty much a joint Franco–Ottoman enterprise, Selim's engineering school, based as it was upon the precedents established by Count Bonneval and Baron de Tott in the earlier engineering and artillery schools, can be considered a French undertaking.

The school and the formidable education it offered must have been highly esteemed. Applicants had to stand for an entrance exam supervised by an Imperial Council. This was designed to prevent privileged families from entering their sons through bribery, favoritism and nepotism. By having the Shaykh al-Islam join the Imperial Council in overseeing the entrance exam, Selim sought to give the school religious credentials without it having anything to do with religion. Quite the contrary. The students were dressed, drilled, and disciplined in the best European military style by European officers. The European staff was assisted by Ottoman officers who had been, to some extent, Europeanized themselves. On the marching field and in the classroom, the students and officers-in-training were constantly being drilled in the principles of duty, obedience and responsibility. The officers-in-training, along with the New Order troops they would be leading, were paid on time and lived in large, relatively comfortable barracks that were distant and hidden from the Janissaries and other traditional military organizations, whose conditions would have appeared wretched in comparison.

A serious flaw on the academic side of the program was the lack of adequate student textual material, not an unfamiliar problem. It had crippled reformist education ever since its beginning under Bonneval. The reformers relied on the press at the French embassy to print up as much reading material for the students as the embassy was able to do. The embassy used its Arabic font to print Turkish translations of French lectures and texts for students to study. Having sufficient texts for students would be a problem in the schools established by all Muslim reformers in the 19th century. There would be no books at all for the students when Egypt set off on its road to reform in the early 19th century. The language of instruction was another stumbling block. Lectures continued to be in French, a language that the students failed to understand adequately until they were about to graduate, if then. As previously mentioned, local French-speaking Armenians were used as translators, when they could be found.

In addition to the Muhendesehane, Selim revived and expanded the naval engineering school that the French had built during Khalil Hamid's vizierate. It had the same basic scientific and technical curriculum followed by the artillery and infantry cadets in the Muhendesehane. The French also renovated the naval arsenal that their compatriots had built during the time of Khalil Hamid. A marine medical school was built inside the arsenal to train naval physicians. Traditionally in the Ottoman navy, the practice had been to leave wounded seamen to die or recover on their own, so there had been no need of physicians.

The problem of student textual material was perhaps less acute in the naval medical school than in the Muhendesehane, probably because someone high in the government favored the school. In 1795, the government generously gave the naval school its own printing press and covered expenses for 15 operators to print Turkish translations of European medical works. The medical books, almost all French, were hurriedly translated to Turkish and printed. Western visitors to the school report having seen up-to-date European medical journals in the school's library, but it is doubtful the students knew enough French, German or English to benefit from them, since right up to the last year of the curriculum European instructors were obliged to communicate their lectures through translators.

When the French were forced to leave in 1798, the military engineering school, which had by then been in existence barely four years, had a total of 80 students, all specializing in one of four fields of military science: artillery, fortification, mining, and engineering. The Muhendesehane ceased growing after that and, like the teaching institutions of previous reform efforts, its student population and course offerings began to shrink. Shortly after the execution of Selim and the blood-drenched end of his New Order, a French traveler visiting the school reported that it had an enrollment of only 40 students.[9] Concomitantly, the initial enthusiasm for the naval medical school, and the government generosity that gave it its own printing press and operators, failed to last. Just before the revolt broke out that ended the New Order, the school had only two full-time teaching physicians and a student body of six, showing it suffered the same decline as the Muhendesehane. This sorry picture was all in the pattern of Ottoman reform. The oft interrupted reforms during the critical century beginning with the Tulip Period deprived the new institutions of continual growth and the chance to sink roots and become institutions accepted as authentic, of sprouting and spreading their own seeds to produce a flourishing indigenous intellectual landscape. As will be seen, Egypt also was to suffer the same halting on-and-off pattern of reform; the same gross underfunding and political whiplashes that would cripple reform in the 19th century, whose seeds of failure would produce bitter fruit for generation after generation of Middle Easterners up to the present day.

Signs of frugality were to be seen everywhere in Selim's schools: in the absence of blackboards and chalk needed by instructors to explain technical diagrams and paper needed by students for taking notes. For all the grand plans of Selim and his ministers, the glacial progress of Ottoman scientific and technological assimilation can be directly related to the inadequacy of government funding. One can imagine the humiliation of Ottoman officials when the French embassy had to offer the Muhendesehane its printing press to print student copies of the Turkish translations of the lectures. Penury was emblematic of the new schools. Levying taxes on the wealthy, reducing palace expenditures, selling off slaves and concubines and other such measures were not options that the sultan and his band of reformers seriously considered as a means to add revenues to the constantly depleted treasury, though there were voices imploring them to do so. Resistance to the reforms was bad enough without having the whole extended royal family and its wealthy friends rise up in revolt.

Most of the funding available for reform went to pay the high salaries of the foreign experts and military advisors. Infrastructure was left to suffer. Selim's government considered making a loan from a European country to help fund the schools, the first instance recorded in Muslim history of such a consideration, but the conservative opposition so aggressively propagandized the idea as being an unacceptable humiliation to Islam – Muslims begging from the infidel West! – that the government quickly dropped it. A loan from an Islamic government was then considered, but upon some investigation it was discovered that all the other Islamic countries were even poorer than the Ottomans.

Desperate to climb out of their financial hole and fund reforms, Selim's government tried to emulate Europe by promoting international trade. To this end the Ottomans moved to terminate those detested capitulations that gave European traders and merchants a privileged legal status and advantageous customs rates over indigenous traders. This met with such fierce resistance by Europe's consuls and ambassadors that the government backed off. Even those countries most sympathetic to the cause of Ottoman reform, France at the front, refused to surrender their capitulatory privileges. "Thus in many ways Europeans in the Empire became as strong defenders of their vested interests and opponents of real reforms as were the most reactionary members of the Old Ottoman ruling classes."[10] Here again was another instance of Ottoman reformers being caught between the anvil of conservative opposition and the hammer of European economic self-interest. The reformers stood alone, resisted from within and without. The consequence was a perennial story of mounting humiliation, territorial loss, debt, and ultimately Ottoman extinction. In the end, both reformers and conservatives went down.

Simply put, Ottoman government and society failed to come up with the money to reform the state out of its sorry plight. Reformers made the best of it by scrounging in every corner for funds to keep the reforms moving, one of those corners being the free cadavers of freshly deceased prisoners in the arsenal's jail that was delivered to the medical school's anatomy classes. After 12 years of scraping funds from her and there, Selim abandoned the common sense of historical experience and in a reckless moment of pecuniary impatience imposed on the Janissary Corps the same drills and discipline given the New Order troops, whose ranks he wanted to swell without having to incur additional expenditures, which could not be covered. This was the beginning of the end. Selim had earlier planned to put the Janissaries down, once the New Order troops were capable of facing up to them. He knew it would not be easy: the Janissaries fought many times more bravely and fiercely inside the city against their own than on the battlefield against the enemy. In the meantime, he mollified their officers with the age-old custom of bonus pay. Thinking to end this wasteful drain on the practically empty treasury and at the same time build up the new troops, he fatefully decided to fold the old into the new. The old refused to fold.

Descriptions of how arrogant and recalcitrant the once-famed Janissaries had become challenge belief, but their unruly insubordination is one of the few things upon which European and Ottoman accounts are in complete accord. Europeans who witnessed the corps' behavior on the battlefield wrote in astonishment of the fear that the Janissaries struck, not in the enemy troops, but in the hearts of their own commanders and grand viziers. They refused orders to engage in battle and acted more like a rioting mob of wild thugs than an army. When discontented, they tipped over their huge soup cauldrons and went on rampages, threatening the government until their demands were met.

This is what they did in 1807, when the sultan imposed the New Order discipline on them. Ordered to dress in the European-style uniforms of the New Order troops and march in step, they broke loose and rampaged through the streets. They were joined by the throngs of discontented and wretched, by everyone who felt anger at the social injustice and who suffered from the plunging economy: religious students clamoring for Islamic justice, unemployed craftsmen whose trades had been undermined by the flood of cheap goods from Europe and dislocated peasants who had lost their land or abandoned it because of the heavy taxes and who had fled to cities and were struggling at the margins of the faltering urban economy. The bloody burst of emotion, building for years, was fed by political factors as well, among them Europe's refusal to renegotiate the capitulations, the French occupation of Egypt and Bonaparte's justification of it by propagandizing Selim as a tyrannical heretic, a traitor to Islam, a charge that the conservative opposition would champion. Once the rampage had gathered force, the reformers were hunted down in an orgy of violence that reads like a replay of the blood bath that terminated the Tulip Period.

One of the first victims was Mahmud Ingiliz Efendi, killed by a Janissary guard for his having ordered the troops to dress in the English style uniforms that he himself had designed. Sultan Selim, uncompromisingly tarred with the brush of treachery by virtue of his reforms that the conservatives claimed served only the interests of Europe, looked to be that, a traitor and a heretic, as proved by the loss of Egypt. French perfidy won the conservatives their case, proving that dealing with and trusting Europe could only be detrimental to Islam and the empire. It was to sleep with the enemy. It was degrading. It was the beggar kissing the hand that held the dagger.

To the conservatives and their traditionalist allies in religion, the French invasion and occupation of Egypt in 1798 at first appeared to be a godsend in that it forced Selim to order the French instructors, engineers, technicians and advisors out of the country, for how could the French invade a province of the empire while at the same time reform the Ottoman military to defend the empire? That presented too much of a contradiction even for the reformers, who had shown themselves willing to accept a great many contradictions in their critical dealings with the French. And so the conservatives had their victory over the reformers. But the victory was illusory. It only opened the way for the British to take the place of the French, both as advisors to the sultan on reform and, a few years later, as occupiers of Egypt: after an Anglo–Ottoman alliance had succeeded in forcing the French out of Egypt, a British naval squadron returned to the shores of Alexandria intending that the British should take the place of the French. The attack failed, but the conservatives in Istanbul could not have asked for a better confirmation of European perfidy. The British were as faithless as the French.

The French occupation of Egypt in 1798 was a political watershed in Ottoman relations with the West. It opened Egypt and Syria to new influences. It led to Egypt becoming an independent power that threatened Ottoman existence as much as the Russians did; and it led to Egypt at long last becoming a possession of the British Empire. The French occupation also brought the arts, crafts, technology, medicine and science of modern Europe right to the heart of the Muslim world.

Notes

1 For origins, scope, financing and structure of the New Order as it related to the military, see Stanford Shaw, "The Nizam-i Cedid under Sultan Selim III, 1789–1807," *Oriens*, Brill, Leiden, the Netherlands, vol.18, 1965–1966, pp. 168–184.

2 When he was seven years old, Yusuf had been enslaved by the commander of the Janissary garrison on the island of Crete. He was later made a palace household slave in the capital, where he received an education and rose up through the administrative ranks to become the grand vizier's personal secretary. S. Shaw, *Between Old and New*:

The Ottoman Empire under Sultan Selim III 1789–1807, Harvard University Press, Cambridge, 1971, pp. 87–88.

3 In England, Mahmud imbibed English culture to the full. He began dressing in the English fashion of a gentleman, adopted English furniture and manners, which was all well and good in London, but keeping to his adopted identity back in Istanbul met with derisive scorn, not the least expression of which was his being branded with a new name, Ingiliz Mahmud. He was the first Ottoman to break decisively with tradition by creating a new kind of personal identity, a break that would not have been unusual a half century later but for his time was decidedly daring, even reckless, as it turned out.

4 For a general account of the high ulema's cooperative and contributive role in reform, see Uriel Heyd, "The Ottoman 'Ulema and Westernization in the Time of Selim III and Mahmud II'," *Scripta Hierosolymitana*, XI, Jerusalem, 1961, pp. 63–86.

5 Shaw, "Nizam-i Cedid Army Under Sultan Selim III, 1789–1807", *Oriens*, Brill, Leiden, the Netherlands, vol. 18/19, 1965/66, p. 171.

6 Uriel Heyd, "Ottoman Ulema and Westernization in The Time of Selim III and Mahmud II," in *The Modern Middle East*, edited by A. Hourani, P. Khoury, and M. Wilson, University of California Press, Berkeley, 1993, pp. 29–60.

7 Shaw, "Nizam-i Cedid Army Under Sultan Selim III," pp. 171–178.

8 Mehmed Esad, *Mir'at -i Muhendishane*, Istanbul, 1896, pp. 9–26, 30–31; A. Adivar, *Osmanli Turk Ilim*, Remzi Kitabevi, Ankara, 1982, pp. 186–192.

9 Charles Pertusier, *Picturesque Promenades in and Near Constantinople*, Richard Phillips and Co., London, 1820, pp. 46–49.

10 Shaw, *Between Old and New*, p. 179.

Part IV

Catching up to the West

Science assimilation in Cairo and Istanbul under autocratic reformers

14 The West's continuing progress

While the miniscule student communities recruited into the Ottoman military schools were being introduced to Copernicus, Galileo, Kepler, Descartes and Newton, science in Europe continued to progress, accelerating from discovery to discovery, leaving Muslims ever more behind. A brief survey of discovery and invention across Europe from the time that Western science first penetrated the Ottoman Empire in the 17th century up to the early 19th century gives an idea of the growing chasm that would work harshly against Muslims, and all the more harshly with each succeeding generation, as reformers struggled unsuccessfully to close the gap, condemning the children of generations to pay in blood, subjugation and humiliation the price of their fathers' failure.

In Scotland around 1663, the astronomer–mathematician James Gregory designed an improved telescope by combining a primary parabolic mirror and a secondary concave elliptical mirror. A few years later, Christian Huygens in Holland discovered the relationship between a pendulum's length and period of oscillation. In Switzerland, Johann Bernoulli (d. 1738) laid the basis of hydrodynamics and advanced the calculus of Newton and Leibnitz. Bernoulli's Swiss student Leonhard Euler (d. 1783) formulated new methods of analysis in pure and applied mathematics and made contributions in practically all branches of mathematics. In the next generation, Daniel Bernoulli (d.1782), Johann's son, carried on his father's work in fluid mechanics and achieved enduring fame for the Bernoulli Equation, a mathematical relationship based on the principle that the total energy of a fluid particle flowing through an orifice remains constant, any change in the kinetic state being balanced by a corresponding change in the potential.

In Sweden, the physicist Samuel Klingenstierna developed lenses for large achromatic telescopes, eliminating the color distortion produced when combining convex and concave lenses for greater magnification. In England, the astronomer Edmond Halley, who it will be recalled was instrumental in Newton's publication of the discoveries he had made during the plague years of his youth, applied Newton's law of universal gravitation to predict the return of the comet that bears his name. In Germany, the astronomer Johannes Hevelius (d. 1687), assisted by his talented wife Elizabeth, made new observational discoveries with an improved telescope. The chemists Stahl and Homberg in Germany, and Boerhaave in Holland, were preparing the way for Boyle, Cavendish, Priestly, Lavoisier and Dalton in laying the foundations of modern chemistry, the latter three, two Englishmen and a Frenchman, sharing honors as the Fathers of Modern Chemistry.

In Italy, where science quietly continued after Galileo's trial, Galileo's assistant, Torricelli, discovered in the 1630s that a column of mercury in an evacuated tube would rise to a maximum height (30 inches in English units), which he found corresponded to the height of 34 feet that an artesian pump could lift water from a well. Pascal in France advanced on this in 1648 by climbing a mountain and experimentally finding the difference in height that mercury rises in a glass column at different altitudes, proving that atmospheric pressure decreases with elevation. The 18th century Italians Giuseppe Compani, Gian Cassini and Joseph Lagrange were among the most productive astronomers in Europe. Cassini's fame earned him an invitation to Paris to found the Royal Observatory of Louis XIV; Lagrange, from Turin, also went to Paris where the opportunities of scientific endeavor were more lucrative and the atmosphere more congenial.

In Denmark, the astronomer Ole Roemer in 1676, aided by advances in telescopic technology, measured the speed of light by determining the difference in time it took light to reach earth when a satellite of Jupiter first appeared from being eclipsed behind the planet at the moment the earth was closest to Jupiter and then when the earth was furthest away from the planet. The speed of light was then computed by dividing the mean diameter of earth's orbit by the difference in observed time between the two appearances of the satellite.[1] Roemer's telescopic determination of the speed of light exemplified the close relationship that was emerging between science and the technology of observation and measurement, an interdependence that was rapidly becoming fundamental to advances in both science and technology.

Improvements in telescopic technology in the 17th century enabled astronomers to detect Jupiter's and Saturn's perturbations, which, applied to Newton's law of universal gravitation (that the force of attraction between two bodies was directly proportional to the product of their masses and inversely to the square of the distance between them) pointed to the presence of a planet whose distance was too far to be observed directly but whose mass and orbit around the sun could be calculated. The unknown planet (Uranus) was observed in 1781 by William Herschel (d. 1822), inventor of the reflecting telescope and the most famous astronomer of his time. His sister Caroline assisted him and also became a recognized astronomer. Women first came into science as family assistants in the 18th century, and by the last quarter of the next century, they were becoming scientists on their own initiative, potentially doubling the gene pool of scientific genius.

Observations of Uranus revealed perturbations, indicating that yet another planet was out there. An English and a French astronomer each independently made the Newtonian calculations of the unknown planet's orbit and mass, while a German astronomer telescopically detected the planet based on their calculations. This one was named Neptune. Pluto, recently having been demoted from planethood, was the last to be discovered. In the course of the 19th century, with the continual development of more sophisticated telescopes, the earth and the solar system were seen to be swimming in a galaxy whose spatial dimensions and breathless infinitude of stars defied imagination.

Medicine was also a beneficiary of the experimental methods and enthusiasm generated by Newton's discoveries in the physical sciences. With the authority of Galen, Razi and ibn Sina still holding, the Englishman William Harvey (d. 1657), who studied at the famous medical school in Padua, applied the empirical method to the human body and, by dissection of cadavers, uncovered the circulatory system of the blood. His discovery was published in 1628 as *De motu Cordis et Sanguinis*. Following in Harvey's tradition, the Dutch physician and chemist Herman Boerhaave, convinced that the old medical theory had to give way to new facts, proposed an empirical philosophy of medicine based on the theory and method of Newtonian science.

Scientific knowledge had been expanding in two interdependent dimensions since the beginning of the 17th century: vertically, with the new discoveries that followed one upon the other, and horizontally, with the communication of those discoveries from scholar to scholar and city to city across the European continent west of the Danube. The printing press, of course, had been an effective means of communication since the late 15th century. Communication was enhanced by the scientific societies that arose. Italy, the leader in so many features of early modern civilization, was the home of three of the earliest and most important of these societies. The first was founded in 1560 in Naples, the Academia Secretorum Naturae; shortly after, in Florence, the Academia dei Lincei, of which Galileo was a prominent member, was founded. The Academia del Cimento (referring to experiment) was also founded in Florence, in 1651, some years after Galileo's trial, another indication that the event did not bring science to a standstill in Italy.

Though Italian in origin, the Cimento included scientists from all over Europe and was transitional to the national societies that arose in the next decade, the leading ones being England's Royal Society, founded in 1662 by Charles II for the advancement of science. (This was in its origin a small society founded in 1645, in London, by members of Oxford's Gresham College who wanted to promote the appreciation of science in the capital city.) Not to be outdone by his English rival, Louis XIV founded the Academie des Sciences and the Observatoire Royale four years later. As had been the case with the medieval universities whose models had been Paris and Bologne, national science societies sprouted up all over Europe modeled after those of England and France, and like them, they published their own journals to broadcast new discoveries and promote further research.

Scientific enthusiasm expanded beyond the institutional professional ranks to the non-professional, where the amateur became a common feature of intellectual life. The society begun at Gresham College was originally founded to bring together in a forum of discussion those who were interested in natural studies but were not necessarily professionally involved in science or even enrolled students of it. The Royal Society's membership reflected the mixed amateur and professional nature of the Gresham group. An important member of the Royal Society was an Irish amateur astronomer residing in London, Thomas Streete, an ardent observer of the heavens who kept his fellow members abreast of astronomical advances made throughout Europe. It was in fact through Thomas Streete that Newton gained knowledge of Kepler's first and third laws. Newton's laws were communicated in turn across the channel to the continent by another amateur enthusiast, one better known than Streete.

Voltaire, a literary writer without formal scientific training, introduced Newtonian science to the continent when he discovered Newton's *Principia* during his stay in England between 1726 and 1729, where he had gone to escape political difficulties in France. He arrived in London in time to attend Newton's funeral at Westminster Abbey and was so impressed by the profundity of Newton's work that he devoted five years to studying the *Principia*, the main part of which he translated into French and had published in 1738 as the *Philosophy of Newton*. The cross-channel communications of Kepler's and Newton's laws give an idea of the time it took in the 17th century for diffusion of new knowledge between Britain and the continent, about half a century in each case. That narrow channel of sea made England a world unto itself. The channel may as well have been an ocean. The effect of it is seen currently in Britain's economic separation from the continent, where communications were much quicker and would become more so in the succeeding century.

The propagation of scientific interest and quickening pace of discovery created a social pressure in the world of science where competition over awards, positions, endowed chairs, prestige of publication, and bitter disputes over priority of discovery that led to lifelong hatreds, such as that between Newton and Hooke, became part of the intellectual terrain, revealing both the greatness and pettiness of scientists.

By Newton's generation, the institutionalization of science had progressed so far and was so deeply rooted in European society that science was able to propagate itself, independent of the storms of religion and whims of kings. Interest in science permeated the whole of literate society, drawing into its orb amateurs from a large range of non-scientific professions. On occasion, amateur interest became consuming, and a scientist was born. Women began making names for themselves in astronomy. Families, societies and institutions were taking up science, providing it with support outside of government and academia. Where in Greco–Roman and Islamic society science had withered because it "did not take root in the general education nor yield the benefits it is capable of to men at large,"[2] science in the West lived by independent societies; it was integrated into the primary and secondary educational curriculum and popularized and identified with health and civilization. Science was recognized as basic and indispensable to the sustenance of life.

So fully integrated into Western society did science become, and so towering the figure of the scientist, that in the 19th and 20th centuries the scientist came to be more respected and to carry more prestige than clergymen, a feature that would astonish Muslim observers, striking some with horror, others with admiration. The magnetic pull of science was irresistible enough that religion recast itself in its image. Christian Science and Scientology live on and have given birth to many offspring that may be bogus as far as science is concerned but are nonetheless spiritually uplifting to many.

Not that science in the West was all that free of its own problems with religious authorities. What society was? Anaxagoras had to flee or was banished for teaching that the sun was a burning stone whose light was reflected by the moon, which was a planet physically no different from the earth. Xenophanes was persecuted; Socrates was executed – and Aristotle took flight to escape a similar fate at the hands of Athenian government. Aeschylus and Euripides voluntarily left Athens because of conservative reactions against ideas of theirs that were seen to be opposed to religious and social norms. Two thousand years later, Luther prohibited publication of Copernicus's book, Bruno was burned and Galileo was forced to recant. When in 1721 Lady Wortley Montague, whose husband was British Ambassador to the Ottomans, brought the technique of inoculation against smallpox from Istanbul to England, the idea was so strenuously resisted by the Anglican clergy that only after the royal family agreed to be inoculated did religious resistance begin to waver. Jenner's introduction of vaccination was also resisted on the religious basis that it was a rejection of God's will, disease being divine retribution. The use of anesthetics in obstetrics was at first opposed in the United States because it was thought that the avoidance of pain was an impious escape from the Biblical injunction of suffering imposed on women for Eve's disobedience in Genesis 3:16.[3] Darwin's theory was welcomed even less than inoculation, vaccination and anesthesia. But the pervasiveness of science and all that had been invested in it, the momentum built by its prestige, the depth of its social roots and the diffusion of its institutionalization had become too great for religious resistance to curb or impose a constricting redefinition of limits.

Following the publication of Newton's *Principia*, scientific discovery became not only commonplace but expected. Fundamental to this was the advancement of technological craftsmanship that went hand in hand with the exploration of nature. The ties linking applied science to warfare and economics go back to Hellenistic times; to the state's support of alchemy and astrology in medieval times; to ballistics and metallurgy during the Renaissance; and no end of instances after that. The reckoning of longitude was one of the earliest and most salient.

As Western maritime powers were creating their commercial empires around the globe and as a few degrees error in computing longitude resulted in loss of ships, crews and treasure, an accurate method of measurement became imperative. Latitude is easily known by sun and stars, but traveling over the undulating featureless ocean and accurately knowing one's position east or west of anywhere was most difficult, especially when measuring elevations of stars on a surface that was anything but placid. The problem presented a challenge that the British Parliament hoped to resolve by offering a huge prize to the inventor who built an accurate chronometer that could withstand violent rocking, humidity, corrosion, severe changes in atmospheric pressure and temperature and all the adverse conditions of the high seas. When John Harrison in 1763 finally produced his marvelous spring-loaded and solidly encased device after two decades of experimental development, during which Parliament did everything to get out of paying him the prize by repeatedly revising upwards its demands on the degree of precision, durability and severity of physical conditions, a reliable means of measuring longitude at last became available. Set at Greenwich meantime, the chronometer measures the hours and minutes, Greenwich being the standard meridian of zero degrees longitude. Knowing local time, one need only subtract it from Greenwich mean time and multiply the difference in hours by 15 (15 degrees/hour multiplied by 24 hours equaling 360, the angular circumference of the earth) to determine local longitude, east or west of Greenwich. Harrison's chronometer of 1763 was to navigation what Herschel's reflecting telescope of 1778 was to observational astronomy: triumphs of technology serving science.

The 18th century revolution in industrial production began with the steam pump, a device that was invented to remove water from the shafts in English coal mines that flooded when the shaft reached a certain depth. At an earlier time, the flooded mine would have been abandoned, but by the mid-18th century the technology of Western society had so advanced the mental frontiers of what was possible that water knee-deep in the shaft simply amounted to a problem to be solved. All was possible. In such a mind-set, the application of the steam-driven pump to the textile industry that initiated the industrial revolution, less than half a century after Newton's death, was no more than a logical step along a path already well-established.

The Industrial Revolution held far graver consequences for the world at large than the Scientific Revolution, though the web of history knits one to the other. One was a revolution of mind interpreting nature through mathematics, the other a revolution of material production, but both were rooted in mind-sets that went far back in time, the former to Socrates, Plato, Aristotle, ibn Rushd, Aquinas and the scholastic cult of reason, the latter to the complex machines envisioned by Archimedes, Hero, al-Jazari, Roger Bacon, Leonardo da Vinci, and the mathematical and experimental refinements of ibn al-Haytham and the Italian filosofo geometra with their dexterous craftsmanship. Capitalist investment in industrial production, protected by the formidable power of the state, and fortified of course by the exigencies of war, made the dream machines of Archimedes, al-Jazari and Leonardo a reality.

The second of the great revolutions accelerated and expanded the technology of society, completing the centuries-long shift from reliance on tradition, authority and continuity of social custom to reliance on private calculation and innovation. In the technical society, specialized technical considerations took precedence over all others and facilitated a change in the patterns of investment of time and money. If social and religious change characterized Western society in the 16th and 17th centuries, purposeful technical innovation characterized it in the succeeding centuries:

> A rationalizing calculativeness, crucial to technical specialization, depended, especially at first, on an expectation of continuous innovation: on encouraging an attitude of willingness to experiment, taking as little as possible for granted what had already been thought and done, rejecting established authority of every sort, and running the inherent risks of error that such rejection entails.[4]

The heightened social expectations of continuing discovery excited Enlightenment essayists to prophesize universal health, education, affluence and general well-being. Progressing science, medicine and technology, nurtured by rationalized government, would eradicate society's evils and solve its problems by eliminating disease, war, famine and hunger, lifting mankind to moral perfection. The Enlightenment's dressing up of science in a rosy philosophy of positive change toward the best of all possible worlds was an optimism science never claimed for itself. Having shed its medieval metaphysical teleology, science was eagerly veiled in a modern one that was formulated not by scientists but by political and social critics. Condorcet's *Sketch for a Historical Picture of the Progress of the Human Mind* was the everyman's bible of religious faith mapped in science as the road to human perfection. Diderot's encyclopedic 17-volume *Analytical Dictionary of the Sciences* popularized science, while Voltaire's 3-volume *Elements of Newton's Philosophy* presented Newton in simple terms to the general public.

Voltaire's Newton was more to the point of what science was, stripped of the Panglossian cheerleading of the *philosophes*. Voltaire's lead was followed in the last quarter of the 18th century by Lagrange and Laplace, who fleshed out Newtonian mechanics and made it accessible to a broad public, thus performing for Newtonian mechanics the service Puerbach and Regiomontanus had for Ptolemaic astronomy three centuries earlier. In his *Mecanique Analytique*, Lagrange systemized differential equations and founded the whole of mechanics on the principle of conservation of energy. In his *Celeste Mecanique*, he translated Newton's *Principia* into the mathematical symbolism of the infinitesimal calculus, borrowing much from the Leibnitz system of notation; while Laplace, in his *System du Monde*, published two years before General Bonaparte set off for Egypt with his floating university, extended his general explication of Newtonian mechanics to include a hypothesis of the nebular origin of the solar system, according to which the solar system evolved from a spinning mass of incandescent gas.

Laplace carried mechanical theory to its logical limit by claiming that if at any instant the momentum and location of all matter in the universe were known, he could mathematically predict the history of the universe. His statement to Napoleon has to stand as the briskest declaration of divorce between science and religion ever recorded. After Bonaparte had abandoned his troops in Egypt and become emperor, he asked Laplace where the place of God was in his nebular theory, to which Laplace replied that God was a hypothesis he could do without. Amused, Napoleon later told Lagrange what Laplace had said. "Ah!" responded Lagrange with a smile. "But that (God) is a good hypothesis. It explains a great many things." Even deism was being elbowed to the margins by the aggressive secularism of the Enlightenment and the French Revolution.

To pious Muslims, who were already at the time being aggressively confronted by the new science, the message was crystal clear: the West had severed science from God and scripture and made it a one-way road to atheism, more lethal, evil and heretical than anything ever posed by Aristotle and ibn Sina. Western science did not even make room for Intelligent Design!

On the institutional level, after the end of Jacobin rule and the Reign of Terror, the leaders of the French Republic energetically set up scientific institutions and medical centers throughout France. The Ecole Normale, Ecole Polytechnique, Institut de France and Musee d'Histoire Naturalle collectively contributed to making France the unquestioned leader in science during the first half of the 19th century. Before the end of the 18th century, the French had instituted the decimal system to standardize measurements and end the competing systems that proved to be a source of confusion in science and technology and still are in some corners of the world. The French Academy defined one meter to be 1/10,000,000 of an earth's quadrant as measured from the North Pole to the equator running through Paris. The measure was marked on an iron bar at the freezing temperature of water, which was standardized as 0° centigrade. One hundred degrees centigrade was the temperature at which pure water boiled at sea level. A kilogram was standardized as the weight of 1/10 of a cubic meter of water at 4°C, whereby a gram was equivalent to one cubic centimeter of water. The system has been almost universally adopted.

Scientific progress measured across the spectrum of natural philosophy was uneven. Astronomy, the queen of the sciences, received abundant attention by legions of professionals and amateurs throughout Europe. Chemistry lagged behind where Paracelsus had left it; optics also stagnated. Both began advancing around the middle of the 18th century.

In chemistry, bases, acids and salts were redefined. Hydrogen and oxygen were liberated by reacting oxides and acid: hydrogen in 1766 by Henry Cavendish (d. 1810) and oxygen in 1774 by Joseph Priestley (d. 1804). Chlorine was discovered. Cavendish demonstrated water to be a compound and not an element. Advancing the discoveries of Torricelli and Pascal in the previous century, the English physicist and chemist Robert Boyle (d. 1691) discovered the inverse relationship between pressure and volume (at constant temperature) that governs all gases. His experiments on the pneumatic character of air opened the way to Dalton's atomic theory of chemistry. Here again technology came to the service of science. The air pump had been invented in 1657 by the German Otto von Guericke and improved by Boyle and Hooke. By pumping air up to pressure, Boyle could feel the compressed air springing against the pump handle, which led him to conclude that air was composed of miniscule corpuscles that when crowded together acted in a springy resistant fashion, the result of the tiny particles of air bouncing off each other.

Antoine Lavoisier (d. 1794), the French Father of Chemistry, set forth a system of chemical notation and performed fundamental experiments in oxidation and reduction. By using carefully measured amounts of mercury and air to produce mercuric oxide and then reversing the reaction by reducing the oxide to produce the original amount of mercury and oxygen, he proved the falsity of the phlogiston theory of combustion. By the end of the 18th century, Cavendish, Boyle and Lavoisier had set the stage for Dalton's quantification of the corpuscular theory, a revolution in chemistry that was as significant for science in the 19th century as the 17th century revolution that mathematically united physics and astronomy. Dalton (d. 1844) was honored as yet another Father of Modern Chemistry.

Progress in optics suffered a long pause following the discoveries of Descartes and Newton. Newton's experiments in refraction led him to conclude light was corpuscular, that a ray of light was composed of tiny disconnected particles traveling in a straight line. Hooke and Huyghens concluded from their own investigations that light traveled in waves, but Newton's authority had reached Aristotelian proportions in the world of science in the late 17th and 18th centuries and so for over a century his views went unchallenged, which in effect froze optical theory in the state he left it.

Toward the end of the century the Englishman Thomas Young (d. 1829) bent himself to the subject and repeated the pinhole refraction experiments in a dark room. After meticulously analyzing and measuring the light and dark areas of the interference patterns on the screen, Young revived the wave theory of Hooke and Huyghens. The debate over the wave-particle nature of light made optics an exciting challenge that attracted researchers to attempt a resolution of the dualism. Young's research and conclusions were not long in being communicated across the channel. The cross-channel exchange registered another triumph for science. Transcending politics, national rivalries and war, the peaceful and productive collaboration of science was recognized when the Academie Francaise made both Cavendish and Priestley honorary members, even with the two countries being at the time locked in battle.

In France, Augustin Fresnel (d. 1827) applied a recently discovered mathematical method of harmonic analysis to the patterns of refracted light. This was a method of analyzing periodic phenomena such as the vibrating waves emitted by drums that had been developed by Fourier, one of the legion of scientists who had accompanied Bonaparte in the French conquest of Egypt in 1798. Fresnel intuited that if the propagation of light was indeed a wave function, then the diffraction patterns could be mathematically determined by Fourier's method of periodic analysis. Accordingly, Fresnel was able to prove that the dark bands in the shadow pattern were caused by interference of waves that were out of phase, showing that light traveled in waves. The authority of Newton's corpuscular theory gave way. Fresnel's confirmation of the wave theory of light presented a new problem that dogged physicists and astronomers through the 19th century. By what medium did the waves propagate themselves in space, where it was assumed that at a certain distance beyond the earth's surface the atmosphere thinned out to nothing, to a vacuum? Waves could not travel in a vacuum. Newtonian mechanics and common sense said so. Some material substance had to carry them. Theory came to the rescue to explain what could not be sensed or detected but had to exist. And so it was assumed that an ethereal luminiferous substance pervaded space and bore the waves of light over the astronomical distances from sun and stars to earth: as phlogiston supported fire, so the ether carried light.

The French remained dominant in optical studies deep into the 19th century. Fresnel improved on Roemer's 17th century value for the speed of light by using the mechanical method of projecting a beam of light through a calibrated system of two gapped gears rotating on the same shaft and set at a known distance from each other. The speed of light could be indirectly determined by measuring the rotation speed that permitted a beam of light to pass through the two rotating gears. It was an indirect but brilliant way to measure the time interval of the beam's passage from the first gear to the second, and hence determine the speed of light.

Armand Fizeau and Leon Foucault, each using an ingenious method, further improved on the value for the speed of light and on Fresnel's mathematical and experimental analysis of diffraction.

Along with advances in chemistry and optics, theories of magnetism and electricity based on experimental research were formulated in the 18th and 19th centuries. It was discovered that the inverse square law relating force to distance in Newton's equation of universal gravitation applied also to magnetism. The research of Henry Cavendish and Joseph Priestley, who was a Unitarian minister and an amateur chemist, determined that electric force was directly proportional to charge. Benjamin Franklin, the first American scientist and an amateur, like so many who at the time were making themselves a name in science, experimentally demonstrated that lightning was electricity. Electricity and magnetism were likened to gravity, forces acting at a distance through ethereal space. Newtonian physics being the central paradigm, scientists looked to it in their search for relationships. But what was it that permitted these forces to transmit themselves and act on distant bodies? Pondering the problem led to the idea of an invisible physical field of force reaching out in all directions, its strength diminishing in proportion to the distance multiplied by itself – an early groping toward a conceptualization of gravitational and electromagnetic fields.

Discoveries in electromagnetism and chemistry had an industrial counterpart. As wind and waterfalls had powered machines in an earlier age, electromagnetism and electrochemistry would power them in the new age. In 1800 the Italian physicist and chemist Alessandro Volta (d. 1827) set up columns of alternate disks of copper and zinc separated by paper soaked in brine to produce the first electrochemical battery. He also produced an electric charge by placing a strip of copper and one of zinc in diluted sulfuric acid. Here in its most primitive form was the source of power that was to turn driveshafts, gears, axles, wheels and propellers. Andre Ampere (d. 1836) carried out more experiments on electricity and found the mathematical relationship for the electromotive force produced by two wires carrying currents. By the middle of the century, the electric motor was taking the place of the steam engine as the source of power to drive the machines of industry.

Advances in chemistry and electricity produced a second industrial revolution. Iron and steel production were revolutionized. Ships that were driven by wind and sail for millennia began in the 1820s to be driven by coal and steam; very early in the next century coal would give way to oil. Science and technology were producing industrial and military power unimagined before 1820, except perhaps for such visionaries as Roger Bacon, Leonardo da Vinci and Condorcet.

In the latter part of the 19th century, electricity and magnetism were discovered by Faraday to be physical aspects of the same force, giving birth to the science of electromagnetism. James Maxwell, who found that light was a manifestation of electromagnetism, supplied the basic mathematical laws. Carrying on from Maxwell's work on electromagnetism, J. J. Thomson, by analyzing the ghostly light produced by passing an electric current into a vacuum tube, discovered the particle he later called an electron. The first glimpse of an electronic age was caught peeking over the horizon.

Thomson's discovery made it obvious that the structure of a simple atom was more complex than had been conceptualized by Dalton; so Thomson hypothesized his own model, a static structure that consisted of negatively charged particles scattered in no particular order throughout a positively charged body: the plum pudding model as it is called. Early in the 20th century, the plum pudding model was revolutionized by Thomson's student from New Zealand, Ernest Rutherford, who put the electrons in orbital motion around a positively charged nucleus of protons and neutrons. In 1905, young Albert Einstein took time out from his job in the Swiss patent office to publish five articles that did not land him a university position but did revolutionize science.

By the dawn of the 20th century, Europe had experienced two industrial revolutions and was rapidly verging upon a second scientific revolution. Muslims were meanwhile left grappling with the scriptural contradictions of a moving earth. The theory of Charles Darwin added another weapon of mass destruction to the arsenal of religious conservatism against science and was wielded with effect when the theory penetrated the Muslim world in the late 1870s.

The rise of science was a success story somewhat more inspiring than other modes of Western success. Unlike those unsavory features related to slavery and capitalism that made possible the economic and political rise of the West to world dominance, science did not conquer, subjugate, colonize, brutalize and dehumanize. Men who did engage in such things could nonetheless be generous benefactors of science, which, through its marriage to technology, empowered the state in protecting and extending the overseas ventures of its entrepreneurs, all the more of whose profits became available for scientific and technological investment. As the Western grip increasingly extended and tightened over the globe in the 17th, 18th and 19th centuries, merchants, industrialists, financiers and colonialist entrepreneurs would endow university chairs, found research centers and institutes of technology and create scholarships, all in the name of "progress:" the logo of higher civilization and code for higher profits. Viewed from the eyes of those upon whose backs progress was made, science could be seen as an accomplice, a weapon in the hands of the enemy oppressor. This would be the view taken by an important section of Muslim society.

Muslim nations had been slowly succumbing to the industrializing nations since the 18th century. In the 19th century and early decades of the next, as Muslims struggled to master the steam age, their countries would succumb at a more rapid pace to a Europe whose industrialized economy was now driven by electricity and electromagnetism. In a few Muslim countries there arose small parties of reformers who advocated the adoption of Western learning and techniques to employ the secrets of Western power and stave off further defeat, occupation and economic servitude. While reform and innovation were being planned and acted on from above, Muslim scholars, from whose ancestors the West had gained in great part the origins of its early science back in the 12th and 13th centuries, and whose ulema had looked askance at science and its underlying philosophy and had done what could be done to undermine the rational study of nature, argued anew whether it was permissible for Muslims to study science and import the technology of unbelievers.

Notes

1 The difference in time he measured as 16 minutes. The speed of light then became 3×10^{11} meters (diameter of the earth's orbit) divided by 960 seconds: 3.1×10^8 meters per second.
2 Benjamin Farrington, *Science in Antiquity*, Oxford University Press, London, 1950, p. 246.
3 John W. Draper, *History of the Conflicts Between Religion and Science*, Appleton, New York, 1875.
4 Marshall Hodgson, *Venture of Islam*, vol. 3, University of Chicago, Chicago, 1977, p. 193.

15 Bonaparte's expedition
Savants, shaykhs and the *Institut d'Egypte*

The French Expedition to Egypt in 1798 is consistently cited as the event that opened Egypt, and ultimately the other Arab provinces of the Ottoman Empire, to Western intellectual influences. In particular, it is the extraordinary scientific dimension of that French enterprise that is claimed to have made the most profound impression on Egyptians and initiated the intellectual revival that came to be known as the *Nahda*. With its many accomplished savants and the scientific equipment accompanying them, the Expedition of 1798 was indeed one of the most remarkable invasions in the annals of military history: a singular intersection of imperialism and enlightenment.[1] But how deeply the scientific component of Bonaparte's transported institute of science and technology impressed and influenced Egyptians is problematic. Egyptian sources indicate that those Egyptians in a position to appreciate the grand display of France's accomplishments in the modern arts and sciences were more confused and dismissive than impressed or inspired by the demonstrations the savants put on for them at the *Institut d'Egypte*. Recall Ottoman ambassador Hatti's reaction to demonstrations of Western science in Vienna.

The French conquest of Egypt had been long premeditated.[2] King Louis IX's crusading incursion in 1249 was a reminder of France's rich historical connection to Egypt. Though the expedition ended badly with the king's capture, imprisonment and ransom, the idea of taking possession of that most advantageously placed country was never far from French political imagination. In 1672, with the bellicose competition of European maritime powers raging over the globe, the mathematician Leibnitz advised another French Louis, the Sun King, to destroy Dutch commerce by seizing Egypt, which was even then seen as the gateway to India, the source of Holland's commercial fortune. Nothing came of the idea until a century later, when Egypt came to be seen as adequate compensation for French losses to Britain in India and North America. Nurtured by French travelers in Cairo and Alexandria, where sizeable French merchant colonies had been established during the 18th century, the idea of France taking possession of Egypt frequently arose in deliberations at the highest levels.[3]

The prize of Egypt became the central theme of a grandiose project written up in 1769 in a report by foreign minister Choiseul-Gouffier, assisted by Talleyrand and the king's ambassador to Istanbul, St. Priest. Egypt had to become French property. Only then could the country's strategic position at the corner of Asia, Africa and Europe be fruitfully exploited. A canal connecting the Red Sea and the Mediterranean would make Egypt a French-controlled gate to Afro–Asian trade with Europe. Under France, Egypt's agriculture and industry could be made to flourish. The marvelous but neglected antiquities could fill the new art museum in Paris, the Louvre. Stately obelisks would adorn the city's squares. Another report submitted to Louis XV proposed transporting Pompey's Column from Alexandria to Paris and surmounting it with a statue of the king.[4] The importance of Egypt to France's imperial destiny, her "mission civilisatrice," transcended all consideration of France's long friendship with the Ottoman Empire in the minds of French expansionists.

As the Ottomans were falling back before the Russian army in the war of 1768–1774 and Istanbul looked about to succumb, French ambitions soared. If destiny and France's civilizing mission failed to convince skeptics, then certainly Russia's occupation of the sultan's capital would justify France's being compensated with Egypt, which at the moment seemed invitingly ready for the taking. The country was in a state of turbulent autonomy under the Mamluk rebel Ali Bey the Great. Having invaded Arabia, Palestine and Syria, he was busily carving for himself an Arab empire out of the seemingly supine Ottoman body. France's conquest of Egypt would be easy and might even be interpreted as a favor to the sultan – an intervention against rebels carried out in the name of Islamic legitimacy and Ottoman integrity. The project, however, was delayed, then shelved when the Revolution broke out, only to be dusted off and refined when Britain and France fell into war. "To truly destroy England," Talleyrand wrote, "it is necessary we take Egypt, and the time is not far off." Less than a year later the long contemplated Egyptian project was at last launched.

The army Bonaparte commanded had originally been intended for England and accordingly named *l'armee d'Angleterre*. Since the landing craft for the invasion had not been prepared and the government was uncomfortable with a huge army left unemployed, General Bonaparte, eager to keep alive his reputation as a conquering hero, convinced the Directoire to send the army to Egypt under his command, whereupon *l'armee d'Angleterre* became *l'armee de l'orient*.

At dawn on May 19, 1798, a fleet of 400 ships carrying 34,000 men (300 of them women smuggled on board dressed as soldiers) and 167 civilian scientists, scholars, engineers, physicians, technicians and artists, sailed from Toulon. Slipping through the heavy fog past Lord Nelson's fleet blockading the port, the French expedition headed out to sea. It was not until the next day when the fog had lifted that Nelson became aware of what had happened. Sure that Bonaparte's destination was Egypt, Nelson sailed off in pursuit. But Bonaparte had the fleet sail first to Malta to liberate the little island of its gold and jewels to help finance the campaign; so while Bonaparte was plundering Malta, Nelson was in Alexandria being haughtily informed by the Mamluk bey in charge of the city that no Frankish fleet had come, and if one ever did foolishly dare to come, the invader would be trampled into the dust under the hooves of the Mamluk's horses, for Egypt was the land of the mighty sultan in Istanbul. With that, Nelson was ordered to leave. Figuring that Bonaparte must have stopped at Malta, Nelson raced his fleet there, but by the time he arrived, the French had finished their rape of the island and were back on their way to introduce Egypt to the modern world.

Like no other imperialist invasion, the French expedition to Egypt gave a semblance of substance to the usual empty, self-serving slogans of state propaganda used to excuse aggression. This was because of the scores of scientists, engineers, artists and scholars, the "savants" as they were called, who had been assembled to accompany the military. Primarily, the savants and their technical support staff were to study Egypt, to learn, record, collect and sketch what they found in Egypt, to produce a written, graphic and physical record of all facets of Egyptian life and culture, from pharaonic antiquity to the present, in addition to building a canal between the Red Sea and Mediterranean; also, the mission was to introduce the arts and sciences and modern civilization to the country. Including a major portion of France's scientific elite, the savants, with their shipload of crates of laboratory equipment and books, formed an advanced scientific institute, a floating university bound for Egypt across the sea: the Enlightenment's contribution to an otherwise mundane scheme of military conquest.[5]

In the words of a memoir of Talleyrand to the Directoire on the goals of the expedition, "Egypt was a province of the Roman Empire, it must now be one of the French Republic. The Roman conquest was a period of decadence of that beautiful country; the French conquest will be one of prosperity." Accordingly, the expedition was framed in the noblest, most altruistic intentions of liberating Egypt from the ravages of the tyrannical Mamluks. The French were to bestow upon the country the unquestioned benefits of modern civilization and the Revolution. Agriculture was to be expanded and modernized, new crops and industries introduced, the people educated in science. A Mediterranean–Red Sea canal would be constructed to make the country a major center of trade. For the young Arabist Edme Jomard accompanying the expedition, France would "carry the arts of Europe to a half-civilized people devoid of industry and the light of the sciences."

The savants were coming with the grand mission of enlightening and transforming the country while preserving for posterity a record of the history and religion, of the social, political and economic organization and the art, architecture, crafts and sciences of Egypt: Pharaonic, Ptolemaic, Roman and Islamic. The geography, topology and geology of the country would be studied, monuments carefully drawn, specimens of the flora and fauna collected and the Nile explored. It was to be a magnificent field trip in scientific investigation, combined with the noble task of educating a once-civilized but now barbarous people in the arts, crafts and science of modern civilization!

To accomplish this, Bonaparte, a military man who considered himself a son of the Enlightenment and who took keen interest in science and delighted in discussing chemistry, physics, astronomy and mathematics with scientists (a year before the expedition he had been elected to the French National Institute), enlisted as his chief scientific advisors two formidable personalities, the chemist Claude-Louis Berthollet and the mathematician Gaspard Monge. He put the two leading scientists in charge of recruiting the best talent they could for the ambitious mission. Both men were in the midst of stellar careers in their professions and had been recognized by the revolutionary leaders as prime assets of the Republic. Berthollet was the recognized founder of physical chemistry and peer of Lavoisier; Monge was the equal of Laplace and Lagrande.[6] Monge had been Lavoisier's assistant and was credited by the great chemist for discovering that water was composed of hydrogen and oxygen.

Monge's career as a researcher had ended in 1787, two years before the Revolution, when the Ministry of War sent him to inspect the iron works at Le Creusot, which still manufacturers some of the world's finest cooking utensils. Principal founder of the prestigious Ecole Polytechnique, Monge served during the Revolution on the Commission of Weights and Measures, which introduced the metric system. He later worked with Berthollet in finding a method to extract saltpeter from the soil to keep the munitions factories going during the revolutionary wars. Two years before the expedition, Monge and Berthollet had been appointed to a governmental commission whose grandiose name, Research of Artistic and Scientific Objects in Conquered Countries, was a euphemistic cover for the organized French plunder of Italy's treasures, presently on display in the Louvre.

In preparation for the Egyptian campaign, Monge returned to Italy, confiscated the Arabic font from the Vatican's printing office and enlisted typesetters to join the expedition. In Rome's School of Medicine he found four Arabic-speaking students, Christians from Syria and Mt. Lebanon, whom he obliged to join the expedition and serve as translators in Egypt. Bonaparte's choice of Monge and Berthollet to staff and organize the scientific wing of the expedition did not disappoint the young general. Between them, they stripped the French Academy of its leading young researchers – mathematicians, physicists, chemists, astronomers, geologists, geographers, orientalists, physicians, surgeons, pharmacists, botanists, zoologists, metallurgists, mineralogists, artists, engineers, historians, architects, economists and antiquarians – later to be called archaeologists. Bonaparte ecstatically complimented them, and of course himself, for their having enlisted a third of the Academy.

He was not exaggerating. Among the scientific catch were, to name a few of the brightest lights: Etienne Louis Malus, a leading scientist and engineer (who would go on to discover the polarization of light), who was appointed head of the mathematical division; going to Egypt with him was Jean-Baptiste Fourier, famous for his discovery of the mathematical series named after him; and the chief zoologist was Etienne Geoffrey Saint-Hilaire, whose research foreshadowed Darwin's in the next century. (Bonaparte had wanted the naturalist Cuvier to join the expedition but Cuvier, a sybaritic socialite–scientist who was horrified at the thought of leaving the plush comforts of Parisian salons and cafes for the heat, flies, scorpions, snakes and mosquitoes of Egypt, respectfully declined the free trip.) The engineer Nicolas-Jacques Conte, famous for inventing the graphite pencil and the first to urge the use of hot air balloons for transporting troops, was recruited. Deodat-Guy-Sylvain Tancredi Gratet de Dolomieu, after whom the Dolomite Mountains were named, was one of the expedition's mineralogists.

The expedition's chief interpreter was the distinguished scholar of Arabic language and literature, Jean-Michel de Venture de Paradise.

No military campaign before or since could have boasted such a splendid array of top scientific, technological and linguistic talent. As one historian of Bonaparte in Egypt expressed it, "never, until the most recent times, were so many excellent people working in such a variety of fields brought together in such close cooperation"[7] – especially considering that the scientists constituted an adjunct of a military campaign. Nor had any piece of laboratory or workshop equipment needed for the scientific mission been forgotten. The instruments and apparatus that would furnish complete chemistry and physics laboratories were crated and put on board along with equipment for a hospital, a pharmacy, a telescopic observatory, a laboratory for natural history and a scientific and technical library of around 600 volumes, not counting the books in Arabic, nor those on Egypt that Monge liberated from the Vatican and Maltese libraries. Also lifted from the Vatican and Malta were two Arabic printing presses. In addition, Malta offered up a Greek and Syriac press, along with their operators.

Remarkably (or perhaps not, considering his broad interests) Bonaparte put as much thought and energy into the scientific side of the expedition as he did the military. In fact, from the point of view of results, the scientific side of the mission was exceedingly better thought out, planned and executed, since militarily the expedition was an unmitigated disaster. When Nelson returned to Alexandria from Malta and found the French fleet anchored unprotected in the open bay at abu Qir, he sent it to the bottom like so many sitting ducks, with all the French silver and much of the loot from Malta that was to finance the expedition, not to mention the scientific equipment that went down with the loot.[8] The loss of the fleet was an unhappy turn to what had begun as a glorious campaign of conquest, though the glory was more in Bonaparte's imaginative reports to the Directoire than in actual battle. Military disaster followed one upon the next.

Weeks after the fleet went down, a costly and bloody revolt against the French unexpectedly burst out in Cairo, further demoralizing the troops who saw themselves far from home, trapped in a hostile land whose people, culture and religion they failed to understand. With the fleet at the bottom of the bay, the invaders had no way out. Lured on by the desert mirage of glory, Bonaparte imagined he saw one. Mindless of the relentless heat radiating off the sands, the raging plague and a British naval squadron patrolling the coast, he marched a part of his now miserably unhappy army through Sinai to Gaza and Palestine, hoping that his ragged, plague-stricken troops would take Syria, Anatolia, Istanbul, and, like Alexander his hero, march on to India. Bonaparte reached no farther than the walls of Acre in South Palestine, where he met a defeat no less humiliating than the sinking of the fleet. What losses were inflicted on the French on the battlefield were more than matched by the losses inflicted upon them by the raging plague. The failure of military leadership was now complete. Rumors were that Bonaparte had his wounded soldiers poisoned by the plague to spare the survivors the added burden of having to carry them back to Cairo (see Plate 9).

The remnant of the army that made it back to Cairo was a pitiable sight. In his report to the Directoire, Bonaparte described the senseless campaign that killed thousands of his troops as yet another great victory. His men knew otherwise. If the loss of the fleet had not made clear that the expedition was doomed, the defeat in Palestine did.

The commander-in-chief was not about to admit defeat or take responsibility for failure. Within a year of his having first arrived in Egypt, he disguised himself and, like a thief in the night, secretly boarded a ship for the trip down the Nile to Alexandria, and from there he sailed back to France. Not even the general who was to assume command, Kleber, knew of his slipping away. The battered and abandoned soldiers, having been reduced by a quarter of their number over the year, were all the more drained of morale and any sense of what they were doing in Egypt. Militarily, the expedition was a total waste of men and treasure. Bonaparte's desertion exposed the campaign for the misadventure it was from the beginning: a pointless exercise of imperialist ambition in a region less known to the French military than the surface of the moon. The desertion of their commander-in-chief depressed the military contingent of the expedition and the savants as well. Many thought Bonaparte should be court-martialed. They too wanted to go back home.[9]

The French managed to fight off popular revolt from within and Anglo–Ottoman pressure from without for two more years before finally giving Egypt up as the lost cause it was recognized to be. The final humiliation came when the haggard troops were stripped of their Egyptian and Maltese loot and transported back to France on British ships. Had it not been for the savants and their accomplishments, the French debacle in Egypt would be no more than a footnote in history on the delusion of empire. Only because of the savants did the expedition become the glorious episode in cultural discovery and scientific scholarship it has for over two centuries now been seen to be. In 1998, the expedition's bicentennial was celebrated in a profusion of scholarly books and articles, but it was the scientific accomplishments that were celebrated, the *mot clé* of the publications being *Lumiers*, Enlightened Ones. Nothing else of the wretched campaign was worth celebrating.

The outlook of the savants was different from that of the military. Their mission was different. Bonaparte's desertion was depressing but not essential to their mission. They were stuck in a country that was hot in summer and socially boring all year. The local populace was not pleased to have the uninvited guests in their country. However, hot or cold, boring or not, leader or no leader, the savants had work to do and they loved doing it. Bonaparte's absconding had not torn meaning and purpose from their being in Egypt, as it had the military's, and they immersed themselves in their monumental pursuit of capturing the whole of Egypt in word, sketch and specimen. The military component had no such compensation and resented the savants for theirs. The military generally, but not completely, resented "les anes" (the donkeys: play on words for les savants) for their upbeat acceptance of Egyptian captivity. Indeed, sustained by the work of science, the donkeys saw it as an experience of a lifetime – capturing and bringing home the life, culture, art, nature and history of this ancient land of fable and mystery.

The expedition's scholarly work was organized and supervised by the Commission of Sciences and Arts. This had been established by Berthollet and Monge, assisted by Bonaparte, before the fleet had set sail from Toulon. Once Cairo had been taken and the Mamluks scattered, the savants were set up in two of the most magnificent of the Mamluk palaces, which now became the *Institut d'Egypte*. Whatever equipment that had not gone to the bottom of the sea was moved into the spacious rooms to become laboratories, workshops, ateliers for artists, libraries, lecture rooms, exhibition halls, all organized on the model of the French Institute in Paris. The Egyptian version was in every sense a research center, whose single but vast subject of inquiry was the country itself.

Bonaparte's order of August 22, 1798 officially established the Institute and declared its principal aims to be the progress and illumination of Egypt through the sciences of France. The order also established the publication of a journal that was intended to keep the public informed of the Institute's research in the natural history, industry and arts of Egypt. The journal also published its research on technical, scientific and economic questions that aimed to make the country as prosperous as possible. It has been called the Muslim world's first scientific periodical, though it was written in French and no Muslim read it.

The Institute was also a window through which visiting Egyptians could view Western civilization. For obvious reasons, not as many Egyptians came as the savants had hoped. If the brutal face of one people conquering another can possibly show a gentle side, nothing could come closer to it than the Egyptian Institute, though even in this instance the inescapable odor of cultural arrogance rose above the disarming smile of good intentions. The few Egyptians who visited the Institute, the learned elite of Azhar shaykhs, came out of curiosity. They did not know they were supposed to be illuminated, and when they gave no outward signs of lighting up, or even astonishment, the savants were disappointed. They were expecting looks of awe and appreciative applause. When the only literates who could possibly have appreciated the exhibits they witnessed expressed neither astonishment nor appreciation, the savants thought them ungrateful.[10] The only demonstrations that managed to bring the visiting shaykhs to their feet were the frightening chemical explosions of Berthollet and the electric shocks given the terrorized shaykhs from the charged brass globe in the physics laboratory. The scientists failed to appreciate the threatened disposition of the defeated and humiliated Egyptian religious intelligentsia who saw the demonstrations of science to be another form of attack on Egypt and Islam. What was on the part of the shaykhs guarded self-restraint in defense of religion was taken by the savants to be mental lassitude and indifference regarding progress or anything new. The savants were simply too absorbed in themselves and their mission to understand the difficult position of their Egyptian counterparts. It was unfathomable to them that the knowledge they brought could be so coolly dismissed.

Many, if not the vast majority, of the savants sincerely believed in their mission. The mathematician Fourier wrote that its purpose was "to restore to the banks of the Nile the sciences that had been exiled for so long time."[11] The savants often expressed their desire to return to Egypt the arts and sciences that the country had created and lost. Even the military, no friends of the civilian eggheaded donkeys pampered by their commander-in-chief, was to a degree infected by the savants' sense of mission: General Andreossy, calling Egypt the cradle of the arts and sciences, wrote that the army would provide Egypt with the means to rectify many errors and reestablish the works fallen into oblivion because of the lapse of time and the barbarity of the rulers who had neglected to preserve them.[12] According to Colonel Jacotin, an engineer and geographer in addition to military officer, Egypt invented geometry to divide the land along the banks of the Nile and deserved that contemporary science and its methods be reestablished in the country: restoring civilization to the people would be a homage for all that Egypt has given.

The crusader-like passion of the savants was embodied in their *Institut d'Egypte*, the cathedral of knowledge that they had brought to Egypt as a gift from the modern gods to their benighted creatures. Modeled on its Parisian forerunner, the Institute was organized in four main sections: mathematics; physics; political economy; literature and art. In addition to research laboratories, the palaces of the Institute housed engineering quarters, workshops, lecture halls, a library, a museum, and exhibition rooms for the specimens collected from Egypt's minerals, animals, fish, birds, reptiles and plant life. Geoffrey Saint-Hilaire wrote that the accommodations that the Institute offered were at least as plush as those of the Louvre.

Most laboratory equipment had to be improvised, since the largest part of it had been lost with the fleet. Fortunately, the savants had a master craftsman in Conté, founder of the Parisian Conservatory of Arts and Crafts. He was able to remake surgical instruments, astronomical lenses, compasses, microscopes, and whatever else was needed. Considering the paucity of technical and scientific materials in Egypt, Conté's was an incredible feat. Bonaparte, a man who usually did not let himself appear overly impressed by other men's abilities, was so amazed at the man's mechanical and scientific ability in replacing the lost equipment that he declared Conté to have recreated the arts and crafts of France from the desert floor of Egypt. It was Conté who designed the hot air balloons that flew men over Cairo in demonstration of French technology. The Cairenes were meant to be flabbergasted at the sight of men flying; there is no record that they were. For the historian al-Jabarti, it was an unimpressive oddity just barely worthy of mention.[13]

The central garden between the two palaces housing the Institute provided a delightful setting for informal gatherings. Every evening 40 to 50 scholars would congregate with military officers and the occasional visiting shaykh. New findings would be discussed. According to Saint-Hilaire, the discussions were "at least as interesting as the seminars of the French Institute," unquestionably the mother of all scientific institutes. The evening lectures in the illuminated garden featured both pure science and practical. Fourier once lectured on the resolution of algebraic equations and another time the complex mechanics of a wind-powered machine for irrigation, a combination of the windmill and water wheel.

The Institute was home of the first printing press to be brought to Egypt. Directed by the young orientalist J. J. Marcel, the press was supplied with the three fonts (Arabic, Greek and Latin) purloined from the Vatican and Malta. In its brief period of operation in Cairo, the press put out two newspapers: one on the scientific side and the other political and administrative. The first was the aforementioned scientific and literary journal that kept the French community informed of the Institute's progressing work. It came out every ten days, hence its name: *La Décade égyptienne*. The second paper publicized governmental policy, new orders and affairs that related to the country. It was intended to keep the troops informed. This was issued every five days and called *le Courier de l'Egypte*. These two papers were the first in the Muslim world and would be models for those first Egyptian journals that began to be published a quarter century after the French had departed.

The savants were justifiably proud of their research institute on the Nile. Modern French historians of the expedition have with good cause inherited that pride. The description given by one of them frames it with feeling if not eloquence:

> What a rare and astonishing spectacle would this vast enclosure of the Institute have offered, what with that Noah's Arc Geoffrey Saint-Hilaire strove to reconstruct in his menagerie and aviary, and close by it the botanical garden where Raffeneau-Delille and Nectoux seed their plants, and the laboratory where Berthollet, Champy and Descotils handle test tubes and retorts, the room of physics and natural history where Dolomieu examines his minerals and Savigny his insects, the study salon where Monge, Fourier, Malus and Corancez have suspended their blackboard, the observatory where Nouet, Mechain and Quesnot inspect the firmament, the library where Marcel, Venture de Paradis and Panhusen translate their Arabic manuscripts, the artists' workshops where Vivant-Denon, Duterte, Casteix and Rigo set up their easels, Norry, Balzac, Protain and Lancret their architectural spreadsheets, and yet still more – those workshops of every sort where Conté, Cecile and their assistants fabricate on order a great part of what the industry of man has learned to produce.[14]

The public lectures and industry of the printing presses could have been added to this glowing portrait of the scientific family happily at work in the faraway land. A project that failed to get off the ground was the French hospital and medical training center Bonaparte and chief physician Desgenettes wanted built. The planned 400-bed hospital was to have been open to Egyptians free of charge. A medical and pharmacy school was to have been annexed to it to train Egyptians as physicians and pharmacists. A preparatory school was to have prepared young Egyptians for entry to the medical college.[15] The grand project was abandoned owing to the adverse political atmosphere that followed the first Cairene rebellion, ending what appeared to the French to have been harmonious relations with the subject population.

The tranquility prevailing during the first few months of the occupation, when Saint-Hilaire could favorably compare the comforts of the Institute to those of the Louvre, was shattered by the bloody uprising that had been fomented by Ottoman agents and religious elements outside of the Azharite establishment. Up to that point, the French and Egyptians had coexisted more or less peaceably. Hoping to win the Azhar shaykhs over and use them as a front for legitimacy, Bonaparte had played to their vanity by having his officers stand up, snap to attention and salute whenever a shaykh came into their presence. Banquets were held in honor of the leading shaykhs; they were put at the head of military parades and celebratory processions. Bonaparte appointed them to a sham administrative council to run the city. To the dismay of his officers, he even tried to pass himself off to the shaykhs as being a Muslim at heart, vowing he would convert at some point in the near future. He was forever professing his Muslim sincerity to the shaykhs and promising them to build the largest mosque the world had seen. His grandiose politics was a theater of absurdity when not just plain comedy. Upon landing in Egypt, he had issued a proclamation printed in Arabic declaring that he had come in the name of Sultan Selim to liberate the nation of Egypt from the barbarous misrule of the Mamluks and restore it to true Islamic rule. He professed himself to be an agent of Ottoman legitimacy and a great believer in the Prophet Muhammad and the Quran. One absurdity followed the next.

After his army had taken Cairo, Bonaparte assembled the shaykhs and told them to write up a constitution and formulate the laws by which they wanted to rule their country. The shaykhs had no idea what to make of it. How could they make law? The Shari'a was their law. It had been for over a thousand years and would continue to be. The idea of a constitution was equally meaningless. And who was this young Christian foreigner telling them he was a Muslim at heart come to save them?

Bonaparte sometimes left his generals as much at a loss as he did the shaykhs. Early during the occupation he had his personal tailor sew up a kaftan and immense turban of his own design which he intended to wear in an administrative council meeting with the shaykhs. His chief officers succeeded in convincing him that the shaykhs would not be impressed. If Vivant-Denon's painting of him dressed in his oriental fantasy bears any resemblance to reality, they advised their little general wisely.

The first several months of the deceptively quiet occupation led Bonaparte to believe the shaykhs, and therefore the people, had accepted him as their new ruler. "Sultan al-Kabir," Great Ruler, was the name people called out to him in the streets, probably in good-natured Egyptian fun, sometimes no doubt mockingly. To Bonaparte, who wanted to believe it, the words must have sounded sweet. The more or less deceptively peaceful 70 or so days that passed between the occupation of Cairo and the first revolt gave reason for the French to believe they may have been grudgingly accepted. French and Egyptians associated freely during those early months of the occupation. Soldiers and Egyptian prostitutes were drawn together. Some soldiers and officers married Coptic girls of various social levels. The daughter of one of the leading shaykhs had taken to dressing French and keeping company in French society. General Menou converted and married a girl from a well-to-do merchant family. Troops lived among the populace and went unarmed through the streets and bazaars, feeling themselves among a friendly people, which the Egyptians were, then as now, though not overly much to military occupiers. The chronicler al-Jabarti writes how free the soldiers were with their money, paying many times more the going price for eggs or whatever it was they would buy, how they paid large amounts of money just to ride on donkeys and how they would race the animals like children through the streets, having a great time terrorizing the populace with their reckless jockeying. The revolt changed all that.

It originated from elements other than Egyptian. When it broke out, a majority of Cairenes and shaykhs were against it. It was not that they accepted French rule, but they knew resistance with clubs and knives was hopeless against muskets and artillery. The French reaction turned out to be everything the populace feared it would be and worse: no doubt just what the Ottoman instigators were hoping for in order to end the seeming amicability. From the heights of Muqattam plateau, French artillery pounded down on al-Azhar Mosque. The leading shaykhs attempted to stop it, and failing, acted to limit the savagery of the French reprisal by appealing to Bonaparte, who had up to the time of the revolt been trying every kind of superficial inducement to win the shaykhs' support, and he seems to have deluded himself into believing he had.

The shaykhs, with possibly one or two exceptions (Bakri and Mahdi) never supported or accepted the French. They suspected the French of wanting to destroy Islam by undermining their faith. Their suspicions of French intentions included the *Institut d'Egypte*. The furthest the shaykhs went was to feign cooperation – serving on government councils, attending Bonaparte's banquets in their honor, acting civilly to their uninvited guests – in hope of protecting the Muslim community, and this was enough for Bonaparte to be deceived. Consequently, he was genuinely shocked at the revolt. Assuming the leading shaykhs had to be behind it, he took the uprising as a personal insult. He had elevated the shaykhs to high (but powerless) positions in government; he had shown them great respect, had reversed the contempt and neglect they had suffered at the hands of their former Mamluk and Ottoman masters. He had showered them with honors, gifts, positions and prestige. And then they betrayed him. He was certain they did. The uprising could not have happened without their knowledge and support. He suspected that some of his favored shaykhs were its very leaders and inspiration.

Lower members of the ulema were indeed involved. They called from the tops of their minarets for the populace to rise up against the French. This was proof enough for Bonaparte: the lower members of the ulema could not have acted without knowledge and permission of the high shaykhs.

Bonaparte completely failed to understand the looseness of organizational hierarchy of the Sunni ulema. His military intelligence equally failed to have any idea of the sentiments of the people regarding the invaders, or of what was going on in the streets outside the quarters directly occupied by the French. Groping in the dark, clueless as to the disposition of Egyptians beyond his miniscule circle of dissimulating Azharites, Bonaparte reacted in typical colonialist fashion: he held the leading shaykhs responsible. He had some of them arrested and incarcerated in the Citadel prison on the Muqattam plateau. A few of the younger ones were later put to death.

From a careful examination of the sources, the chief instigator of the revolt was in fact shown to be an outsider, a popular Maghrabi preacher acting in concert with secret Ottoman agents. To the Ottomans it looked as though the shaykhs were doing too good of a job in working with the French, honored and banqueted as they were, enjoying their prestigious positions on the administrative councils. It was those agents who organized the lower levels of the ulema to incite the poorer classes of Cairenes to rise up. The French were unaware of what was happening around them, even though the planning and organizing of the revolt had started shortly after the French arrival. The high shaykhs must certainly have had an idea of what was about to occur but had little if anything to do with it, according to the best of the few sources on the subject, the Azharite shaykh and chronicler abd al-Rahman al-Jabarti.

Because of wrong assumptions, lack of intelligence and his military's rage for vengeance, Bonaparte turned a deaf ear to the shaykhs' appeals to spare the people. French artillery continued to rain fire and destruction on Cairo and al-Azhar. Destruction of the ancient institution would be the price the shaykhs paid for their supposed treachery. The first two days of violence took the lives of 200 Frenchmen, including a dozen of the savants, some of whom, armed with 40 rifles and 1200 cartridges and led by Monge and Berthollet, defended the Institute and its equipment. Rather than obey military orders to abandon the Institute and join the well-defended French barracks and military headquarters two kilometers away, the savants invested their lives in protecting their work and what they had built. It was perhaps the first time in recorded history that leading scientific researchers took up arms to defend their institution. The young Arabist J. J. Marcel rushed to al-Azhar as soon as the bombardment had begun and, braving the fire, salvaged as many books as he could, one of them a precious 12th century illuminated Quran.

When the revolt had been crushed and the dust cleared, Bonaparte transferred military headquarters and the barracks to the area around the Institute so the savants and non-military personnel would be less exposed in the event of another uprising. The relocation of the military displaced the indigenous inhabitants of the quarter and cut the Institute and its savants off from Egyptian life all the more. The October revolt, which took over 800 Egyptian lives, had sensibly diminished the more or less live and let live atmosphere that had existed in August and September.

Life continued, nonetheless. The administrative council of shaykhs continued to function (discussing and deciding on issues already decided upon by Bonaparte and his chief generals) and the Institute's doors remained open to all. Contact between prostitutes and soldiers continued. Several dozens of soldiers and officers had long-term relations with Coptic girls that ended in marriage; some others had liaisons with Muslim women, a few of which ended in Muslim conversion and marriage. Life went on in Cairo as it had before the revolt but now with a nervous edge. The Muslim dread that any contact with the French was a threat to faith was sharpened, as was French fear of Egyptian "treachery" – the euphemism for national resistance against military occupation.

One of the founding purposes of the Institute was to introduce Egyptians to the modern sciences and their benefits. In the short run, this was no more successful than the French project to build a hospital and medical school. Egyptians failed to flock to the Institute to learn about the modern sciences. They had little interest. They had even less after the revolt. French sources substantiate this. The shaykhs and savants were looking at each other from opposite ends of an eyeglass that not only stretched across a millennium of intellectual separation but was obfuscated by cultural contradiction. French expectations of constructing a bridge of understanding that would convey the principles of the Enlightenment, modern science and the Revolution were as unrealistic as the expectations the shaykhs harbored that the French would convert once introduced to Islam. Bonaparte's demand that the shaykhs write up a constitution and frame their laws left them speechless. Only God made law. The sultan might make a law related to government, structure, taxes and administration, but the ruler's true duty was to protect God's law and the Muslim community. Bonaparte's appeal to their patriotism and national sentiment, translated to them in Arabic, was equally confounding. Arabic equivalents had to be coined for love of motherland, liberty, freedom and patriotism, but the words had no context to become anything more than just words without meaning. The symbolic "Tree of Liberty" Bonaparte planted to commemorate the Revolution in a grand public ceremony in central Cairo, which the shaykhs and notables had been obliged to attend, became an object of gross ridicule. The celebratory toasts of raised glasses by the officers and the savants in salute to the shaykhs – "á la civilization de l'Egypte!' followed by "á l'union de la science et de la force!" – left them smiling politely to their hosts and wondering in confusion what nonsense the French were chattering about now. The French insistence on reason, science and progress, the public reenactment of Louis XVI's beheading, planting the Tree of Liberty, the carnival atmosphere of the Tivoli Gardens that the French built with its drinking, dancing and popular music, men and women embracing in public, the banquets and evening candlelit processions of shaykhs and officers arm in arm in celebration of Franco–Egyptian unity, none of this made any sense to Egyptians.

One of the few Frenchmen to see the futility of the parades, feasts, banquets and public ceremonies that were meant to enlighten the Egyptians and foster their friendship was the chief physician Desgenettes: "Much has been said, even in Europe, about the effect these celebrations have in the minds of Egyptians. I positively can assure you that in spite of all their magnificence they impress the inhabitants of Cairo very little."[16]

Shaykh Jabarti dismissed the Institute's science and laboratories as childish playthings not fit for Muslims, sentiments that mirrored the Ottoman ambassador Mustafa Hatti's contempt of the scientific instruments he saw in the Vienna observatory 50 years earlier. Only two Azhar shaykhs are on record expressing enthusiasm for French learning. One was Shaykh al-Mahdi. Fancying himself a talented story writer, the shaykh admired the printing press at the Institute, and, before the French were expelled, gave the Arabist J. J. Marcel some short stories that he had written, asking him to translate them into French, which Marcel did and had published many years later in Paris under the title *Contes Arabes*.[17] Shaykh Mahdi's literary sensibilities, akin to the flights of fantasy found in *The Thousand and One Nights*, made no room for science and technology. The only shaykh who did have an interest in this direction was Shaykh Hasan al-Attar. Decades after the French had departed he would urge Egyptians to learn their sciences, which he recognized to be the path to strength. Shaykh Attar was the first Muslim figure of any religious stature in the Arab world to advocate such a course. Other than al-Jabarti and al-Attar, few leading shaykhs visited the Institute. Among those who did were Shaykhs Khalil al-Bakri and Sawi, but neither shared Attar's attitude about the learning that could be gained there. In this, Shaykh Attar stood alone. Shaykh Khalil al-Bakri was another who stood alone in his relations with the French, but this had nothing to do with science or learning.

Shaykh Khalil al-Bakri was a complex and ambitious man who bitterly suspected his relatives were thwarting his rise to the top of the religious establishment. The arrival of the French changed everything. Doors that had been closed to the shaykh were opened when he became Bonaparte's foremost Muslim collaborator. In return for the shaykh's cooperation, Bonaparte arranged to have Bakri receive the honorary titles that he had long coveted but failed to get. The tie between Bakri and Bonaparte was a marriage of convenience, each one using the other toward his own end. For the shaykh it was to advance his career, for the commander-in-chief it was to have a religious front legitimizing his position as ruler. Bakri became so closely involved with Bonaparte that he allowed his young daughter to socialize with the French, even to dress and act like French women, a good number of whom, wives of officers mainly, had accompanied the expedition disguised as officers. Zaynab would later pay for this with her life, condemned by her father who testified against her in court to save his own neck for having collaborated too closely with the French, once they were out and the Ottomans were back in.[18]

In spite of his good rapport with Bonaparte, Bakri showed little interest in the Institute. Like the other shaykhs who embodied the educated elite of Egypt, he visited out of passive curiosity. The shaykhs were encouraged by the French to visit and when they did they were received, according to the chronicler Jabarti, with warmth and hospitality that often expanded to good-natured joviality, laughter and demonstrations of kindly affection. Whenever Muslims showed any interest in understanding the scientific and technological exhibitions, the savants "would exert themselves in friendliness and love," so proud of their accomplishments that the least interest fed their vanity.[19] The visiting shaykhs were polite but deliberately phlegmatic, giving no indication of being impressed. Their lack of enthusiasm disappointed the savants, but it was asking much that the religious leaders of a defeated people would publicly applaud the accomplishments of their irreligious conquerors.

This simple norm of human nature and culture was something that seemed not to have dawned on Bonaparte. Bent on awing the shaykhs with French power through science and technology,

> he invited the principal Sheiks to be present at some chemical experiments performed by M. Bertholett. The General expected to be much amused at their astonishment; but the miracles of the transformation of liquids, electrical commotions and galvanism did not elicit from them any symptom of surprise. They witnessed the operations of our able chemist with the most

imperturbable indifference. When they were ended, the Sheik El Bakri desired the interpreter to tell M. Bertholet that it was all very fine; "but," said he, "ask him whether he can make me to be in Morocco and here at one and the same moment?" M. Bertholett replied in the negative, with a shrug of the shoulders. "Oh! Then," said the Sheik, "he is not half conjurer."[20]

To the French, Bakri's question was frivolous, if not stupid. Berthollet, Europe's greatest chemist, could respond only with a perplexed shrug. To the shaykhs, whose religious culture was rich in the mystical lore of the Quranic stories of Muhammad's miraculous transportation from Mecca to Jerusalem where he was in both places at the same time, of his ascension to heaven, and of Sufi saints flying through the air, raising the dead, miraculously curing diseases, and appearing in different places at the same moment, to them Bakri's question had purpose: that the mystical, supernatural power and truth of Islam was superior to the physical power of French science. Refusing to lose their composure in front of the savants as electric sparks flew, gases exploded in flames, bottles burst and liquids magically changed color, solidified or went up in smoke when mixed, the shaykhs denied the non-believing conquerors the awestruck applause that Bonaparte and the savants fully anticipated their magic would elicit from their scientifically ignorant guests. Bonaparte's attempt to seduce the shaykhs through the Institute failed as much as did his inducements of administrative councils, honorific titles and banquets. This is clear from both the French sources and the very few sources on the Egyptian side.

The paucity of Egyptian sources on the French occupation is a measure not only of the country's intellectual depression but also of the difficulty Egyptians of the time had in writing about the traumatic experience. It was a surrealistic interlude the shaykhs would have preferred to forget. Shaykh Jabarti, however, was able to absorb it. He wrote three works on the three-year period of the French in Egypt, two of them relatively short treatises, *Tarikh Mudda Faransis bi Misr*[21] (*History of the French Period in Egypt*) and *Mazhar al-Taqdis fi Idhhab Dawlat al-Faransis*[22] (*Blessed Sight of the French Regime's Expulsion*). His third account of the French period is found in the third and fourth volumes of his four-volume chronology, *Aja'ib al-Athar fi'l Tarajim al-Akhbar* (*The Marvels of the Past*).[23] In this last work the author gives an account of the Institute and the work of its savants. It is the only account of substance from Egyptian sources that offers an insight into the intellectual and psychological impact that French science may have had on those Egyptians who visited the Institute.

Of all the Azhar shaykhs, al-Jabarti would have been the one most likely to be avidly concerned with the marvels the French performed in their institute, not only from the title of his chronology that refers to recording marvels of the past, but for the fact that he was Egypt's leading scholar, the last of a long line of high quality chronologists that reached back to early Mamluk times. In addition to the curiosity he should have acquired by virtue of his historian's craft, his father, Hasan, had the reputation of being Egypt's foremost mathematician and astronomer. A decade before the French invasion, the Ottoman governor of Egypt, who also had an interest in astronomy and mathematics and who reportedly urged al-Azhar to reform its education system, hired Shaykh Jabarti's father to teach him some science. Though Jabarti appears to have inherited very little of his father's scientific knowledge, he did visit the Institute on a number of occasions and took enough interest to describe it. He commented respectfully on its regularly scheduled public lectures that were of a scholarly nature and open to the public. The library and reading room, its well-ordered tables and chairs, the shelves of neatly arranged books, the silence of the room and people who sat reading – features that might have passed unnoticed by a Western observer – made a deep impression on him. But most remarkable to him was that even the lowest ranking soldiers would come to the library and open a book, meaning that common people could read, not just a few officers at the top of the military elite. It was another of those marvels that made the French so incomprehensible. They could be unclean, unmannered and ungodly – but the meanest of them could read! He also marveled at the French artists' realistic renderings of flora, fauna and human figures, so real they looked practically alive. Among the depictions were some leading shaykhs, who were pleased, if not flattered, to have French artists paint their portraits. Forgetting for the moment the Muslim proscription against depicting human and animal forms, Jabarti even expressed his admiration of the lifelike renditions the artists made of the Prophet Muhammad and the caliphs Ali, abu Bakr and 'Umar. The caliphs, he wrote, looked as if they were about to walk out of the frames.

He described the chemical laboratories with their variously colored bottles set evenly in rows, shelf upon shelf, and the wizard-like demonstrations of the chief chemist (Bertholett) who is described in French sources, and by Jabarti, as delighting in performing explosive chemical demonstrations to frighten the shaykhs.[24] One of Bertholett's favorites was having the shaykhs gather round the glass collecting vessel as he produced hydrogen from an acid and zinc. While the unsuspecting shaykhs were inspecting the mysterious bubbles quietly rising up through the water and into the collecting vessel, he would ignite the invisible gas and then laugh as the shaykhs jumped back in fright at the explosive burst of flame. He also delighted in delivering electric shocks to the shaykhs. The shock-producing device especially intrigued Jabarti. He describes the metallic sphere with glass handle that, when turned, generated a flash of fiery light that made a cracking noise as it leapt to make contact with an object some distance away. "When anyone touched or approached it with his hand, his body would shake and quiver, shoulder and forearm contorting in a sudden convulsion, upon which the French savants would giggle in great delight."

Having described in lively fashion the chemical and electrical tricks and the strange clear liquid that preserved dead fish, snakes and insects as if they were alive, Jabarti comments on the many machines and instruments demonstrated to him by the savants, of their many books on mathematics and astronomy and their amazing watches of every size and shape. He was shown instruments he had never seen before. He attempted to describe a musical pipe organ, but with limited success. He expressed admiration for the complex astronomical equipment that allowed French astronomers to calculate the sizes and distances of planets and the telescopes: there was a folding one, and a larger one that was used to observe the starry heavens. Whether he looked through it or not to see those heavenly stars he left unmentioned. It was an odd omission, if omission it is, for the French would have been no less eager to amaze the shaykh with what he could see through their telescope than their predecessors had been to amaze Yirmisekiz 80 years earlier at the observatory in Paris.

Jabarti appears to have been commenting favorably on how French scientists discover natural laws through experimentation and interpretation of their observations; but he then concluded his reflections on the Institute and its savants in a somewhat confusing ambivalence of wonder, dismissal and contempt. The French, he wrote, "have many more things of this nature, and unusual combinations, which produce results that are beyond the comprehension of minds such as ours."[25] The remark seems to indicate how alien science had become to Egyptians. Possibly the author meant to imply that Egyptians and Muslims should not try to understand these alien things, since understanding the infidel knowledge could put their faith dangerously in the balance. God, faith, punishment and preservation of Islamic society are themes woven deeply into Jabarti's writings on the French incursion. The overriding tone of his account presents the occupation as an aberration of history, a curse that has turned the world upside down. The French have come as a plague, a divine catastrophe that will end once God is done punishing the Muslims. Jabarti's regard of the infidel occupiers is most usually one of scorn and contempt: the French are the accursed ones; the names Bonaparte, Kleber, and the French as a whole, are generally accompanied by "may God curse him" or them, as the case may be.

Taken as a whole, Jabarti's response to the French suggests a split-mindedness that reflects the wrenching reality of God's community being suddenly confronted by the superior power and knowledge of unbelievers. While cursing the French as enemies of God, he can grudgingly praise the Institute and Desgenette's pamphlet on the pox (*jadari*) that was printed in Arabic during the terrible plague of 1799.[26] Being forced to exist under a despised foreign occupation while responsibly shepherding the community through the cruel trial produced equivocal behavior. Accursed and despised though the French may have been, Jabarti did not refuse to serve on Bonaparte's central administration council. A good many leading shaykhs served with him: Bakri, Sharqawi, Sadat, Mahdi and Sawi, among others. To serve on the diwan, in the minds of those shaykhs who did, was not to serve the French or themselves (though some did, such as Bakri and Mahdi), but to see Egypt and the Muslim community through this terrible time of troubles. In quest of survival during such confused times of defeat and occupation, thoughtful men of conscience and responsibility were hard put to emerge from the wrenching turmoil unscathed and whole of mind. Not surprisingly, the response of individual shaykhs to the challenge they confronted found wide divergence, even among those who were close friends.

Shaykh Attar, the only Azharite whose response to French science and the Institute was entirely positive, was a close friend of Jabarti's. The difference in views between the two shaykhs, one assimilationist, the other rejectionist, as opposite in mind to one another regarding the West as had been the views of Ottoman ambassadors Yirmisekiz and Mustafa Hatti during the first half of the century, did not injure their friendship. The two shaykhs had much in common to bind their friendship – faith, education, values, interests and a quest for learning. They composed poetry to each other, and, according to Jabarti, they shared a companionship so deep, relaxing and congenial that the two men could discuss any subject at all, however intimate it might be, without embarrassment or discomfort.[27] There is no accounting for Attar's and Jabarti's individual differences in response and reaction to Western knowledge and techniques.

Jabarti fled Cairo with many of the other leading shaykhs after the French defeat of the Mamluks, but he then responded to Bonaparte's call for the shaykhs to return; and while keeping a circumspect distance from the occupiers, he accepted to serve on the administrative council. Attar also fled Cairo after the French defeat of the Mamluks and did not return until early 1800, almost a year and a half after the revolt. In the relatively mild political climate of 1800, he could visit the Institute without incurring resentment from the other shaykhs. Attar did not serve on the administrative council, though he visited the Institute often enough to learn something from the French. It is reported he gave Arabic lessons to French savants in return for lessons in science. Attar left Egypt just after the French were expulsed and stayed away until Muhammad Ali had seized power in Egypt. He may have felt that the returning Ottoman authorities saw him as having been too close to the French. He was wise to have left. Any shaykh that the Ottomans thought had collaborated more than was necessary suffered humiliating punishment.

Once the French had been expelled by the joint effort of a British naval squadron and an Ottoman army in the summer of 1801 and Muhammad Ali had, within a few years, installed a dictatorial government in Egypt, Cairo and Istanbul would follow parallel lines and encounter similar obstacles, inside and outside of religious circles, in the course of their struggle to modernize through the assimilation of Western science and technology. The many internal and external factors aligned against modernization in Egypt would make it a struggle as harsh, and on occasion as bloody, as the Ottoman experience.

On the French side of the balance sheet, the expedition was politically and militarily a loss of men and treasure. The one positive product of significance that resulted from the Egyptian adventure was the work of the savants. A creation of enduring fame, their collective work was published as an Imperial Edition in 23 outsize volumes of narrative and plates between 1809 and 1818. Known by the simple title *Description de l'Egypte*, a second edition of 37 volumes was published between 1821 and 1826. The monumental work introduced the field of Egyptology and provided in detailed description, scholarly narrative and precise drawings, all those previously mentioned aspects of Egypt and Egyptian life that the scientific mission set out to capture for posterity.

Notes

1 For general histories of the expedition see Henri Laurens, *L'expedition d'Egypte, 1798–1801*, Armandi Colin, Paris, 1989; Partice Bret, *L'Egypte au temps de l'expedition de Bonaparte, 1798–1801*, Hachette, Paris, 1998; also by Bret (editor) *L'expedition d'Egypte: une enterprise des Lumiers, 1798–1801*, Acacemiedes Sciences, Paris, 1999;

Francoise Charles-Roux, *Napoleon, Gouverneur d'Egypte*, Plon, Paris, 1936; In English: Nina Burleigh, *Mirage: Napoleon's Scientists and the Unveiling of Egypt*, Harper Collins, New York, 2007; Juan Cole, *Napoleon's Egypt: Invading the Middle East*, Palgrave Macmillan, New York, 2007; Christopher Herold, *Bonaparte in Egypt*, Hamish Hamilton, London, 1963.

2 Francoise Charles-Roux, *Les Origines de l'expedition d'Egypte*, Plon, Paris, 1910 .

3 Anwar Louca, "La renaissance Egyptienne et les limites de l'oeuvre de Bonaparte, *Cahiers d'histoire e'gyptinne*, 7, 1955, p. 104.

4 Charles-Roux, *Les Origines de l'expedition d'Egypte*, p. 33; and more recently: H. Laurens, *Les origins intellectuelles de l'expedition d'Egypte, L'orientalisme islamisant en France, 1698–1798*, Isis, Istanbul and Paris, 1987.

5 For the purely scientific dimension of the expedition: Charles Gillispie, "Scientific Aspects of the French Expedition," 1798–1801, *Proceedings of the American Philosophical Society*, 133, no. 4, December 1989, pp. 447–474. Much of the scientific work that resulted from the expedition was done in France during the years following the end of the occupation of Egypt by the naturalists who accompanied the expedition and continued their studies of the collections they brought back with them. The two most important original scientific studies accomplished in Egypt were Gaspard Monge's optical analysis that explained the mirage and the chemist Claude-Louis Berthollet's experiments on natron.

6 Gillispie, "Scientific Aspects of the French Expedition," p. 449. One of Europe's leading mathematicians and chemists, Monge was Bonaparte's choice for the chief recruiter and organizer of the scientific contingents of the expedition. As a mathematician, Monge had been called the father of descriptive geometry. As a chemist, he ranked with Lavoisier, who might himself have been a member of the savants had he not been guillotined during the Terror.

7 Christopher Herold, *Bonaparte in Egypt*, Hamish Hamilton, London, pp. 27–33; 164–200. In addition to establishing the field of Egyptology, the savants produced at least two original works in pure science: Monge's optical analysis of the mirage and Berthollet's study of the chemistry of natron. Gillispie, "Scientific Aspects of the French Expedition," pp. 450–452.

8 The ship containing most of the medical and scientific supplies escaped Nelson's guns by having struck a reef and going down just before the British fleet's arrival. Gillespie, "Scientific Aspects of the French Expedition," p. 455.

9 Except for the unpopular General Menou who converted to Islam, took the name Abdullah, married a Muslim woman and became commander-in-chief upon the assassination of General Kleber, who was Bonaparte's successor, all the French wanted to go back. Menou would have been happy to remain forever as King Abdullah of Egypt.

10 For a cultural analysis of why the shaykhs found it so difficult to be impressed see André Raymond, "Les Egyptiens et les Lumiers pendant l'expedition francaise," in *L'expedition d'Egypte, une enterprise des Lumiers, 1798–1801*, edited by Patrice Bret, Academie des Sciences, Paris, 1998.

11 The expedition's chief physician, Desgenettes, was of like mind, though he at the same time believed Western medicine could just as well benefit from Egypt since medicine originated in the East, particularly Egypt. Believing some medical wisdom must have survived, he urged the troops to keep an eye open for ancient medicines, practices and remedies that could be beneficial.

12 Roux, *Napoleon Gouverneur d'Egypte*, p. 165.

13 Maria Luisa Ortega, "La 'régéneration' de l'Egypte: le discours confronté au terrain," Chapitre 7 in *L'expedition d'Egypte, une enterprise des Lumiers 1798–1801*, edited by Patrice Bret, Academie des Sciences, Paris, 1998. See Ortega's footnote 16.

14 Roux, *Bonaparte Gouverneur d'Egypte*, p. 172.

15 Roux, *Napoleon, Gouverneur d'Egypte*, p. 221.

16 "On a beaucoup parlé, meme en Europe, de l'effet que produisirent ces fetes sur l'esprit des Egyptiens. J'affirme cependant positivement qu'elles frapperent trés peu

les habitants du Caire, malgré toute leur magnificence." Roux, *Bonaparte Gouverneur de L'Egypte*, p. 186.

17 Mustapha al-Ahnaf, "Chiekh al-Mahdi (1737–1815) Ulema, Médiateur et Business-man," *Egypte/Monde Arabe*, 1, 1999, pp. 115–150.

18 For details on Zaynab's misfortune see: John Livingston, "Shaykh Khalil al-Bakri and Bonaparte," *Studia Islamica*, vol 80, spring, 1994, pp. 125–143.

19 Abd al-Rahman al-Jabarti, *Ajai'ib al-Athar fi'l Tarajim wa'l Akhbar*, vol. 3, Bulaq Press, Cairo, 1879, p. 34; John Livingston, "Shaykhs Jabarti and Attar: Islamic Reaction and Response to Western Science in Egypt," *Der Islam*, 74, 1997, p. 94.

20 Bourrienne, *Private Memoires of Napoleon Bonaparte*, Colburn and Bentley, , London, 1830, vol. 1, p. 279; J.J. Heyworth-Dunne, *An Introduction to the History of Education in Modern Egypt*, Luzac and Co., London, 1968, p. 97.

21 Brill, Leiden, 1975.

22 Bulaq Press, Cairo, 1969.

23 Bulaq, Cairo, 1879.

24 Jabarti, *Athar*, III, pp. 34–36.

25 Jabarti, *Athar*, IV, 2.

26 Jamal al-Din al-Shayyal, *Tarikh al-Tarjama fi-Misr fi 'Ahd al-Hamlat al-Faransiyya*, Cairo, 1951, p. 82.

27 Jabarti, *Athar*, IV, pp. 232–239.

16 Muhammad Ali's militarization of modernization and educational reform

When the French departed Egypt, they took with them everything they had brought: their scientific equipment, library, printing presses and all the results of their research. The two Mamluk palaces that had housed the Institute were left as an empty shell, as if the French had never been. With Egypt having now reverted back to Ottoman control and the situation being what had prevailed before the occupation, it looked as though the expedition and its Institute had been a futile exercise of French power and intellectual pride, and that the three-year colonial episode would sink into the sands of oblivion.

However, neither invasion nor Institute were to be forgotten by any of the countries involved. Mindful of Egypt's importance to both the Ottomans and European maritime powers, and equally mindful of the French attempt to colonize the country, the new ruler of Egypt, Muhammad Ali, would in self-defense strive to modernize the country's economy and military, in pursuit of which the Institute would be revived in one form or another.

Western invasion of the country would also be revived. The French incursion of 1798 jolted the British government, and the East India merchants whose interests it protected, into an awareness of Egypt's strategic importance to imperial communications with India, making Egypt as much an object of European contention as Istanbul with its strategic waterways. When the Suez Canal was built later in the century, Egypt's geopolitical importance would be too great for France and Britain not to want to seize control of the new waterway and the ancient country to which it belonged. As it turned out, a joint Anglo–French armada of invasion in 1882 ended up becoming a unilateral occupation by the British, who would remain in military control of Egypt and the canal zone until 1954, outdoing the longevity of the French colonialist venture by 70 years. The more that Middle Eastern countries strove to modernize in defense against Europe, the more intent European powers were on directly controlling them. The narrowed eyes of imperialism saw in them a source of natural resources, markets for domestic manufactures and strategic way-stations for control of global trade routes. The defensive strength that resulted from the effective modernization of non-European states was not in European interest.

A complex five-year struggle over power in Egypt followed the French evacuation. The main contenders were: the Ottoman central government represented by the newly appointed governor; the commanding officer of the Ottoman garrison; and the Mamluks who were divided into two leading factions, one led by a chieftain backed by the British, the other by a French-supported chieftain. By cleverly playing one Mamluk faction off against the other and at the same time undermining the authority of the newly installed Ottoman governor, the chief officer of the garrison, Muhammad Ali, an able Albanian trained in Selim's New Order, intrigued his way to power. Once he had outmaneuvered his Mamluk rivals, he sent the Ottoman governor back to Istanbul claiming him to be a corrupt traitor to Islam and the empire, precisely of what Bonaparte had accused the Mamluks to justify his own take over. Accepting the reality of the usurper's power, the sultan refrained from being drawn into an unaffordable war against a recalcitrant officer in a distant province and so opted to treat the affair in traditional manner of the enfeebled empire, by recognizing his position as governor while waiting for the first chance to bring him down, whether by assassination or removal by the forces of a neighboring provincial governor.

Taking a more active interventionist stance against the vigorous ruler of Egypt, a British naval squadron landed troops on the shores of Egypt to finish Muhammad Ali off and install the British-supported Mamluk leader in his place. But the small invasion force, a joke compared to the planning and magnitude of the French expedition, was badly mauled by Muhammad Ali's garrison and sailed off in defeat.

Adept equally in warfare and intrigue, the new Egyptian ruler outwitted Ottoman maneuvers to unseat him. Knowing that the sultan and the British would be constant threats to his position in Egypt, and that his only ally against them would be his own military power, Muhammad Ali applied the same New Order reforms that Sultan Selim had undertaken to reform Ottoman government and industry. As an Ottoman military officer, he was a product of those reforms. To succeed where Selim's reforms had failed, he knew he would first have to annihilate or coopt all opposition. Reform required leveling the field for a new foundation. Existing institutions and interest groups would have to be emasculated or uprooted. The most powerful of them would be the first to: Egypt's equivalent to the Janissaries, the Mamluks. As for the Egyptian ulema, they had been readily cowered the moment the Ottomans replaced the French in 1801, as exemplified by the humiliation of Shaykh Khalil al-Bakri and the execution of his daughter by his own testimony. In case the shaykh's fate had failed to deliver the message, the shaykhs were told directly that they had better forget the honor, prestige and administrative positions they enjoyed under the French, those days were over. Indeed they were. Muhammad Ali ruled the al-Azhar shaykhs with the same stern hand he ruled everything in the country. To make the ulema wholly dependent on him he confiscated their chief sources of wealth – the non-taxable charitable foundations, the *awqaf*. Those shaykhs who cooperated with the reforms were rewarded; those who opposed or criticized them were exiled or worse.

Bringing the Mamluks down would require a decisive stroke of quick and well-executed violence. They, or at least their leaders, had to be driven out or slaughtered and their system of slave recruitment obliterated. As Muhammad Ali had only an Ottoman garrison and some Albanian troops who had come to Cairo with him, his meeting the Mamluk beys and their slave forces in open battle was out of the question. Destroying them required cunning. Taking a deception from a well-worn page of history, he annihilated them the way the first Abbasid caliph had the Umayyad princes. It was an old ruse, but the cupidity of men in high position should never be underestimated. All Muhammad Ali had to do was put it out that he was going to richly reward the Mamluk leaders for their services at a banquet in their honor at the Citadel fortress on Muqattam plateau, and they came galloping to their doom. In the middle of the festivities, armed guards appeared from their hidden posts and cut the unsuspecting guests down. With their beys and top officers done in, the thousands of remaining Mamluks fled in terror, either melting into the local population in the city or the countryside or fleeing the country altogether, never to appear or be heard of again as Mamluks, though the names of leading Mamluk houses continue today in Egypt as family names.

In being delivered by Muhammad Ali from one form of rapacious military rule, Egypt fell victim to a new one. The trade-off was that whereas Mamluk rule during the half century before the French invasion was disorderly to the point of becoming chaotic, Muhammad Ali enforced a strict level of order and security that won high praise from even his harshest critic, the chronicler abd al-Rahman al-Jabarti. The historian had at first been strongly in favor of Muhammad Ali's strict and orderly rule but turned against the ruler when, after the slaughter of the Mamluks, he began confiscating the *awqaf* and the wealth of Egypt, showing that he placed himself above the law, custom and tradition of the country.

Having eliminated the Mamluks, Muhammad Ali brought the Sudan under his control. The gold he found there would finance his reforms; the young men captured would be enslaved and trained for the infantry in his new army. Instead of Turkish, Georgian and Circassian slaves, who were expensive and hard to come by with the Russians taking over the region, Muhammad Ali planned to create a black African slave army.

The Africans did not fare well under the harsh military conditions imposed on them in their new Egyptian environment. Many died because of disease and the change in climate; others ran away because of the brutal coercion to make them soldier-slaves. Those of them who survived the conditions and failed to escape did not easily submit to working and fighting for Muhammad Ali, and so the project was abandoned. The ruler then turned to Egypt's peasants. This was revolutionary. Native Egyptians had not been used in the military for two and a half millennia. Turkish rulers considered Egyptians competent in trade, religion and agriculture but not in anything military. However, when drilled, disciplined and trained in contemporary warfare by a handful of French and former Ottoman officers in Muhammad Ali's pay, the traditionally docile peasantry proved their ability by defeating the Wahhabi tribesmen from central Arabia, who, in the name of purifying Islam of its foreign accretions and corrupted leaders, were advancing northwards into Syria. In this, the peasant army performed a great service not only to Muhammad Ali but to the Ottoman Empire, whose Arab provinces were being threatened by the invading Wahhabis.

Inspired by the strictly traditionalist and scripturally literalist legal school of Ahmad ibn Hanbal, the Wahhabis were named after the 18th century religious reformer, Muhammad 'abd al- Wahhab, whose vision of a purified Islam had won over a tribal chief in central Arabia, al-Sa'ud. Inspired by Wahhab's puritanical vision of a return to early Islamic practice and virtue as a source of religious renewal and political strength, Sa'ud's fighting tribesmen came charging northwards out of Arabia into Palestine and Syria, emulating the original Arab conquerors of the 7th century. Too strapped for funds at the time to send an army, the sultan ordered his powerful governor in Egypt to put them down. With his defeat of the Wahhabis, Muhammad Ali added the Hejaz, with the holy cities of Mecca and Medina, to his Egyptian–Sudanese empire. Syria and southwest Anatolia would come a decade later. At the height of his moment of military glory, the conqueror was fond of claiming he was born in the land of Alexander the Great in the same year as Napoleon.

His reorganization of Egypt was centered on building and supporting a strong military. The military was at the heart of his reforms, of the state he created, and of its institutions and administration. The state was an extension of the military, and Muhammad Ali ruled it as a general would command an army. The civil officers running the administration were given military rank and paid accordingly. A military purpose informed everything that Muhammad Ali introduced or reformed during the 40 years and more that he ruled Egypt. His conquests, reforms and institutional structuring emanated from one idea: to construct and maintain a powerful army and navy, the first and last guarantor of his absolutist position in Egypt, which he possessed as fully as an old kingdom pharaoh. The farmlands of Egypt were seized and under his ownership were given to his officers to administer; the new industries he founded were his private possessions; agriculture, trade and industry collectively formed one great state monopoly belonging to the ruler and used to aggrandize the military and its infrastructure. The reforms he instituted were designed toward military objectives, issued as military orders, and executed in military fashion. Muhammad Ali's state was based on a modern army and navy that was intended to be too powerful for the sultan to overcome and powerful enough to deter the French and British from attempting to add Egypt to their imperial holdings.

The military ethos ran through the entire body of Muhammad Ali's reformist state. Nothing went untouched by it. Even his schools were organized and administered as military training camps: the students strictly regimented, given military rank and pay, and gripped in perpetual fear of corporal punishment for the least infraction of the severe rules of discipline they lived under. European teachers, physicians, technical experts and advisors were also given officer ranks and paid commensurately. Nothing escaped the military framework and regimentation imposed by the ruler. The School of Veterinary Medicine was founded for the sole purpose of caring for animals serving the military. The Medical School he founded in 1827 under French guidance was no less of a military installation. It had originally been an annex of the military hospital that the ruler had built around 1812 in the village of abu Za'bal, some 15 kilometers outside of Cairo. Like the hospital, the school was founded to serve the military and was under the authority of the Ministry of War. Graduate physicians, all native Egyptians, were to serve in the military.[1]

The hospital and medical school were military assets, as was the printing press the ruler established at Bulaq:

> Remarkably responsive as he (Muhammad Ali) was to the vistas which modernization unfolded before his eyes, a printing press was not for him a grand cultural gesture, but rather a coolly practical instrument for improving his army, the agriculture of his territory, and the usefulness to him of his subjects, whether Turkish- or Arabic-speaking.[2]

Indeed, if there was anything that gave unity to the patchwork integuments of the ruler's reforms that came into existence over a period of three decades, it was their relationship to the military.

Students in the new state schools, like military recruits, received uniforms, free lodging and board, ate in mess halls, slept in barracks, were given rank and paid accordingly. They were student–soldiers whose duty it was to study whatever they were ordered to study. When it was determined that their studies were finished, they worked at the task assigned to them. Their stipends were scaled to military pay rates, from entering recruits to mid-level officers. Students received 15 to 30 piasters a month during their primary school years; 30 to 50 during their four years of secondary school; and if chosen for specialized training, 100 to 200 piasters as they advanced through the four years of specialized training in science, engineering or medicine. They were drilled and marched in military formation to and from classes. If deemed slackers or insubordinate, they were beaten and thrown in the stockade in the same manner common soldiers were. The bullwhip was commonly applied as a tool of discipline to keep them studying hard and from complaining about their misery. Muhammad Ali was determinedly set on instilling a sense of military discipline in those that he used to build his state and its military organizations, and beating it into them was the norm. His standard of military discipline and severity of punishment favorably impressed European observers.[3]

The task Muhammad Ali set for himself was enormous. From an illiterate, backwater agricultural society steeped in tradition and religion, barren in the fields of science, technology and medicine, with a medieval religious institute as its sole organ of upper education, he set out to create, in the shortest time possible, a modern army and navy with supporting infrastructure of administration, industry and system of medicine and technical education. With little else to build on other than his experience of being trained in Selim's New Order, he started from scratch. The only education in Egypt was religious. The savants of Bonaparte had left nothing behind but memories. Compared to the Ottoman reformers in Istanbul, Muhammad Ali had it hard.

Consider the level of Egypt and Syria around 1800. According to an early 19th century author from Mount Lebanon, Mikhail Mashaqa, people knew little more mathematics than simple addition. Multiplication was done by successive additions of the number being multiplied. Astronomy was a virgin subject no one went into.[4] The chronicler Jabarti wrote that in 1748 the Ottoman Governor of Egypt, Ahmad Pasha, who had been in Istanbul during the Tulip Period and the years Ibrahim Muteferrika's press was operating, 1727–1747, and who himself had a lively interest in science, had a conversation with three Azhar shaykhs on the state of scientific learning in Egypt, the conclusion of which was that no science at all existed among the learned of Egypt, that is, al-Azhar's shaykhs. Egypt, in the Ottoman view, was a scientific vacuum.

Jabarti diverges somewhat from this bleak assessment. He mentions in his chronicle the names of around 30 shaykhs who, during the half century interval between the Ottoman governor's assessment and Bonaparte's arrival, had an interest in science and who wrote scores of books on the subject.[5] He must have had in mind the *miqati* who reckoned the times of new moons, religious festivals and prayer orientation in facing Mecca, in addition perhaps to a few shaykhs who may have had a smattering of mathematics, enough to make them think highly enough of themselves to write on it. Shaykh Jabarti's reckoning stands for something: science in Egypt was not quite at absolute zero. An idea of what that something was may be seen in the scientific collection that existed in Egypt's primary library at al-Azhar. At the time that Shaykh Jabarti was writing, the library contained five books on algebra and 13 each on astronomy and methods of computation. How often the books were read is not known. Jabarti's figure of 30 shaykhs interested in science over a span of half a century helps explain why during the three years of the French Institute only one shaykh, Hasan al-Attar, was recorded to have taken a positive interest in it. Of Jabarti's 30 shaykhs, the leading one was his father, Shaykh Hasan al-Jabarti. According to the chronicler, his father learned astronomy and mathematics from an Indian Muslim who came to Cairo and gave lessons in the mosque he was living in. Hasan invited him to stay at his house and in return received lessons. What this says is that there was some interest in astronomy and mathematics, but so few people in Cairo knew anything about the subjects that the arrival of a traveling Indian who did know something about them was a noteworthy event. Jabarti relates that the Indian made such a scientific scholar of his father that Europeans residing in Cairo came to listen to Shaykh Hasan's lectures on geometry, which enabled them, when they later returned to Europe, to invent windmills and load-drawing machines. Hasan's experiments with sundials and his reform of the system of weights and measures gave him such fame, according to his son, that his name reached the sultan, who rewarded the shaykh with gifts.[6]

Despite the reputation that al-Jabarti believed his father had as a scientist and mathematician, and the contribution his father supposedly made in providing ideas to Europeans for mechanical inventions, he could still write dismissively of the *Institut d'Egypte* as a French circus for the savants' amusement. Their scientific knowledge he regarded as being unsuitable for Muslims. Other than those brief references in Jabarti's chronicle, there is scant evidence of any scientific interest in the country.

Syrian Christians were less restrained than Egyptians in social and intellectual intercourse with the French occupants and in general more open to Western learning.

In Damietta, a few years before Bonaparte's arrival, a Muslim shaykh taught astronomy to a Syrian Christian merchant from Mt. Lebanon. The shaykh, Muhammad al-Sabbagh al-Miqati, who was reputed to be learned in astronomy (his title *Miqati* indicates he knew something about it) taught the subject to Butrus Anhuri, an uncle of the author referred to above, Mikhail Mashaqa. Mashaqa cites his uncle Butrus as having said: "Then, when Bonaparte came . . . I did not waste my time but worked hard to learn the language of the French, among whom were many scholars. From them I learned the latest discoveries in astronomy, physics and geography."[7]

A decade after the French departure, Basili Fakhr, a Syrian Christian who was acting French consul general in Damietta, helped make Arabic translations of French books on physics and astronomy, one of them being Jerome Joseph Lalande's *Abrege' d'astronomie*, published in 1795. Another source of intellectual contact that Egypt had with the West before the French incursion was also Christian in nature, namely the Arabic-speaking Franciscans, and to a lesser extent, Jesuits who taught in the missionary schools that had been set up in Luxor, Aswan and Asyut to instruct boys in the Coptic community. The instruction was strictly religious, with little if any science entering the mission curriculum. However, the most promising of the graduates were sent to Rome to complete their religious studies and, it was hoped by the Church, to become proper Catholics and go on to become priests and missionaries in their Coptic communities back home. A major consequence of Rome's missions to Egypt came in 1736, when the first printed book entered Egypt, the *Missale Copto – Arabicum*, compiled by a Coptic convert to Catholicism, Rafa'il Turki. This was just about the same time the printing press of Ibrahim Muteferrika started turning out books on geography, astronomy and history, giving an idea of the difference in content and interest between Cairo's and Istanbul's interaction with the West in the 18th century. Muslims in Egypt did not see a printing press until Bonaparte came with one, and even then there was no Egyptian enthusiasm to exploit the invention.

The Egyptian Christians who studied in Rome may have learned something of European science during their years of religious study there, but nothing of it seems to have been transferred to the larger community. Also, as if the Ottoman capital and Cairo existed in different worlds and were not major cities in the same empire, there seems to have been very little if any intellectual transfer coming to Egypt from Istanbul. Mamluk beys and Ottoman pashas in Cairo did not receive from the European consuls general the books of astronomy and science that the sultans and grand viziers in Istanbul received as gifts from European kings and ambassadors in the 17th and 18th centuries. Egypt had no Tulip Period, no Yirmisekiz, Ibrahim Muteferrika, or Khalil Hamid, and no printing press until Muhammad Ali established his government press in 1826, a quarter century after the French had left and a century after the founding of the Tulip Period press. On such unpromising ground did Muhammad Ali hope to transform Egypt to a modern state within his lifetime, from water buffalo and Archimedes screw to steam engine and electrodynamic: a road that took Europe over seven centuries to travel.

Muhammad Ali was not sure how to go about reaching his goals. The cunning but untutored soldier had little idea of all that was involved in creating a system of education that would produce the expertise needed to build and run the modern industries, arsenals, agriculture and state administration to support the military institution he knew he must have to survive as the ruler of Egypt. He never contemplated following Peter the Great's example of going to Europe to see what had to be done and how, nor in all fairness did any other Muslim ruler. He was forced to rely on those who did know, first the Italians, then the French, and to a lesser extent the British, none of whose interests were by any means congruent with his own. His dependence on foreigners rendered him incapable of even getting rid of the detested Capitulations.

The more problems that arose in his quest for a modernized state, the more the foreigners were needed, which contradicted what he wanted. The system was intended to train specialists to take the place of Europeans whose presence would be gradually phased out. The problem was that the privileged Europeans did not want to be replaced. Highly paid and protected by Capitulations that gave them what approximated to ambassadorial status, they had a princely life in sunny Egypt and did what they could to keep it. Consequently, the relations between the foreign faculty and the students who were out to take their positions were as frigid as those prevailing among the bickering French, Italian and British directors and faculty who were supposed to be working together to train the Egyptians.

The new education

The main pillar of Muhammad Ali's industrial, agricultural and military reforms that were to make Egypt into a modern state were the translators of scientific, technical and medical books. The books were printed by Muhammad Ali's new printing press and used in his system of education. The printing press was introduced in 1822 and housed in Boulaq, a section of old Cairo along the Nile. The running of the press offers a glimpse of the corruption that crippled the dictator's reforms. Copies of many of the books that were translated and printed were sent to Istanbul, Izmir, Salonika and other cities of the empire where they could be sold for three and four times their price in Cairo, with the people running the press pocketing the difference. This explains why in the 1840s catalogers of the books published by the press were unable to find any copies of many of the works published by the press.[8] Such scams were endemic. Undermining the reforms, they were simply one more charge added to the price a society burdened by one-man rule had to pay.

Between 1822 and 1842, over 240 works were published. More than 50 were Arabic and Turkish translations of French books on mathematics, mechanics, geodesy, military science, gunpowder, medicine, surgery, physiology, pharmacy, veterinary medicine, naval technology, hospital administration, sanitation, agriculture, natural history and botany. The remainder pertained to Islamic subjects: treatises on Arabic, Turkish and Persian grammar and lexicology, theology, jurisprudence, rhetoric, logic, metaphysics, history, mysticism and poetry.[9]

Muhammad Ali got his technical and scientific books for translation and printing from wherever he could – Europeans working in Egypt, consuls general, European publishing houses that responded to orders for specific books. Except for the last source, the selection of books that were translated and printed was a happenstance affair. The ruler bought three presses from Milan in the earliest stage of his educational initiative. Ink and paper came from Livorno and Trieste, respectively. This was when Muhammad Ali was employing mainly Italian experts and instructors. Later, when he turned to the French, he bought more presses from Paris. Eight presses were reported to be in use in 1831.[10]

The translations were made by Egyptians in the Translation School. Translators had to invent Arabic and Turkish equivalents for the technical and scientific terms that were foreign to the language. The Arabic translations were passed on to Azharite shaykhs who edited and corrected them to assure they were in good Arabic. The shaykhs who did the editing were called *muharrir*; those who did the correcting, *musahhih*. The *muharrir* corrected the translation, checked the technical terms and advised on their suitability based on standard Arabic usage. The *muharrir* then submitted the translation to a *musahhih* who made sure the composition was in acceptable literary Arabic. The changes that the shaykhs suggested would then be made by the translator. Once the translation was accepted, it was delivered to the press. In this way, the translation and printing procedure brought the shaykhs into the secular business of modernization. Those editors and correctors were at the lowest level of the translation system, though their work was absolutely essential. The shaykhs were the only literates around who could guarantee that a text was written in sound Arabic. Their contribution in creating technical terms and new phrases "must be considered one of the most important contributions to learning during the whole of the Muhammad Ali Period, if not the nineteenth century."[11]

Like all of Muhammad Ali's reform programs, his system of education took practically the whole of his reforming life to come into being. It was not until 1840, the beginning of the end of Muhammad Ali's expansionist policy and reforming career, that something approaching a system of education had come into place, innovation by innovation. The unplanned rush to modernize is apparent in the piecemeal fashion in which the system came together over a quarter century, without any apparent overall template or logical order of progression from one level of education to the next. Muhammad Ali wanted immediate results and began sending students abroad for advanced specialized study soon after he had cleared the field with his destruction of the Mamluks. It was not until almost a quarter century later that serious attention was given to preparing students before sending them on to advanced technical and medical study, which usually would be in French, the language of texts and lectures. Without the foundation in language, the attempt to educate was short-circuited. The system of advanced education that was at last developed was in itself sound enough in overall design, but outside of itself it was dysfunctional in its inheritance of flaws that were transmitted from the lower levels to the upper levels, undermining the whole process of educational progress which the overall system was intended to ensure.

The system was structured on four levels: primary, secondary, specialized professional and overseas advanced graduate study. The specialized institutes of higher training were the schools of engineering, medicine, language and translation.

The engineering school was designed to turn out military and industrial engineers, technical experts, factory managers and teachers of science and mathematics. These would be the experts who would oversee Muhammad Ali's military, industry and agriculture. The medical school was to supply military doctors for the military hospital; the language and translation school was to provide language teachers for the secondary and specialized schools, and translators to translate scientific and technical books into Arabic and Turkish to be used as school texts.

Graduates of the specialized schools who were sent to Europe for higher studies in their specializations were to replace the many highly paid Europeans who had been brought in as military advisors, trainers, teachers, technical specialists, industrial managers, physicians, and directors of the specialized schools. The Europeans who advised, organized, planned, designed, taught, staffed and administered the reforming institutions formed a privileged class. The large majority of them being French, they may be seen as a second invasion of savants, very highly paid ones this time, not at the cutting edge of their science as had been Fourier, Berthollet, Monge, Malus, Saint-Hillaire, Conte; they were hardly as motivated by Enlightenment ideals that had inspired many of the savants accompanying Bonaparte's expedition. The second invasion was invited and went for the money.

Muhammad Ali paid foreigners handsomely. But the mind behind the grand projects of the hired savants faced obstacles almost as formidable as had the grand designer of the projects of their predecessors, the Enlightenment savants: for what greater contradiction could a Muslim leader fall into than hiring Europeans to help him and his country to become strong, the very people whose robust merchants, industrialists, financiers and governments were now more than ever becoming a threat to Egypt and the Ottoman empire, to Islam, Asia, Africa, to whatever patch on the globe that had not yet come under European colonization and could be exploited for a shilling or a slave? Relying on the West could be a costly trap. Did Muslim rulers have any choice but to learn from the West? Where else to go for introductory lessons in industry, economic management and defense? But then, would the West allow any Muslim state to modernize and become strong enough to escape dependency, that is, lift itself from a captive market to a competitor, if it could be prevented?

Muhammad Ali thought he could do it by using European rivalries to his advantage and keeping out of debt. Britain and France were the greatest threats. Egypt, at the corner of Asia, Africa and Europe, with the Mediterranean to the north and the Red Sea to the east, would be a prize addition to either of their growing maritime empires, both of which had already shown themselves invasively covetous of the country. Muhammad Ali went first to northern Italy for help. Italy was not as modern as the maritime powers, but it was divided, weak and not even a nation state. The Italians would be in no position to threaten Egypt. Also, the Italians had long experience in Egypt. A few of them had powerful political connections. The merchant and industrial entrepreneur Carlo Rosetti had been an advisor to Mamluk rulers ever since the early 1760s and lived long enough to be an early advisor of Muhammad Ali. Italian trading colonies, especially the Venetian, had existed in Egypt for centuries, as had Italian missionary schools serving the Coptic community. It was from Italy that the first non-religious foreign teachers in Egypt had come, and it was to Italy that the first Egyptian student missions went. It was Italian in which were written the earliest text books used by students in the nascent military, engineering and medical schools.

As many as 105 Italian doctors and pharmacists were employed in Muhammad Ali's military hospital. From 1816 to 1825, somewhere between 30 and 40 of Muhammad Ali's elite Albanians, Turks, Circassians, Armenians and Greeks who had come from all over the empire to take up service with him in Egypt were taught Italian and sent to Pisa to study industrial technology and military science. Italian advisors also predominated in Muhammad Ali's administration during this early period.[12] Accustomed as they were to enjoying almost a monopoly on Ottoman reform, the French were not pleased to be denied the same in Egypt.

Halfway through Muhammad Ali's reign, the few French advisors and military officers working for him at the time, jealous of the favored position of the Italians, were at last able to convince the ruler to turn to France to support his modernization programs. The French had been trying to win Muhammad Ali over ever since his annihilation of the Mamluks, as it was a faction of the Mamluks that the French had relied on to protect their interests. That same year, 1811, Jomard, the geographer who had accompanied Bonaparte on the Egyptian expedition and who as the editor of the monumental *Description de l'Egypte* had kept close contact with Egypt, submitted a plan of modernization to the French consul in Cairo to be forwarded to Muhammad Ali. Jomard's plan called for sending Egypt's best students to France to complete their advanced study. It was several years before Muhammad Ali acted on it, but when he did, Frenchmen began taking the place of Italians.

By 1825, Frenchmen were rapidly replacing Italians as teachers and directors in the medical, engineering, military and veterinary schools; more and more student missions were being sent to France. Within a decade, the French had taken over the Italian position. A few Italians were retained in order not to give the French a monopoly: only Muhammad Ali had monopolies. British technical and military experts were also hired, and a few small student missions sent to Britain. The British had defeated the French at Waterloo and so to the victor went several prestigious posts in the emerging complex of European-staffed educational institutes and government bureaus, to the chagrin of the French. Their running of the Military Engineering School, the Medical School and abu Za'bal Hospital complex and the Artillery School at Turah was not enough. The French desired a total monopoly in modernizing Muhammad Ali's Egypt. Technocratic colonization would substitute for the military one Bonaparte had failed to make.

The two most important specialized schools were medicine and engineering. The director of the abu Za'bal Medical School was Dr. Antoine Barthelemy Clot. He had been elevated to the rank of bey by the ruler and was known as Clot Bey. Muhammad Ali respected him as a physician and admired him for his devotion to improving the country's health but probably more so because of the stern discipline, bordering on cruelty that he inflicted on his medical students. He took over and expanded abu Za'bal where the Italians had left off.[13]

Clot Bey designed a three-year medical school program that enlisted 100 new students a year. The first hundred came from al-Azhar by order of Muhammad Ali. In the following years the number of recruited students was much less than a hundred. A few recruits were taken from the secondary (*tajhiziyya*) schools of Cairo and Alexandria, but mostly they were drawn from al-Azhar, and some from even as low a level as the rural madrasas, where it was hopefully thought that the minds of the young boys had not yet been cast hard in the mold of rote memorization of long texts of the Quran. Wherever they came from, the students were all woefully unprepared to study medicine. The principal consideration that guided selection was that the students had to be of rural stock. Farm boys were thought by Muhammad Ali to be more pliable and obedient than young Egyptians from the urban areas, who were considered too corrupt, devious and obdurate to be made into obedient servants of his state.

Before entering the medical school, the students were obliged to prepare themselves in geometry and the natural sciences, with an emphasis on physics and astronomy. These were exercises that would presumably wean their young minds from the rigid tradition of textual memorization they had been religiously subjected to in the village schools. Math, physics and astronomy would prepare them for the logical rigor and precision of thought demanded by medicine, and would "stimulate interest in observing the natural world" and foster "the spirit of critical inquiry."[14] Unfortunately, very little of that was done. In a hopeful attempt to help bridge the huge disconnect between the traditional and the modern studies of incoming students, Clot Bey had the walls of the hospital study and library decorated with geometric figures and diagrams of astronomical phenomena, as if by osmosis the meaning of the wall decorations would permeate the atmosphere, giving the students the knowledge they needed but did not get in al-Azhar or the *tajhiziyya* and madrasa schools. In the same vein of wishful intent, the walls of the anatomical amphitheater were written over with names from the era of Islamic scientific greatness – Khwarizimi, Razi, ibn Sina, Biruni, Haytham, as if to say that classical Muslim science legitimized cutting up dead bodies.[15]

For all of his overbearing demeanor and assumed severity, Clot Bey appears to have been a diplomatically adept operator, if his own testimony is to be credited. He claims that he was able to overcome the suspicions of the Azharite shaykhs and religious students who considered Western science and certain medical practices, such as using cadavers in anatomy classes, to be repugnant, if not heretical, to Muslims. Clot did not meet the opposition head-on, or obsequiously patronize the shaykhs as had Berthollet and others of the savants 30 years earlier. His method of enlisting their support was to invite them to give sermons to the medical students and staff. When they had finished, he would then take the occasion to extol the past greatness of Muslim science and medicine, tying in the old with the new and assuring students and shaykhs alike that Egypt's medical school would revive that greatness and shed new glory upon Islam. His conversion to Islam, followed shortly by his being elevated from bey to the high rank of pasha, could not but help his mission as an ambassador of modern science and medicine. He boasted that he had convinced the shaykhs of al-Azhar that it was permissible to dissect cadavers for the sake of medical knowledge and that he had obtained a secret agreement from the rector of al-Azhar allowing students to practice dissection in the anatomy classes. What he probably meant by secret agreement was that the shaykhs associated with the medical school as translators and proofreaders would wink at the practice and not let word of it get outside of the guarded walls of abu Za'bal, as few people outside the school knew what was going on inside, inasmuch as it was built alongside the military hospital, a good distance from Cairo and surrounded by high walls and guarded by soldiers on all sides to keep snoopers out and students in.[16] As for the higher shaykhs who had problems with what they suspected or knew was going on at abu Za'bal, they were bought off, threatened or exiled to the distant provinces. Those who bravely opposed the ruler were eliminated, as Shaykh Jabarti had been, while those who accepted to join the new education and cooperate with the ruler were rewarded, like Shaykhs Attar, Tahtawi and Arusi.

The students entering the medical school were woefully unprepared. This was the story in all Muhammad Ali's specialized institutes, but it was particularly true of medical students, mainly because they had been drafted from village madrasas and were therefore very young, or they were recruited from al-Azhar and were therefore older but steeped in religion and mentally frozen in textual memorization. Students coming from the two secondary schools were still young, but their minds were less religiously rigid. From wherever the students came, the problem of unpreparedness was so severe in the medical school that Clot advised the ruler to build a preparatory school for students entering the medical school. Muhammad Ali agreed. The school was called the *Madrasat al-Maristan*, or Hospital School. It was an annex to the Medical School, which was itself an annex to the Military Hospital, whose graduates did not have far to march in going straight from their medical studies into the military.

Much was expected of the *Madrasat al-Maristan*. Its curriculum was designed to accomplish much: in four years the school was to teach the students what they were supposed to have learned in the secondary schools but did not; or for students coming from al-Azhar, it was to transform graduates of religion into students ready for medical study. Students studied logic, geometry, algebra, trigonometry, general science, biology, chemistry and descriptive cosmology, the last including Newtonian mechanics and the new astronomy. Also included were ancient and modern history and French and Arabic languages.[17]

Entering students usually ranged from 10 to 14 years of age. The ones who performed best in the Preparatory School were admitted to the Medical School to study either medicine or pharmacology. Most students entering the Medical School would have been between 14 and 17. The younger the student the better. Clot Bey put the upper age limit at 30. The minds of ten-year-olds were still fresh and open. A boy who had not been introduced to new ideas and taught to think independently by his early teens would have been mentally devoured by his unquestioning acceptance of tradition and the mind-numbing exercise of Quranic memorization.[18] This was the perception shared by observers from both the East and the West. In his book *Growing Up in an Egyptian Village*, Father Henri Ayrout, a French-trained Egyptian Jesuit who spent his life teaching young boys in rural Egyptian villages during the middle of the 20th century, writes of the light in the eyes of his young students and of their curiosity, their eagerness to learn, comparing them favorably to Europeans of the same age; but once the deadening hand of rural tradition and rigorous religious memorization was mechanically drilled into their young heads by the village shaykhs in the mosque schools, their eyes dimmed, their eagerness left and creative learning came to an end.

The quality of education in the *Madrasat al-Maristan* was plagued by the same problems that beset all of Muhammad Ali's schools: shortages of teaching staff, reading material, blackboards, chalk, paper: a lack of supplies of every kind. Conditions improved somewhat when in 1841 the school was taken over by the School of Languages. This gave it a more professional level of instruction in French. The School of Pharmacology that was added to the Medical School also benefited the Hospital preparatory school by making the pharmacological teaching staff available to it. An unintended consequence of the ad hoc method by which Egypt's system of secular education arose is that the Medical School's incoming students, thanks to the *Madrasat al-Maristan* and Clot Bey's influence with Muhammad Ali, ended up being even better prepared than were students entering engineering, the school that had been given top priority over all the specialized schools, and for whose service the two secondary schools had been created.

The top-down sequence in which the hospital and its two annexes were founded over the years, from military hospital to medical school to preparatory school, reflects the way reforms were drafted. Orders were sent down from the top without feedback from below, without an initial overall plan to turn non-French-speaking religious students into medical students and physicians by way of French instruction. Attempts to meet the problems that soon appeared were made only after they could no longer be avoided or concealed in some manner of denial. The result was that the system was for the most part self-defeating in terms of preparing students for the next level of study. Tardiness in establishing primary and secondary schools and the medical preparatory school undermined the efficiency of the much more costly specialized institutes and foreign student missions.

The Military Engineering School was encumbered by the same drawbacks as the Medical School. Engineering, the jewel of the specialized schools, was the discipline that in the ruler's mind most immediately impinged on military modernization. Equating political power to military might and might to technology, Muhammad Ali gave engineering top priority. Altogether, six schools of engineering were created in the 1820s and 30s. The schools of medicine, languages and translation, whose students were most usually native Egyptians, were effectually annexes in support of the more prestigious engineering schools. Graduate engineers were to provide the expertise for building roads, bridges, dams, barrages, canals, irrigation works, factories, railways, public works, and most important of all, the military industries of mining, steel mills, artillery foundries, gunpowder plants and arsenals. One of the large construction projects in which several graduates from the Bulaq Engineering School had important positions was the Mahmudiyya Canal (named in honor of Sultan Mahmud II) connecting the Nile and the Red Sea, a precursor to the Suez Canal.[19] The engineering schools also produced the mathematics and science teachers who were to replace the Europeans in the specialized schools.

The embryo of what would become the first engineering school was established in the Citadel, near the ruler's palace where, a few steps away and a few years earlier, the Mamluks had met their end. The Citadel school was an early step in Muhammad Ali's creation of a modern military institution to replace the unruly medieval one whose members he had massacred. In preparation of its being expanded to an engineering institute, the Citadel school was transferred in 1821 to Bulaq, a quarter of Cairo along the Nile where Muhammad Ali had recently founded his printing press and where the National Library of Egypt (*Dar al-Kutub*) stands today. The school's name was changed to *Dar al-Handasa*, literally House of Geometry, and was a version of the Ottoman Muhendeshanes established the century before in Istanbul by Bonneville and de Tott. Over the next decade the Bulaq school expanded and became known as the Bulaq Muhendeshane, the first of the specialist (*khususiyya*) institutes. Another engineering school was founded in 1833 in the countryside at Qanatir al-Khayriyya, where a barrage was being built to expand agriculture. Several other engineering schools were built in Cairo over the next few years, most notably the one at Qasr al-'Ayni. The sixth and last was built in Alexandria. These were the years when Muhammad Ali's power was at its height, ruling over an Afro–Asian empire that included the Sudan, Palestine, Syria, Mt. Lebanon, southwest Anatolia and the Hejaz. As his empire expanded, so did the ruler's engineering schools and industrial projects. In spite of their many flaws in priorities and planning, Muhammad Ali's reforms appeared to be lifting Egypt to the ranks of a developing industrial state.

The flagship of all the engineering institutes was the Bulaq *Muhendishane*. Its director was the French engineer Jean Henri Lambert, who with other French Saint Simoneans had come to Egypt in the mistaken belief that Muhammad Ali was a progressive, humanist-minded leader who would lead Egypt and the Arabs to a socialist enlightenment. The Bulaq School of Engineering suffered Muhammad Ali's parsimony the least of all the specialist schools. It had the best faculty, mostly French, and was reorganized by Lambert along the lines of the famous Ecole Polytechnique in Paris.[20] Bolstering its prestige was an observatory, built at the same place that the savants of the *Institut d'Egypte* had built theirs. More prestige came with the school's participation in a European research project to measure the earth's magnetic moment. Requiring the purchase of sophisticated and expensive magnetic measuring equipment, the project was the only occasion Muhammad Ali ever laid out money for pure research.

Students for the engineering schools came from the favored "northern" nationalities, Turks, Circassians, Albanians, Bosnians and Armenians, most of them being the sons of officers and former Mamluks who had survived and were later to join Muhammad Ali's state-building program. These were men who came from the racial stock that Muhammad Ali considered would make good officer material and whose minds were capable of mastering Western technology and organization. There were exceptions. Between 1820 and 1825, the years of Italian influence and educational directorship, a few shaykhs who had learned Italian were enrolled in the Qasr al-Ayni School. This was before it had been expanded to include engineering. The shaykhs were sent to study mathematics and scientific subjects. This was not because they were being prepared to become engineers but because they had studied Italian. Since the teachers at the school were Italians lecturing in Italian, they could understand the lectures without need of an interpreter. The intent was that these shaykhs would become general science and mathematics teachers lecturing in Arabic to students entering their first year at the Bulaq Engineering School. In the late 1830s and through the 1840s, some native Egyptians who graduated from religious schools were at last enrolled to become engineers. Several of them became famous, not only as engineers but as leading voices of reform. Amin Sami and Ali Mubarak were the most notable exceptions to Muhammad Ali's dictum that native Egyptians were fine for medicine, language and translation, but not engineering: an engineering diploma meant an officer rank in the military and no Turk, Circassian or Albanian considered native Egyptians as officer material.

The French director ran the Bulaq Engineering School more like a military institution than even the notorious disciplinarian Dr. Clot did his abu Za'bal Medical School. The director and department chiefs of Bulaq were high ranking military officers who lorded over the students as if they were the lowest of lowly recruits, which they were. The department heads never let them forget rank. A *bin bashi*, or commander of a thousand, which in Western terms would equate to a major, headed each department. The engineering school's director of mathematical translations also held the rank of *bin bashi*. The rank appointed to the heads of departments in the medical, translation and language schools was only *yuz bashi*, commander of a hundred, or a captain, which gives an idea of Muhammad Ali's militarized priorities in his approach to education. Directors of translations of history, geography and literature were mere lowly *malazims*, lieutenants. Upon graduation, engineers became officers. They were not directly assigned to military related projects, but like all the graduates of specialized schools they had first to serve a period in translating the foreign texts they used in their studies. Post-graduation translation duty slowed the pace of development, but the old soldier in Muhammad Ali wanted to exploit his student recruits for all he could and save on translation costs.

The major weakness in Muhammad Ali's system of specialized education was the woeful unpreparedness of the entering students who graduated from the poorly staffed and underfunded secondary schools. Language became a tremendous problem once students advanced to the higher schools without having the Italian, or later the French, in which the lectures were delivered and the texts written. Instead of Italian or French, secondary schools taught Arabic, Turkish and Persian, even though a fair number of non-Arabic-speaking Italians were teaching technical, mathematical and scientific subjects in the secondary schools. So even at the secondary level, a language barrier existed between the students and some of their teachers. Arabic, Turkish and Persian were important for Islamic literature and cultural continuity and had a place in Islamic learning comparable to Greek, Latin and modern European languages that at the time made for a sound humanist curriculum in Western education, but the emphasis on Muslim languages did not advance the preparation of students for specialized technical studies.

The problem was how to give secondary students the languages they needed to engage in their own culture while giving them the language skills they needed for engineering, medicine and science. It could have been solved by introducing Italian and French instruction at some point in the primary school curriculum and intensively continuing it through the secondary level. However, part of the deal Muhammad Ali had made to enlist the shaykhs as teachers in his secondary schools was to let them teach the three primary Muslim-speaking languages. Instruction in these languages was not an expression of cultural aestheticism on the part of the military ruler but of his effort to staff his secondary schools with Azharite teachers who cost him next to nothing. Rather than sending students to Europe to learn a language in preparation of their being translators or language teachers in the primary and secondary schools, Muhammad Ali tried to do it on the cheap by having the specialized students act as language instructors and translators, thus fragmenting the professional work and academic concentration of his doctors and engineers. The frugality slowed the process of assimilation. The dictator further undermined the education reforms by staffing the secondary schools with shaykhs, their predominant numbers having the unwanted effect of transforming the curriculum from a secular to a religious emphasis. Consequently, the two secondary schools that were supposed to prepare students for technical studies were able to offer little in the way of science and mathematics.[21]

The shaykhs in the secondary schools tended to be less educated and more oriented to Shaykh Jabarti's rejections, as opposed to the shaykhs at abu Za'bal and the other specialist schools who, sharing an assimilationist attitude similar to Shaykh Attar's, had been chosen for their willingness to associate with Europeans in learning and teaching science, technology, medicine and languages. The strict conservatism of the secondary school shaykhs is reported to have so intimidated the Italian teachers that they dared not introduce courses in Italian and science, defeating the purpose for which the schools had been created. In effect, the secondary schools ended up preparing students more for religious study at al-Azhar than for engineering or medicine at the *khususiyya*.

Each level of schooling undermined the one above it. The primary was mired in religious tradition where the Quran was the sole book of instruction. Hamont, the director of the veterinary school, painted a dismal picture. At neither the primary nor secondary level were there texts for the students. Blackboards, chalk, paper and writing materials were not provided. Slates, tin plates and whatever was at hand were used for writing, and the boys, having been given nothing to mark them with, had to wet their fingers in their mouths and make the marks as best they could on whatever they had provided themselves to write on. However, since in the primary schools there was not much to write, the time being taken up by memorization of the Quran, the problem was not felt to be all that acute. As in the mosque schools, the primary school religious instructors bobbed their heads back and forth as they read to the children, their heads rocking in rhythm with their teachers, a style that was general throughout the Ottoman Empire and Islamdom. When the students graduated, those who went on to the secondary level knew well the reading and recitation of the Quran but barely a smattering of arithmetic.[22] The best thing about the primary level may have been the yearly two shirts and pair of shoes the government allotted the students.

The shortcomings of the primary and secondary levels became critically apparent at the specialized level, where much of the time that would ordinarily have been devoted to specialized study had to be used to teach the students what they should have been taught at the secondary level. The deficiency was transmitted upwards. Students sent abroad for study had to learn what they should have learned back during their years in the specialized school. It will be recalled that the abu Za'bal Medical School was obliged to create its own preparatory school to compensate for the failings of the secondary level. The engineering schools had no such equivalent. Director Lambert of the Bulaq School proposed to improve the secondary level schools by having graduate engineers do a tour of duty teaching mathematics and science in them, but the proposal came to nothing, probably because the shaykh secondary school teachers opposed it. In any case, the specialist school graduates would have been too busy translating their textbooks into Arabic or Turkish to have time to teach the secondary students.

One can only wonder at the shock of incoming engineering students suddenly confronted with a five-year course of study that included advanced geometry, algebra, trigonometry, descriptive and analytical geometry, differential and integral calculus, physics, chemistry, statics and dynamics, fluid mechanics, astronomy, Newtonian mechanics, topography, structural analysis, civil engineering and industrial processes. Not only did the students have no preparation, they had no knowledge of the language their teachers were lecturing in. Even by the third year they were unprepared to digest the material. In fact, at least half of the five-year curriculum was spent tutoring the students in subjects they should have got in secondary school introductory mathematics, physics and chemistry.

A step was taken in 1837 that was thought might ameliorate the failings of the secondary schools. Students who were considered mentally or physically deficient were sent to work in the hospitals and factories, but all that this accomplished was to reduce the student body in the two schools from around 400 to 176.[23] It did nothing to change the faults in the system. One of the consequences of the wretched state of secondary education was that the most cherished of Muhammad Ali's specialized schools, engineering, fell furthest from meeting expectations. Not even public whippings, 30 days in the stockade and the most dire threats from the ferocious French mentors and their Turkish and Circassian acolytes were able to terrorize the students enough to meet the challenge of their coming to calculus without any algebra, geometry or trigonometry. By the time the students who had been selected for advanced study abroad reached Europe and saw his curriculum he must have been on the verge of breakdown – unless he was a privileged relative or favorite of the ruler, or a son of his Turko–Circassian and Armenian elite caste. These would have had private tutors to bring them up to standard.

The advanced study missions abroad, in addition to all the flaws they inherited from the preceding levels of training, were beset by their own inconsistencies and contradictions. Beneath the many reasons for all these problems was the stark reality that no bond existed between the ruler and the people of the country he wanted to modernize, nor between the ruler's government and the people. And therefore, none between the reforms and the people. This goes to explain much. The students Muhammad Ali sent to Italy and France were not young Egyptians but military and administrative officers in his service, his followers and their sons, a miniscule elite of Armenians, Turks, Circassians and Albanians. The ruler had no interest in educating native Egyptians. He thought it a waste of resources. Muhammad Ali's contempt of the indigenous people of the country that he ruled was legendary. When Jomard suggested sending native Egyptians abroad to study, Muhammad Ali is said to have scoffed at the idea, declaring that Egyptians were too ignorant to learn anything from Europe. The ruler's contempt was shared by his Turko–Circassian followers. The effect was to eviscerate leadership of the power to inspire: only tyranny to depress, humiliate and crush was seen to come from the leaders. Obviously, the psychological and structural complexities that bedeviled reform in Egypt were many times more formidable than those confronting reformers in Istanbul. This arrogance of power to humiliate also explains, in some part, the huge difference in achievement between the Middle Eastern and Japanese reach for modernity.

Not until the mid-1830s when the French were directing the specialized schools did primary and secondary education come into serious consideration, a full quarter century after the ruler's destruction of the Mamluks and beginning of modernization. In 1836, 50 primary schools (*ibtida'iyya*) were built throughout the country, serving a total student enrollment of 5,500 young boys. Only two secondary schools (*tajhiziyya*) were founded, one in Cairo, the other in Alexandria, which gives an idea of the miniscule base of Muhammad Ali's educational reforms. It took a long while before he considered using Egyptians as anything more than peasants and fodder for his military. Muhammad Ali was not thinking in terms of educationally regenerating a whole nation from the bottom up; his vision was focused no wider than on producing a cadre of technical experts to run his industry, staff his administration and train his officer corps to lead an army and navy that would deter Western and Ottoman designs to oust him. Reformist Egypt was all about Muhammad Ali, not Egypt or Egyptians.

His political narcissism doomed his reforms. The Military Engineering School and Medical School and foreign study missions were not designed to give root to a nascent culture of science and technology. Muhammad Ali had no desire to produce such a culture. Indigenous science and technology were long-term endeavors. His needs and his means of meeting them were immediate, practical, political and framed in a military regimen. He wanted trained experts, engineers, managers, specialists to run the specialist schools, technocrats who would follow orders and not think beyond the limits of the assigned task. He wanted the industry, technology, administrative organization, managerial skills and high caliber officer corps that made a modern state, but not the initiative, independence of mind and entrepreneurship that came with it. He wanted his men to serve him, not work for themselves. He is reported to have said that Egyptians (and this included his entourage of Turkish, Circassian, Albanian, Greek and Armenian followers) should become accustomed to industry but not profit from it.

Small wonder no entrepreneurial class developed in Egypt until long after Muhammad Ali's reign. The ruler wanted the power of the modern industrial state without the human development that sustains it. For fear of the dissatisfaction it would create if there were more educated Egyptians than there were jobs to employ them, he was hesitant to introduce general public education. For what he wanted to achieve he did not need it. It was an expensive frill that would take resources from the military. Muhammad Ali's goals and the methods he used to reach them provided poor soil for science to take root.

Education, or rather specialized training, was limited to the direct needs of the state as determined by the ruler. There was no emphasis on pure science or creativity, of opening minds to thoughtful reflection and exploration. Nurturing of creative and inquisitive minds was the last thing Muhammad Ali wanted from his schools. Independent thought had no place. The schools were set up to produce servants of the state, of the ruler. Creativity, independent action and any individual thought openly expressed were acts of insubordination tantamount to military disobedience. The military character of the schools was meant to instill discipline, order and obedience. Engineering and medical students were treated like cadets in an officer training school; the grounds and buildings of the schools were more like military camps than campuses. Students had little independence or free time. Even when they graduated and became professionals serving in the state industries or, in the administration or hospitals, they were treated like low-level officers in a chain of military command.

Students for the primary and secondary schools were rounded up in the same way as military conscripts. In fact, the system of primary school selection was an extension of the military conscription system. Provincial mudirs, or governors, were issued student quotas from Cairo and would go into the villages looking for fit-looking boys between 7 and 12. Because villagers feared for their children in the hands of a hated and ruthless government, and because the boys were needed for farm work in helping the family survive, few families volunteered their children to go to the state schools. Consequently, many were taken coercively from the madrasa religious schools. Once seized, the boys had no choice but to go. In the eyes of the peasants, a boy being drafted to go to school was as terrible as being forced to serve a lifetime in the army.[24]

They had reason to see it that way. Students wore military uniforms and many graduates did indeed end up serving in the military, since failing in school was punished by having to serve in the army for not less than 20 years. To save their boys, mothers were reported to have mutilated them. To keep them out of the army they cut off their trigger finger, blinded them in the right eye so they could not aim a rifle, or pulled out their teeth so they could not open a powder cartridge. What precisely it was that mothers did to mutilate their children to keep them from being conscripted into primary school was not reported, but those found guilty of it were to be drowned in the Nile.[25]

Peasant horror of the state schools, fueled by dark stories that had permeated the countryside, had a basis in hard reality. The school buildings were walled in and guarded by troops to keep the students from running away. Students slept, ate and studied behind the walls of the schools, were never allowed out, day or night, and for long periods they were prohibited from visiting their villages. Incoming students were grouped into battalion-like formations, each commanded by an upper-level student acting as its drill sergeant. The students remained in their groups throughout their course of primary and secondary education, each group competing as a team against the other. It was no different in the specialized institutes. Keeping to the military character of the system, teachers received an officer's rations according to rank. A primary school teacher received the pay of a lieutenant and the school director received a captain's.[26] European teachers were ranked and paid accordingly, but on a higher pay scale.

The Europeans are reported to have taken to their disciplinary duties with militant zeal. Clot Bey is reported to have ruled over the students more fiercely than a Prussian general over his soldiers. Sentencing medical students to 30 days in the military-like student stockade for the least infraction of discipline was Dr. Clot's usual prescription.[27] When questioned on this he simply claimed to be following the same disciplinary methods inflicted on him as a student and doctor in Marseilles. On one occasion, finding the hospital's preventive care remiss during a cholera epidemic, Doctor Clot had the Ministry of Education (*Diwan al-Madaris*) order the hospital's whole body of students, teachers and staff (Europeans excluded of course), incarcerated for a month in the stockade.[28]

Dr. Clot's standard was ardently emulated by his fellow countrymen. In fact, French directors of the other specialist institutes seem to have surpassed him in inflicting bodily punishment and incarceration on their Egyptian students. No less enthusiastic disciplinarians were the Turkish and Circassian officers that Muhammad Ali placed in charge of academic departments. One need not imagine the cold fear of impending brutality and pain that terrorized the students. The historical record is clear enough to fill the picture. Failed assignments or absences from class were considered as serious as being AWOL, punishment for which was the bastinado: lashing the bottoms of the feet with thin wooden whips until the feet were so bloody and swollen the miscreant would be unable to walk for weeks. Permanent damage often resulted. Any medical student who failed his exams was expeditiously delivered to the army as a lowly private or made a miserably paid lifetime orderly in the hospital.[29] The brighter side of the abu Za'bal school was that a successful medical student gained a life of possibilities beyond any he could have experienced or imagined in the village – though few young men could have perceived that bright light at the end of the dark tunnel during their years of misery in medical school, the end of which, they were sure, would be their graduation into the dreaded military. All too often their worst fears were realized. Being a medical student at abu Za'bal was as good as being in the army already.

Language and translation schools

One of the most serious of the many problems dogging Muhammad Ali's system of education was the chronic lack of Arabic and Turkish translations of European books for students to study as course texts. French or Italian texts were of no use since students entered their medical and engineering studies without knowing the languages, and by the time they did learn them they were ready for graduation. In addition to the students not knowing the languages they were being lectured in, those few texts in science, technology and medicine that had been translated to Arabic were not available in sufficient copies to be of much use to students in the engineering and medical courses, even as late as the 1840s. The Language and Translation Schools were meant to alleviate these problems. However, the School of Languages (*Madrasat al-Alsina*) was not founded until the mid-1830s, by which time, given the exigencies of funding, the problems were too great to be resolved.

The Language School was housed in a former Mamluk palace, Bayt al-Daftarder, in Azbakiyya. By training students in a foreign language so they could take over from the Europeans in translating both books and lectures, the Language School served the School of Translation. Some of the students were sent abroad to perfect their language skills before becoming translators and language teachers. At first the school drew its students from the religious *makatib* schools, but after the state primary schools had been in operation several years, the students were then drawn from them.[30] The Language School did not graduate its first class until 1839, a quarter century after Muhammad Ali sent the first batch of students abroad for specialized education. The long interval was a primary source of problems plaguing specialized state education.

The school taught its students either French or Italian in preparation for them to serve as translators and language instructors. Graduates of the school taught the students in the engineering and medical schools the language they needed to know in order to understand their lectures and read their texts, if students were fortunate enough to have any reading material at all. Some were not. Priority of translation was given to books and manuals on science, technology and medicine that were urgently needed by students as study and reference texts. It was considered to be quicker to translate European texts into Arabic and Turkish than it was for students to learn to read and understand French or Italian. But since the translations and printing them took so much time, the students were left for the most part of their studies without the reading material they needed for their courses. And since the large majority of the students could not understand what their French and Italian teachers were saying, they were forever at a loss in the lecture hall. Having no reading material for the subject and hearing lectures delivered in an unknown language would challenge even the brightest student. A stopgap measure was to have the lectures translated into Arabic for the medical students and into Turkish for those in engineering. But again, the translation and its printing took time. By the time copies of the translated lectures became available – and for some reason there were never enough copies – the French and Italian teachers had progressed several lectures down the line of the syllabus. Because the punishment and humiliation for failure was so severe, the stress on the beleaguered students sometimes resulted in nervous breakdowns, especially among native Egyptians, young boys who had only a few years before been in the lap of familiar village life along the Nile.

The challenge the students faced is not easily imagined. Classroom transfer of knowledge from teacher to student was close to a total loss. The professors proceeded steadily through the material, not repeating, clarifying, inviting discussion or asking questions, which in any case the students would not have understood, so accustomed were they to uncritical memorization. Discussion was out of the question. The diagrams offered another problem. Often the biological, mathematical, chemical and engineering diagrams that were chalked on the board were erased all too quickly for the students to get down, even when they were supplied with paper and pen, which mostly they were not. The students had no idea what the diagrams were about.[31] It is remarkable that out of this mind-boggling muddle any progress was made. Some was, but not until the late 1830s, after the Language and Translation School had been in operation a few years.

Anyone knowing French or Italian in addition to his native Arabic or Turkish, usually someone who had graduated from the Language School, would be used to translate the lectures of the French or Italian lecturer. The translators, in their local Eastern garb, would stand next to the Western-dressed European lecturer at the front of the class (the two styles of dress emblematic of the cultural division that was just beginning to form) and, sentence by sentence, translate into Turkish or Arabic what the lecturer was saying in Italian or French. If an engineering student happened not to know Turkish, as happened with those few native Egyptians who were allowed to study engineering, then both lecture and translation were lost on him.

The difficulties were endless. The translators of the lectures often knew nothing of the subject they were translating. This made for many problems. Since the European lecturers did not know Arabic or Turkish, they were of no help to the translators. It was later decided to have the lecturer and translator work together in producing a clear translation by having the translated lecture translated back to the original language by another translator. The lecturer would then read the retranslation to make sure that the Arabic or Turkish translation had captured the ideas that the original lecture in French or Italian had expressed. Once this procedure had been established, the Arabic translations were then given to Azhar Shaykhs, who were employed as proofreaders and correctors of grammar, syntax and phraseology. Turkish translations received no such favored treatment. Once a more or less reliable translation had been produced, the students would then cluster around and copy it out in their notebooks, if they had one, and commit it to memory. This was considered a good substitute for the more expensive route of printing copies of the translations for all the students.

Muhammad Ali did not wish to waste resources or spoil students. Making reliable translations of lectures was time consuming, but over time it produced a lexicon of Arabic equivalents for modern technical, scientific and medical terms, resolving one of the stiffest problems students had in learning the new material.

In extremely rare instances, the horrendous adversities of communication plaguing the teaching process disappeared, as when a student happened to know French, Italian and Arabic. Such was the case of the Syrian Christian translator and medical student Rafa'il. He was sent abroad to study physiology and years later became an assistant to the French professor of psychology. Doctor Clot Bey thought Rafa'il to be as intelligent, skillful and efficient as even a European doctor. What higher praise could a European give a Syrian?[32] Clot admitted there were even others like Rafa'il!

Recalling the vital part played by their coreligionists in the Caliph Ma'mun's House of Wisdom exactly a thousand years earlier, a community of skilled Syrian Christians was employed in Muhammad Ali's School of Translation, one of the most prodigious of them being Yusuf Far'un, a translator of medical texts.[33] Another noteworthy Syrian doing translation work in Egypt was Yuhanna Anhuri, nephew of the previously mentioned Butrus Anhuri who was in Damietta before Bonaparte's arrival. Yuhanna knew Italian, French, Arabic and perhaps Turkish. He had also studied some mathematics and physics, no doubt from the French books of science acquired by his uncle. Praised by Clot Bey as a skilled translator, Yuhanna Anhuri was one of the pioneers in finding or creating Arabic equivalents for words in modern mathematics and physics.[34] Evoking the memory of Hunayn ibn Ishaq and his contribution to creating an Arabic scientific and medical taxonomy during the translation of Greek texts a millennium before, Yuhanna's fellow translators expressed their respect for his skills by using the diminutive form of his name, Hunayn, Little John, when referring to him.

Whereas Syrian Christians performed the task of translating from beginning to end of the translation period, trained Egyptians were eventually able to take over from the Syrians in Muhammad Ali's School of Translation. Several of them are worth mentioning. Ali Izzat was a graduate of mathematics in the engineering school who translated a two-volume physics textbook from French. It unfortunately was not published until 1854, after the death of Muhammad Ali and the closure of the engineering schools by his successor. Salih Majdi (d. 1888) was the most renowned of the Egyptian translators. His emergence as a scholar and modernizer is revealing. A country boy, he graduated from one of Muhammad Ali's 55 rural primary schools and went on to secondary school in Helwan, a village some 20 miles south of Cairo, where he is reported to have been introduced to science. The school in Helwan had been built a decade after the two secondary schools in Cairo and Alexandria and may have benefited from their mistakes. From Helwan, Salih went to the School of Languages where he learned French and studied the methodology of translation under the leading light of modern reform in Egypt, Shaykh Rifa'ah Rafi'i al-Tahtawi. Possessing some rudimentary knowledge of science from his secondary schooling, he was sent to the Bulaq Engineering School to translate texts on military science. Salih Majdi blossomed as a writer and reformer during the generation that came to maturity following the death of Muhammad Ali, as did many other Egyptians educated in the schools the ruler had founded. In the 1860s he was translating books on astronomy, mathematics and physics to replace the French texts still being used in the engineering schools, which had by then reopened after a period of closure.[35]

Europeans employed in the professional schools were also enlisted in the enterprise of translation, but they were highly paid and Muhammad Ali was impatient to be rid of them since he could pay Egyptians next to nothing for the same work. This was another flaw in the multi-flawed system of translation. The pittance he paid Egyptians in his service bred inefficiency, carelessness, demoralization and deception – and not only among the Egyptian translators but Europeans as well. Corruption became endemic. The attitude was that since the ruler was cheating everybody in the country why should they not cheat him, and so some of the highly paid Europeans stretched translations out to years that should have taken only months.[36]

In 1841, a Bureau of Translations was founded. The Bureau was an extension of the School of Translation and was organized and staffed predominantly by graduates of the Language School. Here at least was a semblance of planning: a school in existence long enough to supply trained graduates to a newly established government bureau designed to carry on the work of modernization through translation as a step toward assimilation of applied science, medicine and industrial technology. Drawing its trained staff from the School of Translation, the Bureau, at its peak in the late 1840s, employed about 50 translators working in four different departments: three departments for subjects in Arabic and one for subjects in Turkish. The difference between the Bureau and the School was that the latter, in addition to making translations, took care to train translators. As a school, training was its primary function. The Bureau translated books required by the Engineering and Medical Schools. Following the military ranking employed by the School of Translation, the Bureau was organized and staffed as a military unit. An officer with the rank of major headed the department specializing in translations of mathematical texts; a captain was in charge of the department translating physics and medical texts; and a lieutenant was in charge of the department translating books on history, fiction, law and geography. The fourth department oversaw translations into Turkish. The department heads were all officers who had been students in either the School of Medicine or Languages and so therefore they had some experience in what they were responsible for, which was not always the case in Muhammad Ali's Egypt.

Military regimen reigned as rigorously in the Bureau of Translation as it did in the Schools of Engineering and Medicine. Translation School graduates were automatically employed in the Bureau of Translation and not paid until they had each translated a book in his field of study. The translation had to meet the department head's satisfaction, and until it did, the graduates were kept locked up in an isolated room at the Citadel. That the room overlooked the military stockade was intended to speed up the translations.

Students returning from a tour of advanced training in Europe were treated to the same fate as Translation School graduates entering the Bureau. It must have been with some dread that students returned from their tours of foreign study. Met by military officers at the port of Alexandria as soon as disembarking after several years in Paris or another European city, they were marched to the Citadel where they were imprisoned and forced to translate one of their European textbooks. They lived in a small cell, were given their meals in it and at the end of each day, their pages of translations were checked to make sure they had translated no less than the minimum number required for them to finish the book within a set number of weeks or months. Those who fell short had their rations shortened accordingly. Only after having produced a satisfactory translation were the students released. No one escaped translation duty. All students became indentured servants in the Bureau, whatever their fields: engineering, medicine, language, chemistry or industrial mechanics. Even commissioned officers sent abroad for specialized study in military science had to serve their sentence. Not even Azhar Shaykhs escaped. Those who accompanied the educational missions as spiritual guides and chaperones of the students in Europe were equally obliged to do their translation duty. It was a coercive, fear-inducing system. Nonetheless, with all its consequent inefficiency and corruption, it did manage, over a period of 30 or 40 years, to put a part of Egypt on the road to modernity, a road from which, for better or worse, there was no turning back. However uneven, pocked, unsurfaced and detoured the road was and continues to be, there could be no turning back. Progress down it was slow and sometimes halted, but always resumed. Political survival depended on it.

It was no coincidence that by the time the new institutions had achieved a semblance of logical organization, and Egypt was moving just a little bit faster on the road to industrial modernization, the British pulled the rug from under one of Muhammad Ali's major manufacturing industries: textiles, which were threatening to undersell Britain's in the local markets of the Middle East. The next step was to finish him off militarily, which Britain did several years later. As in the case of the French government's refusal in the 1780s to give up its capitulatory privileges as a part of Ottoman economic reform, Britain's crippling of Muhammad Ali showed once again that no European power would tolerate Muslim progress to the point that it intersected European interests. How could reform anywhere in the Muslim world not do so if its purpose was to defend against a European stranglehold? Reforming Muslim states were locked in a 19th century imperialist Catch 22. However minimally successful reform might be and however much any European power might clamor for Muslim reform, the moment reform gave any sign of putting the power's strategic or economic interests in question, that power was quick to take action to cut progress short. Consequently, industrial progress in Egypt was slow and then ultimately aborted.

Muhammad Ali's economic and imperial projects came tumbling down around him just as the translation of technical, medical and scientific texts was gearing up. Carried by the momentum built up since the early 1830s, the productivity of translation reached its apogee during the last years of the 1840s.

Although founded late in the course of reform, the Schools of Languages and Translation served the state well when the wheels began at last turning in the right directions. The Language School continued functioning right to the end of the reformer's reign. In 1844, it was expanded with the addition of a newly founded School of Administration, whose students learned French as a vehicle to their study of administration in preparation of their becoming civil servants in the police department, and in the bureaus of government and public utilities that were being introduced from Europe during the middle decades of the century. In 1846, a School of Administration was founded and put under the auspices of the School of Languages, which was then renamed the School of Languages and Administration, a seemingly awkward coupling of fields. Directed by a Frenchman, the new school followed the precedent of the abu Za'bal Medical School, whose translators composed an Arabic medical dictionary. The School of Languages and Administration in its turn composed a technical dictionary of Arabic, French and English terms for professionals working in the crafts and industries.[37]

Translations continued after Muhammad Ali's demise. Egyptian translators, most notably Shaykh Tahtawi and Salih Majdi, were joined by a European contingent of scientists, physicians and orientalists, predominately French but joined by a good number of Italians and several Germans, who had come to Egypt primarily as teachers but ended up translating texts in their professional specialties. European Arabists working with European scientists, and assisted by the leading Egyptian translators in the School of Languages, made for a collaboration that in the long run was more fruitful than the translation efforts of Muhammad Ali's time.

Women also made contributions working in the School of Translation. One of them, a remarkable Ethiopian slave named Jalila Tamirhan, learned Arabic, studied midwifery in a school attached to the military hospital, then learned French and made an Arabic translation of a book in French on that subject. Jalila's *Muhkim al-dalala fi a'mal al-qibala* (*Guiding Principle in the Operations of Midwifery*) was published in the Egyptian medical journal, *Majalla al-Ya'sub*. The most salient success of the reform period under Muhammad Ali was the nursing program from which Jalila had graduated. It began as a course in midwifery under the auspices of the abu Za'bal Hospital Medical School. This was one of Clot Bey's prized projects. Knowing that a corps of trained midwives would reduce the high Egyptian mortality birth rate that took the lives of so many women and weighed so heavily on the people, he convinced Muhammad Ali to fund a program for such a school and in 1832 the *Madrasat al-Hakima* opened its doors.[38] The name could be taken to mean School of the Female Doctor. *Hakima* literally means wise woman, but in the vernacular, *hakim*, the masculine form of the word, is used for doctor, an association that goes back to antiquity and the relationship between the philosopher being the guide to a healthy soul, and the physician the guide to a healthy body.

The problem faced by the *Madrasat al-Hakima* was finding girls or young women to study in the school. Egyptian families were averse to having their daughters leave the home for a hospital filled with men. The idea was too alien, the imagined threat of female dishonor too strong. Their insecurities were cast in a pretentious disdain that the work was beneath them. So then where to find the students? Clot Bey procured the school's first class by purchasing ten young, strong and intelligent looking girls at the slave market. Five were blacks from the Sudan and Nubia and five were Abyssinians. They were taken to the school next to the hospital and carefully warded by eunuchs. His ten young slave girls were precious to Clot Bey, for upon these prospective students depended the success of his project. As he did in the abu Za'bal Medical School, for which the Marseilles medical school was the model, he set the same courses for his student nurses as French nurses took in the Paris nursing school. For two years they learned to speak, read and write Arabic. After that, they entered the 4-year study program Clot Bey had adopted from the maternity school in Paris. In addition to the curriculum followed by nursing students in Paris, he had his nurses take courses in anatomy, physiology, general medicine, obstetrics, pharmacy, women's and children's illnesses and also had them assist in surgery.

Except for obstetrics, the courses were supervised by the professors in the School of Medicine. Obstetrics was supervised by an Arabic-speaking French woman who had trained in the maternity school in Paris and come to Cairo. During the demanding six-year course of study, four students died, but the success of the six slave girls who graduated so excited the wives of the hospital functionaries that the *Madrasat al-Hakima* had 22 voluntary students in 1837 and 60 in 1846. Clot Bey's long-term goal was to have a first year enrollment of 100, with an annual graduating class of 25.

The graduate nurses assisted in vaccinations, obstetrics, consultation and examining ill women. As their numbers grew, they were assigned to work in the provinces where they did their rounds on donkeys: "doctoresses a dos d'ane," Clot Bey affectionately and proudly called them. But no exceptions were made for them in Muhammad Ali's Egypt. When the nurses graduated from the *Madrasat al-Hakima* they advanced in military service and received the pay of their rank as second lieutenants. This was almost two decades before Florence Nightingale and the Crimean War that saw the first women serving as military nurses in the West.

Among all of Muhammad Ali's many projects, the *Madrasat al-Hakima* stands out as a most accomplished success, all credit going to Antoine Barthelemy Clot who started it and to the slave girls he bought into medicine who made it work.[39]

The school's symbol of that success, and the most memorable of its accomplishments by one of its graduates, is the contribution to the practice of midwifery made by Jalila Tamirhan and her translation.

The teaching of language and the execution of translation brought the culture and knowledge of Europe to Egypt, and in doing so introduced ideas far beyond anything the ruler and his Turko–Circassian coterie had envisaged, intended, or wanted. The effects of this intellectual influx were not to become apparent until decades after Muhammad Ali's demise.

Notes

1 Clot Bey, *Compte rendue des travaux de l'ecole medecine*, Victor Masson Libraire, Paris, 1849.

2 Richard Verdery, "The Publication of the Bulaq Press under Muhammad Ali of Egypt, *Journal of the American Oriental Society*, vol. 91 (pt I), 1971, p. 132.

3 Puckler Muscau, *Travels and Adventures in Algiers and Other Parts of Africa*, 4 vols., Wilson Publishers, 1831–1832, vol. I, p. 203, pp. 271–273.

4 Mikha'il Mashaqa, *al- Jawab 'ala Iqtirab al- Ahbab*, translated by W. Thaxton under the curious title *Murder, Mayhem, Pillage and Plunder*, State University of New York, New York, 1988, pp. 95–97.

5 J. Heyworth-Dunne, *An Introduction to the History of Modern Education in Egypt*, Luzac and Co., London, 1939, pp. 78–83.

6 Jabarti, *Athar*, III, p. 397.

7 Wheeler Thaxton's translation of Mashaqa's *Jawab ala Iqtirab*, p. 97.

8 T.X. Bianchi, "Catalogue generale des livres arabes, persanes imprimes a Boulac en Egypte depuis L'introduction de L'impimerie dans et pays," *Journal of Asiatique*, 2, July–August 1843, pp. 24–61.

9 T.X. Bianchi, "Catalogue generale des livres arabes, persanes et turcs," Bulaq Press, Cairo, pp. 28–29.

10 J. Heyworth-Dunne, "Printing and Translations under Muhammad Ali of Egypt: The Foundation of Modern Arabic," *Journal of the Royal Asiatic Society of Great Britain and Ireland*, 3, July 1940, pp. 325–349.

11 Heyworth-Dunne, "Printing and Translations," p. 338, 342. Some *muharrirs* and *musahhihs* closely cooperated with translators in bringing a translation to completion. Shaykh Imran al-Hirrawi worked with the Syrian Christian translator Anhuri on medical translations. Shaykh Mustafa Hasan Kassab, a *muharrir*, worked with a Syrian Christian named Fira'un. Shaykh Muhammd Umar al-Tunisi was the best known of all the editors and correctors for his expertise in finding Arabic equivalents to scientific terms, or finding a better one to replace an older one. The work of translation was on occasion contributed to by Europeans. A bright young orientalist named Koenig helped create Turkish terms for scientific and technical words that did not exist in the language. When he came upon a technical term in French for which there was no Turkish equivalent, he referred to the main workhorse at the Bulaq printing press, Uthman Bey Nur al-Din, "a first-rate all-around man." Uthman Bey would search Arabic for an equivalent, or create one, which was then taken over as the Turkish equivalent. Arabic was the ultimate source. Any Arabic term could be used in Turkish or Persian and be accepted as part of the language without any notice being taken.

12 P.N. Hamont, *L'Egypte sous Mehemet Ali*, 2 vols., Leyautey and Lecointe, Paris, 1843, vol. II, p. 89; Ahmad Izzat, *Ta'rikh ta'lim fi 'asr Muhammad Ali*, Maktaba al-Nahda al-Misriyya, Cairo, 1938, p. 9.

13 Clot's education came at a time in the French revolution when medicine was undergoing profound changes, the new sweeping away the old like an ancienne regime that had lived its life and was standing in the way of progress. In spite of the prestige and the technical innovations associated with science, medical tradition stood its ground, holding onto old concepts such as contagion and denying newfangled contraptions such as the microscope. Beginning as a lowly assistant to a barber-surgeon in Grenoble – even as late as the 1790's surgery was considered to be on the level of shaving and cupping – Clot entered proper medical studies in Marseilles when he was 23. Three years of study in Marseilles was followed by another year in the famous medical center of Montpellier. He received his doctorate in surgery, which had by then been elevated to a respectable field of the medical profession. He was employed as a surgeon in a hospital in Marseilles and was having some problems with his medical colleagues that seemed to be related to his ambitious character. The chance to go to Egypt on a 5-year contract as chief surgeon in Muhammad Ali's army was his way to a new life of untold possibilities. He was 31 in January of 1825 when he sailed from Marseilles to Alexandria. Within a few years he was chief of chiefs of Egypt's military physicians, hospitals and medical schools. Daniel Panzac, "Medicine revolutionaire et revolution de la medicine dans Egypte de Muhammad Ali: le Dr. Clot Bey," *Revue du monde musulman et de la Mediterranea*, 52–53, 1989, pp. 95–110.

14 La Verne Kuhnke, *Lives at Risk: Public Health in 19th century Egypt*, University of California, Berkeley, CA, 1966, p. 39.

15 La Verne Kuhnke, *Lives at Risk*, pp. 34–35.

16 Puckler Muskau, *Travels and Adventures in Algeria and Other Parts of Africa*, vol. I, London, 1847, p. 227.

17 Izzat, *Ta'rikh ta'lim*, p. 258.

18 Hamont, *L'Egypte sous Mehemet Ali*, II, p. 92.

19 Amin Sami, *Taqwim al-Nil wa asma man tawallaw amr Misr wa muddat hukumihim*, *matba'ah al-amiriya*, vol. 4, Cairo,1836, p. 583.

20 Izzat, *Ta'rikh ta'lim*, p. 362.

21 Izzat, *Ta'rikh ta'lim*, pp. 228–231.

22 Hamont, II, pp. 192, 319–322; Heyworth-Dunne, "Printing and Translations," pp. 213–214.

23 Izzat, *Ta'rikh ta'lim*, pp. 366–367.

24 Madden, *Egypt and Muhammad Ali*, Hamilton, Adams, London, 1841, p. 76; Heyworth-Dunne, "Printing and Translations," p. 215.

25 Izzat, p. 37.

26 Heyworth-Dunne, "Printing and," pp. 325–349.

27 Izzat, *Ya'rikh ta'lim*, p. 37.

28 Izzat, *Ta'rikh ta'lim*, p. 286.

29 La Verne Kuhnke, *Lives at Risk*, p. 38.

30 Izzat,*Ta'rikh ta'lim*, p. 51.

31 Izzat, *Ta'rikh ta'lim*, p. 369.

32 Clot, *Compte Rendue*, p. 44.

33 Jacques Tajir, *Harakat al-Tarajima fi Misr khilal al-qarn al-tasi' 'asara*, Cairo, n.d. p. 24.

34 Clot, *Compte rendue*, p. 45.

35 Tajir, *Harakat al-Tarajima fi Misr*, pp. 99–100.

36 Izzat, *Ta'rikh ta'lim*, p. 339.

37 Tajir, *Harakat al-Tarajima fi Misr*, pp. 85–86.

38 For the *Madrasat al-Hakima* (and a finely researched account of Egyptian medical modernization in the 19th century) see: LaVerne Kuhnke, *Lives at Risk: Public Health in Nineteenth century Egypt*, University of California Press, Berkeley, CA, 1990: pages 122–133 for women health officers.

39 D. Panzac, "Medicine revolutionaire et Revolution dans l'Egypte de Muhammad Ali: Le Dr. Clot-Bey," *Revue des Mondes Musulmans et de la Mediterranee*, 52/53, 1989, pp. 105–106.

17 Foreign missions

Over a 35 year period, a dozen and more student missions of varying size were sent to Europe, the large majority to France. The students chosen for these missions were intended to be an elite selection comprising the brightest and most promising graduates of the specialized schools, though in many instances an inside connection had more to do with selection than brilliance. Many pioneers of the *Nahda*, as the intellectual and literary revival is called, were products of these missions: Shaykh Tahtawi, Ali Mubarak, Mahmud Hamdi al-Falaki and Amin Sami, to name but a few of the most well-known. The missions gave students the opportunity to master, at the European cultural sources, the science, medicine and industrial technology that they had been introduced in the Egyptian specialized schools. Some students went outside of the syllabus of their specialty to imbibe other things that Europe had to offer. A few students learned things that Muhammad Ali would have preferred they did not: philosophy, literature, economics and political systems, ideas and institutions that had been transforming Western society since the 12th century, back when the two sibling civilizations of Islam and Latin Christendom had been joined from hip to head.

Back then, Latins had trekked to Toledo and Palermo to mine the Muslim intellectual heritage, quite as in the 8th and 9th centuries the progenitors of what became Muslim science, medicine and philosophy had searched the old Hellenistic lands for Greek and Syriac manuscripts. The learning process was now reversed. Muslims were coming to the revolutionized West for knowledge. But unlike their Latin predecessors who had ventured forth in search of learning on their own, Muslim students came in tightly controlled groups, missions that were organized, delivered and supervised by absolutist modernizers who decided what would be studied, who would study it, where and for how long, and where and what they would work at when they returned. The missions were based on a narrowly defined set of objectives designed to provide the ruler with the technical specialists he wanted. They were not meant to be journeys in search of knowledge.

The first Egyptian mission, in 1813, went to Italy. The first to France was five years later. The Arabist Jomard was influential in convincing Muhammad Ali to direct his students to France, the fount, as he claimed, of science, medicine and engineering. Of the 12 and more missions sent to Europe between 1825 and 1833, three quarters of them were sent to France, the rest to Italy, Austria and Britain.

One went all the way to Mexico for some obscure reason – some Egyptian prince wanted to see the pyramids. The early missions, those between 1813 and 1825, tended to be small, usually around six students, never more than 20. A total of about 28 or so students were sent to France during that period, and little more than double that to various cities in Italy. The numbers were modest, no more than 100 students in over a dozen years. An average of eight students a year does not suggest a deep investment in preparing a cadre of specialists to manage the modernization of the military, industry, agriculture, government, public administration and technical training schools that the ruler was founding.

The students were mainly of Albanian, Turkish, Circassian, Armenian or Greek origins, most usually the sons of favored officers that Muhammad Ali had surrounded himself with since his rise to power in Egypt. The Albanians, Turks and Circassians studied military science. Greeks and Armenians were usually chosen to study the printing craft. Native Egyptians who were selected for overseas training generally studied medicine.[1]

With the establishment of the Ministry of Schools (*Diwan al-Madaris*) in early 1826, the missions became larger and more organized. Compared to the previous missions, more is known about them. A considerable body of information was recorded for the Paris mission in 1826, the largest and most famous of all the missions. Numbering 44 students, the mission gives an idea of the areas of government and technologies Muhammad Ali was aiming to develop. Four students were sent to study government administration; four to study military administration; three students each of naval administration and military engineering; two for artillery technology; two for forging, casting and arms manufacturing; one student each for mechanical and hydraulic engineering; four for chemistry; two for metallurgy; one for natural history; two for printing and engraving; two for medicine and surgery; two for agronomy; two for political science; and one to learn the art of translation. Of the 33 students listed to pursue specialized technologies, 26 were directly or indirectly related to the military. The study of chemistry, metallurgy and surgery was for military purposes. The seven students who had not been assigned subjects at the time of departure were most likely selected because of inside connections.[2]

The mission included several native Egyptians, two of them al-Azhar shaykhs. One of the shaykhs was to study medicine; the second was added as an afterthought: at the last moment it was decided the students needed an imam as a spiritual guide and chaperone. The position fell upon a young, recent graduate of al-Azhar, Shaykh Rifa'a al-Tahtawi, who would become famous as a writer, translator, proponent of reform and founding father of the Egyptian Awakening, guaranteeing that the 1826 mission would be considered the mother of them all. The students' ages ranged from 15 to 37, the older ones being Turko–Circassian and Albanian favorites of the ruler sent to study government and military science, the youngest being sons of the favorites. Eighteen students were Muslims born in Egypt; 12 were Muslims born outside of the country. Eighteen had Ottoman origins, six of whom were born in Egypt.[3] Four Ottomanized Armenians who had retained their Christianity were included in the mission. Two of them, the brothers

Estefan and Artin Sarkis, studied political science and government administration, respectively. Both would achieve outstanding success, earning the prestigious title of bey, the former as a member of the Advisory Council for Schools, the latter first as Director of Education and then as Muhammad Ali's Secretary of Foreign Affairs.

Other successful members of the mission included a future Director of Schools, a Director of the Ministry of Religious Schools, and a Minister of the Navy. Two native-born Egyptian students, Muhammad Bayyumi and Bahjat Efendi, would become respected mathematics professors in the Bulaq Engineering School where they taught descriptive geometry, mechanics and surveying.[4] Another Egyptian, Mazhar, whose grandson Ismail Mazhar would translate Darwin's *Origin of Species* in the early 20th century, specialized in engineering. He would go on to become Muhammad Ali's chief engineer for the construction of a new port in Alexandria, then of the barrage at Khanatir al-Khayriyya, and crown his career as Minister of Public Works.

Because of this mission's contribution to the ruler's policy of replacing Europeans in the military and specialized schools with his own men and native Egyptians, the 1826 group has been given special attention by historians of Egyptian reform. Portrait sketches of a dozen or so students that were made before the mission departed for France have in fact been preserved. They are shown as slim young men with fine features, serious expressions and elegant turbans, their wide eyes staring eagerly out in expectation of their grand adventure in another world. Uniformly handsome, the students of this first big mission to France may possibly have been chosen in part for their physical attributes. No sketch of young Shaykh Tahtawi was made, most likely because he was chosen just as the mission was about to depart.[5] Native Egyptians chosen for the mission were graduates of either al-Azhar or the School of Medicine. Later missions would include graduates of the medical school's annex, the *Madrasat al-Maristan* or Hospital School. When the *tajhiziyya* schools (secondary) were founded in the 1830s, some of their native Egyptian graduates who went on to graduate in engineering at one of the *khususiyya* institutes were selected for foreign missions to study engineering, physics and chemistry, rather than the usual medicine, language and translation; though already in the 1826 mission at least three native Egyptians had been appointed to study chemistry and engineering, proving there were exceptions to everything in Muhammad Ali's Egypt. As previously mentioned, native Egyptians were rarely appointed to study military science or anything directly related to military and government. These fields were reserved for the Turko–Circassian and Armenian Ottoman favorites of the ruler and their Egyptianized sons. Muhammad Ali was too cautious to let the sons of the indigenous people master the arts of warfare and government. He wanted a modern state but he wanted it on his own terms as an absolute ruler, with no trouble coming from below. Another significant point is that no Egyptian Christians from the Coptic community were selected for this mission or any others. In his pursuit of absolute power, Muhammad Ali had determined to put an end to the traditional monopoly the Copts had in being appointed government accountants and chief customs officers and set about confiscating the wealth they had accumulated during their control of these positions over the years.[6]

By the time the 1826 mission reached Paris, French Arabists commissioned by the Egyptian Ministry of Schools, the *Diwan al-Madaris*, had established the "Egyptian School" there. The school had begun as a residence for arriving students during the early years of the missions to France and then developed year by year into something larger, with the addition of classrooms, a dining hall, dormitory, and a teaching staff conducting classes in what could be called an acculturation program designed to introduce students to French language and history. Before they were transferred to French institutions to commence their formal studies, students were prepped in subjects they should have studied in Egypt but may not have, such as geometry, algebra, chemistry, physics and engineering principles. In 1828, another mission was sent to Paris. This numbered less than half of the 1826 mission and included a Sudani and an Ethiopian, in addition to the usual majority of Albanians, Turks, Circassians, Georgians and Armenians. Five members, about a quarter of the whole mission, were to study mathematics, physics and engineering. The others were assigned to naval science, veterinary medicine and the manufacture of surgical instruments. The students assigned to physics and mathematics were intended, once they had finished their studies, to replace European teachers in the specialized schools. A large mission of 58 students was sent the following year to various European countries: 34 to France, 20 to England and four to Austria, with the emphasis of study on arms manufacturing, textiles, industrial tools, medical instruments, watches and, surprisingly, candles. The 20 students sent to England studied industrial mechanics and cannon manufacturing.[7]

This was followed by a medical mission consisting of 12 graduates from the abu Za'bal School of Medicine's first graduating class. The 12, personally selected by Clot Bey, were sent to Paris for further study in preparation of their taking over teaching positions in the medical school. These 12 physicians were expected to become the nucleus of the medical school's teaching staff, whose Arabic lectures would do away with the lectures in French and the necessity of translators. Clot Bey had defined six areas for their years of study in Paris and assigned two students to each: natural history; physics and chemistry; pharmacy and materia medica; anatomy and physiology; pathology and internal medicine; and surgery.[8] As if their lives as student–prisoners in the regimented medical complex at abu Za'bal had not been punishing enough, the young physicians-in-training were treated like personal servants by the French physicians in the Parisian hospitals to which they were assigned, though they may have felt themselves fortunate for having escaped the bullwhip lashings and Dr. Clot's standard 30 days in the stockade.[9] Five of the students returned in 1836 with their French medical degrees and joined the abu Za'bal faculty. The seven others remained in France for another two years to complete their internships and residencies. When they did return, abu Za'bal was able to give its courses in Arabic and do away with the interpreters.

These French-trained Egyptian faculty members appreciably increased the rate of translations. A total of 24 medical texts had been translated from European languages to Arabic between 1820 and 1840; between 1840 and 1849 another 31 were translated. Slow progress, but the pace was picking up. Had it been sustained, and the same methods applied to the engineering schools that were applied to the abu Za'bal Medical School, the overall effect of the reforms may have approximated a systematic process of scientific assimilation and technical modernization. A few decades of continuous application might have produced an indigenous scientific and technical culture. But nothing was sustained once Muhammad Ali left the scene. Even while he was in power, his reforms slackened off once the British had driven his army out of Syria and forced him to disarm. When his expansionist foreign policy met defeat and the British successfully pressured the sultan to have Muhammad Ali reduce his army to 16,000 men, the heart and soul of his reformist purpose evaporated. The heart of his reforms had been the military. With that taken away, he no longer felt the urgency to invest in reform. The energy for it had been sucked out of him by aggressive British defense of its imperial interests.

Yet, carried by the momentum of reform that had been built up over a quarter of a century, even after the British had forced the Egyptian army out of Syria and Anatolia and emasculated the military, student missions continued for a while. A large mission of 70 students was sent to Paris in 1844. Because it included six of the ruler's family, it has accordingly been named by Egyptian historians "The Mission of the Nephews and Grandsons." Several of the six princes studied engineering at the prestigious Ecole Polytechnique. The others studied military science, armaments and industrial technology.

Around this time, Muhammad Ali had his French advisors open an Egyptian military academy in Paris. Under the jurisdiction of the French Ministry of War, the academy was known as the "Egyptian Military School in Paris" and was totally separate from "The Egyptian School in Paris." This also was a mission that was carried by the momentum that had been built for a cause that turned out to be lost, as far as it concerned building the Egyptian military. But as a special academy in Paris for the children of the ruling family and its supporting elite, it served a purpose that was still a century from being lost: the ruling dynasty itself. In the cause of dynastic longevity, the school was a concession to those officers whose support he needed in ruling the country. As the students selected for the Paris military school were for the most part young members of the ruling family and, primarily but not exclusively, sons of the ruler's Turkish, Circassian and Albanian officers, they were given ultra-special consideration and ranked as, in ascending order, effendi, bey or amir, depending on the position the student's father held in Muhammad Ali's military. A most unusual exception was Ali Mubarak. Ali was the first and only native Egyptian to be sent to the Egyptian Military School in Paris to study military science and engineering. Much will be said of him later.

After 1847, by which time Muhammad Ali, defeated and weary, had given authority to his chief general and son, Ibrahim Pasha, the missions became more purely scientific and academic. The last military-oriented mission was in 1848–1849. The Paris Military School closed the next year.

The final mission was in 1851, during the reign of Abbas, Muhammad Ali's son and successor. It was a purely scientific mission, and very small: three students were sent to study astronomy at the Paris Observatory. The head of this little mission was a native Egyptian, Mahmud Hamdi, an 1830s graduate of the Citadel Engineering School. Hamdi was another exception to Muhammad Ali's reported bias against Egyptians studying engineering. Mahmud Hamdi's career offers an insight to the extent to which science had been assimilated at mid-century by at least a few of the graduates of Muhammad Ali's specialized schools. At the time of this 1850 mission, Mahmud Hamdi was assistant director of the Bulaq Observatory and a teacher of mathematics and astronomy at the Bulaq School of Engineering. By the end of his nine-year tour of foreign study, he had become a practicing scientist, Egypt's first, and without doubt the most scientifically knowledgeable person in Egypt, if not in the Arabic-speaking world. This was primarily due to his research related to the calendar and the earth's magnetic field. Mahmud Hamdi became known as al-Falaki Efendi Pasha, Lord Astronomer, as it were. His reputation as a reformer of education and advocate of science ranks with the leading figures of the Egyptian intellectual revival. He regarded himself as accomplished enough to warrant publishing his memoirs, in which he describes his studies, his research and the various scientific academies he studied at and visited in Europe.

His scientific activity ended with his being appointed Minister of Education. The appointment was an honor, with prestige and salary that he understandably must have found hard to resist, but his accepting it deprived Egypt of a science teacher who could have contributed to nurturing an indigenous scientific culture through his training of dozens of young men to follow in his path. This appointing of accomplished men who were scientists, or well on their way to becoming scientists, or were effective science teachers, to non-scientific government positions would occur again and again. These men were capital assets of the education programs necessary to initiate, sustain and advance the momentum of an autonomous scientific culture. Taking them from their fields of expertise was self-defeating. Instead of being used to train graduate students and multiplying the fund of expertise, these bright lights were being taken from the classroom and laboratory for the glitter of high office and all the seductive perks that went with it. By stripping the rudimentary scientific community of the talent it needed in order to develop toward the creative stage, the government was defeating itself and society.

In sum, like all Muhammad Ali's reforms, his education missions were a product of the moment's need. Rather than a comprehensive plan to educate students in increasing numbers over time to nurture and sustain an independent culture of science and technology, the missions were sporadic, of greatly varying size, with each one concentrated on a specific subject. In 1847, for example, a mission of five al-Azhar students was sent to Paris to study French law of contracts and claims. The purpose was no doubt to satisfy a need of the moment for contract law in regulating agreements between the ruler and Europeans. Muhammad Ali's ideas in the largest sense went in the right direction, but they were found wanting in regard to organization, continuity, maintenance, financial investment and overall planning. Conditions did improve over the years, but time ran out for Muhammad Ali when Britain decided he had gone too far and denuded him of his military. Being relieved of the burdensome expense of a large military would have been a boon to the kind of ruler-reformer who saw that a country's strength depended on developing a sound educational system deeply invested in science and technology, but such rulers exist more in hope than history.

Notes

1 For details of the missions: George Douin, *Une Mission Militaire Aupre de Mohamed Aly*, in *Societe Royale de Geographie d'Egypte*, Cairo, 1923, p. 40; Umar Tusun, *al-Bu'ath al-'ilmiyya fi 'ahd Muhammad Ali wa Abbas wa Sa'id*, Alexandria, 1934; Zaki Salih and Mahmud Mursi, *al-Bu'athat al-'ilmiyya fi qarn al-tasi' 'ashar*, Cairo, 1959, pp. 9–12.
2 Alain Silvera, "The First Egyptian Student Mission to France Under Muhammad Ali," in *Modern Egypt*, edited by E. Keddourie and Sylvia Hayem, Frank Cass, London, 1980.
3 Edme-François Jomard, "L'Ecole Egyptienne de Paris," *Nouveau Journal Asiatique II*, August 1828, p. 105; Ahmad Izzat, *Ta'rikh ta'lim fi 'asr Muhammad Ali*, Maktaba al-Nahda al-Misriyya, Cairo, 1938, p. 435.
4 Izzat, *Ta'rikh ta'lim*, p. 364.
5 Salih and Mursi, *al-Bu'athat*, pp. 16–20.
6 Salih and Mursi, *al-Bu'athat*, pp. 16–20.
7 Izzat, *Ta'rikh ta'lim*, p. 446.
8 La Verne Kuhnke, *Lives at Risk: Public Health in 19th century Egypt*, University of California, Berkeley, CA, 1966, p. 41.
9 Salih and Mursi, *al-Bu'athat*, pp. 61–64.

18 Assessment of Muhammad Ali's reforms

Muhammad Ali's educational reforms were seriously underfunded and severely, in some instances brutally, executed. Mismanagement and inefficiency were rampant. During 35 years of reform, only around 216 students or so were sent abroad, nowhere near the number of trained personnel that would have been required to take over the positions of the many Europeans that the ruler had brought in to modernize the military and administration, to create and manage modern industry and agriculture, to build and maintain armaments factories, to organize and staff a modern system of education and develop a self-sustaining pool of scientific and technological expertise to keep the machinery of modernization going.

The graduates of the few specialized schools in Cairo displayed the same deficiencies as the students sent abroad: too few of them and too poorly trained to effectively assist in this immense undertaking of birthing a modern state from Egypt's medieval precedents. Consequently, Egyptians had great difficulty taking over the managerial reins from their European guests. In fact, following the reign of Muhammad Ali, Europeans became increasingly more ubiquitous. Before the French invasion in 1798, fewer than 1,000 Europeans resided in Egypt. By the 1830s the number had risen to 10,000. Within another 40 years, the European population would be ten times that. In trying to strengthen the country from external military aggression, Muhammad Ali in effect set a precedent of inviting a small army of highly paid European experts into the country whose numbers rapidly multiplied. Under less capable rulers, particularly Muhammad Ali's grandson, Ismail, Egypt would eventually become a European holding. Rather than Egyptians taking over from Europeans, as Muhammad Ali's plan called for, it was more a case of Europeans taking them over, lock, stock and barrel. Once Muhammad Ali had invited them in, they kept coming.

t Muhammad Ali put a Frenchman in charge of the naval yards in Alexandria and, under him, a corps of French engineers and technicians. Another French Officer, Colonel Seves, was made head of the Egyptian War Office. A Swede, Joseph Bokti, in collaboration with the French Consul, Jumel, built the textile industry. An Englishman, Thomas Galloway, assisted by English engineers and technicians, built the foundries and installed the steam engines. A Frenchman supervised the public works projects. Frenchmen were directors of all but one of the specialized schools. The paltry number of Egyptians trained at the highest level of Muhammad

Ali's schools and sent abroad for more advanced professional training had no chance of replacing the Europeans.

None of Muhammad Ali's foreign legion of advisors seems to have laid out for him a coherent program of producing physicians, technical craftsmen, engineers and educators from the ground up. Why should they have? Waste, corruption, redundancy and inefficiency produced the advisors' bonuses: bukshish on top of their princely salaries. The condition was generic. The ruler's European directors were at odds among themselves in their jealous rivalries and ran their institutions as their own despotic monopolies patterned on the way Muhammad Ali ran the country. Each institutional monopoly was a jealous rival of the others; with all of them striving to be the chief monopolist's favorite, intrigue flourished. Cooperation was not the common currency of the directors who designed and ran the reformist machinery that came to be assembled piece by piece.

Improvised solutions to problems were concocted as they arose. The lack of infrastructure and overall planning made for a constant barrage of problems, some of which had no solution, or had to wait years for even a partial solution. It was not until late in Muhammad Ali's reign that something as basic as a lithograph press was introduced to print some texts in order to alleviate the dire lack of student reading material. The geographical separation of the schools that were spread across and outside the city, from abu Za'bal to Azbakiyya, to Bulaq, to the Citadel and Qasr al-Ayni, shows the uncoordinated and piecemeal fashion in which the system of education had come into existence, the product of so many afterthoughts hitched tenuously together over the years. Had they been located in one quarter in the city or outside of it as a single complex of buildings, the teaching staff could have been coordinated and used more effectively to the benefit of the students in all the schools. Four separate engineering institutes bear testimony to the scattered and wasteful nature of the system that came into existence in fits and starts.

The system's impairments and disconnects were exacerbated by discontents among faculty and students. The miserly pay of Egyptian teachers and graduate doctors drained enthusiasm. All the Egyptian graduate physicians worked for Muhammad Ali, who paid them as little as he could. He wanted to build up his army, not waste money on educating Egyptians. If some of the locals needed training in order to create the infrastructural requirements for a powerful military, then that was a necessary loss of resources that just had to be tolerated. Often the pay of Egyptian teachers and physicians was in arrears. Never with the Europeans. They could leave if their pay was not delivered on time, but not Egyptians. They could not leave the country or retire from government service to set up a private practice. The only schools available for Egyptians to teach in were state schools. It was the same for Egyptian engineers, physicians and other professionals. No private industry or hospitals existed to offer alternate employment to them. No Egyptian physician had a chance to go into private practice. Private practice in Egypt was monopolized by the better trained European doctors. The technically and medically trained Egyptian was locked in a dead-end, career-shunting cage built by Muhammad Ali.

Entrepreneurship was an alien idea. Europeans practiced it in Egypt, not Egyptians. Egyptian professionals were confined to state service and remained forever subject to the will of the ruler. Once a student graduated into professional government service, his incremental increase in pay was practically insignificant. The ruler's closefistedness when it came to funding education and paying professionals made for conditions verging on the absurd. Something as inexpensive as leeches had to be purchased through a requisition issued by the Ministry of War's executive council. The same with purchasing candles used by the interpreters and correctors who worked through the night on the Arabic translations of classroom lectures and French technical and medical manuals. Funds were short and patience long. The time from requisitioning to receiving of such simple items as chalk and candles was months. Expensive equipment could take years. The Russian consul general in Cairo wondered why Muhammad Ali bothered to set up a medical school, "for which the country cannot offer even the basic requirements."[1]

The huge differential in salaries between European and Egyptian professionals created another rift in the system. On the Egyptian side, the differential resulted in envy, uncooperativeness and intrigue. The unhealthy atmosphere infected both sides. The French knew that the intention of the foreign missions was to train Egyptians to take over their positions and they sorely resented it. That they could possibly be replaced by locals was an insult to their professional expertise, to their being European. When Clot Bey convinced Muhammad Ali to send the 12 abu Za'bal medical graduates to Paris in 1836 to study, train and work in French hospitals and obtain French medical degrees in order to Egyptianize abu Za'bal Hospital and Medical School, the French physicians under him did not like it. Egyptianization threatened their comfortable, well-paying positions. They openly expressed their contempt toward their younger Egyptian colleagues and did what they could to undermine their confidence and self-esteem as professionals.

Hamont, director of the School of Veterinary Medicine, venting his frustration and insecurity, wrote that the Egyptian medical students who trained in France returned having added the faults of Europeans to those of their own original Egyptian faults: they neither shed the bad nor gained the good. Believing them educated and giving them his confidence, Muhammad Ali was contributing to the decline and deterioration of all he was trying to accomplish. The majority of Egyptians educated in Europe, Hamont continued, were far from possessing a solid knowledge of their field of study:

> They learn French, which some speak well enough, but all of them return from Europe with highly elevated pretensions. They say to the Turks: "We will be more valuable than you, because we know the language of the country and we join to it the sciences of Europe, neither of which you know." And in addressing themselves to Europeans they say: "We have acquired in your country the sciences and we add to that the intimate knowledge of the customs and practices of Egypt, of which you know nothing." These "nouveaux venus" are the enemies of Europeans employed in Egypt, and either we must submit to them

or be replaced by them . . . Muhammad Ali knows how to create, but not how to preserve. . . . The ruler bestows his favors upon a recommendation, not upon insightful knowledge.[2]

Hamont's plaintive criticisms are echoed by the Austrian prince Puckler Muscau, who was traveling in Egypt during Muhammad Ali's time. Writing of Mukhtar Pasha, one of the students who had returned from a tour of study abroad and had risen to the rank of pasha in the ruler's service, the Austrian prince seems to have borrowed Hamont's bitter words straight from the Frenchman's mouth. Mukhtar, he wrote, returned from his study abroad having learned only the vices of Europe, which he added generously to his own: "Prompted by stupidity and inflated by arrogance, he constantly was trying to persuade Muhammad Ali to be rid of the foreigners since they, the Egyptians, had learned all they could from them."[3] Culturally biased as these negative sentiments of European observers may certainly have been, they nonetheless point to the difficult relationships that hampered educational reform. The bitter pill of swallowed Egyptian pride for their having to be tutored by Europeans, whose arrogant contempt was freely expressed for all to hear and see, made for a poisonous atmosphere that only added to the grave problems already encumbering education in the new schools. After years of having to suffer the bastinado and Clot Bey's standard 30 days in the stockade, Egyptian students then graduated to suffer the open arrogance and insolence of their European teachers. If, as the European sources say, the graduate Egyptians showed spates of arrogance to their foreign guests, it is clear they had excellent models to emulate.

Low payment was an additional source of resentment that drained Egyptians all the more of whatever enthusiasm and dignity they may have acquired for themselves and their professions during their years of study at home and abroad. The lack of respect liberally shown to Egyptians deprived them of any chance of professional pride. Muhammad Ali made no secret of what he thought of local Egyptians. The professionally trained Egyptians were subject to the same cursory treatment by the whole of the Turko–Circassian ruling elite. They were treated as servants, especially by the princes of the dynasty who used the technical experts for their own private use to decorate palaces and repair what needed repairing. The servitude lasted until the end of a professional's career. Joseph Hekekian, a graduate of the specialized schools who was sent abroad for further training and rose to a high position in government, tells of his being in the middle of an important meeting in 1844 to discuss the school system when suddenly a palace messenger entered the chamber and ordered him to appear at once to advise the tailors on the material to be used in covering the cushions for a divan on the yacht of one of Muhammad Ali's sons.[4] If a Christian Armenian in a high government position could be so unprofessionally treated, what of a native Egyptian doctor or engineer of rural origin?

The French may have had a point in claiming that students returning from tours of European study were not up to replacing them, but this was more because of the flaws in the educational system, right from the primary *ibtida'iyya* up to the specialized *khususiyya* level, than it was the fault of the students who suffered through the system, coming out of it as they did so unprepared for advanced study abroad. Professional training was sorely deficient at every level. Personnel were not trained sufficiently, or in sufficient numbers, to maintain the machines and factories imported from Europe. Expensive steam pumps bought from England broke down for lack of maintenance and were never repaired. The jealousy with which Muhammad Ali controlled his industrial monopolies precluded indigenous entrepreneurial initiative. Preclusion of private ownership dampened any sense of personal responsibility, which to foreign observers appeared as indolence and incompetence.

Traditional attitudes stood fast against any pressure exerted on them to adapt to a social outlook amenable to modernization. In the words of one of the Arab world's leading analysts on science and technology transfer from the West to the Muslim world:

> It appears safe to conclude that there was little science, technology, discipline, or education imparted to the population and a great deal of coercion, force, drill and blind obedience. So Muhammad Ali who knew next to nothing of science and technology and a great deal about power and control almost succeeded in his political ambitions by merely importing engineers and technicians from Europe, harnessing them to build an army strong enough to defeat all other Middle Eastern armies. Most of the industries established between 1818 and 1836 collapsed before 1841 for lack of technology and basic operating techniques.[5]

An exception or two to that harsh but generally correct judgment should be noted, for exceptions can highlight the rule with which they must contend and are certainly more interesting than the non-exceptions. About 1840 or shortly thereafter, the education system that had been cobbled together so fitfully over the previous quarter century began turning out a promising handful of young men sufficiently competent to teach basic chemistry, physics, mathematics and medicine. One of the bright lights was an Armenian, Artin Hekekian, the aforementioned Joseph's brother. An 1837 graduate of the Bulaq Engineering School, Artin returned from a tour of advanced study in France to take up upper-division teaching duties at the Bulaq School under the French director, Charles Lambert. In 1849, by then having risen to the rank of effendi, Artin took over the position of director when Lambert was appointed head of administrative affairs under Muhammad Ali's successor, Abbas. This was what the system was designed to do, replace the European directors with locals, though in this case the director was given a higher position rather than being replaced. Artin's rise was nonetheless a success story, a case where the system worked.

The success, however, was cut short when, a few years later, Abbas shut down the school, along with the medical and language schools. He simply let the schools dry up and die from attrition by cutting off funding. This came just as the system, ramshackle and inefficient as it was, had started producing more exceptions to the rule: skilled native Egyptian graduates like Ibrahim Nabrawi, Muhammad al-Baqli and Ali Mubarak. Ibrahim Nabrawi rose to become the ruler's personal physician. Another medical graduate, Muhammad al-Baqli, wrote an important treatise on cataracts. Both Nabrawi and Baqli had been sent as students on missions to France to perfect their medical expertise upon finishing their training at abu Za'bal and returned from France with prospects of promising professional careers as teaching physicians. Again, the contributions these graduates could have offered the Medical School were aborted when Abbas let it die along with the other schools and the foreign missions.

Educational reforms in Egypt were destined to suffer the same confusions and discontinuity that scarred Ottoman reform through the whole of the 18th century, from Mezzomorto to the New Order. Instead of producing yearly batches of graduates from the specialized schools and sending the best of them to Europe for two or three years of further education and training, thereby building up a body of scientific and technical expertise that could supply competent science and mathematics teachers to the secondary and specialized schools and provide physicians for public health, the enrollments in the schools started small and failed to increase, while the foreign missions, right up until they were terminated, remained too few, too sporadic and too miniscule to make enough of a difference in catching up to the West, as Japan was just beginning to do.

What was needed was a generation of men that included a Shaykh Attar, a Shaykh Tahtawi, a Shaykh Arusi, a Hekekian, an Ali Mubarak, an Amin Sami, a Nabrawi and Baqli, each one of these multiplied by ten every five years, to produce in human terms the critical mass for a self-sustaining and dedicated reformist movement, complemented by the reflective historical and religious thought needed to gain the supporting consensus of al-Azhar's quiescent leadership, which, together with the secularly educated community, could communicate the spirit of reform from the top of government to society at large. Creating a competent school system for a large number of pupils required a cooperative ulema and, on the part of the leadership, decent funding, determination, consistency and unity of vision, none of which characterized Muhammad Ali's leadership or his reforms. The ruler perceived clearly enough the relationship between the powerful army he desired and the academic training and industrial economy that would make possible his having it, but his vision fell short of the need to continually build and maintain the supporting infrastructure, such as providing for a ready supply of things like chalk, blackboards, books, spare parts and repairmen. An account left behind by Ali Mubarak, one of the few to write of his experiences in Muhammad Ali's schools, offers an idea of the penury and wretched conditions the students were forced to endure, and at the same time of the possibilities that opened to a student who successfully survived the conditions.

Born deep in the rural countryside, Ali attended a village mosque where, in addition to memorizing the Quran, he learned to read and write. From there he entered al-Azhar. He would have been only 12 or 13 at the time. At 14, he was selected as a student for Muhammad Ali's new Qasr al-Ayni state school in Cairo. The pain and shock of going overnight from a traditional religious routine to a secular setting was vivid in Ali Mubarak's memory many decades later when, having become well-known and Muhammad Ali having been long dead, he wrote freely about it in his memoirs.[6] Mubarak described the Qasr al-Ayni school as being hardly more than a bare structure. The students, housed in crude military barracks, were given a cheap military blanket and a mat. The food was so bad Ali learned to live on bread, cheese and olives. Morning to noon the students studied military subjects. Other than that there were no real studies.

They were beaten, treated with contempt, insulted and made to feel they were in a military prison rather than a school. Unremittingly mistreated, the miserable students had no recourse to outside help; those who mistreated them were accountable to no one. Even the fathers of students could not visit them. As much as he wanted, Ali Mubarak dared not try to escape. Students who had tried it were hunted down, mercilessly whipped, their parents arrested, shackled and humiliated. As he put it, the authorities held the sharp edge of the sword of destruction at the neck of the parents to keep the students submissive. Some became broken in spirit by the constant stress of their inescapable condition of imprisonment. Ali Mubarak was one of them. The horror of the place, combined with his fear of bringing punishment to his parents and shame to himself if he did not succeed, broke him. He fell into a delirium. He was haunted by the conviction that the school was a punishment for some unknown crime he had committed. The physical and mental anguish affected him to the point he had to be hospitalized. The hospital was an even crueler prison than the school. Scabies started to cover his body. He was sure that he would never leave the hospital alive. His affliction reached a critical point. And then, just as he thought he would succumb to the affliction, it began to ease, for what reason he did not know. His relief from the affliction gave him strength to endure. His mental suffering eased. By the time he left the hospital his strength had grown enough for him to accept the school as his destiny but no longer necessarily the agent of his death. As he put it, the power of his deeply ingrained religious conviction had overcome his guilt, revulsion and self-hate. His acceptance of God's will gave him the stamina to survive. He was then able to apply himself to his studies and never felt sick again.

When Qasr al-Ayni was made into a medical school, the students there were transferred to Clot Bey's medical preparatory school at abu Za'bal. It was there that Ali Mubarak began what he calls his "real" studies: geometry, physics and mathematics. The problem here was his lack of preparation. He could not understand a thing. He recalls, with perhaps some exaggeration and humor, the diagrams and equations that struck him as being diabolic, the geometrical shapes being some kind of talisman or dark magic, and the incomprehensible lectures that accompanied them a sorcerer's incantation.[7] Incompetent instruction compounded with lack of student preparation made for futility. But in the middle of that first year at abu Za'bal, a new teacher replaced the first one and the fearful night of magic, talismans and incantations turned to simple, precise, methodical explanation, clear as the light of day, as if an angel had descended to disperse the darkness. Writing his memoire 50 years later, he praised the secondary school teacher who made all scientific and mathematical subjects comprehensible. The teacher's name was Ibrahim Rif'at. He inspired Ali to do well. Determined to reward his tutor's investment in him, Ali drove himself to excel. The effort resulted in his being selected to attend the Bulaq Engineering School rather than the abu Za'bal Medical School, where Egyptian country boys chosen for specialized education were usually sent. Engineering was a big step up.

During his five years in engineering school, Ali studied mechanics, statics and dynamics, machine technology, advanced algebra, descriptive geometry, hydraulics, physics, chemistry, optics, geology and cosmography. He studied optics and differential calculus under Mahmud Hamdi al-Falaki, that other native Egyptian who was one of the exceptions to Muhammad Ali's prejudiced policies. Ali Mubarak graduated first in his class, winning the praise and respect of the director. The subjects were difficult, but it was not just the nature of the subjects but the lack of texts:

> Since there were no printed books for the subjects, the students would fill their notebooks with as much of the lectures as they could get down, and in those days the teachers raced with all their energy to get through as much material as they could, so it was rare that a student got it all, especially the diagrams and drawings.[8]

Toward the end of his engineering study some books were printed with lithography, alleviating the problem but far from solving it.

His performance in engineering school earned him a place in the 1844 "Mission of Princes" (composed mostly of nephews of Muhammad Ali) that was sent to France for the study of military science, making him the first native Egyptian sent abroad to study military subjects.[9] Even this honor had, like every stage in Ali Mubarak's education, its stumbling block. In Egypt it was first his mental and physical health, then it was lack of texts; in France it was language. Whether the students knew the language or not, the lectures were given in French and no Arabic translations were permitted. Since a couple of the princes were already competent in French, Ali Mubarak and all the other students who were lost in the lectures would meet with the princes after lectures and try to pry out of them what had been said. Unwilling to surrender the advantage that made them appear smarter, the princes refused to divulge. Complaints to the French instructors were of no avail. In frustration the students stopped attending classes. When word of this reached Muhammad Ali's agents in Paris, they came and put the students under house arrest and wrote the ruler what had happened. Muhammad Ali's return letter ordered that the arrested students should be released to continue their studies, but if there was a repeat performance they should be immediately returned to Egypt in chains. Ali Mubarak was so terrified at being put in chains and returned home in humiliation, and above all, causing unknown suffering to his poor parents, that he finally set himself to learning French on his own. For months he worked through the whole night on children's books in French, memorizing vocabulary, deciphering syntax, declensions. From there he advanced to French books on history and geometry. By the end of the third month he had progressed enough to make sense of the lectures.

The students who had the drive to improvise and learn for themselves had a chance of making up for the faults in the system and succeeding, but it was emblematic of the system that responsible educators and state officials could let students who did not know the host country's language go abroad for a tour of expensive study. It might at first also seem inexplicable that students did not apply themselves to learning the language the moment they knew they were going abroad, but it is not hard to understand why they did not. The lack of individual motivation was a natural product of the reformist state's totalitarian monopoly on decision-making, planning and executing. So much had been assumed by the state in determining the students' lives that individual responsibility and action became alien to them and, if ever considered, terrifyingly dangerous. Lack of time, guidance, money for introductory books to a foreign language or tutor, and the constant drumbeat of obedience that was drilled into the conscript students, froze out personal initiative. Thinking and acting on one's own could be a punishable offense. One never knew. If the state, which decided everything, had not thought of something and ordered it to be done, dare a student presume to take over the state's responsibility and risk the consequences? Better to do nothing and wait until ordered than risk punishment for acting on one's own thoughts. For this reason Ali Mubarak kept his autobiographical thoughts closely to himself until long after Muhammad Ali had departed from the scene.

Ali Mubarak did well in his studies abroad once he knew the language, well enough to receive praise from the Egyptian commander-in-chief. While Ali was in Paris studying, Ibrahim Pasha himself had come to France to study firsthand the military and industrial institutions that gave European nations such power over the world. While in Paris, Ibrahim visited the students at the Egyptian School and took the occasion to reward Ali Mubarak with a copy of Maltebrun's famous geography and atlas. The gift figures large in Mubarak's memoires. It was owing to Ibrahim Pasha's favor that Ali Mubarak went from Paris to the French military school at Metz, where for two more years he studied artillery, fortification, mining and engineering. By the end of that tour, he had firmly fastened his life to the military and engineering, though the arc of his career was neither smooth nor continuous, reflecting as it did the jagged trajectory of Egyptian reform.[10]

When Abbas Pasha came to power, Ali's studies came to an abrupt end. Called back from Metz, he was given the military rank of *yuz bashi*, literally commander of a hundred troops, equivalent to first lieutenant, and appointed instructor in an artillery school in the outskirts of Cairo. It was then that he married the daughter of his dear teacher at the abu Za'bal school, Ibrahim Rif'at, whose lectures had taken geometry and physics from the scary realm of magic and sorcery to the sweet sunlight of reason. His marriage, he confesses, was one of self-imposed duty in payment for all that his old teacher had done for him, for he had died while Ali was in France, leaving behind a poor unmarried daughter who had no means of support. And so the young officer, without conferring with his parents, honorably took it upon himself to provide for her.[11]

He had been away from his village for 14 consecutive years without ever having returned even once for a visit. "Foreign travel," his commanding officer told him, "erases half your identity," and, perhaps realizing the depth of meaning in those few words, the officer gave him leave to have a few days home to recapture his lost half. Leaving his new wife in Alexandria, he traveled to his village, arriving late at night. When he knocked on the door of his home and his mother answered, she failed to recognize who it was. It was dark and he was dressed in his French officer's uniform with the ceremonial sword at his side. "Who are you?" She asked, and he replied, taken aback: "Your son." The woman peered uncomprehendingly out into the night at the strangely dressed figure at the door with a sword hanging at his side and repeated, "Who are you?"[12] Many times the former village boy must have asked himself that very question during all those years of suffering and studying in Muhammad Ali's wretched schools in Cairo, followed by a decade in exotic France.

Abbas Pasha gave Ali Mubarak a medal and put him in charge of solving the problems plaguing the barrage that had been built at Qanatir al-Khayriyya during Muhammad Ali's time. The problem was that the barrage was not collecting water from the Nile for agriculture as it was designed to. Ali saw that the sluices were not being operated correctly and reported what had to be done. The order came back: whatever had to be done, do it, but in the cheapest way possible.[13] Best was not a word in the government's vocabulary.

Abbas was pleased with his native Egyptian engineer. The new ruler preferred Egyptians over Europeans. He did not trust Europeans. Europeans were out for what they could get, which included the country, if they ever had half a chance to seize it. Under the new ruler, the talents of Egyptians were beginning to be recognized. Ali Mubarak was one of the first. Abbas appointed him administrative director of all the engineering schools, with authority to set curriculums and designate the books to be used. The position came with a fine house in the engineering complex.

Ali Mubarak was proud of his tenure as director, brief as it was. He writes glowingly of his having increased the school's scientific and technology library collection to 60,000 volumes, of having badly needed textbooks printed on the school's lithograph and font presses and of running the school like a proud father runs his home and family. He made atlases for the students to help them in their geography and history courses; he kept close supervision over courses and instruction, and he himself gave courses in structural engineering. He claimed that under his guidance students learned French as well as those who studied it in Paris. "And to sum it up, in regard to the students, my goals were fatherly, looking to all of them, both teachers and students, like a father looking after his children."[14]

In his memoires he writes, with perhaps some exaggeration but showing clearly he realized what the problems were, of his having transformed the schools. He stopped the abuse and the curses that rained down on the students, and he almost eliminated the practice of student imprisonment and corporal punishment. He claims to have eliminated the corruption, bribery and favoritism that pervaded the administrative system of education and to have improved the food, drink, clothes, health care and instruction of the students. In place of fear, rivalry and rancor, he instilled a sense of harmony and affection among the teachers and students and he planted the seeds of progress, honor and self-respect so that both students and faculty excelled.

Ali Murbarak writes proudly of his success and sadly of the false rumors and accusations spread about him by those who were envious. Envy and intrigue were rampant in an environment where high positions were up for grabs and the only rules were those made up by the ruler who doled out the positions. As long as Abbas ruled, Ali Mubarak was safe from the slander. In 1854 Abbas died and was succeeded by Sa'id Pasha. The new ruler may have believed what was being whispered against him by "corrupt teachers," since he sent him to fight with the Ottomans against Russia in the Crimea, putting a temporary end to his career as an academician. Knowing he was soon bound for Crimea to fight, and perhaps recalling the language problem he had had as a student in Paris, Ali learned Turkish in four months of concentrated study in Istanbul before going off to the Crimea with an Ottoman detachment. Upon returning to Egypt at the conclusion of the war, he was aimlessly shunted from position to position. After two months in the Ministry of Commerce he was transferred to the Ministry of Interior as an inspector of engineering projects in the provinces.

It was during this period that he gladly accepted the somewhat low-level task of teaching mathematics to some Turkish officers in the Egyptian military. It was as a teacher of Turkish officers that he discovered that things had not changed all that much from his own student days at abu Za'bal, when the chronic problem of pitifully underfinanced schools made a mockery of state education at every level. Conditions seemed no better in the late 1850s than they had been a generation earlier, and in some ways they were worse. Not only did the Turkish officers not know any mathematics, they incredibly did not know how to read and write. This well could have been true, but it could also be that Ali Mubarak meant they did not know Arabic, though the script was the same in both languages. In any case, his teaching assignment looked impossible. There were no chairs, desks, blackboards, no classrooms, books or writing materials. He had to teach the officers wherever he found a place where they could assemble – in their tents, in the mess hall, or sitting on the ground outside in the shade. If outside, he would scratch circles and triangles in the sand with a stick. If inside, he would use a piece of charcoal and a slate or a slab of tile or a piece of tin, anything he found that could be used to write on. Sometimes he used ropes and sticks to represent geometric figures. If these were the conditions of education at the officer level, one can hardly imagine the conditions at the lower-level schools.

Blessed with the patience of Job, he resorted to the fallback method employed back in Muhammad Ali's day, of using students who had some understanding of the lesson to tutor those who had none. "The bright one or two students who learned the lessons would then help the others to learn, and eventually things became easier and progressed as the circle of learning widened."[15] To help his officers learn geometry, science and construction engineering, he reworked his lecture notes into a readable text, copies of which he had printed on the lithograph press. This was later published (1873) as an engineering science text.[16] Remarkably, neither as an engineering student nor many years later as a mathematics teacher did Ali Mubarak have the benefit of textbooks. He reports it humorously, but his light heartedness says much about the lack of serious financial regard Muhammad Ali and his two immediate successors, Abbas and Sa'id, had for reform and science education. His low-paying position as a military teacher obliged him to find outside work to survive. At one point, he went into low-level trade, selling whatever was at hand. He describes his life at this point in terms that could be used to characterize Egyptian reform: a series of intermittent advances and retreats.[17]

With the demise of Sa'id and rise of Ismail, Ali Mubarak's career took off once again. The new ruler appointed him as supervising engineer of the Qanatir al-Khayriyya Barrage. He was now high enough in the technological realm of government administration to be included in the ruler's 45-day mission to France, the purpose of which was to give Ismail the chance to parade down the Champs d'elysee in his gilded carriage while contracting another of those killer loans that was to sink the country and for Ali Mubarak to study the French school system, which he admired as an educator, and the Parisian sewage system, which he admired as an engineer.

Back from Paris, Ismail elevated Ali to *mir miran*, commander of commanders, equivalent to a four-star general. This was more honorific rather than a real military position. The position he was given was in the national railway administration (railways being one of the reasons for Ismail's ruinous European loans). This position was soon followed by his appointment as director of schools in the Ministry of Schools, the *Diwan al-Madaris*. In this capacity he worked toward two main goals, to centralize secondary level education and to get notables, merchants and ulema together to contribute to school expenses in provincial cities like Asyut, Minya, Beni Suwayf Benha and Tanta. It was important, he believed, that primary and secondary schools should be supported in part by the people whose children were educated in them. This was not just to get the public involved in secular education but to compensate for the ruler's unwillingness to lay out good money for educating the young of the common people. Ali Mubarak repeatedly emphasized how his centralization of secondary level education and his other innovations came with little additional cost. Cost was the anchor that held education to the reef.

Combined with his work in the Ministry of Schools, he continued on in his other positions as supervising engineer of the Qanatir Barrage and chief project engineer in the Ministry of Public Works. He writes of being very busy.[18] At the height of his career he became Director of the Ministry of Education and founded what is now the National Egyptian Library. He held the post of Minister of Education until the British occupied the country. As imperialists remove what they cannot control, the new rulers replaced Ali Mubarak with one of their own favorites.

The medical analog of Ali Mubarak's engineering success was Ibrahim Nabrawi. Nabrawi was another early figure signaling that native Egyptians were beginning to take a place in the secular world Muhammad Ali had unwittingly created in the midst of their medieval society. Nabrawi was a mere boy when his impoverished parents sent him from his peasant village with a little money to set up a watermelon stand in Cairo, the profits of which he was to bring home. Ibrahim was soon scammed of the money in the big city and, too afraid to return home, sought refuge in the mosque of al-Azhar. He was not long hiding there before he was discovered and, thought to be illiterate, obliged to enroll as a low-level student of religion. He learned to read and write and did well enough that upon graduation he was more or less coerced to enroll in the abu Za'bal Medical School. He continued to do well and was then included in a student mission to advance his medical training in Paris. He earned a French medical degree there, married a French woman, became fluent in French and began to translate French medical books into Arabic. What had begun with lost money for watermelons culminated in a stunning medical career. Nabrawi's translations were so highly regarded that upon returning to Egypt Muhammad Ali made him his personal physician. From there he rapidly advanced in rank, from effendi to bey to pasha and finally emir (prince or commander). Bathed in glory, wealth and status, Ibrahim Nabrawi nonetheless remained faithful to his professional training and continued translating books on natural philosophy to become one of the most important members of the highly educated and Western-ized group of reformers inching Egypt toward that distant goal of becoming a modern secular state – inching and distant only because of the minimal human and material resources that were invested in meeting the challenge of that momentous goal, which meant overcoming the formidable resistance that Egypt's traditional society posed to Westernized change.

The undercurrent of resistance that minimized the results of Muhammad Ali's decades of reform came from many sources. In addition to the antipathy between ruler and ruled, underfunding, and the dubious commitment of the Europeans upon whom Muhammad Ali relied, not to mention the total lack of an indigenous edu-cated class that could willingly relate to science and technology, serious structural flaws in the foundation supporting his complex of reforms undermined the whole edifice of his New Egypt. It should also be noted that the miserly pay that graduates in engineering and medicine received from Muhammad Ali's government did little to energize an already flagging will.[19] The perennial lack of funding at all levels, but especially at the primary and secondary, went against any chance of the flaws being worked out in time. In addition, the system was undermined by the contra-dictory vicissitudes of despotism as, after Muhammad Ali's death, one ruler fol-lowed the next in the dynasty with repeated reversals in policy and aborted educational initiatives.

If internal faults crippled the reforms, external factors cut them off at the legs. Muhammad Ali proved correct in believing that the West and the Ottomans would do everything to bring him down. In the middle of the 1830s, just as the primary and secondary schools were being established, Britain, in alliance with the sultan, was undermining Muhammad Ali's industrial projects. A few years later he was forced to evacuate Syria and reduce his military, thus cutting away the raison d'être that had motivated his reforms. The precipitous drop in primary school enrollment to a meager 780 students reflected the ruler's dashed dreams. As far as it concerned Muhammad Ali, without a big military and a burgeoning industry supporting it, there was no need for schools. Consequently, Egypt's attempts to reform in the 19th century were stifled by the same debilitating reversals that had crippled Ottoman reforms in the 18th century: discontinuity and inadequate funding. Nonetheless, as it had been with the Ottomans, the experience of reform during one generation lived on to be inherited by the next generation. Languages were learned, books translated and schools created. Egyptians had seen the West, had been trained in science, mathematics and medicine; and although the schools and reforms fell short, the experience was indelibly there, awaiting the opportunity for it to be employed. Beneath the tumultuous surface of the government's inefficacy, the experience and knowledge of several thousands of men that had accumulated over a period of almost two generations flowed quietly on. Reforms in education, though alternately succumbing and being resuscitated during the century from the first student missions of Muhammad Ali to the end of the First World War, did in fact give birth to a small, secularized intelligentsia that viewed the West, through whatever refractive lens – nationalist, socialist, communist or capitalist – as the model to be emulated for a future in which Muslim countries might regain their lost leadership. But it was a small, fragile, thin layer of society composed of writers, academics, lawyers, doctors, journalists, engineers and intellectuals that could be, in moments of crisis and defeat, all to easily swept away and driven underground by the storms of military and religious revolution.

Defeat of Muhammad Ali, 1840–1841

As usually happens when a regional power strikes an independent path counter to the interest, or perceived interest, of a world power, the lesser is taught a lesson that can range from boycott and embargo to subversion, invasion and regime change. In Egypt's case it was regime diminishment, but so diminished it amounted to change. The lesson was clear: a regional power must know its place and keep to it. It was a lesson not easily learned. Britain would have to teach it to Egypt again in 1882, this time by invasion, occupation, regime change and direct military control.

Muhammad Ali's undoing was first his textile industry that was competing with Britain's. Then ultimately, it was the brilliant military ability of his adopted son, Ibrahim, who marched from conquest to conquest: Sudan, Greece, Palestine, Syria and to the heart of Anatolia. Muhammad Ali was at the height of his power when the Greek revolt against the Ottomans broke out in 1821. The Janissaries having humiliatingly failed to crush it, Sultan Mahmmud II called on his governor of Egypt to do it, in return for which service Muhammad Ali was to be given Crete. Ibrahim Pasha's defeat of the revolt drew France, Russia and Britain together to save, as they grandly proclaimed, "The Cradle of Civilization from Turkish Barbarity." In 1826, at the battle of Navarino Bay, a joint European armada destroyed the French-built Egyptian navy, depriving Muhammad Ali at one blow of both fleet and Crete. Sultan Mahmud reneged on Crete when the Greek revolt flared up again and owing to Western intervention, particularly Russian, was successful.

Rejecting the sultan's loss of Greece as an excuse for him not to be given Crete, and also in compensation for his lost navy, Muhammad Ali sent Ibrahim Pasha at the head of the army into Palestine, Syria and Mt. Lebanon. Handily defeating the Ottoman Army, Ibrahim Pasha crossed over the Taurus Mountains into Anatolia where he crushed another Ottoman army, laying open the undefended road to Istanbul. It seemed Egypt, not Russia, would be the country to swallow the empire. Panic overcoming humiliation, Sultan Mahmud II appealed to the West for help. With France now shifting to a pro-Egyptian stance and Britain at the time indisposed for internal reasons, the sultan turned in desperation to the Ottoman's foremost foe, Russia, who eagerly offered help. The turnabout was an unimaginable contradiction of Ottoman history. For three centuries, relations between Ottomans and Russians had been defined by war, territorial loss, peace treaties and more war and loss. On more than one occasion the Russians had been on the verge of avenging the fall of Byzantine Constantinople. Catherine's imperial dream that Russian Orthodoxy's defeat of Islam would avenge Greek Orthodoxy's loss to it, once again loomed on the historical horizon. Catherine in paradise was waltzing with the angels.

The tsar extended his saving hand and Mahmud, the energetic reforming sultan, grasped it. To explain the mind-boggling turn to Russia, Ottoman historians resort to the Turkish proverb: a drowning man grabs the tail of the snake. The snake saved the drowning man from the Egyptians by sending its navy and 30,000 troops to the Bosphorus, the sight of which must have drained the last drop of imperial spirit from what was left of Ottoman pride. Seven years earlier it had been a ragtag army of Greek peasants inflicting a humiliating defeat on the Janissaries. Then it was Egyptian peasants overrunning Mahmud's reformed army. And now it was satan's slavic soldiers rushing to rescue the caliph and sultan of God's true children. Russian infidels protecting the once-mighty ruler of Islam from a Muslim army of Egyptian peasants created by an Albanian adventurer, a middling Ottoman officer of Selim's New Order! What was left to be salvaged of dignity?

While the Ottomans were cast into the depths of depression by having to be saved by their number one foe, the British were utterly horrified. The sultan had willfully surrendered Istanbul and the straits to the tsar, who was not only the sultan's greatest enemy but, in Central Asia, Britain's greatest threat to imperial communication with its empire's prize possession, India. The specter of Catherine's dream of a resuscitated Byzantine empire under Russian tutelage was more than the British could bear. In the name of liberalism and free trade, and to satisfy British industrialists who had already rallied their government against Muhammad Ali over his monopolistic industries that threatened the British textile market in the Middle East, Britain's foreign secretary, Lord Palmerston, set himself to get the Russians out of Istanbul and the Egyptians out of Anatolia. Also, as the ruling elites of empire never fail to spot in the distance an evil menace or two salivating covetously over the far-flung possessions that are by self-declared legitimacy rightfully theirs, the British ruling class looked with disfavor on Muhammad Ali's strong army and navy as another threat to Britain's India.

Once the threat of the Egyptian army marching on to Istanbul had been checked by the Russians, a treaty between the tsar and the sultan put the empire under virtual Russian protection: the fox guarding the hen house. It was with a deep sigh of Ottoman and British relief that the Russians lived up to the treaty and withdrew their troops. But Ibrahim Pasha and the Egyptian army were still in control of Syria and southwest Anatolia, and the tsar still enjoyed his recognized protectorship of Ottoman sovereignty, guaranteed by treaty. As they had 30 years earlier when Britain and the Ottomans had come together to drive the French out of Egypt, they now awaited their chance to drive the Egyptians out of Anatolia and Syria – and then cancel out the Russian protectorate.

Sultan Mahmud II had destroyed the Janissaries in 1826 and then had proceeded, with European help, to build a new army along the lines set by Selim's New Order. As we have seen, Mahmud's new army had failed against Ibrahim Pasha's generalship of the Egyptian peasant army. Ten years later, in 1839, Mahmud II risked another go at the Egyptians camped to the south. Again Ibrahim Pasha was victorious. The Ottoman army was smashed. For many Ottoman leaders this was too much: a sultan who had delivered the empire of Islam to Russian protection and whose twice-reformed army suffered defeat upon defeat from provincial Greek and Egyptian peasants. Giving up on the sultan, the grand admiral of the Ottoman navy sailed the fleet to Alexandria and surrendered himself and the newly built fleet to Muhammad Ali, the only ruler in Islamdom left standing. What was too much for the Ottomans was too much for the British. Lord Palmerston decided the time had come to cut Muhammad Ali down to size once and for all and protect British interests before Russia and Egypt ended up partitioning the Ottoman empire between them. A British fleet was sent to the eastern Mediterranean and an ultimatum issued to Muhammad Ali: retreat to Syria within two weeks and Syria would be under Egyptian authority for the duration of the ruler's life, with Egypt belonging to the ruler's family in perpetuity, or suffer Britain's determination to drive the Egyptians out of Syria by force. In either case, the Ottoman fleet had to be returned to the sultan. Unable to face defeat, Muhammad Ali let the deadline of the ultimatum pass, at which point the British went to war. Only when the firing began did the Egyptian ruler realize he had no choice but to surrender Syria. Ibrahim Pasha was ordered to withdraw to Egypt with the army. Muhammad Ali's dreams of a regional empire had foundered on the shoals of Britain's global empire.

The following year, a meeting in London of the European powers, joined by a representative of the sultan (who, hardly more than an observer, was invited to the gathering only to cloak the treaty with international legitimacy), drew up what was pompously called the London Treaty of the Pacification of the Levant. The treaty internationally established the Muhammad Ali dynasty by granting the hereditary right of ruling Egypt to Muhammad Ali and his descendants. To render him "moderate," meaning emasculated, his army was reduced to the level of a police force, from over 100,000 troops to 16,000. Another crippling clause, this one economic, obliged Muhammad Ali to reduce the customs duty on British textiles imported to Egypt. This, and the huge reduction of his military, effectively terminated his purpose in reforming and industrializing the country. As if the economic influence that the Capitulations gave European powers in the Ottoman Empire was not enough, the Treaty of the Pacification of the Levant tacitly legitimized European authority in its capacity to arrange political settlements within the empire and its recalcitrant provinces. It was all but over for the Ottoman Empire.

France in 1798, Russia in 1828, and Britain in 1840 had proved decisively that the last word concerning the affairs of Egypt, the Ottoman Empire and its Arab provinces, as they related to European self-interest, belonged to one or another of the powers, or several of them acting in concert. In 1860, the French fleet would come to Beirut, to the hypothetical rescue of fellow Christians and Maronites. French marines would occupy coastal Syria and Mt. Lebanon, whose autonomy from Ottoman rule would be legitimized by international treaty – international in this instance meaning Britain and France forcing a settlement that the Ottomans could not refuse. This simple geopolitical reality would be demonstrated by Western powers over and over again, and ever more forcefully and brutally into the 21st century.

At some point between 1840 and 1882, when Britain at last occupied Egypt and by which time the British had long been in control of India with its tens of millions of Muslims, the Russians were in control of large parts of central Asia with its tens of millions of Muslims, a conglomeration of European industrialists, financiers and entrepreneurs was taking over Iran and Afghanistan was being contested by Britain and Russia, the gloomy reality took hold of Muslim leaders everywhere that there was no escape from the iron grip of direct or indirect Western domination. The rivalry between Istanbul and Cairo in the 1830s had opened another fissure into which the European powers could wedge their levers of control to advance their self- interests. This was precisely what the reforms of Muslim states were meant to prevent.

The disintegrative heat generated by the social friction of modernizing reform and reformers grinding against conservatives, was heightened all the more by the continual advance of European economic interest and control in the region; then rendered again all the more explosive by the socially degenerative acid of another influence from the West penetrating the empire's body politic: nationalism. Reform, conservative reaction, Western penetration, and finally nationalism would be leading features in the relations between the West and Muslim states, dividing and wasting the powers of the latter, while enhancing those of the former in their control of the internal affairs of the region.

The geopolitical reality of power relations between Islamdom and the West had been long in the making. Precedented by the centuries-old Capitulations, Muhammad Ali's experience made it all the more obvious that the West would henceforth wield a heavy hand in the internal affairs of Muslim countries. Muhammad Ali is reported to have said that he knew Britain would one day seize Egypt to protect the route to India, and it was for that reason he refused to consider the French proposal of having a canal built to link the Red Sea and Mediterranean, as it would suck Egypt all the more into the vortex of European rivalries by heightening the country's desirability to be possessed by one or another of the European maritime powers, Britain being the lead candidate. How right he was. Muslim rulers who failed to take careful account of Western interests, who gambled on a course of action inimical to those interests, would have fates far worse than Muhammad Ali's: Khedive Ismail, deposed and exiled in 1879; Ahmad Urabi, exiled in 1882 and Egypt occupied; King Faysal of Syria, exiled by the French in 1920; the shah of Iran, deposed and exiled in 1943 by the British, who also did the same to the prime ministers of Iraq and Egypt during the war years. The line of deposed and crippled Middle Eastern leaders following Muhammad Ali is impressive. Any Middle Easterner or Muslim reflecting on it can only be depressed and angered at the humiliation of supine impotence. When the Japanese defeated the Russians in 1905, while Muslims were being squeezed ever tighter by Western economies and empire builders, the humiliation of failure in the face of Japanese success deepened the sense of Muslim helplessness. That their leaders were comical puppets and playthings of the West was an open theater for the world's amusement.

In retrospect, Muhammad Ali's reforms, and the concomitant ones of Sultan Mahmud II, may have been Islam's last chance to close the scientific and techno-logical gap. Regardless of the many sound reasons explaining their minimal suc-cess in modernizing, the consequences were no less fateful. In the 1830s, 1840s and 1850s, Europe was already going through a second industrial revolution with the new technologies of steel processing, metallurgy, chemical manufacturing, railroads, telegraph, steamships and electric power, while Egypt, the Ottomans and Iran were struggling with the rudiments of the first one. Success in narrowing the gap required far more vision, determination, planning, organization, leadership, social unity, money and sacrifice than Muslim leaders and their societies were willing to expend.

Notes

1 La Verne Kuhnke, *Lives at Risk: Public Health in 19th century Egypt*, University of California, Berkeley, CA, 1966, p. 43.
2 Pierre N. Hamont, *L'Egypte sous Mehemet Ali*, 2 vols., Paris, 1843, vol.1, pp. 440–441. A Frenchman employed in Egypt calling Egyptians "nouveaux venus" gives a clear enough idea of the colonialist sense of proprietorship.
3 Puckler Muscau, *Travels and Adventures in Algeria and Other Parts of Africa*, 3 vols., London, 1847, vol. I, p. 250. Puckler Muscau, undoubtedly informed of it by Hamont, accused Mukhtar Pasha of an intrigue that aimed at having Muhammad Ali dismiss Hamont from his position as head of the veterinary school.
4 A.B. Zahlan, *Technology Transfer and Change in the Arab World*, Pergamon Press, Oxford, 1977, p. 7 note 17.
5 Zahlan, *Technology Transfer*, p. 14.
6 Ali Mubarak, *Hayati (My Life)*, Maktabat al-Adab, Cairo, 1989.
7 Mubarak, *Hayati*, pp. 13–14.
8 Mubarak, *Hayati*, p. 16.
9 Yusuf Ilyas Sarkis, *al-Matbu'at al-Arabiyya wa'l-Mu'arriba*, vol. 2, Cairo, 1928, pp. 1367–1369.
10 Mubarak, *Hayati*, pp. 17–20.
11 Mubarak, *Hayati*, p. 21.
12 Mubarak, *Hayati*, pp. 22–23.

13 Mubarak, *Hayati*, pp. 24–25.
14 Mubarak, *Hayati*, pp. 26–28.
15 Mubarak, *Hayati*, pp. 36–37.
16 Mubarak, *Hayati*, p. 38.
17 Mubarak, *Hayati*, p. 39.
18 Mubarak, *Hayati*, pp. 40–45.
19 Reid, *Making of Modern Egypt*, Cambridge Middle East Library, Cambridge, 2002, p. 127.

19 Azharite shaykhs and modern science

A great misconception is that the defenders of traditional religion stood united against change, modernity, science and Western innovation in general. There was no such thing as a unified religious opposition to Muhammad Ali's reforms. As much diversity existed among the ulema regarding reform as existed within society. Betting, as had Bonaparte, that the yeast of payment and prestige would leaven the loaf of cooperation, Muhammad Ali found in the shaykhs amenable supporters of his programs. He employed them in various capacities in all his specialized schools. The most controversial of these schools, from a strict religious viewpoint, was the Medical School at abu Za'bal, if for no other reason than the school's teaching and practice of dissection. This was clearly against traditional religious practice. But distance, isolation and guarded walls removed abu Za'bal from the public eye, relieving the shaykhs working or studying there from the pressures of their more conservative peers. Shaykh Rifa'a Rafi'i al-Tahtawi, famous as translator, reformist writer and director of the School of Translation, was the first director of the Medical School's preparatory annex, the *Madrasat al-Maristan*. Shaykh Arusi, another leading Azharite exponent of reform, followed him as director when Tahtawi moved on.

Arusi believed that the benefits of the new medicine would help reduce any qualms that the more traditionalist shaykhs and religious students had in accepting science in general. Modern medicine had also to be cloaked in the mantle of religion. To boost acceptance of the modern, he encouraged the rector of al-Azhar to introduce a course on Islamic medicine (*al-tibb al-shar'i*): medicine that had by tradition originated with the Prophet and had come to be considered sacred (also called *al-tibb al-muhammadiyya* or *al-nabawiyya*: *muhammadan* or prophetic medicine). Analogous to Clot Bey's depictions of the great Muslim physicians on the walls of abu Za'bal's dissection theater, Shaykh Arusi's idea was that teaching the sacred and the new alongside each other would engender a sense of religious legitimacy, one sanctifying the other by transference through proximity. This was not to happen. The reformist shaykhs were more effective in contributing to change outside the formidable walls of al-Azhar than inside. Every indication is that when distanced from the main body of shaykhs and the authority of al-Azhar, individual shaykhs were willing to contribute and were capable of performing an active and important part in the reforms. Azharite scholars contributed by working as "correctors" of lectures translated from French into Arabic in the medical and engineering schools. Some shaykhs became students in the Language and Translation School in preparation for employment as language teachers or as translators and proofreaders of the scientific and technical texts being translated into Arabic. As translators of books and lectures, these shaykhs were among the first to provide Arabic equivalents for European medical terms, accomplishing in some part what Hunayn ibn Ishaq had done for medicine in Arabic a thousand years earlier when translating Greek and Syriac texts, or what Gerard of Cremona had accomplished for medicine in the Latin world when translating Arabic texts.

Young Azharite students also contributed to reform. The transformation of many young Azharites from religious student to graduate physician and reformer took place within the grim walls of abu Za'bal. Azharite students who had been selected to study at abu Za'bal, and then gone on to Europe for advanced specialized training after graduation, returned to join the small but growing community of educated elite coming out of Muhammad Ali's institutes. Other Azharite students who were educated at the medical school but did not graduate went on to teach the science and mathematics they had learned as medical students.

It was these Azharites, both shaykhs and students, who formed the first generation of reformers opening the way to Egypt's stressful passage to modernity. While the shaykhs themselves were recruited to be trained as translators, correctors and proofreaders of lectures, their students provided many of the recruits to the state specialist schools, the medical school in particular. Egyptian medical students were reported to be expert dissectionists. A British physician visiting abu Za'bal in the late 1830s wrote that cutting up cadavers was more accepted in Egypt than in London. In Istanbul, where Sultan Mahmud II was at the time vigorously pushing reforms parallel to Muhammad Ali's, dissection was being practiced in the Galata Saray Medical School as freely as at abu Za'bal. Ottoman students were at first obliged to practice on wax dummies, until the government reached a compromise with the religious opposition and issued an imperial decree permitting the use of cadavers. Clot Bey had been able to provide his abu Za'bal students with cadavers a decade earlier.

During the first half of the 19th century, Muhammad Ali's educational innovations introduced modern medicine and science to several thousands of students, many of whom had been students in Islam's most venerable institution of advanced religious study.[1] But none left any record describing his reception of the new sciences or of problems he faced in squaring religion with principles of the new knowledge. They may not have thought about it, just as they may not have grasped Clot Bey's intended implication when seeing the images of al-Razi and ibn Sina on the walls of the dissection theater. And if they did, they may have seen no reason for the subtlety. After all, back when the French were occupying the place, Rigo's depictions of the prophet Muhammad and the early caliphs elicited praise from the shaykhs for the artistic realism of the portraits, mindless of the Muslim proscription against depicting the human form. Except for the students who had studied seriously and graduated from al-Azhar, their knowledge of religion may not have gone beyond fasting, prayers and the rest of the five pillars. Knowing little of theology, they would hardly have been aware of the metaphysical problems modern science presented to belief. Reconciliation would not have been an issue. The scholastic dialectics of *kalam* were beyond what they accepted to be their religion. Those who might have been aware of contradictions made no fuss. Reconciliation was not an issue because the science taught in the schools was not science. Education in the specialized schools was strictly limited to the overall goal of creating a technologically skilled workforce that would eventually take over the European-run military–industrial complex, along with its schools and training annexes. Anything beyond that was irrelevant. Science existed in a vacuum, wrapped in the most immediate practicality of technical craftmanship. The science the students learned was purposefully practical, unhinged from theory and cosmological principles. Also, any mention of contradictions in private conversations could have come across as criticism of Muhammad Ali's projects and no student wanted to be reported for that and hauled off in chains, his family punished and humiliated. The heavy hand of authority in Muhammad Ali's state would have suppressed student comment. The ruler's order to learn was reconciliation enough.

Muhammad Ali did not tolerate the divisions that came from religious debate, or debate of anything. The single most important exception to this dearth of student commentary comes with Ali Mubarak, whose literary and teaching career did not begin until the more relaxed reign of Ismail, in the 1860s and 1870s. In any case, the work of reconciliation between modern science and Islam would properly have been restricted to al-Azhar's shaykhs, specifically those who had been brought into the new system of education and who would have supposedly been exposed to the new science. Voices of theological conciliation or critical rejection would be expected to have arisen from among those scholars. And it was indeed from among them that the voice arose.

In the half century from the days of the French *Institut d'Egypte* to the end of Muhammad Ali's reign, only two shaykhs, Hasan al-Attar, and to a much greater extent his student, Rifa'a Rafi'i al-Tahtawi, publicly exerted themselves to argue a place for science in the Muslim community as a bridge between secular and religious education. On the contra side of the ledger, Shaykh 'abd al-Rahman al-Jabarti wrote critically that Muslims had no business studying the science demonstrated by the savants of the French Expedition. His criticisms of Muhammad Ali's tyrannical policies appear to have cost him his life. No other shaykh dared speak out against the reforms or the ruler. There were those who served and praised, and those who did not and kept silent.

Shaykh Attar served and praised. His highest service was as chief editor of the government journal founded by the ruler, *al-Waqa'i' al-Misriyyah*. As a shaykh and an important member of the reformist government, he was acutely aware that religion was being marginalized in the state created by the ruler he was serving. Attar's criticism was directed not against the current of innovation but against the shaykhs who stood by watching the current sweep past them. To avoid the growing split in education that was bypassing al-Azhar and threatening to leave society morally and spiritually impoverished, he called for the shaykhs to prepare themselves for an active role in the new state by becoming educated in modern science, for only as active participants in the new institutions could they preserve religion's place in society. The shaykhs would either have to become teachers of science and relate it to religion or see themselves become irrelevant as the chasm between science and religion deepened. Their remaining outside the new system, passively or actively resisting, would deform both society and religion. If al-Azhar as an institution could not reform itself to bring the modern sciences into its curriculum and turn out graduates versed equally in religion and science, then individual shaykhs had to do it on their own within the state structure and outside al-Azhar. Only thus, Shaykh al-Attar believed, could a vital place for religion and the shaykhs be assured in the reforming state.

He offered no details on how religion and science could come together as parts of a unified system of knowledge, but he did go so far as to say, in referring to astronomy and anatomy as empirical sciences, that God's greatness is manifested through the lights of science, the fading of which brings the dusk of ignorance.[2]

> We have been dissuaded from the pursuit [of the empirical approach] because it has been considered an obstacle to God's law and to faith, and an impediment to His mercy and charity which dwells in our hearts. But for this reason it might be said that there exist two sciences, the careful study of which leads to a growing certitude concerning the wonders of creation and the masterful skill of God. These two fields are *'ilm al-hay'a* and *'ilm al-tashrih* [astronomy and empirical anatomy]. And I say that these two fields transform the mind, but that precision in the knowledge of them is concealed from the masses. For that reason you will see that the majority of distinguished people in our time have not been influenced at all by the two fields. In fact, these fields are almost nonexistent now, except among a very small number of people, who save us from ignorance.[3]

Clot Bey had a high estimation of Shaykh Attar. The French physician–director of abu Za'bal considered him to be an important religious support in the medical school's practice of dissection in the anatomy classes. Attar contributed in a modest way to Arabic scientific literature by composing a treatise on the construction of the astrolabe and a few others on medicine, geometry and astronomy.[4] In Damascus, where he had gone for a while after the Ottomans had returned to Egypt upon the French evacuation, he may have studied the medieval reform astronomy of the Maragha school. But of the new Copernican astronomy he makes no mention to all, though he must have learned of it from his association with the savants. His not writing about it was most likely because of his sense of caution. As a scholarly shaykh versed in theology, he would have seen the problem that a moving earth would give to his Azharite colleagues, and so Shaykh Attar played it safe. Playing it safe was in his nature. Just as he avoided the controversial pitfalls that Muslim theology posed to science, he avoided the anti-collaborationist wrath of the Ottoman authorities by departing for Damascus after the expulsion of the French and their savants with whom he was known to associate. He knew when to leave and when to return. He had earlier fled Cairo when it became apparent the

French were going to occupy the city in the summer of 1798 and only returned in early 1800, after the smoke of the revolts had dissipated.

In the relatively mild political climate of 1800, Attar could visit the Institute without arousing resentment on the part of his fellow shaykhs, a number of whom were cooperating with the French by serving on Bonaparte's administrative councils. The resentment that the shaykhs harbored for any of their own who associated too closely with the French was directed against Shaykh Bakri, the star collaborator. Also, Attar did not advertise his positive feelings for French learning until long after the conquerors had been forced out and Muhammad Ali was in power, with the French back again as employees rather than conquerors.

Shaykh Attar, according to the historical record, was the first Egyptian whose eyes were opened to the importance of the science that the French brought with them. Muslims could only resist the West by possessing its knowledge. He is said to have offered the savants Arabic lessons in return for instruction in science; then, after the French had left, he urged his students and colleagues to put their minds to science and technology:

> There is no choice but for our country to change and renew itself through the new learning that has not yet made its way into our country. It is amazing what the French have accomplished.[5]

Muhammad Ali provided Shaykh Attar a secure place in government service, in exchange for which the shaykh sang the ruler's praises. The rumored accusations that the ruler had ordered the death of his best friend, Shaykh Jabarti, who had also praised Muhammad Ali for the security his government gave the country but also condemned him for his un-Islamic rapacity in confiscating the wealth of the people, imposing illegal taxes and trampling old traditions, did not hinder Attar from continuing to serve him to advance his own career. In this he was eminently successful, first as the chief editor of the ruler's newly founded government gazette, *al-Waqa'i al-Misriyya*, and later as Rector of al-Azhar, the highest religious position in Egypt. On the other hand, serving the tyrannical reformer presented to Shaykh al-Jabarti a moral problem no less conflicting than had been the question of his cooperating with the French. Jabarti, along with most all high shaykhs, grudgingly cooperated with the French, but with Muhammad Ali he refused. Very few shaykhs did that. Jabarti was an exception. He believed a good Muslim may be coerced to accept and live under tyranny but that did not mean he had to cooperate with it. With the French, he brought himself to cooperate. After the battle of abu Qir and the sinking of the fleet and Bonaparte's humiliating defeat in Syria, the French became a fleeting interlude soon to be swept away, an incursion without roots or sustenance to hold it. Muhammad Ali, on the other hand, as an Ottoman Muslim, presented a home-grown illegitimacy that might not so quickly disappear. Jabarti found no room for compromise with the Muslim tyrant's rule. This was not the case for Shaykh Attar. Attar, al-Jabarti's closest friend and soul-mate, chose to serve and praise the ruthless autocrat, even after Muhammad Ali was suspected of being behind the murder of al-Jabarti, whose severe political and religious criticisms the insecure ruler could not tolerate. Shaykh Jabarti lived and died by his principles, which included rejection of what the Institute and Western knowledge had to offer. Shaykh Attar did not let politics or religion stand in the way of either his advocacy of science and modernizing reform or the upward trajectory of his career in serving the ruler who was suspected of having assassinated his dearest friend.

The first reformist to address the question of the religious legitimacy of Western science in Arab Muslim society was Attar's student, Shaykh Tahtawi.[6] Ottoman writers and translators had been commenting on the compatibility of science and Islam in a most general manner since before the Tulip Period, but figures such as Ibrahim Muteferrika and Hajji Khalifa did not have the credentials to act as spokesmen for what was permissible or not in Islam. In Muslim society, the task of determining what is within religious legitimacy and what is outside of it falls to the men of religion, those who are professionally educated and employed in religion, the shaykhs of al-Azhar in Cairo; the imams and Shari'a judges, the chief *mufti* or shaykh al-Islam in Istanbul; the mujtahids, marja's and ayat Allahs of the Shi'i theological seminary of Qum in Iran; and their equivalents for the Muslims in India, the Caucasus and Central and Southeast Asia. Any other Muslim, however devout, religiously educated and scientifically knowledgeable, avoided the subject like poison, and wisely so, for he had not the authority to comment on such matters and could find himself in deep trouble with those who did if he stepped over the line. Religious commentary and definition were monopolies of the leading shaykhs and imams whose *fatwas* were recognized as having the authority to condemn and prohibit, or commend and legitimize, unlike in the West, where men outside of the religious establishment had since late medieval times argued for or against the harmony of science and religion. In Islam, if the case was to be made, it would be the learned elite of the high ulema who would have to make it. But for fear of condemnation by their peers for taking a stance on science, something they knew little about and even suspected as being heretical, few had a mind to risk it. Advocating science education presented a heavy psychological burden to the ulema. Modern science, the offspring of those Greek rational sciences that had been warned against by the great theologians of the past, was a product of the West, millennial enemy and perverter of the good name of Islam and its Prophet. For a shaykh to praise and encourage the study of something from the West was fraught with danger. The science that Shaykh Attar called for consisted of technical craftsmanship, cut off from the core of theoretical, philosophical and theological implications that would give trouble to whatever traditional belief could not bear. Shaykh Tahtawi went deeper into it than that. He was better prepared to do it.

Tahtawi lived and studied in Paris for five years. At the newly established "Egyptian School in Paris," where he and his student charges resided, he learned French and read French literature and was tutored in astronomy, physics, mathematics, geography and European history.[7] By some unpredictable trait of openness in the young shaykh's character and intellectual curiosity, he turned out to be the most articulate product of all the missions. His Parisian experience endowed him with the required leverage in seeking and finding a balanced position regarding the question of orthodox belief and scientific theory. In addition to his years of study-ing science and reading French literature, he translated 12 books from French to Arabic and composed a treatise on astronomy that introduced him to the physics of Descartes and Newton, which he accepted at the time without question. Yet even with this, the positive position that young Tahtawi took on science, with all his cautions limitations, qualifications and equivocations, would be abandoned toward the end of his career, possibly because no other shaykhs had come forward to help hold the line he had established. Tahtawi's writings reflect the struggle on a per-sonal level between the strong hold of traditional religion and the perceived neces-sity to reform. After 35 years of reformist writing, with no scientifically endowed scholastic conciliation welling up from the ranks of the ulema, the aging shaykh, growing timid and more conservative, recanted what he had allowed during his younger, more hopeful years. He had begun his career as a reformist writer with bright hope for the future.

His career had been launched by his teacher, Shaykh Attar, who recommended Tahtawi, a 25-year-old recent graduate of al-Azhar, to Muhammad Ali for the posi-tion of imam and spiritual guide of the 1825 student mission to France. At the last moment Muhammad Ali had decided that the students going to Paris needed a shaykh to keep an eye on them so they would not venture into Parisian life and pick up infectious political ideas that they might bring back with them. He asked Attar to choose a shaykh and he chose his brightest student.[8]

Tahtawi returned from his five years in Paris versed in the principles of modern science, but also with ideas he learned from his reading in political philosophy and the tutorials he received from his French mentors, the Arabists Caussin de Percival, Sylvestre de Sacy and E. F. Jomard, the same Jomard who as a young Arabist had assisted Venture de Paradis on Bonaparte's Egyptian expedition. The science that failed to penetrate Egyptian thought through the *Institut d'Egypte* was now, a quarter century later, entering it through its successor, the Egyptian School in Paris. Tahtawi's witnessing of the 1830 liberal revolution in Paris was itself an education in political action giving reality to the ideas he absorbed from books, Rousseau's *Social Contract* being one that especially appealed to him. At the end of his second year of study in the Egyptian School, his French tutors, with customary flair, reported he was destined to become "one of those who would render the highest services" to his country.[9] This Tahtawi did. For a generation, his was the clearest voice calling for change and arguing that Muhammad Ali's reforms not only con-formed to the basics of Muslim belief but were what true Islam demanded.

Tahtawi's book on his experiences in Paris and impressions of French life, and his books that followed, created the first body of Arabic literature devoted to legiti-mizing reform and laying the intellectual groundwork that claimed it to be a reli-gious duty for Muslim society to adopt the natural sciences as a modern extension of that lost heritage of Islam, when Muslim scientists, philosophers and physicians led and enlightened the world.

Publicly, he was a spokesman for the ruler, but beneath his expressions of sup-port, the conservatively reared and educated shaykh of rural Egypt had some seri-ous problems with the science he learned in Paris. Some of the stumbling blocks were the same old medieval ones of Ash'ari theology. Some were new. This new science of the Europeans was far from the natural philosophy Muslims and their non-Muslim colleagues had excelled in during earlier centuries of Islamic history. The latter had been an amalgam of Aristotelian physics, Ptolemaic astronomy, and Galenic medicine refined by Muslim scientists and philosophers to accommodate the presence of a Quranic deity's guiding hand. The science that had come back to Islam was unrecognizable. Galileo, it seemed to some, had divorced God's caring hand from the operations of the universe. Mathematics, Galileo had declared, was the truest knowledge of God. Newton, though no less than Galileo a deeply devout believer in a divine creator, had gone a step further and framed no metaphysical hypothesis of a final cause, unwittingly reducing the universe to a self-operating system of parts in a wonderfully assembled machine. A century later Laplace would take the mechanistic logic to its end, declaring himself to have no need of that hypothesis when Napoleon asked him where God fit into his cosmology. God, whose name and verifying scripture liberally graced every Muslim author's scientific and philosophical text, was redundant in the modern version. As Tahtawi came face to face in Paris with the implications of this new science that posed a moving earth and a mechanistic universe working in accordance with mathematical law, he had second thoughts. In Paris, he appeared to have accepted the implications of science being a set of unbreakable laws. But later, when back in Cairo, he wavered. Sentences he wrote in Paris on an earth in motion and a universe structured by mechanical forces that acted in accordance with absolute laws bonded in causality and mathematics were crossed out in Cairo. Where was God in such a world? Were God's law and natural law consonant? Then where did God's miracles come in? Science made God irrelevant.

As materially beneficent and mathematically convincing as modern science was taken to be by Tahtawi, he could not openly embrace it. His problem with science was no less a dilemma than the one he faced as an employee of a state that was continually diminishing the social importance of the religious studies he so valued. Tahtawi's was a dilemma of double complexity: an Azhar Shaykh advocating a system of secular education that replaced religious study by a modern science whose metaphysical foundations his religion could not accept. He advocated that science be studied while rejecting its principles. It was a quandary: adopting a philosophically eviscerated science that became meaningless as a set of principles integrating a system of nature? He recapitulates the tradition affirmed by al-Ghazali of there being no harm in the study of mathematics, logic, or any of the natural sciences so long as no religious inferences are drawn from them and as long as their underlying principles, that is, the metaphysical foundations giving unity to a universal system of nature and existence, are rejected. He affirms the other side of the coin, that a society which on the basis of religion rejects the mathematical sciences and workings of logical reasoning will lose its best minds and end up a society of ignorant believers. The difference in objectives between al-Ghazali and Tahtawi for holding to this, however, was as wide as the seven centuries separating them. While al-Ghazali was writing emphatically for a widened Sunni embrace of an indigenous mysticism that was already an accomplished fact in that Sufism was at the time firmly placed in the religious mainstream of Muslim society, Tahtawi spoke equivocally for a widened embrace of a natural science that was extraneous and far beyond the mainstream. What united them across the divide of time was their fear of the acid of critical reason spilling into scripture and Shari'a, thereby loosening the moral glue that bound as one the individuals in a society as pages in a book. Religion was that glue. Nothing should be tolerated that might dissolve it.

What mattered for Tahtawi were the benefits of science, not science. The technology of science held the promise of the material regeneration of society, and this required universal education of both sexes. With material regeneration would come spiritual regeneration, as long as religious study accompanied scientific study. For these beneficial ends, the material fruits of science could be legitimately taken, but not their philosophical seeds. His equivocal position made consistency impossible. The mental cleavage would take many forms in Muslim society during and after the 19th century.

It did seem for a moment that while he was in Paris even the seeds were swallowed. In the book he wrote on his experiences and impressions of Paris, everything is positive, or at least neutral. The institutions he saw there appealed to him; his admiration of the libraries, museums, observatories, and scientific institutions comes through clearly.[10] He originally wrote the book as a graduating composition for his French tutors, a summing up of his education and reflections after five years in Paris, and so of course the tone would be appreciative; he was a guest writing for his hosts. But there is no reason to suspect he was not sincere: the same positive tone remained when he edited the book for publication in Cairo to be read by Egyptians. In Paris, he tells us, he read Voltaire, Racine, Condillac, Condorcet, Rousseau and Montesquieu – whom he calls the ibn Khaldun of the French, reversing the joke of his Orientalist French tutors who referred to ibn Khaldun as the Montesquieu of the Arabs. Tahtawi's study of ibn Khaldun's *Muqaddima* under his tutors gave him a Muslim philosophy of the interrelationships of law, justice, social cohesiveness and economic affluence that complemented his understanding of the Enlightenment's vision of science in its social context.[11]

Having imbibed the cocktail of ibn Khaldun's social justice and Enlightenment progress, he could not help but get political. Science, he writes in Cairo some years after the publication of his Paris book, does not exist in itself but is the result of a set of ideas realized only in certain social and political conditions and is possible only in the condition of freedom. Science and freedom are causally related to progress and material well-being. The production of science involves a process that, once underway, creates a self-sustaining system whereby freedom, science, and social regeneration feed into one another.[12] Underlying the scientific discoveries that impel social progress is the principle of freedom, and underlying the condition of freedom is the inquisitive mind liberated by literacy and education, education through understanding not memorization. For this, one must have access to books. The presence of books was an expression of the general public's literacy and intelligence, twin necessities for freedom and science.

Of the many features of Parisian life that impressed him was the presence of books, books everywhere, in every home, in every Frenchman's hand. Even women read. This amazed Tahtawi in Paris as much as it amazed al-Jabarti in Cairo to see common soldiers reading in the library of the Institute. Clarity of thought and writing were also essential for science. This was another feature of French culture that impressed him. French writing was clear, simple, precise and direct; French books needed no gloss or elucidating commentary (*sharh*). Arabic books, on the other hand, were laden with hidden meanings and obscure words. It was a highly regarded literary device in Arabic to keep meaning veiled in symbol and metaphor so that different levels of interpretation were possible; in Arabic, the florid eloquence of words and their patterned sounds and rhythms were more prized than brevity and clarity. On the other hand, Tahtawi was amazed that French had so few synonyms, in which regard Arabic was a much richer language.[13]

To Tahtawi, coming from a society where only a small fraction of the people could read or were educated, literacy and education looked to be widespread in France. Common people went to school, girls went to school. Men, women, young and old, everybody read daily journals that informed them of events that affected their lives as a social body.[14]

It was literacy, he concludes, that enabled French society to advance and the state to become powerful.[15] Literacy and the knowledge of science: such were the products of education, and such would it take to make Muslim society strong and prosperous like the West. But this could not be without freedom. To be meaningful, science had to be cultivated in a free environment; otherwise, the exercise would be a hollow imitation, a degraded mechanical mimicry as culturally perverse and deadening as authoritative *taqlid*. To emphasize this, Tahtawi produced in a later book a verse of poetry: "All things having a goal strive to attain it. Freedom makes its goal the attainment of the highest."[16] Freedom is fundamental. There can be no science without it and no civilization without science. Freedom implies justice; both are aspects of the same thing, in the same way that science and civilization are aspects of the same thing, one unable to exist without the other.[17]

Political freedom was more important to Tahtawi than the theological freedom that would accept science in its entirety, taking nature as acting autonomously within the Creator's cognizance. In none of his writings, except for the instance referred to in his Paris book, did Tahtawi pay much attention to the theological obstacles in the way of science. Intellectual renewal took precedence over theological consistency. In the process of renewal that Egypt has now entered, the shaykhs of religion have a dual responsibility: assist the state in universalizing science education and stand as guardians to preserve the holy law, the Shari'a, that guarantor of social justice without which there can be no strength in a state or a people. An enlightened guardianship of religious scholars acts as a ministry of justice and education in the government of an enlightened ruler.[18] Tahtawi carefully certifies Muhammad Ali's enlightenment credentials by comparing him to earlier Egyptian autocrats who brought the glories of civilization to Egypt, rendering them enlightened.

In his works that followed the publication of the book on Paris, he appeals to Egyptians to accomplish what the French have accomplished. To do so would not be blind imitation of the French. Whatever wisdom or knowledge a people lost in the past remains their property and it would be legitimate for them to repossess it. A lost heritage is still a legitimate heritage, even if reworked and incorporated into the civilization of an alien people with an alien religion. Old texts and recorded history of past accomplishments are the title deeds authorizing repossession. Ancient Pharaonic learning was Muslim Egypt's legitimate heritage. The science of Ptolemaic Alexandria is essentially Egyptian, allowing Egypt to claim a fair share of Hellenistic and Abbasid science and, of course, to all Fatimid Egypt's natural philosophy.

Egyptian science went back to the earliest times; in fact, it could be claimed that Egypt originated science. The wisest of the Greeks visited Egypt: Pythagoras, Solon and Plato. It was those visitors who brought Egyptian medicine, mathematics and other sciences back to Greece with them. True, the Pharaoh Psamettichus had brought Greeks to Egypt to teach Egyptians, but as Plato had said, the Greeks had originally learned their science from the Egyptians.[19] By taking back what the Greeks had taken from them, ancient Egypt was merely reclaiming what was its own. The Pharaoh Ramses, he claims, had invited Greeks into Egypt; he appointed a Greek as admiral of his navy and had Egyptian youths learn Greek so they could become translators. As a result, Egypt entered a period of prosperity and civilization based on agriculture, trade, the rise of new cities and public works – exactly what Muhammad Ali was doing to make Egypt great again.[20]

Tahtawi's praise of Egypt's pre-Islamic greatness was as original in Muslim thinking as was his praise of political freedom, the former as disturbing to many Azharites as the latter should have been to Muhammad Ali. Tahtawi had no intention of disturbing the ruler, nor did he, or he would have been exiled, as were other shaykhs, and as he himself was when Muhammad Ali's successor came to power. Muhammad Ali could not have read his chief translator's work closely enough. On the surface, the autocrat is praised, but beneath it, autocracy is condemned.

Tahtawi's historical examples of ancient Egypt's power, glory and prosperity convey a justification of Muhammad Ali's reforms. Ptolemaic Egypt was a period in which Greeks and Egyptians worked together to make Egypt the brightest light in the world of medicine, mathematics, astronomy and the mechanical sciences. Greeks and Egyptians had built a great navy and maritime service, which put the country at the center of international trade, producing great wealth in the same way that the French and Egyptians were now doing in collaboration.[21] In response to those Muslims who were complaining that Christian foreigners had no right to educate believers, Tahtawi points to the caliph Ma'mun's House of Wisdom that was staffed and directed by Christians for a century, and to the part played by such Christians and pagans as Hunayn ibn Ishaq, his son Ishaq, Qusta ibn Luqa, and Thabit ibn Qurra, all of whom collectively brought into Arabic the mathematics, astronomy, optics, chemistry, medicine and philosophy of the Greeks.

The historical example that should inspire al-Azhar to take up modern science was to be found in its very origin as an institution of advanced Fatimid studies, where religion, astronomy, mathematics, natural philosophy, physics, and medicine flourished within an integrated Islamic cosmology.[22] Al-Azhar had no choice but to become the vanguard of social regeneration if Egypt was to regain the position that it and other countries of Islam lost when they fell under the heavy rule of Mamluks and other Turks.[23] The shaykhs should be the first to know that the science of the West is no more than an advanced version of that same body of Islamic natural study that was transmitted to the West several centuries earlier.

The most acute issue for Tahtawi was what was to be the place of religion in this new state Muhammad Ali was bringing into existence with all of his reforms. The pace of secular education threatened to reduce al-Azhar and its scholars to irrelevant vestiges of the past. It was an ironic situation in that many of the leading shaykhs, himself included, were contributing to their own institutional irrelevance by serving at all levels in the ruler's new schools. Conjoined to this was the poverty of Egyptian society, which Tahtawi could measure in a new light after his years in Paris. Science and technology could redress the problem of poverty, but if the secular state was left to monopolize science and technical education, where would this leave al-Azhar as an educational institution or the charitable institutions of Islam that traditionally gave succor to the poor? Without a vital al-Azhar, Islam would wither and society would disintegrate in a confusion of opposing factions fighting like packs of wild dogs.

Religion had to contribute toward social progress in cooperation with the state to produce a reformed society framed in an Islamic Enlightenment. This meant al-Azhar must forge a central place in education. To gain that place science had to become an important part of the curriculum. Otherwise, science and technology would be taught in the spiritual vacuum of state schools. The state would thus be shaped, and its citizens educated, by people torn from their religious moorings. Muslim society would be deprived of its best minds and moral exemplars, as was happening every day that the ulema and al-Azhar remained outside the sphere of change. Shaykh Tahtawi saw it happening before his eyes. Indeed, by the time he returned from Paris, Muhammad Ali's program of modernization had gone far in relegating al-Azhar to a marginal role in a social order dominated by a new class of technocrats composed of military officers trained in European methods of warfare and weaponry, and just beneath them a legion of Westernized specialists – doctors, industrial and agricultural engineers, administrators, shipbuilders, factory managers, metallurgists, typesetters and translators – and in their wake an indigenous corps of young specialists being trained in the ruler's militarized technical training programs in the preparatory (*tajhiziyya*) and specialized (*khususiyya*) schools. The Azhar shaykhs and students who received training in the specialized schools and in Europe were being alienated organizationally and spiritually from al-Azhar and religion. Muhammad Ali was having the best students from al-Azhar reeducated and remade to serve his state. Having torn away the land and waqf that supported the ulema, he was now intellectually impoverishing al-Azhar. How to reverse it? Shaykh Tahtawi lamented that the ruler had not forced reform on al-Azhar so that the power of religious acceptance could have helped foster a scientific culture by enlisting the shaykhs and their students to learn and teach science. But he could have had no illusions that the intellectually timid shaykhs would add science and the mechanical arts to theology and jurisprudence in order to make al-Azhar the mother institution of reformist education.

He claimed to be disappointed with al-Azhar's disinclination to embrace the new science, but he could not seriously have thought the venerable institute of advanced religious study had any intention, or capability, of transforming itself into a polytechnical university by expanding its curriculum to accommodate imports from the West at the call of a nonconformist whom the Azharites generally disliked because of his Paris book and the favor that the ruler showed him. It is reported that the shaykhs were so jealous of Tahtawi's relatively high position in Muhammad Ali's service that they spread rumors of his having drunk wine, eaten pork, and cavorted with French women while he was in Paris.[24] The leading shaykhs, far from young men, were not apt to encourage the study of foreign sciences or energetically partake in reforms that were pushing them to the margins. Only the state could have forced such a change in al-Azhar. Toward the end of his life, Tahtawi blamed the shaykhs, but above all Muhammad Ali, who was by then dead, for al-Azhar's failure to modernize. He blamed the shaykhs for their conservatism and Muhammad Ali for his having failed to force them and al-Azhar to change, as he had forced everything else in Egypt to change.

Muhammad Ali's successors were even less inclined to impose change on al-Azhar. The most liberal of them who ruled before the British seizure of the country, Ismail (1863–1879), as Tahtawi would lament, "has been unable to spread the lights of these diverse fields of knowledge in al-Azhar and to attract its scholars to perfect their minds with the natural sciences that are so undeniably useful to the homeland."[25] The split society of secular and traditional at war with itself that Tahtawi had feared for Muslims in the 1830s and 1840s had by the late 1860s become reality. Having worked on both sides of it, he was an intimate witness of its evolution. His writings mirror the struggle within himself as a leading figure and victim of those forces that drove Egypt to its fate of existing in two separate and contradictory worlds.

Indeed, al-Azhar was the one institution that Muhammad Ali and his immediate successors left untouched. Rather than reform and use it, he chose to bypass it and build new schools. He used individual shaykhs in his reforms but not the shaykhs as a collective body within al-Azhar. By leaving al-Azhar as he found it, Muhammad Ali isolated it from the currents of change. The result was that Egypt had two rival educations, one totally alien to the other. Muhammad Ali's state schools produced a technically educated class that was set apart from traditional religious education. In terms of numbers, the state-educated students were only about half of those who went to the traditional religious schools. In 1836, for example, a total of 10,715 students were enrolled in all the primary, secondary and specialized schools, while double that number attended the religious schools, but it was the people from the state schools who were running the government and shaping a dominant secularized society in which little place was left for al-Azhar graduates, the mentors of the majority. The intellectual bifurcation was fraught with disaster if left unbridged, for the two kinds of education could not have been more different in method, content and world outlook.[26]

Tahtawi's fears were prescient. The two minds and worlds of Egypt that emerged from the educational divide that horrified him in the 1850s diverged further and further over the succeeding decades and generations, each world living in itself with its own set of values, dressing in its own style, guided by its own principles of what is legitimate and what is permissible, one providing an audience for secular Westernized reform, pluralism and democracy, the other for political Islam, the Shari'a and Islamic justice.

As a shaykh of al-Azhar, Tahtawi had an uneasy conscience serving an absolutist who ruled with secular tyranny.[27] It was not just the religious problem with secularization. In Paris, he had witnessed the free exchange of information and the freedom of expression in journals, newspapers, and scientific societies. In the political parties, he witnessed popular participation in government, and in the uprising of 1830 he witnessed its corollary, constitutional limits to authority. Returning to Egypt, he found himself committed to serving the absolutist ruler who had sent him. Reflections of the inner conflict between his vision of a liberalized Egypt and the actual condition of the country under the man he was serving, and whose actions he was at the same time justifying, can be seen throughout his writings. This is particularly true in the works that followed the publication of his Paris book. In the later ones, he dares to express, though ever so cautiously, from subtly veiled criticism to gently coaxing counsel, that the ruler be guided by a more liberal policy of enlightened despotism. Then, in almost the same breath, he justifies the ruler's reforms and all he is doing.

Spokesman and velvet critic in turn, but always the visionary looking back to the early days of pristine Islam, Tahtawi sees Egypt remade in the image of an Islamic Enlightenment, wherein religion forms a legal and constitutional framework limiting and giving moral guidance to the power of a benevolent despot, with science acting as the intellectual driving force that creates cultural renewal. It is the moral, just, and enlightened ideal society that, as Muslim reformers have claimed ever since, existed in early Islam. The classical Abbasid caliphate of the 9th century, with its science and philosophy and Shari'a, approaches Tahtawi's romanticized image of the ideal. For Muslim society to be truly Muslim, the benevolent caliph's absolutist authority would have to be limited by the Shari'a and informed by wise ministers.[28]

Within the New Egypt of a law-abiding benevolent despot taking counsel from wise ministers, science and science education would take their place alongside religion and religious education, for no state could be strong and healthy without both. Religion informs science as science does religion. Informed by moral guidance and the principles of religion, science is the firmest foundation of a state's health, wealth, and strength because it awakens a people's social consciousness and their love of country, *hubb al-watan*.[29] This in turn energizes the people to strive for their own and their country's improvement through dedication to the arts, crafts, and applied sciences, producing progressive civilization.[30] Because science is such a vast and expanding field, science education and specialized study must be prime concerns of the state, since only the state has the resources to educate its people in science and technology. The state must take the lead in this, enlisting the efforts of the ulema in providing teachers and students for the state schools.[31] The state must plan, organize, and build a modern educational system from primary to advanced levels and must create the industry and economic institutions to sustain the system. It falls to the state to initiate economic revival through international commerce, agriculture, and industry. Of these, Tahtawi argues, the first is most important because overseas commerce opens a country to new ideas that are the building blocks of science and progress. Egypt is a special case, thrice blessed: by the Nile and its sunny climate for abundant agriculture; by its long historical tradition reaching back to the pharaohs when Egypt was rich in science and the arts; and by being located at the crossroads of three continents and two seas, at the very hub of world trade.[32] A good beginning to Egypt's maritime trade has already been made by Muhammad Ali's shipbuilding projects and schools of naval science. Tahtawi compares Muhammad Ali to Peter the Great, praising the Egyptian ruler for building his own ships and, ignoring here the ones he acquired from France, not buying them ready-made from Europe.[33]

The greatest obstacles that Tahtawi sees to Egypt's achieving science and industry are the country's aristocratic intellectual attitudes and its deeply rooted patterns of traditional life. Educated people tend to be elitists in regard to applied science and the technical crafts. They put these beneath pure thought and knowledge.[34] The second obstacle, tradition, has been sanctified by religion and so has become identified with religion. He assures his Shari'a-oriented readers that change can be made within strict Islamic limits. The legitimacy of change depends on what change itself is measured against and how it is defined. Measured against the intellectual history of Islam, change in terms of the adoption of science turns out not to be change at all, since Islam was the light of the world for a millennium and science was one of its beacons. Now it is the West carrying the light, and it is from the West that the torch of Islam must take fire if Muslims are to regain their leading position, just as the torch of the West took fire from Islamic knowledge to become the torch-bearer that it presently is. Taking is not imitating. It is claiming one's own, as the West would only be returning what it had taken from Islam centuries earlier. Science is not a European creation. Nor is it an Islamic creation. Science is a gift going back to the beginning of civilization in Egypt and Mesopotamia.

Tahtawi is the first modern Arab Muslim writer to sweep away the traditional wall separating the *Jahiliyya*, the Age of Ignorance, that existed before the coming of Muhammad, from the Islamic period. He posits an intellectual continuum from people to people, age to age, civilization to civilization. In pre-Islamic times, Syrians, Egyptians, and Greeks each built upon the accomplishments of its predecessor. The Syrians were the inventors of reckoning, *'ilm al-hisab*, and were the first to use numerical tables and account books. (Here Tahtawi is referring to the trade registers of the Phoenicians.) Others who came after them built on their achievements.[35]

Tahtawi recognized the cosmological and metaphysical contradictions between Islam and Western thought and discussed them at different stages of his life in brief passages in three books, but he was too divided, too "deeply rooted in his inherited convictions," to produce a reasoned harmony. Preferring to stand aloof of the Ash'arite metaphysics, he accepts the pursuit of science for what it could produce. Science was not a valid field of metaphysical speculation. Freedom of thought was fine in the worldly political sphere, but not when it came to understanding God's power or mind. There were some aspects of what the West has done to science that should be rejected, for they contradicted the foundations of Islam. Natural law did not pervade the heavens as justice did the Shari'a.

Tahtawi did not go beyond that in pointing out or resolving contradictions between Islamic theology and Western cosmology. He had no interest in reformu-lating theology to agree with the perceived logic of the universe. He did not speak for theology, nor did he bend it to the specifications of modern science. He was an apologist, a reformer, a pragmatist, a man of action helping to religiously inform the modernizing state through education because he believed that the comforting familiarity of religion would temper the harsh edges of modernity. When he pon-dered heliocentricity, Cartesian rationalism, and the mathematical relationships expressing a system of natural law that governed the physical structure of the universe, his enthusiasm for Western science chilled. The very heart of science was suspect. Belief in natural law was an arrogance; the claims of science were an affront to God's power. The metaphysical foundations along with the mathematical positivism of demonstrable laws were inimical to the conservative spirit that existed at all levels of religious consciousness, from the lowest level of the ulema to the highest level of the Azhar shaykhs. He was equally repelled by the arrogance of those cocksure French scientists who would straightjacket God's almighty power with their universal system of self-contained causality. They may have been the best scientists in all of Europe, as they claimed they were, but their insufferable arrogance misled them to believe that science and scientists were more important than religion and prophets:

> They deny the breaking of nature's custom [miracles], they believe that the course of nature cannot at all be negated, and that religions arose to guide man in doing good and avoiding its contrary, and that civilization and progress in morals have taken the place of religion, political affairs thus being elevated to a kind of shari'a. One of their repulsive beliefs is that the intellects of their philosophers and scientists are greater than that of the prophets. They have many odious beliefs, such as the denial among some of them of divine predetermination.[36]

To Tahtawi, this was loathsome; though he regarded it in part as a problem of personal attitude and perception and not wholly a problem between science and religion. Properly married, the two could live together in harmony. Those who believed that scientists were above the prophets were typical products of the false and dangerous pretensions that the arrogance of scientists could lead to when religion was cast off. The theory of heliocentricity presented another problem. It was a completely different sort of problem but one just as threatening to belief as natural law. Tahtawi was sure it was a problem, but he was unable to decide whether the theory was religiously compatible or not. So it may not have been a problem. His position shifted from book to book. He studied modern astronomy in Paris and appeared at one point to accept the sun-centered universe when he wrote that "there is no harm in a Muslim's believing that the earth moves or is stationary:" *La yadurru i'tiqad taharrukihi wa sukunihi.* Both systems are equally valid. It had only been three years before Tahtawi's arrival in Paris that the Church had issued a public declaration that the sun was indeed at the center of the universe and the earth and other planets revolved around it, reversing the Church's stand that had condemned Galileo. Now it was for Islam to come to terms with it. But the shaykh was not sure. The science seemed right, but what of scripture? He decided it was best to leave it up in the air and make no decision. What did astronomy have to do with religion?

Incredible as it may seem, the earth's sphericity was at the time also a question of debate, among Ottomans as well as educated shaykhs of Egypt and North Africa. Tahtawi described a lively discussion he had with two shaykhs from Tunis on whether the earth was flat or round and whether or not it moved. The exchange occurred during the period of quarantine he was required to spend in Marseilles before proceeding to Paris. Tahtawi told the two shaykhs the earth was round and gave proofs.[37] He knew the earth was a sphere without having to go to Paris to find out. He felt no threat in holding to sphericity. No Greek, Syriac or Muslim astronomer or philosopher, nor anyone even remotely familiar with the subject had ever believed otherwise. A motionless spherical earth at the center of a spherical world was a hallowed Aristotelian–Ptolemaic fixture in Islamic cosmology. But concerning the discussion of earthly motion that he had with the Tunisians, of this Tahtawi related nothing, possibly because there was no disagreement. Being that he and his interlocutors were firm believers in the Ptolemaic system, the idea of a moving earth was dismissed as so much European folly. But then he reintroduces the subject later in the book, in a section on French mathematics, *'ilm al-hisab*, where he describes the various categories of mathematical reckoning, one of which was, as the French termed it, "mathematical geography," *'ilm al-hay'at al-riyadiyya*, which, he explains, is mathematical astronomy and insists that it "must be mentioned." There is nothing forcing him to mention it except his compulsion to give it public airing. He perhaps goes into it to measure the temperature of rejection or acceptance, hoping to arouse public discussion of the sensitive subject:

> The Franks divide the heavenly stars into fixed and moving bodies, and bodies moving around moving bodies, and comets. They consider the sun as fixed and the earth as moving and the moon as moving around it. This system they call the system of the Austrian [*sic*] Copernicus [*madhhab al-Kabarniq al-Nimsawi*]. Some recent astronomers among them have discovered several planets that the ancients did not know of because they lacked the instruments that these Franks have. For this reason, the known planets among the Franks have reached 11, not counting the sun and the moon. From their nearness to the sun, the order is: Mercury, Venus, Earth, Mars, Jupiter, Saturn, Uranus,

plus the four stars of Jupiter. The new planets are difficult to observe because of their smallness and great distance.[38]

Tahtawi concludes this section by saying that the French hope to discover more planets in the future. In the manuscript he wrote for his French teachers in Paris there is an additional critical passage that he deleted when editing the manuscript for publication in Cairo. The deleted passage reads:

> A certain French scientist says that the earth's motion in revolving around the sun and rotating on its axis is not contradictory to what is described in the holy texts. That is because the holy texts mention these things as exhortations and the like in accordance with the way the sun and earth appear to the common people, not according to the niceties of natural philosophy. For example, it occurs in the holy text that God, may He be extolled, stopped the sun. But what is meant by the sun's having been stopped is that its setting was seen by eyewitnesses to have been delayed, and this would produce the illusion of the earth being stopped . . . And so he [the French scientist] renders the text allegorically.[39]

The passage essentially restates Galileo's argument for the allegorical interpretation of God's stopping the sun that he wrote in his famous letter to the Grand Duchess Christina. The reason for Tahtawi's striking it out in this book could have been that he learned of Galileo's troubles with the ecclesiastical authorities when he was studying science in Paris and feared that the argument he attributed to the French scientist might be taken as his own and used against him as a case for heresy by conservative Azharites, as the Church did Galileo. Even when praising French scientists for their highly precise astronomical equipment and proofs of the dual motion of the earth that he says would be most difficult to refute, he covers himself by impugning certain of their philosophical beliefs, hoping that he has left more than he has taken back:

> Whoever reflects on the state of the sciences and literary arts and craftsmanship will see that it is in the city of Paris that humanist studies [*al-ma'arif al-bashari-yya*] have reached their apogee. No other Europeans compare to the scholars of Paris, nor any of those from the past. Most of the sciences are well known among them; but they have some philosophical beliefs that are beyond the laws of reason in relation to their people, although they falsify them and strengthen them so that they appear true, as for example in astronomy, which they are certain of, and in which they are the most knowledgeable of all people because of their instruments and renowned inventiveness. It is well known that knowledge of the secrets of instrument-making is the strongest aid in the arts, which the French have, and although their sciences are filled with error opposing the holy books of revelation, they are based on proofs whose refutation would be most difficult.[40]

Taken together, the passages cited above highlight the issues around which Muslim religious scholars would debate the legitimacy of science for the following half century, until Darwinism took center stage. Tahtawi's view of the issues that religion has with modern science is fraught with equivocation. It is a view of denial, perhaps a half-hearted one, of natural causality, universal law and a moving earth, with the last one coming off as possibly harmless to the believer. Four years after the publication of his Paris book, Tahtawi published another one, *Kanz al-mukhtar fi kashf al-aradi wa'l bihar* (*The Choice Treasure in Quest of Lands and Seas*). A geopolitical handbook of sorts in question and answer form, the book appears to be a reworking of the treatise on astronomy that he mentioned he wrote as a student in Paris and which introduced him to the world systems of Descartes and Newton.[41] In this little book he gives the names of countries, their capital cities, names of rulers, forms of government, and the names of the seas and oceans. He also offers bits of current and recent history, including news of the New World, the 13 colonies and the American Revolution. The conclusion of the book is entitled, "On Earthly Sphericity." In the 15 pages of this chapter (129–143), Tahtawi has come unequivocally to terms with a spherical earth that rotates on an axis inclined 23 ½ degrees to the plane of the ecliptic while orbiting the sun in a period of 365 ¼ days. The inclination and dual motions produce the four seasons, solar and lunar eclipses, phases of the moon, and the earth's bulge because of increased linear velocity at the equator as determined by the English philosopher–astronomer Isaac Newton – the first and only time Tahtawi mentions the name.

The book goes on to present a brief and simple description of geographical astronomy that was well-known to Muslim astronomers for more than a millennium but which is now seen as the result of a revolving earth and a stationary sun and firmament. All this is exposited in a simple question and answer dialectic explaining the underlying astronomy without any hedging, equivocation or pious circumlocution. The sphericity of the earth and its dual motion are accepted as certainly as the existence of the marvelous new world across the western ocean that is described, perhaps, for the first time in Arabic literature. Tahtawi was again testing the intellectual climate. The little book provoked no reaction. Tahtawi was not chastised for publicizing such outlandish ideas, nor was there any follow up commentary on such a revolutionary concept by other writers in favor of reform and science in the following decades. Instead of causing ripples, the book sank into oblivion without a trace, a sure sign that there were no other reformist writers between 1830 and 1860 with the intellectual stamina to keep the discourse alive and move it forward. This could explain why, 33 years after the book's appearance, during which time no Arabic commentary on Copernicus came to life, Tahtawi's tough-minded acceptance of modern astronomy was sucked up into the vacuous silence. In the twilight of life, when he was 68 with but five years left to him, he published his large book on the history of Egypt from Pharaonic times to the Islamic conquests, *Anwar tawfiq al-jalil*, in which he gives his last word on modern astronomy:

But when in the West came Copernicus who devised a new system and established proofs that the earth moved around a stationary sun, Europeans followed the system [*madhhab*], claiming that the sun's movement is only apparent, which mathematically is the same [as a geocentric system]. Copernicus's system is an ancient one going back to Pythagoras and his followers, as stated by Muhammad ibn Muhammad al-Qazwini [d. 1283] in his book *'Aja'ib al-Makhluqat wa Ghara'ib al-Mawjudat*. Among the ancients who followed Pythagoras were those who said that the earth is in constant circular motion, and that the motion of the stars is actually the earth's motion. Some said the sun is stationary and the earth is attracted to the sun and for this reason it does not tend to go off in any direction, the force of the parts being in balance. The force of attraction between sun and planet is like the attractive forces between a magnet and iron. This force maintains bodies in their place and in balance.[42]

This is followed by a description of Descartes's theory of vortices that emanate from the sun and whirl the planets around it periodically. "There are those who said that it [the sun] is stationary, rotating in place at the center, and it is this solar turning which causes the orbiting of a heavenly body around it centrifugally at a certain velocity." To clarify the picture, he likens the earth and planets to particles of dirt or stones in a round jar (presumably containing water) that, when turned rapidly, causes the suspended particles to whirl about the bottle's axis of rotation, the centripetal force concentrating the suspended particles toward the center. The narrative up to this point sounds like an endorsing explication of modern astronomy, but suddenly the author shifts into reverse:

Although Copernicus negated the geocentric Ptolemaic system and reverted back to heliocentric Pythagoreanism, which was accepted and practiced by the Europeans, there is no cause for despair, for the progress of intelligence over a long period of time will bring the *Frangis* back to the Ptolemaic system after an extended period equal to the period which has elapsed from the time of Ptolemy to the time of Copernicus. There is nothing strange in this, since the Ptolemaic system still continues to be current upon their tongues. Hence the *Frangis* say that the sphere of the sun moves such-and-such a distance in the heavens [sun rising, sun setting], viewing it as apparent motion. With respect to purely mathematical observation, both systems [Ptolemaic and Copernican] are identical. The difference is in causation. Both possess powerful signs of God's might, be He praised and exalted. God in every motion and stillness brings witness and in everything there is a sign revealing He is one. The only reason for mentioning the orbiting motion of the earth is not as an exposition of religious beliefs which are dependent on the textual clarity of exalted Qur'anic verses that state the sun goes its own course as was established by God, but to expose the account of Copernicus and Pythagoras and those astronomers and geographers who practice their systems. It is incumbent upon the Sunni community to believe in the moving sun and to follow the God-ordained system followed by our ancestors. We do not follow the way of allegorical interpretation, and we do not say that the moving sun is apparent, originating from the earth's motion according to the system of those who claim it to be so, such as Pythagoras.[43]

Ali Mubarak, who knew Tahtawi well and collaborated with him on a scientific and religious journal for secondary students in the early 1870s, believed that his five years in Paris had no influence whatsoever on his beliefs, character or habits.[44] Of those five years, Tahtawi was actually in Paris only a few hours a week, on Sundays for a closely escorted tour. The rest of the time he was in the Egyptian School in Paris, leading his student charges in prayer, studying under his tutors and breathing the Egyptian air that came with the School.

As early as his Paris days Tahtawi fell back on traditional arguments, asserting religious authority in limiting scientific knowledge to those whose religious belief was strong. Recalling al-Ghazali's caveat to Muslims who would delve into mathematics and logic, he warned that "the books of philosophy in their entirety are filled with innovation." Consequently, anyone wanting to immerse himself in anything in the French language that included philosophy should be well fortified by Quran and Sunna so not to be seduced and let his belief slacken. In other words, those of fragile faith and mind should leave science to minds steeped in religion and girded by unshakable faith, namely the leading shaykhs. This of course throws into question the freedom he asserted was required for science and civilization, hallmarks of the healthy society. The dilemma comes vividly alive in one of the poems Tahtawi wrote in Paris:

Is there anywhere a country the likes of Paris?
The suns of knowledge never set on it,
Nor does the night of unbelief have any morning.
Is not this and your truth amazing!

As his reversion to Ptolemaic astronomy shows, Tahtawi remained through his life "deeply rooted in his inherited convictions and to him what is clear is the contradiction between the two [science and religion], not their possible reconciliation."[45] A synthesis that wove the warp of modern science's metaphysical fundamentals to the woof of a reinterpreted Quran was a task too abstract for his pragmatic interests. Also, such a work would have reduced religion in the measure that science was ennobled by association with the divine. Apart from his book on Paris, the one other work that could have served as a stepping stone to an extended rethinking of Islam in the light of modern science, *al-Kanz al-mukhtar fi kashf al-aradi wa'l bihar*, was itself a simple practical handbook on political and natural geography and physical astronomy as it directly affected the earth. Unfortunately for the cause of a reinvigorated return of Muslims to science, the book stood so alone it might as well not have existed.

If the idea of a moving earth was such an endless source of equivocation for a leading reformer like Tahtawi, a bright young man who studied five years in Paris and became honored as Father of the Egyptian Renaissance (*Nahda*), it could be expected that the Renaissance was in for some hard times.

Notes

1 From 1832 to 1849 around 750 doctors graduated from the Medical School, a good fraction of whom went into other services: Panzac, "Medicine revolutionaire," p. 105. During the same period, the engineering schools graduated several times that number.

2 Ahamd 'Izzat 'abd al-Karim, *Tarikh Ta'lim fi'Asr Muhammd Ali*, Cairo, 1938, p. 25; Ali Mubarak, *al-Khitat al-Tawfiqiyya*, 20 vols., Bulaq Press, Cairo, 1880, vol. 4, p. 38.

3 Peter Gran, *The Roots of Islamic Capitalism*, University of Texas Press, Austin, 1979, p. 5.

4 Peter Gran, *The Roots of Islamic Capitalism*, pp. 197–208.

5 Ali Mubarak, *al-Khitat al-Tawfiqiyya*, 20 vols., Bulaq Press, Cairo, 1888, vol. 4, p. 38.

6 J.J. Heyworth-Dunne, *An Introduction to the History of Education in Modern Egypt*, Luzac and Co., London, 1968, pp. 154–155.

7 From the algebra questions on the mathematics exam he was given in 1828, the Egyptian School in Paris appears to have been equivalent to a modern liberal arts preparatory school. The curriculum, study hours and daily routine of the students were strictly regulated. Students were allowed out a few hours on Sundays for a group tour of the city, under escort. Except for that brief respite, they devoted their time to study. They were never permitted to go off by themselves. Tahtawi, *Takhlis al-Ibriz fi Talkhis al-Bariz*, Bulaq, Cairo, 1958, p. 319; Alain Silvera, "The First Egyptian Student Mission to France Under Muhammad Ali," in *Modern Egypt in Politics and Society*, edited by E. Khedourie and S. Haim, Cass Publishers, London, 1980, pp. 1–22.

8 Zaki Salih, *al-Bu'athat al-ilmiyya fi qarn al-tasi' 'ashar*, Cairo, 1959, pp. 9–20; Izzat, *Ta'rikh ta'lim*, p. 435.

9 Edme-François Jomard, "L'Ecole Egyptienne de Paris," *Nouveau Journal Asiatique II*, August 1828, pp. 96–114.

10 Tahtawi, *Takhlis*, pp. 172, 211, 302–303, 307.

11 Tahtawi, *Takhlis*, p. 124. For a detailed explication of Tahtawi's ideal of liberalism see chapter 3 of Israel Altman, "The Political Thought of Rifa'a al-Tahtawi, a 19th century Egyptian Reformer," unpublished PhD. Dissertation, University of California, 1976.

12 Tahtawi, *Manahij al-albab al-misriyya fi mabahij al-adab al-'asriyya*, Cairo, 1912, pp. 19–21; Tahtawi, *Murshid*, pp. 80, 125; see also Hijazi, *Usul al-fikr*, pp. 113–118.

13 Tahtawi, *Takhlis*, pp. 206–207.

14 Tahtawi, *Takhlis*, pp. 211, 328–329. One of the criticisms his French tutors made of the long essay Tahtawi wrote for them at the end of his studies in Paris was that the author's generalizations were too complimentary and off the mark, the result of his experience in France being limited to an upper sector of Parisian society.

15 Tahtawi, *Takhlis*, pp. 328–329.

16 Tahtawi, *Manahij*, p. 23.

17 Tahtawi, *Murshid*, pp. 80, 125; Hijazi, *Usul al-fikr*, pp. 117–118.

18 Tahtawi, *Takhlis*, p. 334.

19 Tahtawi, *Anwar tawfiq al-jalil*, Bulaq, Cairo, 1869, pp. 117–118.

20 Tahtawi, *Manahij*, pp. 15–19, 185–197.

21 Tahtawi, *Manahij*, pp. 201–205.

22 Tahtawi, *Manahij*, pp. 248–259.

23 Tahtawi, *Manahij*, pp. 205, 248–249. This idea of a dark age of Turkish rule, first broadcast in Egypt by Bonaparte, was popular with the French Orientalists Tahtawi studied under in Paris.

24 Heyworth-Dunne, *Education in Modern Egypt*, p. 297; Gran, *Islamic Roots of Capitalism*, p. 163. For more substantial accounts of this animosity see Husayn Fawzi al-Najjar, *Rifa 'a al-Tahtawi*, Dar al-Misriyya, Cairo, 1966, p. 93; and Ahmad Badawi, *Rifa 'a Rafi' al-Tahtawi*, Bayan al-'Arabi, Cairo, 1959, p. 140.

25 Tahtawi, *Manahij*, pp. 247–248.

26 Izzat, *Ta'rikh ta'lim*, p. 566.

27 Tahtawi's position between the Azhar shaykhs and the political rulers is analyzed in Israrel Altman's "The Political Thought of Rifa'ah Rafi' al-Tahtawi, A Nineteenth century Egyptian Reformer," unpublished doctoral Dissertation, University of California, 1976, pp. 33 ff.

28 The fullest account of Tahtawi's political thought in English is Israel Altman's previously cited unpublished doctoral dissertation, "The Political Thought of Rifa'ah Rafi' al-Tahtawi, A Nineteenth century Egyptian Reformer." A brief and more recent published account is Juan Cole's *Colonialism and Revolution in the Middle East*, Princeton University Press, Princeton, NJ, 1993, pp. 38–52. In both works Tahtawi is seen as having rejected the ultimate underlying assumptions of the Enlightenment's political philosophy. For example, to be in keeping with the absolutist tradition of Islamic government he inverts Montesquieu when arguing that a limited absolutism is the true form of Islamic government and that the legislative, judicial and executive powers of government emanate from the central power of the ruler.

29 Tahtawi, *Manahij*, p. 15.

30 Tahtawi, *Takhlis*, p. 124; *Manahij*, pp. 19–24, 34.

31 Mahmud Fahmi al-Hijazi, *Usul al-Fikr al- Arabi al- Hadith 'inda al-Tahtawi, al-Hay'a al-Misriyya al-'Amma l'il Kitabr*, Cairo, 1974, pp. 126–127.

32 Tahtawi, *Manahij*, pp. 7–18, 201–205.

33 Tahtawi, *Manahij*, p. 246.

34 Tahtawi, *Manahij*, pp. 10, 20–21.

35 Tahtawi, *Takhlis*, p. 339.

36 Tahtawi, *Takhlis*, p. 124.

37 Tahtawi, *Takhlis*, p. 112.

38 Tahtawi, *Takhlis*, p. 392, n. 29.

39 Tahtawi, *Takhlis*, p. 36 ff.

40 Tahtawi, *Takhlis*, p. 206.

41 Tahtawi, *Takhlis*, p. 319.

42 Tahtawi, *Anwar tawfiq al-jalil*, p. 225 ff.

43 Tahtawi, Anwar *tawfiq al-Jall fi-Akhbar Misr wa Tawthiq Bani Isma'il*, Bulaq Press, Cairo, 1285 Hijri (1868), p. 231. See also Youssef Choueiri, *Arab History and the Nation State*, Routledge, London, 1989, pp. 15–16.

44 Mubarak, *Khitat*, vol. 13, p. 54.

45 Albert Hourani, *Arabic Thought in the Liberal Age, 1798–1939*, Oxford University Press, London, 1962, p. 82.

20 Intensification of Ottoman reform under Sultan Mahmud II

In the race to modernize there were to be no Islamic winners. Japan would be a success story beyond Muslim emulation. The Muslim rulers who ran the race ran it with one leg tied to the ball and chain of Western interests, the other to impecunious funding. It was not religious traditionalism that sandbagged reform, as is so readily put forth to explain the unsteady performance of Muslim states; nor was it their conservative opposition that resisted having a layer or two of its wealth peeled away. No class willingly surrenders a part of its wealth except by coercion, and reformist governments did not force the issue as a means of financing modernization.

By the same token, Western countries most involved in the reform of Muslim countries were also not about to sacrifice any part of their interests that were so bitterly symbolized in the eyes of Muslims by the Capitulations and the right that European powers arrogated for themselves to protect the religious minorities. Those interests, compounded with Europe's desires to control or possess certain provinces, and its balance of power diplomacy, which acted to check those desires, made for inherent contradictions in any Muslim government's attempt to acquire modern industry and build a strong military in tutorial association with a Western power. Just as Muslim governments allowed themselves to be dependent on European nations in their quest to modernize, individual Muslims left it to government to assume initiative and leadership in any innovative undertaking. The best laid plans to have Muslims take over from European instructors, engineers and managers after a period of on-the-job training never quite worked out to the point of their reaching technological and educational independence.

Obliged as they were to take, learn from, and cooperate with the West, modernizing Muslim rulers had the odds heavily against them in their struggle to even the playing field in a contest invented by the West, whose nations were forever sharpening their carving knives for the funeral banquet. The Ottomans had seen Austria, Prussia and Russia devour the whole of Poland piece by piece in three partitions at the end of the 18th century and feared they were the dessert. And so they were, in slices large and small. Modernization became a race for survival that Muslims had little chance of winning, but one they could hardly afford not to run. By running it they hoped to have a chance of surviving, of meeting the West on its own terms, enough at least to ward off political and cultural extinction by making it too costly for Western powers to invade and rule. But in order to finance the schools and industries that would have been necessary to build a credible deterrent force, a radical appropriation of social wealth would have been required, a socio-economic revolution of sorts, and this was beyond what Ottoman sultans and their reformers dared do. Muhammad Ali was successful in expropriating the wealth of Egypt but unsuccessful in its careful application to education and industry.

To those who believed renewal and strength had to come from within instead of bringing in Europeans and borrowing their institutions, modernizing Muslim leaders looked at best like lackeys of the West, at worst as traitors. Sultan Selim and his westward-looking reformers, branded traitors and heretics, had paid the price in blood by the reaction from within. Muhammad Ali, who either coopted or destroyed internal opposition in the course of his reforms, and then monopolized the Egyptian and Syrian market for the manufactures of his own industries, survived until reduced by reaction from without. Another pitfall awaiting Muslim modernizers of the late 19th century was the seductive loans offered by European financial institutions. Muslim potentates came to believe they could borrow their way to modernization. Sultan Mahmud II was the last Ottoman reformer to remain free of European debt, one of many lessons he learned from Muhammad Ali. The rulers who followed them were courteously escorted to the European loaning institutions for a five-star cleaning.

Sultan Mahmud II followed Muhammad Ali's reforming example in several ways. First, all opposition had to be crushed. Mahmud knew well from his own harrowing experience that the Janissaries had to be wiped out before reforms could be put in place. Only his wits had saved him from certain death at the hands of the Janissaries during the Patrona Revolt that ended the life of his father and the New Order. Since he was the only surviving member of the Ottoman family who could contest the sultanate with the prince that the revolutionaries had put on the throne as Mustafa IV, Mahmud knew the purpose of the contingent of Janissaries that he saw from his chamber window heading at a quick march through the courtyard toward his apartment in the palace. Minutes before they stormed into his rooms he was out the window and over the rooftop, clambering up a wall, then over more roof tops and walls until finding refuge with surviving supporters of his father's reforms.

He remained in hiding a year, until Mustafa IV was assassinated by a secret coterie at court loyal to Mahmud and the memory of Selim. As the only surviving prince, Mahmud ascended to the throne without opposition. Like his father, he knew that if the empire was to survive, reform was inevitable, but how to avoid the horrific fate of the Tulip Period and New Order reforms? By stealth, cunning and secrecy. The young sultan proceeded cautiously, keeping his thoughts to himself and feigning docility and weakness of character year after year, as he awaited the right moment to strike. He studiously observed the career of his autonomous governor in Egypt, his slaughter of the Mamluk chieftains, his reduction of the wealth and authority of the ulema, his student missions to Europe, his foundations of a new system of education, his industrialization, his resuscitation and expansion of Selim's New Order reforms on a grand scale, his creation of a powerful military that enabled him to seize the Sudan and Arabia, and his triumph as uncontested master of his Egyptian Empire.

One eye on Egypt, the other on courtly politics, Mahmud dissembled for 18 years, slowly and inconspicuously gathering around him, one by one, a body of trusted men whom he sent to serve as ambassadors and secretaries in the embassies his father had established in the capitals of Europe. Mahmud wanted his men to be as familiar with the West as they were committed to modernization. When they returned from their tours of overseas service, Mahmud inconspicuously appointed them as ministers in his government. By proclaiming the reforming governor of Egypt a bona fide Muslim to whom God had given victory over the un-Islamic fanatic Wahhabi Arabs, he cast Muhammad Ali as the embodiment of the true spirit of Islam: a Muslim directed by God in the defense of religion. With God having manifested favor on the Egyptian reformer, and therefore also on his reforms, the sultan was able to convince the Shaykh al-Islam and leading ulema that the reforms he was about to introduce were purely Islamic since they originated with a good Muslim ruler. Since many of the high ulema were as aware of the empire's weakness as was the sultan, a *fatwa* was signed by the Shaykh al-Islam legitimizing Mahmud's intended reforms. Once armed with this, Mahmud issued his army new uniforms and weapons that were similar to those of the Egyptians and ordered that their drills and training be followed. To make his reforms look all the more genuinely Islamic, Mahmud named his army "The Triumphant Soldiers of Muhammad's People."

As Mahmud fully expected, the Janissaries, still unbowed by the humiliating defeat dealt them by the Greek rebels, refused to wear the uniforms or tolerate the drills, or even consider using the cowardly bayonet. Having carefully plotted his course in detail, Mahmud went back to the Shaykh al-Islam for another *fatwa*, this one declaring the Janissaries to be rebels against Islam since they refused to abide by the very reforms that the Shaykh al-Islam had earlier declared to be in accordance with the law of Islam. The trap had been set. Unable to avoid contradicting himself, the Shaykh al-Islam signed the *fatwa*, possibly having little idea of its momentous implications: lawful destruction of the Janissaries. Muhammad Ali had had no need of such legalizing delicacy, since the Mamluks lacked the powerful courtly, religious and social support that the Janissaries enjoyed.

While he had been maneuvering to deprive the Janissaries of ulema support, Mahmud was secretly moving a large number of artillery pieces and a mountain of shells to the palace storehouses in the darkness of night. When all had been prepared, the *fatwa* signed, the artillery in place and ready to be wheeled out and concentrated on the Hippodrome across from the palace where the Janissaries customarily assembled, Mahmud gave the early-morning order that the Janissary corps dress in the new European-style uniforms that had been issued to them. Predictably, they refused. In angry protest, they made their customary sign of revolt by tipping over their big soup kettles in front of their barracks. In anticipation of the events that were unfolding just as Mahmud had foreseen, the sultan read the Shaykh al-Islam's *fatwa* to an assembly of leading ulema and ministers who had been ordered to appear and to hear. Many of them Janissary supporters, they realized too late that they had just been duped into being the unwitting witnesses of the legal authorization of annihilation of their favored and legendary infantry corps. The sultan was commanded by religion to execute what he was about to do.

Minutes after the *fatwa* had been read out, the order was given and the merciless bombardment commenced. Known in Ottoman history as the "Auspicious Event," an unremitting torrent of artillery fire rained down on the Janissary encampment all morning and afternoon. When the smoke of the holocaust finally rose and dispersed, the Janissaries, for centuries the terror of Europe, and then of the sultans, were no more. Those of them who survived threw off their uniforms and melted into the populace, as had the surviving Mamluks in Egypt 17 years earlier. Mahmud and his loyal band of reformers then pressed the Shaykh al-Islam to issue a *fatwa* declaring that the Bektashis, a Sufi order closely affiliated with the Janissary Corps, held beliefs and practices contradictory to true Islam. The Bektashis, who had been effective in getting the lower-level ulema and religious students out in the streets to join the Janissaries during times of protest against the government, were then disbanded, their leaders arrested and exiled, leaving Mahmud and his men free to carry out the reforms that had been so long in the planning.

Among the chief reforms was the establishment of a formal system of imperial diplomacy that, going beyond what Selim had instituted in this regard, was designed to weave the foreign affairs of the empire into the fabric of European diplomacy. The Greek Revolt triggered two fundamental actions that impinged on the radical reform that was about to begin. One was the failure of the Janissaries to put it down, leading to the "Auspicious Event." The other was the establishment of an Ottoman bureau of diplomatic correspondence. Before the revolt, the Ottoman government had let the Greeks of Istanbul handle diplomatic correspondence. The logic was simple and direct: Greeks were Christian, Europeans were Christian, let the empire's Christians do the empire's work of communication with European Christians. This not only relieved Muslims of having to demean themselves by having contact with inferior unbelievers, but it spared them the effort of having to learn Europe's heathen languages. The Greeks who handled Ottoman diplomatic correspondence belonged to the well-to-do merchant community of the Fanar quarter of Istanbul, the Lighthouse section, hence the name they are known by, Phanariot. Because of their international trade and travels, which gave them knowledge of the ways and languages of Europe and their being recognized as capable and loyal subjects of the sultan, the Phanariots were put in charge of diplomatic correspondence and awarded the responsibility to mediate the Ottoman government's affairs with the Western nations. This convenience ended with the Greek revolt. The Greeks could no longer be trusted. The point was brutally made when at the outbreak of the revolt Mahmud had the Greek Patriarch in Istanbul publicly hanged as a traitor in collusion with the rebels and their Russian supporters.

This bloody act signaled the undoing of the millet system of communal semi-autonomy. The communal religious modus vivendi that the millet system had provided was also done in by the growing Muslim hostility to the Capitulations, and, even more, by the infection of nationalism that spread through the Christian communities in the empire's European provinces during the 19th century. The Greek revolt was the first violent expression of the fiery passions of nationalism brought east by General Bonaparte and his citizen armies with their revolutionary war cry of Patria!, breaking down the Ottoman *millet* system and its cushioning effect that heretofore had enabled the many national and religious communities of the empire to coexist more or less amicably. Throughout the 19th century, peoples in the empire's Balkan provinces, encouraged and supported by one European power or another, rose up in nationalist rebellion against "the tyrannical Turk dripping in blood," as the European press was prone to describe the violence in its simplest form, accompanied by caricatures of wild-eyed pashas, daggers in hand, cruel, dark, fanatical and bloodthirsty, seizing terrified Christian maidens. Against the power of the Western press, the Ottomans had no defense, and suffered a poor image in Western public opinion. This was not helpful to a people whose government was taking Europe as a model.

To fill the diplomatic vacuum left by the expulsion of the Phanariots from government service, Mahmud created the Bureau of Translation (*Tercume Odasi*) to train young Ottoman officials in European languages. From the Bureau came the men who served as secretaries and translators in the European embassies, some of whom rose to become ambassadors. Translation became the training ground of the great architects of 19th century Ottoman reform. Because of their knowledge of Europe and its languages, men whose careers originated as trainees in the Bureau became the sultan's avant garde, several becoming grand viziers. In an earlier period, the *Devshirme* had been the institutional springboard for a low-born slave to rise to the highest offices. After its demise and the destruction of the Janissaries following the Greek Revolt, the springboard was the *Tercume Odasi*. It was also the training ground for those liberal constitutionalists who would become the severest critics of the autocratic reformist grand viziers.

Supported by these new men, Mahmud composed a government that brought the provinces under central control, with the notable exceptions of Egypt, Syria, Greece and Algeria, the first two being under Muhammad Ali, Greece being lost to a European-supported nationalist revolt, and Algeria to the sultan's duplicitous diplomatic ally and supporter in reform, France. Territorial loss and reform progressed apace, but in those lands that remained under Ottoman control, Mahmud and his ministers made the central government's authority felt as it had never been before. Adopting Muhammad Ali's method of impoverishing the ulema to a condition of dependency, Mahmud's reformist government seized the *evkaf* (*awqaf* in Arabic) or charitable pious foundations that, traditionally non-taxable, afforded the ulema its economic autonomy. With this, the religious establishment was reduced to a department in the imperial bureaucracy. The sultan then quickly proceeded to enact reforms in a pattern similar to those laid down by his Egyptian rival, though both leaders were in fact following a broadened version of the path of organized innovative reform set by Selim's New Order. In 1827, less than a month after Egypt's medical school at abu Za'bal had opened, Mahmud founded an Ottoman medical school, the Galata Saray Medical School. This was followed by a language and translation school, a school of surgery and veterinary medicine, an artillery school, and a naval school.

European teachers, military officers and technical experts were hired and brought to Istanbul to train Ottomans in the modern methods of military science, construction, industry, agriculture, administration, economics and systems of education. Student missions were sent to Europe and, as in Egypt, military rank, rations and pay were given the students, who were treated much less severely than Muhammad Ali and his agents treated students in the Egyptian schools and missions. Mahmud's first student mission was organized a year after his destruction of the Janissaries. It was sent to Paris, the preferred destination of both Egyptian and Ottoman missions. Ottoman missions tended to be several times larger than their Egyptian counterparts. The mission of 1835, composed of graduates of the new medical school, had 135 students, almost half of all the students Muhammad Ali would send to Europe.

Mahmud paid special attention to the Galata Saray Imperial Medical School. Staffed with French physicians, it was, of all the professional schools founded during his reign, the one that received priority in funding. On the occasion of its opening, it was the sultan who presented the inaugural address. Though God and the ulema may have forgiven him for slaughtering the Janissaries, Mahmud knew that his permitting Ottoman students to be lectured to by Frenchmen in French was more than heaven could bear. Well aware of conservative sensitivity on this point, he related modern medicine to the Muslim scientific heritage in his inaugural address. The language of knowledge was incidental, he said. What was essential was the subject matter, not the language it was expressed in. In substance, modern medicine was nothing more than today's edition of Muslim medicine. Since the French and other Europeans had acquired their medical knowledge from Arabic sources, and then had added to them, the quickest route to modern medical knowledge was through French. Learning French to study medicine would be more expedient than mastering Arabic to study the Arabic texts before going on to the European discoveries based on those texts, or than bringing the combined knowledge of Arabic and European sources into Turkish. That would have taken too long, 16 or more years, and the need for good doctors could not wait. The sultan concluded that the smartest way to go was for students to learn French in order to study medicine and then create an Ottoman medical tradition by translating French texts to Turkish. A community of well-trained and experienced physicians writing their own Turkish manuals would allow medicine to be taught in Turkish.[1] It was a tortured argument, but one that had to be made, because creating a modern school with a European language as its mode of instruction was not easily swallowed by even a cooperative upper ulema, and not at all by the middle and lower ranks who spoke for triumphant Islam and the proud people of a once-powerful empire.

Mahmud's emotional and intellectual investment in the medical school was not entirely shared by the Ottoman populace of Istanbul. The school had an unsavory public reputation. Respectable families had no wish for their sons to be covered in filth, gore and blood, touching diseased people and cutting up dead bodies of both sexes. Engineering was preferred. It was clean. It was honorable. Engineering meant military science and officers in handsome uniforms.[2] Medicine was unmanly. A scalpel could not measure up to a sword. Medical students therefore had to be drawn from Istanbul's lowest class, sons of boatmen, porters and water carriers.

In 1838 the school was reorganized, expanded, renamed and given a revised academic program. This was under the directorship of a young and highly respected Viennese medical professor who found the time to learn Turkish and compose medical texts designed for Ottoman students. The Ottoman government provided enough funding to the medical and engineering schools so that they did not have to face the horrendous problems of lack of texts and supplies and fractured infrastructure that crippled Muhammad Ali's specialized schools.

Housed spaciously in the Galata Saray palace on the shore of the Bosphorus, the Imperial Medical School was adequately supplied and smoothly run by the predominantly French teaching and administrative staff. Problems were resolved amenably; sometimes, however, an independent show of strong support by the reformist government was needed to get over a hump. Because of religious objections over defiling the dead, anatomy students were at first obliged to practice on wax dummies. In this respect, Clot Bey's abu Za'bal school was more advanced. Finally, in 1841, when the Viennese director complained that students could not learn dissection with wax dummies, the government stood up to the religious opposition and, without seeking the traditional *fatwa*, issued an imperial decree permitting the use of cadavers. A compromise with the ulema was then worked out, according to which the only cadavers that could be used were those of Nubian slaves. Not long after that, medical students were reported to be performing autopsies with professional expertise and without supervision. When asked by a visiting Englishman of a conservative turn of mind if using cadavers was not against Islam, the student laughed in apparently genuine surprise and replied in perfect French, "Eh! Monsieur, ce n'est pas au Galata Serai qu'il faut venir chercher la religion!" The Englishman had not come to the medical school looking for religion, but his being answered in French was no less upsetting than witnessing cadavers being cut up.[3]

The serious attention Mahmud paid to the medical school may be evident in the English visitor's description of it, where it is reported that the latest medical advances made in Paris, London and Vienna were being practiced there. The school had a physics laboratory furnished with electrical machines, galvanic batteries, hydraulic presses "and nearly every machine and adjunct necessary to teach, or to experimentalize in the physical sciences. . . . There was also a tolerable chemical laboratory." In addition, the school had a small but well-kept botanical garden with a good collection of colored botanical engravings from Paris and Vienna, a museum of natural history "with a collection of geological specimens . . . a very sufficient medical library," whose books, the visiting Englishman added disapprovingly, were nearly all in French. Not only were the books nearly all in French, they were, to the English visitor's added horror, from the Revolutionary Period! "It was long since I had seen such a collection of downright materialism." He was even more nonplussed coming upon a student who was reading Baron d'Hollbach's *Systeme de la Nature* – "that manual of atheism!" which gave to nature and science the heavenly place of God.

Other students at Galata Saray were reading even worse. One of the students had translated the spicy passages in Voltaire's *Dictionnaire Philosophique* into Turkish and was now translating *Candide*! The English visitor's most shocking encounter with Ottoman atheism came when he found lying open on a table yet another copy of *Systeme de la Nature* that had been much used by the worn look of it, with many of the passages proving that God did not exist heavily marked and underlined. A Turkish physician who saw him holding the book enthusiastically exclaimed in French, "Great book! He's a great philosopher! He is always right!" The Englishman admits to being struck speechless.[4] It was not until 1866 that Turkish had replaced French as the medical school's language of instruction, to the satisfaction of the conservatives, ulema and English visitors.

With the Janissaries gone, the ulema did not give the reforming government a great deal of trouble. The elimination of the threat of reactionary revolt from below gave more leading members of the ulema courage to compromise with the sultan and his reformers. Compromise was not difficult. The Shaykh al-Islam and leading ulema tended to come from the upper strata of society and were as aware of the existential crisis as the political elite. Cooperation on that level became the norm. Also, with the passing of the Janissaries, the high ulema knew that if it did not compromise, the headstrong sultan and his viziers, in their zeal to reform, would only go ahead and do what they wanted, regardless of religion and tradition. On some issues the ulema held fast to tradition. Permitting the use of cadavers had taken much time and negotiation. Mahmud's introduction of the fez was another difficult issue. The same chief *mufti* who had signed the *fatwa* that was used as a legal cover to destroy the Janissaries adamantly refused Mahmud's appeal to allow Ottoman officials to wear the tasseled red headpiece imported from Fez in Morocco. In Mahmud's day, the fez was considered modern apparel and symbolized a break with the past for fast forward change. A century later, when Mustafa Kemal outlawed it for the European brimmed hat, what had once been a racy emblem of modern dress and life had come to stand for tradition and backwardness. Eventually the Shakyh al-Islam allowed it, but only for officials, not the ulema, who retained their voluminous white turbans. What one wore on one's head was no trivial matter, for it signaled what was inside it. It identified one's religion and status. For Mahmud, the fez was a sign of being on the official team for progress.

On the few occasions when the religious institution resisted Mahmud, the reformers simply steamrolled over it, but only as a last resort. Usually it was not necessary. A French observer in Istanbul in the early 19th century considered the ulema to be not at all as inimical to modern science as they had been generally described by Westerners. He found the Ottoman shaykhs able to speak about religious questions with an openness and tolerance not found among Jews and Christians in the orient or among the al-Azhar shaykhs in Cairo. Ottomans would have agreed. They thought their ulema superior in every way to the ulema of al-Azhar. The French observer estimated that the intellectual horizon of the Istanbul ulema and of Ottomans in general regarding scientific matters was where it was in France, England and Germany during the 12th and 13th centuries: in the heart of the scholastic period.[5] That would have made their horizons broad indeed if included in that period he meant the scholastic thinking of Albertus Magnus, Roger Bacon and Siger of Brabant.

How extensively diffused in society was the scientific knowledge that had come into the empire during the century and a half between the translations of Noel Durret's and Jansoon Bleau's works and the time Mahmud's schools were up and running? The scant evidence available would suggest scientific knowledge was not widely diffused, that even into the first half of the 19th century knowledge of the new science did not go far beyond the confines of the classrooms of the new schools. General von Moltke as a young Prussian officer in Istanbul on a mission to train Mahmud's new military spoke of educated Ottoman officers who accepted only out of politeness his statement that the earth was round, an observation later confirmed by an Ottoman writer commenting in the 1850s on the strong religious opposition that the idea of a moving earth produced when it was taught in the new secular schools.[6] "It was claimed that teaching the roundness of the earth or the heliocentric system was contrary to dogma, and it became necessary to seek evidence in dogma to support the truths which have been proven by observation and reason."[7]

The reforms, as they were continuously legislated from the late 1820s, had an unsettling effect on society. As the reforms grew more radical and displaced venerated tradition, the ulema splintered. In the middle were those who grudgingly accepted innovation. At diametrical opposites stood those who accepted it as beneficial and necessary and those who fiercely opposed it as alien, treacherous and heretical. Strong opposition to the innovations and to the reformers who introduced them came from the *softas*. These were undereducated madrasa students drawn from the poorer classes. The madrasa students, for the most part, were young, poor, unmarried and unhappy, coming as they did from the section of society that was the most unsettled by the changes imposed from above, and the most easily aroused to take to the streets, where, in belligerent defense of religion and tradition, they were joined and led by that lower splinter of ulema. The haunting fear of lost livelihood was a strong factor in sending people to the streets in anger. Religious opposition was motivated by fear that the reforms were going to abolish the traditional positions of the ulema as judges, teachers and imams, just as the Janissaries had feared that the military reforms of Selim and Mahmud were threatening their position.[8] Decades of reform during Mahmud's reign, and the succeeding Tanzimat period of even more radical reform, pushed the diminishing ranks of the *softas* and lower ulema to the social margins of populist protest in a gradual, less dramatic version of the Auspicious Event, but not without creating a deep chasm in Ottoman society.

In the half century between the destruction of the Janissaries and the promulgation of a liberal constitution in 1876, Ottoman urban society became progressively secularized and divided. Ottomans educated in the new government schools came to disregard the ulema. Their secular education and sense of belonging to a sacred brotherhood dedicated to state, reform, progress and strength bolstered a contempt for the ulema when some of the conservative scholars openly opposed French being taught and used in lectures, condemned cutting up cadavers, claimed science was irreligious and denied the earth was round and moved around the sun because the Quran stated otherwise. In Istanbul as in Cairo, a small but influential secular-ized class of government officers, technical specialists, writers, school instructors and physicians was drifting away from religion and traditions that defined their civilization, while being only partly familiar with the Western civilization from which government secular education derived. Bereft of firm values and deeply rooted beliefs, they became a lost and disoriented generation in search of identity. "Students," it was claimed by an Ottoman critic at the time, echoing the criticism made by the French in Cairo regarding Egyptian students who returned from study-ing in Europe, "went to Europe to study but acquired only vice and expensive tastes, returning home more syphilized than civilized."[9]

A telling difference between the parallel reform programs taking place in Istan-bul and Cairo was a result of the comprehensive guide and vision that informed Mahmud's projects, something Muhammad Ali's complex of schools and projects never had. Though the "Board of Useful Affairs" came late in Mahmud's reign, its general prescriptions were effective. Established in 1838 under the skilled director-ship of a high religious official, Mulla Mehmed Esad (d. 1847), who was the intel-lectual overseer of Mahmud's reforms, the Board of Useful Affairs served as the structural and philosophical template of reform. Except for some of the passages in Shaykh Tahtawi's early books, a stronger statement in support of social progress through science than the one in Mulla Mehmed Esad's report to the government on the Board's purpose and goals would be hard to find in the writings of a member of the ulema anywhere in the Muslim world before 1880. His report reads like a document from the Enlightenment with a strong dash of the Revolution: science is the foundation of human perfection, of the arts and trades. Just as astronomy is the mother of navigation, upon which commerce depends, mathematics is the par-ent of military organization, tactics and technology, upon which depends arma-ments, industry, trade and national wealth. Without science people cannot know the meaning of love for the state and fatherland (*vatan*).

Mulla Esad urged that new schools be built to teach science as a first step in revitalizing the country:

> In discussing every project for the recovery of agriculture, commerce and industry, the Board has found that nothing can be done without the acquisition of science and that the means of acquiring science and remedying education lie in giving a new order to the schools.[10]

The intellectual roots of Mehmed Esad's report go back to Ibrahim Muteferrika and the Tulip Period, whose spirit of embracing Western science had been driven underground on occasion, only to sprout anew when the climate above ground was favorable, similar in a way to the recurrent revival of the scientific spirit in Muslim societies between the 12th and 15th centuries. It should not be too astonishing that the author, a member of the religious establishment, unhesitatingly encouraged Muslims to claim science as their own. Members who belonged to Mulla Esad's high level of the ulema, with a few exceptions, either actively encouraged and contributed to reform and science education or in passive resentment held their peace, with a neutral middle layer remaining silently and often sullenly undecided on science, one way or the other. Those few exceptions in high positions who actively opposed it encouraged those who would otherwise have remained silent to join the chorus of opposition and, more importantly, excited to action the many lower-level ulema who were adamantly, even violently, opposed to secular reform and science education. These exceptions never had their way, but they were loud, bellicose and active enough to make sure reformers like Mulla Esad did not have a free run.

Rather than leading to a systematic program of action, his program was deliberately mired by a handful of ultra-conservative obstructionists in a bottomless swamp of debate until it sank out of sight. The upshot was that rather than beginning secular education and introducing students to science on the primary level as Mulla Mehmed Esad had advised, the government chose to avoid religious controversy and opened only secondary schools (*rushdiyye*), leaving the primary schools imbedded in the traditional education of Quran memorization and religion. This resulted in a condition of scientific unpreparedness similar to what existed in Muhammad Ali's primary and secondary school system, though the Ottoman *rushdiyye* schools seem to have prepared students for specialized training more than did the *tajhiziyya* in Egypt. The first *rushdiyye* school was not opened until 1840, at the very end of Mahmud's reign. The year before, two special secondary schools for promising boys were opened which graduated several students who would become prominent voices in both the reform movement and the secular literature that came with it after mid-century.[11] The going was slow. Fifteen years after crushing the Janissaries, the score for education reform was three secular secondary schools graduating a few dozen students.

Mulla Mehmed Esad had hoped for a smooth transition from one level of education to the next by having an eight-year long secondary school program that took students from the primary religious schools and immediately introduced them to the sciences. Science and religious study were to live side by side for a time, with science coming more and more to the fore as religious study receded. That did not bother the Mulla. Religion would always be there. It would come from places other than schools. Religion, the earliest teaching, would not easily depart.

However, too much opposition from too many directions kept this temporal cohabitation from becoming a legal marriage. Deprived of its violent militant arm of the Janissaries but still able to voice effective dissent, a few ulema at the top, joined by the secular conservatives at court and in government, were able to put up enough resistance to keep Sultan Mahmud from enforcing Mulla Esad's plan. Equally vociferous in their opposition were the religious heads of the various Christian millets, who appeared even more traditional than the conservative ulema in regard to religious education and fear of science. That decided it. The government wanted to avoid any and all religious squabbling, especially as it involved their Christian communities, since such issues never failed to be used as a pretext for Western intervention, costing the empire either another lost province or more loss of authority over the religious minorities. Consequently, the secular and scientific system of education Mehmed Esad envisioned and planned was ultimately aborted, with the Mulla's liberal program remaining a vision rather than a living reality.

The justified fears of Ottoman leaders reveal just how complex the negotiating of reform was in the face of so many sources of resistance: religious conservatism of the populist level ulema; regional nationalisms; preservation of Europe's capitulatory privileges; minority religious opposition; Europe's assumption of its right to protect the minorities; and most formidable, the annexationist ambitions of the great powers. Whatever area of Ottoman life that reform might impinge upon, there was sure to spring into action one or more fronts of opposition.

The consequence of Mulla Esad's aborted program was a divided system of education where understanding of a mathematical system of nature on one level, and of socio-legal change on another, clashed head-on with belief in God's untrammeled will and the changeless law of God and Quran in human affairs. As the new "transitionless" secular schools became established, a mental rift arose that necessitated a redefining of old words and creation of new ones. One of them was the word for science. *'Ilm* (plural *'ulum*), the common word for knowledge, had been used for natural science in classical Abbasid times and ever since. It was also the word for the religious sciences. Knowledge had been and continued to be divided into two primary categories, the religious (*al-'ulum al-diniyya*) and the rational (*al-'ulum al-'aqliyya*). To keep from stepping on the ulema's toes and to make sure that the spheres of knowledge were kept clearly separate, Ottoman reformers found a new word for the natural sciences: *fenun* (singular *fenn*, from *fann* in Arabic, plural *funun*.) Owing to the enhanced career opportunities offered by the government schools, *fenun* became more prestigious than *'ulum*. As students were being drawn away from religious studies, madrasas began closing down during the last half of the century.

A similar advance in secularization was going on in Cairo, though the prestigious al-Azhar stood its ground and maintained a high student enrollment. In both capitals, Cairo as much as Istanbul, the traditionalist religious side of the intellectual divide was being pushed more and more to defend its diminishing place in society and education, resulting in a more and more divided society.

In spite of the obstacles, Mahmud's reforms made for some notable changes in Ottoman society. His predecessor had been overthrown and assassinated for bringing in European officers and instituting reforms, and here was his son uprooting old institutions, sending students to Europe, having them learn French and study science and medicine from European teachers in a foreign language and even making a royal lecture about it at the new French-directed hospital. Under his orders, medical, science and engineering textbooks were translated to Turkish, bringing into the language a modern technical and scientific vocabulary, paralleling what was happening in Cairo with respect to Arabic. Within a decade of his having overthrown the Janissaries, Mahmud's institutional reforms had brought the Ottomans up to the level reached in Egypt under Muhammad Ali, whose reforms had started 15 years earlier.

Of the many common features of the innovations, institutions and individuals that make for the parallelism between the paths of the two modernizers, both advancing along the trail blazed by Sultan Selim III, one of the most striking is the similarity in the ideas and careers of their respective chief spokesmen, Mulla Mehmed Esad in Istanbul and Shaykh Rafi'i al-Tahtawi in Cairo. Both were members of the ulema, both were appointed chief editors of the new government journals, the *Waqa'i al-Misriyya* (Egyptian Events) founded by Muhammad Ali in 1828 and the *Takvim-i Vekayi* (Survey of Events) founded by Mahmud two years later. Both journals published French translations of each issue in order to keep the many Frenchmen and other Europeans in the employment of the Ottoman and Egyptian state abreast of new laws, political events, policies, regulations, appointments, new projects and changes in government. Intended to inform and build support for government policies, the journals presented the goals of the modernizers to the public in their best light. Mehmed Esad published his report of the Board of Useful Affairs in an 1839 issue of *Takvim*.

Shaykh Tahtawi's role as religious spokesman, or state propagandist for modernization, was in many ways practically identical to that of his Ottoman counterpart. Both expressed the most progressive ideas held by any of the members of the ulema, both were spokesmen for the religious legitimacy of the reforms being instituted by the rulers they served, both argued for science education and a corpus of study that allowed religion and science to coexist and both edited their respective government journals before becoming ministers of education. An important difference was that Tahtawi feared to express himself as openly and forcefully as Mehmed Esad, who went so far as having science take precedence over religious education in the later years of the secondary school curriculum. It was not only Tahtawi's peasant roots in the rural traditions of Egypt that made him waver and equivocate before the dicta of traditional theology but also the vulnerability of his position. He did not have in Muhammad Ali the full political support that Mulla Mehmed had in Mahmud. Nor was al-Azhar so incapacitated it could not intimidate any of its overly liberal members. If al-Azhar shaykhs were ever to gang up on Tahtawi for going a step too far in interpreting religion and early Muslim history to justify modernization, the ruler was not going to get himself embroiled in religious controversy. The skin of no government servant, that is, Muhammad Ali's servant, was worth the ruler's time or trouble. Shaykhs were to be used not protected.

A comparative analysis of the public positions taken by Shaykh Tahtawi and Mulla Esad writing on science in the 1830s and 1840s offers a good measure of the difference between al-Azhar's upper ulema and its counterpart in Istanbul with regard to the place religion gave to science, both as a field of knowledge and a set of disciplines to be studied in school by young Muslims. Al-Azhar comes out as the more conservatively oriented, but that did not stop Muhammad Ali from enforcing his reforms. There was no council for debating what was and was not acceptable. His word was law, whatever the Azharites thought. Mahmud, on the other hand, dared not ignore or roll over the members of the ulema who dissented against Mulla Esad's program, however more liberal the collective ulema leadership in Istanbul was compared to its al-Azhar equivalent. The sultan allowed dissent, listened to it, and often felt obliged to accommodate it, even when it meant cutting back the designs of his chief reformer.[12]

Sultan Mahmud followed Muhammad Ali's path of reducing opposition against innovative reform, but not as despotically. He expanded the control of central government and increased tax revenues, but he did not make himself sole land-owner, as did Muhammad Ali. The Egyptian ruler was able to. The geography and demography of the long fertile ribbon of the Nile Valley that was Egypt made it possible, once the Mamluks had been destroyed. Mahmud would have faced countless revolts. The Greek disaster was enough. Another difference between the reform initiatives of the two rulers was this, and it made all the difference: serious Egyptian reform ended with the premature death of Ibrahim Pasha, while Ottoman reform continued with even more intensity after Mahmud left the scene, with 50-member student missions being sent from Istanbul to Paris in 1848 and 1856 and the following year an Ottoman school in Paris being founded. Although it was closed in 1865, students continued to be sent to Paris, 93 of them, between 1864 and 1876.

Continuity made the difference between standing on wobbly legs and not stand-ing at all. A generation after Muhammad Ali and Ibrahim had passed from the scene, Egypt fell to the British. The Ottoman Empire would succumb 35 years later, but the men who built on the ideas and institutions that had been created by Selim and Mahmud, and who carried the reformist logic forward to produce Tanzi-mat reformers, Young Ottoman liberals, and the Committee of Union and Progress constitutionalists, passed onto another generation the mental, organizational and adrenal stamina to drive out the European armies occupying the Ottoman corpse at the end of World War I. This mercifully allowed the former Ottomans, now called Turks, to escape the humiliation and deforming psychoses of foreign occu-pation and to establish a secular republic from the rubble of Ottoman defeat and collapse. Egypt and the Arab provinces were not as fortunate. Foreign occupation would have a negative effect on how the Arabs viewed the West, its institutions, its ideals and even its science. As for the Turks of the Ottoman Empire, their two centuries of reform did not bring them anywhere near success in reaching the Western level they hoped for in their economic, cultural and scientific productivity, but their long, difficult and repeated forays in reform did at least enable them to escape the wretched fate of their former Arab provinces that fell into Western hands.

Hoja Ishak Efendi

Administered by a corps of dedicated reformers, translators and educators, Mahmud's reforms were, in the mid-to-late 1830s, on their way to surpassing the reforms taking place in Egypt. This would appear to be contradicted by the poor performance of Mahmud's reformed military. The Egyptian army, led by Ibrahim Pasha, decisively defeated the Ottomans on two occasions, the first bringing a Russian army to the banks of the Bosphoros, the second a British fleet to the Syrian coast, both Russian army and British fleet coming in defense of the faltering empire. The difference in military performance may have simply been Ibrahim's superior generalship. Mahmud's achievement was however not in his new army's performance, but in translation and education, notably in physics, mathematics and medicine. Though not enough to create an autonomous scientific culture, the Ottoman investment in education and translation showed itself in the superior facilities available to students and the texts they had at their disposal, compared to what students in Egypt had at theirs. Mention has been made of Mahmud's special interest in the Galata Saray Medical School. His investment in the Imperial School of Engineering was also enough to spare the faculty and students the impossible problems their counterparts in Egypt had to endure. In reference to student texts for the mathematical sciences, between 1826 and 1834, ten different works in 13 volumes were translated, mainly from French, and published in Istanbul. The most important of these was a four-volume encyclopedic text on the fundamentals of the mathematical sciences. This was the accomplishment of an Istanbul Ottoman, Ishak Efendi (d. 1834), whose father was reported to have been a Balkan Jew who converted to Islam.

Ishak Efendi's four volumes were translated and printed for student use within a few years of the engineering school's opening. Of the technical and scientific handbooks, treatises and text books translated during the decade following the destruction of the Janissaries, Ishak's *Compendium of the Mathematical Sciences* (*Majmu'-i 'Ulum-i Riyaziye*) stands out as a major educational contribution in bringing science to Ottoman students. The prestige and academic status he gained from its publication added the honorific '*hoja*' (teacher) to his name. Ishak Hoja was born around 1774, in the midst of that disastrous war with Russia and the humiliating peace treaty imposed by Catherine the Great that bode so ill for the Ottoman future. Whether his father converted or Ishak himself did is unclear, but his origins are reported as Jewish and Balkan. Little is known of his early life other than that he went to a madrasa, was a devout Muslim and performed the pilgrimage to Mecca when middle-aged. He was said to have known many languages, Turkish, Arabic, Russian, Greek, Latin, French and Italian, providing an excellent passport to a promising career in Ottoman service during the reigns of Selim and Mahmud. A student in Selim's Military Engineering School, he graduated just before the New Order met its horrific end. Under Mahmud, Ishak became a translator of =Western scientific and military texts in the Bureau of Translation, then advanced to instructor in the new Imperial Military Engineering School. It was an important position. Translators were many, Ottoman instructors in the prestigious engineering school were few. His receipt of the promotion came directly from the head of the school, a man whose importance in the transmission of science to the Ottomans in the 19th century became second only to Ishak's: Husayn Rifki Tamani (d. 1817).

Husayn Rifki, born in Taman in the Crimea, was one of the first Muslims in the modern period to be scientifically informed to a significant degree. He shared this pioneership with his friend and colleague, Yahya Naji Efendi. But it is Tamani who through his translations and his own original scientific compositions would become the more highly regarded of the two. Both started their careers as students in Baron de Tott's *Muhendishane* artillery school and then became instructors in Sultan Selim's Imperial Naval Engineering School. The careers of both men can be considered important long-term payoffs of those institutions and provide a thread of continuity to the interrupted reforms from the time of Baron de Tott in the 1770s to Sultan Mahmud II in the 1830s. Ishak Efendi, who most likely took Tamani's courses as a student in Selim's engineering school, is another thread in the strand of continuity.

Tamani became head instructor (*bash hoja*) in Sultan Selim's engineering school. This was in 1793. Four years later, having learned French from the French staff as a student and instructor, Tamani began translating scientific and technical books to be used as student texts; another four years after that, he was giving a course to his fourth-year engineering students that included conic sections, differential and integral calculus, mechanics, astronomy, engineering sciences and their application to military science. Some modern Turkish historians of science consider Tamani to be not only the first Ottoman to have mastered science, technology and French – well enough to render competent Turkish translations from the language – but the first Ottoman who could be considered an active scientist. The claim is based on an original work on logarithms and several books he composed for his students. There is a difference, however, between translating scientific treatises, teaching and writing textbooks on science and being a practicing scientist. Tamani's real importance is that his work laid the foundation that would eventually produce Turkish scientists. His assimilation of science and the contribution he made based on it require qualification: not once in his writing on physics or astronomy did he mention a moving earth. Nor was anything said of it in the book on geography he wrote with Ishak when they were colleagues. It was a topic he seemed studiously to avoid.

Among Tamani's other published translations are a treatise on the elements of geometry, a technical compendium for engineering students, a treatise on determining elevations, artillery tables and a handbook on plane trigonometry, enough to make him director of the Engineering School. It was while he was director that he used his influence to have Ishak transferred from the Translation Bureau to the Engineering School, which launched his academic career. As colleagues, Tamani and Ishak collaborated on an introductory geometry text, composed mostly by Tamani, with Ishak doing the work in preparing it for publication. Essentially a student textbook in geometry, it nonetheless contained some modern science related to geometry, such as the geometry of the heavens, but it included nothing of heavenly or earthly motions. Tamani may not have been sure what to make of the unsettling concepts of a moving earth and of laws of nature handcuffing God's ability to strike up miracles. Like Tycho Brahe, a foot each in the old and the new, he compromised by holding to the geocentric model while presenting the new in the framework of the old to avoid the unpleasantness of appearing to support revolutionary ideas. "Let it be known" he wrote, "that the universe is a sphere and its center is the Earth."[13] One might suspect that the impressive Muslim achievement in medieval science, rather than serving as a catalyst of rediscovery and creative rebirth, was a cage from which even some Muslim scientists in the making could not free themselves to embrace the modern.

The pace of assimilation was not quickened when a man very much of Tamani's divided mind-set succeeded him as the Engineering School's head instructor. Seyyid Ali Bey had also started his career in the Translation Bureau, but he only translated works of the old classical science from Arabic, nothing of the new science that Selim's and Mahmud's schools were created to teach. Toward the end of his career as chief instructor, he in fact wrote a book, *Mirror of the World (Mir'at al-Alam)*, in which he described the superiority of the Ptolemaic system over the Copernican and Tychonian systems. Kepler and Newton were given no place. This reluctance to abandon the old and embrace the new was finally abandoned when in 1830, by order of Sultan Mahmud himself. Ishak Efendi succeeded to the chief instructorship, with Mahmud's express command that he either succeed in reforming the school or – employing Dr. Clot's and Muhammad Ali's style of striking terror in their schools – go to prison. That same year, the first volume of his *Compendium of the Mathematical Sciences* was published. The book was the beginning of a tremendous contribution to science education and assimilation. By 1834 the succeeding three volumes had come out, giving Ottoman students an encyclopedic text of the mathematical sciences. In the broad context of secular reform, Hoja Ishak's 2000-page work, much of it translations from a number of French books, was as important to science in Ottoman education as Mahmud's Rose Chamber Rescript declaring religious equality was to Ottoman law.

Seeds that had been sown in the 17th and 18th centuries were at last taking root in the resistant earth. A majority of older Ottoman officers may have believed the earth was stationary, but with Ishak Efendi's *Compendium*, a new generation of engineering and military science students now had, in their native language, access to the rudiments of science based on the collective work of Copernicus, Gilbert, Galileo, Descartes, Kepler, Newton, Laplace, Lavoisier, Young, Cavendish, Dalton, Ampere and Volta, and the astronomy, calculus, universal gravity, mechanics, hydrodynamics, static and dynamics, optics, magnetism, electricity and chemistry that came with them. If the Galata Saray medical students had Hollbach's *Systeme de la Nature* in French, the engineers now had Ishak Efendi's four-volume scientific compendium in Turkish. For many years this was the chief source anyone in the Ottoman world would go to who wanted to learn the rudiments of modern science but could not read a Western language. The work was reported to have been printed in Cairo between 1841 and 1845 for use by Muhammad Ali's engineering students, but no reference to it appears in Egyptian sources.[14] It was not translated to Arabic. Nothing like it or close to it was published in Arabic in the 19th century. Not even the Lebanese Christian Butrus Bustani's Encyclopedia in the 1860s approached Ishak's detailed, mathematical and mechanical explication of the exact sciences. Aside from Tahtawi's little treatise, *Kashf al-Buhar*, which was a translation and only briefly described the earth's motion and oblate sphericity, not a single book on Newton's physics or modern astronomy had been translated into Arabic during the Muhammad Ali era of reform.

Ishak himself was not completely sure how far he could go in positing the solar system as the true system of the world. In comparing the strengths of the various systems – Ptolemaic, Copernican and Tychonian – he clearly leans toward the Copernican, but then adds, as cautious as Shaykh Tahtawi writing in Cairo at the time: "mistakes can happen." Writing of an earth in motion was a touchy subject in the Muslim world, even until the late 1870s in the Arab region. But in Istanbul that began to change as early as the late 1830s. In the 1840s, as the new science and its astronomy came more and more to displace the old, the once venerated name of Ptolemy began to be misspelled and forgotten.

In addition to his compendium of mathematical sciences, Ishak published nine other works, which like his compendium were translations and adaptations. These were chiefly on military subjects: casting cannon, fortifications, trigonometric surveying and one on naval mines based on a French version of Robert Fulton's *Torpedo War and Submarine Explosions*. As chief instructor, Ishak took seriously Sultan Mahmud's threat that he turn the Engineering School into a more productive institution or go to prison. He sacked a good number of the deadwood faculty and rescheduled classes so that students were obliged to attend five classes daily. As memorization remained the backbone of the Ottoman learning process, learning science was a simple matter of memorizing Ishak's four-volume compendium: a volume each year.

At the end of his first year as head instructor, during which the first volume of his encyclopedic work was published, he submitted a petition to Sultan Mahmud requesting that a medal of honor be awarded to him for his work. Since he had not been sent to prison, he presumed himself to have successfully reformed the engineering school and wanted recognition. A medal of honor from the sultan would do nicely. He had, after all, dedicated the book to Mahmud. Ishak must have been indeed proud of himself: to his petition he attached his own personal design of the medal he wanted. Just any old off-the-shelf medal of honor would not suffice. His had to be special, an Ottoman precedent of the Nobel Prize for science, as it were. And this for what was essentially a science teacher's guide and student text consisting of an eclectic assemblage of translations from different sources in French. The sultan might be excused for not taking Ishak's petition and academic pretensions of grandeur too seriously. Greece had just been lost, the Egyptian army was in occupation of the southern half of the empire, threatening to march on Istanbul, while the Russians threatening from another direction as saviors of the Ottomans. No surprise that the beleaguered sultan denied the teacher his medal. But Mahmud was large enough to grant that Ishak's compendium be printed on the royal press and to reward the author with 250 gold pieces. This was not a princely sum, considering that this was all that Ishak was ever going to earn from his prodigious effort. His experience might help explain why it was that legions of young men were not rushing out to devote themselves to the hard work of science.

Ishak's volumes filled the dire need for a comprehensive engineering student text. The book was meant to be more than that. Its full title in translation is *Compendium of the Mathematical Sciences for the Engineering Teacher*, and it was to provide both teacher and student with a text containing the basic knowledge of science required in the technology of modern warfare. The Compendium was the textual spinal column of the Military Engineering School. Meant for future military officers and engineers in the defense industries, it was prefaced accordingly. Written by Hoja Ishak, the preface related the new science to the old ideas of *jihad* and *ghazwa*. The holy war against the infidel was now to be waged by science, the new sword of conquest and defense. Science was to be the modern *jihad* and Ishak its first ghazi. He was later more soberly known as "The Father of Turkish Technology."[15]

The first volume of his *jihadi* Compendium covers plane and solid geometry, algebra, analytical geometry, conic sections, topography and geodesy. The second volume stood on its own as a text in differential and integral calculus. The third was on the principles of physics, motion, force, gravity, mass, velocity, acceleration, the mathematics of Newtonian mechanics, hydraulics, properties of gases and the gas laws and the mechanics and design of pumps, barometers, hydrometers and devices to measure relative humidity. Also covered in this volume was the science of light. This included an analysis of Newton's prism experiment to produce the spectrum; light mathematically defined in terms of power and intensity; the inverse square law of light as a paradigm of gravitational attraction; determinations of the speed of light; immensity of the universe and the earth as an infinitesimal drop in the ocean of space. Also included in this section was: absorption of light; reflection; the physics of color; the action of light impinging on the eye; the physiology of sight; and the rainbow – which Ishak explained was another phenomenon uniting earthly and heavenly physics, implying that here again Ptolemy and Aristotle were proved wrong, which by default left Copernicus and Kepler to be assumed right.

Like the third, the final volume surveyed in some depth a large number of quite different subjects: the physics and mathematical interrelationships of heat; electricity and magnetism; spherical trigonometry; astronomy; acoustics; biology; botany; zoology; anatomy; mineralogy; geology; chemistry; mechanics; and the design of electrical machines.

The section on astronomy, almost a book in itself at 235 pages, came toward the very end of the final volume. Did Ishak fear that the new astronomy would arouse religious antagonism and so deposited it behind 3 ½ volumes of densely written mathematics and physics, which few of the ulema would have had the mental stamina to struggle through? He must have thought there could be a problem, for at the very start he assumed a defensive posture by launching into a long digression proving the superiority of the system developed by Copernicus, Kepler, Descartes and Newton over those of Ptolemy and Brahe. Ten pages (231–239) reviewed the same litany of arguments that European defenders of Copernicus used centuries earlier against their Aristotelian critics: a stationary earth meant the stars in their daily rotation around it had to go at an absurd speed; there was no visible parallax because the stars were so distant from the earth; the new astronomy was supported by neat and simple mathematical relationships; and scripture was meant to call people to obedience and had therefore to use popular language that described natural events in a way that uneducated people could grasp without getting lost in the complexities of scientific reality that went against simple sense perception.

In the course of explaining the laws of Kepler and Newton and the role of the telescope in supporting their mathematical laws of planetary motion, and why it is that the earth's motion is not sensed, he used the standard analogy of people on a calmly moving ship watching objects along the shore pass by and their imagining that it was the shore, not the ship, that was moving. Tycho Brahe's hybrid system was dismissed as a poor compromise that the Danish astronomer made in order to avoid contradicting the literal word of scripture. "These mathematical sciences," Ishak writes in conclusion,

> have been made by God as a blessing for those who study them and as a way to make life more enjoyable and comfortable and not as a manacle for believers or to seduce the believers from the true path of faith. For the good of believers, it is best to preserve these mathematical sciences free of error and neglect, and protect them from the arrows of those who aim to do them harm.

To provide a small loophole for himself in case this failed to disarm a hostile ulema, he adds that there was always the possibility of error, that the new astronomy might after all be wrong.

Ishak's encyclopedia gives the mathematical principles of celestial mechanics in detail for the first time in a Muslim language. Telescopic observations prove that the square of the period of planetary revolution is proportional to the cube of a planet's distance from the sun, and that planets sweep out equal areas in equal times in the plane of their elliptical orbits. The telescope also reveals that the stars show parallax, another indication of an orbiting earth. The telescope has shown that the sun has spots and also turns on its axis, one full rotation every 27 days, 12 hours and 20 minutes. Ishak, or the source he translated from, discusses the effect of the refraction of light on telescopic observations. At the end of the section he compares Descartes's vortex theory with Newton's theory of universal gravitation in accounting physically for planetary motion. But the divine hand is not left out: the power of gravitational attraction originates in God's eternal will. From the Creator's act of structuring the universe by simply willing it comes the force of gravity and its governing laws that preserve the moving planets and all heavenly bodies in their steady orbits, preventing them from crashing into each other or rushing off into space.

But once the system of the universe has been presented as a divine gift and blessing, God is no longer mentioned. The narrative returns to the forces of nature acting as nature; the principle of gravity returns to being the mechanical foundation of all motion, celestial and earthly, understandable in itself. The gravity that pervades the solar system and reigns throughout the universe of stars is the same force that governs an object's rate of fall to earth and the same that governs motion of the moon around the earth and the motion of all planets around the sun. Gravity is what preserves the sphericity of earth and of all heavenly bodies. It gives weight to objects; the closer a body to the center of the earth the heavier it will be. Beyond the earth's gravitational pull, a body would be weightless. The moon's gravity pulls the earth just as the earth's pulls the moon, the former causing the tides. The force of gravitational attraction between two bodies is directly proportional to the product of their masses and inversely to the square of the distance between them. Weight, or mass, has nothing to do with the rate of fall of objects. In a vacuum, cotton and lead fall at the same rate.

A large number of explanatory diagrams are found at the end of the volume. Diagram 86 shows the positions of the planets from Mercury to Uranus orbiting the sun. The narrative related to the diagram gives detailed descriptive explanations of the seasons, eclipses, solar apogee and nadir, and apparent retrograde motion.

Following the sections on the science of the heavens come the earth sciences: geology and its evolutionary transformations that give rise to mountains and volcanoes and earthquakes that churn up pools of molten gold, silver, lead, copper and iron deep inside the boiling earth to form solidified veins nearer the cooler regions, just under the hardened crust of the surface. This is followed by the science of meteorites and electricity, which in heaven is lightening and on earth can be generated by man-made machines that transmit it with such power through a single metallic wire that it can be just as lethal as a bolt of lightning from heaven, hot enough to burn a strong man to a crisp. Modern chemistry is included in this concluding section, grouped with geology, biology and zoology.

Ishak borrowed liberally from French sources in bringing together his encyclopedia. It more or less amounts to a translation. His career had begun in making translations and as a teacher and textbook writer he used what he translated. The sections on arithmetic, geometry, mechanics and navigation came largely from the first five volumes of a French encyclopedic work on science for military purposes, Etienne Bezout's (d. 1783) prestigious nine-volume *Cours de mathematique a l'usage de gardes du Pavillon et de la marine*, repeatedly published up until 1870. Ishak's survey of chemistry relied greatly on Lavoisier's *Traite elementaire de Chimie*, directly or indirectly, and to a lesser extent on Valmont de Bomare and J. Brisson whose works on chemistry were available in the *Muhendishane* library.[16] But Ishak's cursory treatment of the subject would have left it wanting as an introductory text for a first year university chemistry course, probably because chemistry was associated more with medicine than engineering and taught as an independent course in the medical school. It was some years later before chemistry was recognized to be an indispensable science in modern warfare and not just pharmacology. Not until 1834, with the publication of Ishak's fourth volume, did chemistry become a required course in the Military Engineering School. His chemistry section did nonetheless serve as an introduction to the nomenclature of the time for the elemental symbols that were coming into common use, the chemical categories of acids, bases and salts and the emerging shorthand of symbols to express the various types of reactions (reduction, oxidation and double displacement) in the form of a chemical algebra. For the first time in Ottoman and Muslim scientific literature, the names, discoveries and critical experiments of Priestley, Cavendish, Black and Lavoisier were discussed.[17] Not until 1848 did the first book devoted solely to chemistry appear in Ottoman Turkish. This was Mehmet Amin Darwish Pasha's 400-page *Usûl-i Kimya* (*Principles of Chemistry*). Printed by order of the sultan on the Imperial press, Mehmet Amin's book superseded Ishak's contribution. Among other advances, it expanded on Ishak's list of chemical terminology and provided names and symbols for the new elements that had been discovered since Lavoisier's day. Mehmet Amin's book was the first book purely on chemistry printed in any Muslim language.

Only five hundred copies of Ishak Efendi's four-volume *Compendium on the Mathematical Sciences* were printed, the expense covered by the military treasury. The specter of bankruptcy was forever present, hence the paucity of copies. The Ottoman government's method of covering expenses in the 19th century amounted to a kind of bureaucratic kiting, revenues appropriated for one bureau being shifted at the last moment to cover expenses in another in a fiscal frenzy of musical chairs. In more critical situations, payments would simply be delayed or paid with promissory notes. With so few copies, the set became precious in no time, putting it beyond the reach of students to have their own. However, sets of the volumes were made available at the school so students could write out their own copies for memorization.

Ishak Efendi's work encapsulated the science that had been sporadically trickling into Istanbul in one form or another from the West during the 170 years since Kose Ibrahim Tezkireci's translation of Durret's cosmography in 1665. Within a strictly Ottoman context, his service to Islam in its pursuit of science as a measure of civilization can be related to Hunayn ibn Ishaq, Qusta ibn Luqa, Thabit ibn Qurra and all those non-Muslims and converts who translated Greek science to Arabic. Having lost touch with their impressive scientific heritage, Muslims were now obliged to go through it once again, translating from new masters. But now being on the defensive and frightened, former top dogs that had lost their teeth, they were obliged to learn the languages and do the translations themselves. Before the time of Sultans Selim and Mahmud, knowing a foreign language was like a blemish on an Ottoman's character. With their reforms, it became a certified guarantee to a professional career in translation, education and, possibly, a high government position.

The assimilation of science and technology that was taking place in Istanbul, and in Cairo to a slightly lesser extent, was concomitantly occurring in several other areas of the Muslim world, though on a much reduced level. In Tunis, which like Egypt was autonomous but legally an Ottoman province, the ruler Ahmad Bey (d. 1855) undertook to modernize the military and founded a polytechnical school at Bardo to train officers in the science and technology of modern warfare. Emulating his Egyptian neighbor and Ottoman suzerain, Ahmad Bey imported instructors from France, Italy and England. His Bardo Polytechnic was modeled on its predecessors in Istanbul and Cairo, along with their schools of European languages and translations. On the other side of the Muslim world, science was being introduced to Muslims in the private and government schools that the British founded in India. Hindus were the first to attend the schools. Muslims held themselves back for religious reasons, but by mid-century, Indian Muslims were following the footsteps of their Hindu rivals, writing treatises on the positive harmony between Islam and natural science, and founding a college to propagate precisely that very message as an opening for Muslims to engage in science. The movement of ideas related to science and religion in Muslim intellectual history in 19th century India was of great importance to Muslim thinkers throughout the world of Islam and will be treated in its proper place in a separate section.

Another port of scientific entry to the Muslim world was Azarbayjani Baku on the Black Sea, a region of predominantly Turko–Persian culture that had fallen under Russian military occupation in the course of Russia's conquest of the Caucasus. In Baku, a poet, historian and philosopher of local fame wrote in Persian a cosmology that was informed by anonymous Russian sources on modern science but which bore an innocuous sounding traditional title: *Secrets of the Divine Heavens* (*Asrar al-Malakut*). Its author, Abbas Kulu Aga, (1794–1846), known more popularly by his poet's and scholar's nom de plume of Qudsi, is reminiscent of Ibrahim Hakki, the Sufi of Erzerum whose semi-mystical *Marifetname* explained and supported the Copernican system a century earlier. One might suppose that Western science would be the last thing to attract a Sufi and a poet, but nature, the heavens, God's creation of the material universe and its relation to God's other creation, man and his immortal soul, were and continue to be popular themes in Muslim mysticism.

How modern science reached Qudsi in Baku is not so curious as it is in the case of Ibrahim Hakki in Erzerum. The Russian occupation had brought modern ideas, which Qudsi discovered as a soldier in the Russian army of the Caucasus. Son of the Khan of Baku, he and his several brothers joined the Russian military. He learned Russian, read Russian literature, and served as a military translator, having studied Persian and Arabic as a student. His military service took him to Russia, the Ukraine, Lithuania, Latvia and Poland. Through his travels he learned enough science to compose a book on cosmography, in which he stated that of all the proposed systems of the world, the Copernican was the most successful and precise in solving astronomical problems. He had no religious problem with this. Nor did the Ottoman translator of the book who was familiar enough with the new astronomy to add astronomical material to Qudsi's account that the translator must have learned from other works he had either translated or knew from one source or another, one of them probably Ishak Efendi's recently published *Compendium*.

How Qudsi's book reached the Ottoman court and received a Turkish translation is a lesson in the vagaries of scientific circulation in the Muslim world during the 18th and 19th centuries. Much was owed to non-specialist but literate Muslims receptive to new ideas. Toward the end of his life, while on his way to Mecca to perform the holy pilgrimage, Qudsi visited Istanbul and, in 1846, having by then become a known writer and poet, presented an Arabic translation he had made of his *Cosmography* to Mahmud's successor, Sultan abd al-Majid. Qudsi died shortly after that in Arabia, at the end of his pilgrimage. The sultan's reformist grand vizier had his book translated to Turkish, thus making Qudsi's *Cosmography* available in all three major Muslim languages. Another vagary, this in the lack of diffusion of modern science across the language zones of the Muslim world, is that Hoja Ishak's *Compendium* never made it out of Turkish. An Arabic translation was reported as having been made in Cairo under Muhammad Ali, but no one involved in science and science education in Egypt ever mentioned it and one searches in vain for any reference to it in the Arabic sources or any of the catalogues of the books translated and printed during or after Muhammad Ali's reign.

By mid-century, modern science, accompanied by its social and political resonances for change, was entering the Muslim world from many directions and, as Qudsi's book and its Ottoman translator's comments indicate, the heliocentric theory was becoming accepted by the educated of the empire, with the singular exception of the generality of the Ottoman and Arab ulema; of which, as always, there were exceptions to the exception. Qudsi reveals that even as late as the mid-19th century, anyone commenting positively on science was obliged to use caution and placate the ulema, even in those states where the institutions supporting the economic autonomy of the ulema had been battered and impoverished by reformist government, for it was the ulema who had the hearts and minds of the common people, and a good number of the not so common as well. Toward the end of the introduction to his *Cosmology*, Qudsi claims that religion has nothing to do with astronomical systems. But then, in case there are those who would think otherwise, he concludes that the Quran and Hadith do indeed support the Copernican system. And even if they did not, he says, one system is the same as another with respect to religion, which is of faith and tradition, not science and reason:

> Some Muslim scholars of the new astronomy who compared the Copernican view with the rules of reason and observation defended its correctness on the basis of Quranic verses and traditions of the Prophet. They were surprised to see how the Ptolemaic view, which did not conform to the principles of science and observation, continued to be well known for such a long time. I realized that the Copernican view conforms to the clear and definitive proofs deduced from geometry, and moreover, to the Quranic verses and the traditions of the Prophet; as for the Ptolemaic view, it is the opposite.[18]

Familiarity with science was strikingly uneven across the Muslim world in the 19th century. Iranians were light years behind Ottomans and Egyptians. It would be deep into the 19th century, 1875, before Copernicus became a topic of literary discussion in Syria and Lebanon; the 1890s in Iraq; and even longer in Iran where the social power of the Shi'i ulema would not wane as it did in the Sunni regions of Muslim society but would wax stronger as the state of the Qajar dynasty fell more and more under the economic influence of European powers. Muslim India presents a special case because of the British presence, and also because of Hindu scholars who, being relatively free of the religious hang-ups mentally sandbagging Muslims in their regard to Western knowledge, were eager to learn the mathematics, science and philosophy of their foreign ruler.

Regardless of how far modernizing rulers had secularized society and education by the end of the first half of the 19th century and how fiscally and educationally squeezed to the margins the ulema were by the secularizing forces from above, wherever one looked in the Muslim world, copious references to Quran and Hadith were made by those reformers who argued the religious legitimacy of the new science coming from the West. The *fatwa*-issuing ulema in charge of granting the visas that made entry legal held fast in their fortress mosques at the heart of society, guarding religious orthodoxy, and making sure Quran and Hadith, as interpreted by them, would remain the final authority of the crossing guard.

The upper level of ulema could be counted on to cooperate with reform, through conviction, offers of high position, gifts, bribery, coercion or exile. It was the lower levels that held out on what was permissible and what not, and it was this part of the ulema that was most connected to the people and that spoke for them on matters of belief and legitimacy. Secularization of society had hardly gone deeply enough to reduce these lower ulema to anywhere near irrelevancy. Government investment in science had not been generous enough to give birth to an indigenous culture of science that reached into society, which would have forced a revaluation of belief, tradition, custom and culture because of economic concerns of career, prestige, and social advancement. The drag of mass illiteracy was not to be overcome without heavy and intensely focused investment, which would have meant severe cutbacks in expenditures in some places and increased taxation on those who could pay. The sultans and their reformers felt incapable of such strong action. Hence, the old continued on, living alongside the new, with the unreformed ulema of the underprivileged believers remaining in tenuous opposition to the government reformers and those who had been educated and who had benefited by the reforms.

By mid-century the division was clear. The social fissure deepened and widened as imposed reforms became more radical, keeping pace with the mounting fear of extinction engendered by Europe's growing economic domination and open support of the revolt-minded Christian minorities in the empire's Balkan provinces. If the traditionalist ulema and their flock feared that the imported innovations of Sultans Selim and Mahmud might wrench society out of joint, what came next must have seemed like the end of the world.

Notes

1 Berkes, *Development of Secularization in Turkey*, McGill University Press, Montreal, 1964, p. 113.
2 Charles MacFarlane, *Turkey and Its Destiny, The Result of Journeys Made in 1847 and 1848*, 2 vols., John Murray Press, London, 1850. See vol. 2, p. 266.
3 MacFarlane, *Turkey and Its Destiny*, vol. 2, p. 165.
4 MacFarlane, *Turkey and Its Destiny*, vol. 2, pp. 181–184; Berkes, *Development of Secularization in Turkey*, p. 118.
5 Charles Pertusier, *Promenades Pittoresques dans Constantinople*, H. Nicolle Press, Paris, 1815, vol. I, pp. 238–239. Uriel Heyd, "Ottoman Ulema and Westernization in The Time of Selim III and Mahmud II," in *The Modern Middle East*, edited by A. Hourani, P. Khoury, and M. Wilson, University of California Press, Berkeley, 1993, p. 76.
6 Helmuth von Moltke, *Briefe uber Zustande and Begebenheiten in der Turkei aus den Jahren 1835 bis 1839*, 2nd edition, Mittler Press, Berlin, 1876, p. 411, cited by Roderic H. Davison, *Essays in Ottoman Turkish History*, University of Texas Press, Austin, 1990, p. 166.

7 Wayne Vucinich, *The Ottoman Empire: Its Record and Legacy*, Huntington Press, New York, 1979, p. 158.

8 Heyd, "The Ottoman Ulema and Westernization."

9 Davison, *Essays in Ottoman Turkish History*, p. 174.

10 Berkes, *Development of Secularism in Turkey*, p. 105.

11 Berkes, *Development of Secularization in Turkey*, p. 107.

12 Though indeed, Mahmud created a few of his own men as high ulema, deposed and exiled some, coerced others and dismissed Shaykhs al-Islam. He also cajoled high ulema and brought them into government at high positions and included them in his Council of State, commanding an extraordinary degree of control over the higher ulema compared to earlier sultans. See Heyd, "Ottoman Ulema and Westernization," pp. 76–83.

13 Ekmeleddin Ihsanoglu, *History of Ottoman State, Society and Civilization*, 2 vols., Research Center for Islamic History and Culture, Istanbul, 2001–2002, vol. 2, p. 428.

14 Ihsanoglu, *History of Ottoman State, Society and Civilization*, vol. II, p. 429.

15 J.J. Heyworth-Dunne, *An Introduction to the History of Education in Modern Egypt*, Luzac and Co., London, p. 336.

16 Feza Gunergun, "Ondokuzuncu Yuzyil Turkiyesinde Kimyada Adlandirma," in *Osmanli Bilim Arastirmalar,* Istanbul University Press, Istanbul, cilt 5, sayil, 2003.

17 Reports of Priestley's discovery of oxygen in 1774 and Cavendish's of hydrogen in 1776 had found their way into Ottoman literature by 1803; Ihsanoglu, "Bashoca Ishak Efendi," p. 63.

18 Ekmeleddin Ihsanoglu, "Introduction of Western Science to the Ottoman World," in *Transfer of Modern Science and Technology to the Muslim World*, edited by E. Ihsanoglu, A.H. de Groot, Istanbul,1992, p. 104.

Index

Note: Italicized page numbers indicate a figure on the corresponding page.